Teaching Science
in Elementary
and Middle School

A Project-Based Approach

Third Edition

Teaching Science
in Elementary
and Middle School

A Project-Based Approach

Third Edition

Joseph S. Krajcik • Charlene M. Czerniak

LEA Lawrence Erlbaum Associates
Taylor & Francis Group

New York   London

Lawrence Erlbaum Associates
Taylor & Francis Group
270 Madison Avenue
New York, NY 10016

Lawrence Erlbaum Associates
Taylor & Francis Group
2 Park Square
Milton Park, Abingdon
Oxon OX14 4RN

© 2007 by Taylor & Francis Group, LLC
Lawrence Erlbaum Associates is an imprint of Taylor & Francis Group, an Informa business

Printed in the United States of America on acid-free paper
10 9 8 7 6 5 4 3 2 1

International Standard Book Number-13: 978-0-8058-6206-5 (Softcover)

Library of Congress Cataloging-in-Publication Data

Krajcik, Joseph S.
   Teaching science in elementary and middle school : a project-based approach / author/editor(s) Joseph Krajcik and Charlene Czerniak. -- 3rd.
      p. cm.
   Previously published: Boston: McGraw-Hill, c2003.
   Includes bibliographical references and index.
   ISBN-13: 978-0-8058-6206-5 (alk. paper)
   ISBN-10: 0-8058-6206-4 (alk. paper)
   1. Science--Study and teaching (Elementary)--Methodology. 2. Science--Study and teaching (Middle school)--Methodology. I. Czerniak, Charlene Lochbihler. II. Title.

Q181.K73 2008
372.35--dc22
                                                  2007013599

Visit the Taylor & Francis Web site at
http://www.taylorandfrancis.com

and the LEA Web site at
http://www.erlbaum.com

To Ann, whose encouragement, support, and love have allowed me to develop as a person and scholar, and to my children, Michael, Paul, and Ellen; my daughter-in-law, Erin; and my grandchildren, Juliet and Elise, who continuously provide me with the energy to keep young and energetic.

J.S.K.

To my parents, who helped inspire my love of science, and to my husband, David, who is always supportive of my professional endeavors.

C.M.C.

Contents

Preface

Overview of the Book

One of the primary goals of science education in schools today focuses on helping all children learn about their world. In *Teaching Science in Elementary and Middle School Classrooms: A Project-Based Approach,* we focus on helping you learn how to teach science to elementary-age and middle school-age children.

To accomplish this goal of helping you learn how to teach science, we present an exciting science teaching method that we refer to as project-based science. This teaching approach engages all young learners—regardless of culture, race, or gender—in exploring important and meaningful questions through a process of investigation and collaboration. Throughout this dynamic process, students ask questions, make predictions, design investigations, collect and analyze data, make products, and share ideas. As a result, students learn fundamental science concepts and principles that they apply to their daily lives.

The roots of project-based learning are found in the writings of many distinguished educators, including John Dewey, Jerome Bruner, and Robert Karplus.[1] Elements of this approach also can be found in other forms of science teaching: inquiry-based learning, which focuses instruction on students' using science process skills, such as observing and interpreting data; science, technology, society (STS), which focuses instruction on issues related to societal problems; and problem-driven science, which focuses on science issues. Project-based science stands apart from these other forms of science teaching, however, in that it situates the learning of science in questions that children find meaningful. As a result, it shifts the responsibility of learning to the child.

Support for this approach can be found in the works of contemporary educators, including Ann Brown and colleagues (1994), Wolff-Michael Roth (1995), Phyllis Blumenfeld and colleagues (1991), and Richard Roup and colleagues (1992). This

---

[1]John Dewey was the founder of the progressive education movement that encouraged students to be active learners. In the 1930s, the Lincoln School of Teachers College and other schools used real situations to help students ground their learning. Jerome Bruner, an education theorist, helped establish curriculum reform during the 1960s. His book *Toward a Theory of Instruction* set the stage for much of the modern direction of curricula today. Robert Karplus was a physicist who cofounded The Science Curriculum Improvement Study, a pioneer elementary science program of the 1960s. The program exists today in a third incarnation of the "hands-on, minds-on" movement of that era. We discuss the influence of these individuals in more detail in Chapter 2.

approach requires dynamic teaching methods that match the guidelines of today's major science education reports, including *The National Science Education Standards* (National Research Council, 1996, 2000), *Project 2061: Science for All Americans* (Rutherford & Ahlgren, 1989), *Benchmarks for Science Literacy* (AAAS, 1993), and the National Science Teachers Association recommendations for elementary science (NSTA, 1991). At the end of this chapter, we will discuss more thoroughly how project-based science works with these national efforts.

Although it has a solid foundation in educational theory, project-based science represents a fundamental shift in how to teach science in most classrooms today. While it is already being used successfully in many schools, you may find this kind of science instruction quite different from what you have experienced in the past. Learning to teach in a way that is different from what you have experienced can be very difficult. As you work through this book, most likely you will be changing the way you think about science and science teaching. More important, you'll be changing the way you help students learn science. We will assist you with these changes by providing strategies for helping children engage in inquiry that involves important and worthwhile questions.

Although this book does not provide teaching activities to help you learn science content, content understanding is critical to effective science teaching. When appropriate throughout the chapters, we will refer to other references that cover science content or children's science activities.

Chapter 1—"Teaching Science to Children"—introduces you to project-based science. It also answers the question "What is science?" and discusses why it is important for young children to learn fundamental concepts and principles, as well as processes of science. This chapter also presents important information about current national goals in science education and illustrates how project-based science matches these initiatives.

Chapter 2—"How Children Construct Understanding of Science"—focuses on the characteristics of children and the connection of these characteristics with learning science. It pays close attention to factors that influence students' construction of understanding, including prior experiences, social interactions, and teachers.

It is important to make science relevant to children. Driving questions initiate, implement, and sustain inquiry and are, thus, central to project-based science.

Chapter 3—"Establishing Relevance to Students' Lives"—explores characteristics and issues of driving questions. It provides answers to the questions "What are driving questions?" "Where do they come from?" and "What makes a good driving question?"

Investigation is an essential element of project-based science. Chapter 4—"Developing Scientific Investigations"—explores how to help students engage in investigations and find solutions to questions that are of interest to them. It examines the various components of investigations, such as asking and refining questions, designing experiments, analyzing data, and presenting findings. It also discusses ways that a teacher can support inquiry and overcome the challenges of implementing investigations.

Chapter 5—"Making Sense of Data and Sharing Findings"—introduces ways that teachers can help students make sense of data, construct scientific explanations, and make conclusions, key components of engaging in science. It describes ways that students can construct explanations using evidence from their investigations. The chapter also pays attention to ways teachers can work with students in sharing their ideas. Finally, it discusses ways to support students' implementation of investigations and outlines criteria for assessing the value of an investigation.

New technology tools can enhance learning. Chapter 6—"Using Learning Technologies to Support Students in Inquiry"—examines the value of using new learning technologies and explores how teachers can use new technology tools to help students investigate questions important to them.

Collaboration in science teaching is frequently much more effective than individual learning. Chapter 7—"Collaboration in the Science Classroom"—discusses types of collaboration among students, teachers, and members of the community. It presents ways that a teacher can create a collaborative environment, build the social skills students need to sustain collaboration, hold students accountable during collaboration, and overcome challenges that might arise while implementing collaborative groups.

Chapter 8—"Instructional Strategies That Support Inquiry"—focuses on techniques that teachers use to teach science lessons. It presents a variety of types of instructional strategies ranging from direct, teacher-directed strategies to independent, student-directed strategies. The chapter also discusses important skills that teachers need to develop to be successful in engaging students in inquiry.

Assessing students' understanding lies at the center of the educational process. Chapter 9—"Assessing Students in Science"—discusses the purpose of assessment in a science classroom. The chapter also describes characteristics of assessment. We explain what to assess and when to assess. Finally, the chapter presents ideas for using technology tools to examine assessment.

Many teachers today face the realities of high-stakes assessments. Chapter 10—"Assessing Student Understanding"—focuses on the process of assessment and presents numerous ideas about how to assess students' understanding and skills. Finally, it discusses advantages for teachers, students, and parents in using various assessment strategies.

Managing an elementary or middle grade classroom presents challenges for many teachers, especially new teachers. Chapter 11—"Managing the Science Classroom"—discusses classroom climate, classroom organization, and management skills. It presents many practical strategies that teachers can use to manage students successfully as they engage in science learning.

Chapter 12—"Planning a Project-Based Curriculum"—focuses on the important task of planning lessons. We introduce lesson planning formats and provide helpful suggestions for thinking about planning in a project environment. Throughout the chapter, we illustrate the process of developing a project by using the question, "When do various insects appear on our playground?" Finally, we discuss the importance of integrating across subject areas and present ways that project-based science help integrate the curriculum.

Finally, Chapter 13—"Next Steps"—discusses the benefits and challenges of a project environment. The book concludes with a reflection on the principles and ideas explored throughout this book.

Each of these chapters includes learning activities to help you engage in learning this exciting way to teach science. First, each chapter starts with several scenarios that are designed to help you envision various classroom scenes. Throughout the text, questions encourage you to stop and think about ideas. Other questions ask you to think about your own experiences as a student to help you construct an understanding of how to teach science. There are activities throughout each chapter that are designed to help you construct meaning from what you are reading.

Acknowledgments

The ideas in this book would not have been possible without the innovative thinking of faculty colleagues and graduate students at The University of Michigan and The University of Toledo. In particular, we would like to acknowledge Carl Berger, coauthor in the first two editions, for his contribution of ideas. We would also like to thank the many K–12 teachers, who extended our ideas on enacting Project-Based Science in the classroom. Undergraduate and graduate students at The University of Michigan and The University of Toledo, who used the first two editions, provided us with invaluable feedback that was used to enhance this edition. We were encouraged by colleagues in our profession to publish the third edition; and our current editor, Lane Akers, is credited with encouraging us to write the first edition of the book and, being a strong supporter, motivating us to complete this edition.

About the Authors

Joseph S. Krajcik, a professor of science education and associate dean for research in the School of Education at the University of Michigan, works with teachers in science classrooms to bring about sustained change by creating classroom environments in which students find solutions to important intellectual questions that subsume essential curriculum standards and use learning technologies as productivity tools. He seeks to discover what students learn in such environments, as well as to explore and find solutions to challenges that teachers face in enacting such complex instruction. In collaboration with colleagues from Northwestern University, American Association of Science, and Michigan State, Joe, through funding from the NSF, is a principal investigator in a materials development project that aims to design, develop, and test the next generation of middle school curriculum materials to engage students in obtaining deep understandings of science content and practices. He has authored and coauthored over 100 manuscripts and makes frequent presentations at international, national, and regional conferences that focus on his research, as well as presentations that translate research findings into classroom practice. He is a fellow of the American Association for the Advancement of Science and served as president of the National Association for Research in Science Teaching. Joe codirects the Center for Highly Interactive Classrooms, Curriculum, and Computing in Education (hi-ce) at the University of Michigan and is a coprinciple investigator in the Center for Curriculum Materials in Science and The National Center for Learning and Teaching Nanoscale Science and Engineering. In 2002, he was honored to receive a Guest Professorship from Beijing Normal University in Beijing, China. In winter of 2005, Joe was the Weston Visiting Professor of Science Education at the Weizmann Institute of Science in Rehovot, Israel. Before obtaining his PhD in science education, Joe taught high school chemistry for 7 years in Milwaukee, Wisconsin. He received a PhD in science education from the University of Iowa in 1986.

Charlene M. Czerniak is a professor of science education at The University of Toledo. She received her PhD in science education from The Ohio State University. A former elementary teacher for 10 years in Bowling Green, Ohio, she now teaches classes at The University of Toledo in grant writing and science education. She has authored and coauthored over 50 articles. Her publications appear in the *Journal of Science Teacher Education, Journal of Research in Science Teaching, School Science and Mathematics, Science Scope,* and *Science and Children.* Charlene has contributed chapters to books and has illustrated 12 children's science education books. Most recently, she authored a chapter entitled "Interdisciplinary Science Teaching" in the *Handbook of*

Research on Science Education, published by Lawrence Erlbaum and Associates. She has been an author and director of numerous grant-funded projects in excess of $13 million that targeted professional development of science teachers. She is currently the director of a $6 million dollar grant from the United States Department of Education entitled UToledo, UTeach, UTouch the Future (UT³), which focuses on recruiting, better preparing, and retaining science and mathematics teachers for urban schools. She makes frequent presentations at national and regional conferences that focus on her research interests on teachers' beliefs about teaching science, professional development for elementary and middle grades teachers, science education reform, and school improvement. She is an active member in the Association for Science Teacher Education (ASTE), the National Association of Research in Science Teaching (NARST), the School Science and Mathematics Association (SSMA), and the National Science Teachers Association (NSTA) and reviews manuscripts for the journals associated with these organizations. She is currently Editor-in-Chief of the *Journal of Science Teacher Education*, the professional journal of ASTE. She has served on numerous committees for ASTE, NARST, SSMA, and NSTA. Charlene Czerniak was the president of the School Science and Mathematics Association for 2 years, and she is now president-elect of the National Association for Research in Science Teaching.

Chapter 1

Teaching Science to Children

Introduction

When you think about the prospect of teaching science to children, many questions probably come to your mind: What characterizes science in elementary and middle grades? How should science be taught to young learners? How can I motivate children to become interested in science? How can I help them learn about science in their everyday world?

In this chapter, we introduce you to project-based science teaching. **Project-based science** (PBS) is an approach to teaching science that focuses students on engaging in inquiry to learn important ideas and practices of science. As such, PBS aligns with the National Science Education Standards (NRC, 1996) that explicitly state that inquiry should be the primary strategy for teaching science. Later in the chapter, we explore what is meant by the nature of science and how it relates to project-based science. We then examine why young children need to learn science, and we review the goals

Chapter Learning Performances

- *Describe the primary features of project-based science.*
- *Compare and contrast project-based science with reading about science, direct instruction, and process science.*
- *Explain how project-based science reflects the nature of science.*
- *Justify why young learners should learn science.*
- *Summarize the primary features of our national science education goals.*
- *Explain the value of using project-based science to meet national goals and standards, particularly goals of encouraging females and minorities in science.*

CONNECTING TO NATIONAL
SCIENCE EDUCATION STANDARDS

SCIENCE TEACHING STANDARDS

Inquiry into authentic questions generated from student experiences is the central strategy for teaching science (p. 31).

of science education. Finally, we discuss in greater detail how project-based science matches today's science education goals.

First, however, to challenge your thinking about science and science teaching, we start this chapter by encouraging you to reflect on your own past experiences as a student of science. This reflection will help you examine your personal views of science teaching and learning. Take time now to complete Learning Activity 1.1.

As you reflected on your own experiences, you may have found that you did not have many memories of learning science or that you had good memories of science class. Maybe you did hands-on activities. Perhaps your teacher exhibited excitement about teaching science. By studying science, you may have learned about important questions related to your world. However, your memories may not have been positive ones; science may have been presented as dull, boring, or difficult to learn. Perhaps your out-of-school memories about science learning, such as that second-grade field trip to the zoo, were the strongest and most positive. The literature shows that many of us did not experience learning science in a dynamic and active manner that included asking questions, collaborating with others to find solutions, and designing investigations (NRC, 1996; Stake & Easley, 1978). As a result, many of us do not have good models of teaching science to use in our own classrooms.

Let's examine several models of science teaching. As you read these scenarios, contrast them with your memories of elementary and middle school science.

Learning Activity 1.1

WHAT ARE YOUR ELEMENTARY AND MIDDLE
SCHOOL SCIENCE EXPERIENCES?

Materials Needed:

- A sheet of paper
- Something to write with
  A. Think back to your elementary and middle school days. Do you recall your teachers teaching science? What do you remember about learning science in early elementary grades? Middle grades? What kinds of topics did your teachers cover? How did they teach science? Did you take field trips to planetariums, zoos, or science museums? Did you conduct "experiments"? Were you required to complete a science fair project?
  B. What science learning experiences are most vivid in your mind? Do you remember stories, such as about Newton "discovering" gravity when an apple fell on his head? Write a short paragraph about your most vivid memories of science learning experiences.

Scenario 1: Reading About Science

Maybe your class was like this: Mrs. Patterson[1] said, "Okay, boys and girls, let's turn to page 37 in the science book. Don't forget to write down the bold print science words for your spelling list for the quiz on Friday. Al, would you please read the first paragraph?"

Al sat in the middle of the classroom. You all had figured out the order of reading and who would read next, and you sighed a little relief as you counted the paragraphs and found that yours was beyond the last page of assigned text. At the end of each paragraph, Mrs. Patterson wrote new words on the board. If it was after Wednesday, they would be on next week's spelling list. After looking at the colored pictures in the book and daydreaming a little, you heard Mrs. Patterson say, "Now turn to page 41 and answer the first four questions. Make sure to use complete sentences and check your spelling."

Sound familiar? We call this *read about science,* **an expression we will sometimes use in upcoming chapters. Reading is an important part of science teaching and learning. However, teachers often focus primarily on vocabulary words and facts in the textbook, rather than using the reading materials as support for helping students understand phenomena. Although reading about science is one important strategy for learning, teachers need to avoid using it as a substitute for experiencing and doing science. We discuss appropriate reading strategies in Chapter 8, "Instructional Strategies that Support Inquiry."**

Scenario 2: Direct Instruction

Perhaps your class was like this: Mr. Velasquez stood in front of your class by the hot plate he kept near his desk. Normally, he had his coffee on the plate, but today there was a soda can resting on it. Mr. Velasquez said, "This soda can was empty, but I rinsed it out; and I added about 2 centimeters of water to it. You can see the steam coming out of the top of the can, and some of you can hear the water boiling on the inside. I'm going to take the can off the hot plate with these hot pads and quickly turn it over into this icy pan of water. I want you to watch what happens and try to figure out what is going on." You thought it might explode; after all, it was boiling inside and very hot. As you watched, Mr. Velasquez set the can in the icy water. Almost instantly the sides of the can crumpled inward as though it had been crushed by a giant force. "Well, what do you think?" he asked. As usual, Bobby Wilson's hand shot up. "Yes," said Mr. Velasquez. He didn't wait for anybody else to think. You had some ideas, but they had not quite formed in your mind. Bobby Wilson blurted out, "It's the suction; when the can cools down, and something on the inside is sucking the can in!" "Well," said Mr. Velasquez," it does have to do with the can cooling, but you see, as the can cools, the steam turns back to water and that takes up less space. Actually, it is the air on the outside of the can that is pressing in. It is the air pressure that does it." To this day, you remember this dramatic demonstration.

We call this kind of teaching *direct instruction.* **It occurs when a teacher provides the direct answers, sometimes after a demonstration. Demonstrations can be powerful teaching tools because they allow students to experience phenomena that science concepts and principles explain. However, in this scenario, the teacher merely told the students what they had seen. Students were expected to understand the concept told to them simply because they had witnessed the demonstration.**

Scenario 3: Process Science Teaching

Maybe your class was like this: Mr. Haddad said, "Today we're going to find out how high a ball bounces when it is dropped and if the kind of ball makes a difference. This is part of the science process of prediction, and you will be graphing your results. Each pair of you has a ball with a letter on it, a meter stick, and a sheet of graph paper. First, drop the ball from a height of 100 centimeters onto the floor and record the height of the bounce. After you have done this four times, drop it from a different height four times and record your results. Then answer the questions on the board."

You looked up at the board and the questions were as follows:

1. Did the ball bounce back to the same height for each height it was dropped?
2. Graph the height of the bounce on the y- axis and the height from which it was dropped on the x-axis. What is the pattern?
3. What would the graph look like if you dropped the ball onto a carpeted floor?

You and your partner didn't quite understand Mr. Haddad's questions, but you had a meter stick and a rubber ball, so there would be a lot of things you could try. Anyway, if it got too bad, Mr. Haddad would come around and show you what to do. Graphing always presented two difficulties. One, the last point you would try to plot would always fall just outside the edge of the graph paper and, two, you would always do the whole thing in ink and then, when you'd make a mistake, you'd have to start all over. By the time you finished graphing the data, you had forgotten what you were supposed to find out. Oh well, science was fun and you did some interesting activities.

This kind of science teaching we label *process science teaching.* **The primary purpose of the lesson is for students to use science process skills, such as observing, predicting, and graphing. Process science lessons can be used to help students learn various skills associated with science teaching, such as using a ruler or recording data, but processes should not be presented as separate stand-alone lessons. Rather, processes**

need to be connected with important concepts. For example, in the lesson described, the processes could have been connected to learning about forces and motion. Then, students would have understood why they were doing the activity.

Scenario 4: Project-Based Science

Maybe your class was like this: You and a couple of friends were looking at the pet rabbit in the cage in your classroom. Normally, the rabbit was eager to eat the carrots you gave him. "Maybe he's sick," you thought. You and your friends questioned why the rabbit wasn't eating. With your teacher's encouragement, you and classmates formed teams to investigate the sudden change. You had other pets in your classroom (including hamsters, gerbils, a snake, and fish), and your teacher encouraged you to investigate the question "What do pets need to stay healthy?"

Each day, *teams of students* from your class visited one of the classroom pets and provided it with several different foods. You gave the rabbit foods such as carrots, celery, oats, alfalfa, and rabbit pellets purchased from the pet food store. Some teams used computer *technology* to obtain information from the World Wide Web about the needs of various pets. When you heard that one of the members of another team had called the local pet store, and the manager was interested in your class investigation and would come to talk about the needs of pets, you were surprised that members of the *community* would *collaborate* with you in your investigation. You knew you had to come up with some great questions.

After *several weeks* of investigating what pets need to stay healthy, you and your classmates *shared results* in graphic form. A couple of teams included photographs of animals and models of healthy environments in their presentations. You found that different pets need different habitats, special kinds of food, a clean environment, and veterinary care to fight against diseases or infections. In fact, you were able to change a few things in the rabbit's diet to entice him to eat. You still had a question about the needs of your pet iguana at home, but you knew that the classroom rabbit was happy and healthy.

This last scenario is an example of what we call *project-based science*. The primary purpose of the project was for students to collaborate for a substantial length of time in the investigation of an important question that was interesting to them. As they explored their questions, students learned important science concepts and principles linked to national standards, used technology, and developed products.

The scenarios you just read are, in some respects, stereotyped. Although we will reflect throughout this book on the science teaching techniques illustrated by these scenarios, you shouldn't completely dismiss any one kind of science teaching based solely on these scenarios. There is no one best way to teach, but there are ways that can promote student understanding and motivation and do more to develop lifelong interest in science. Although project-based science is, for many reasons that will be elaborated on throughout this book, the most effective overall approach to teaching science, you will find times when teacher-directed activities are not only acceptable, but preferred. Imagine that, as part of answering a question about water pressure in the school's drinking fountains, you have students investigate how powerful the air pressure is in the water lines. For safety purposes, you would not want an elementary student boiling water in a soda can to demonstrate how air pressure can crush the can. Instead, you might decide to demonstrate the activity yourself so that students could develop an understanding of air pressure and apply it to their investigation of pressure in the water pipes. Demonstrations such as this can provide important information needed by students.

Science teaching has gone through several revolutions—or evolutions—in the last 30 years. So, amid all this change, why has project-based science emerged as such an important way to teach science? In the next section, you will explore the features of this approach and learn why it is so important.

An Overview of Project-Based Science

Project-based science helps students and teachers find solutions to questions about the world around them. Investigating real-world questions that students find meaningful has long been touted as a viable learning structure; the roots of the idea go back to John Dewey (1991), who is often described as the father of progressive education. Dewey promoted teaching strategies that helped students actively engage in learning about topics relevant to their lives. Because it focuses on students and their interests, project-based science is sensitive to the varied needs of diverse students with respect to culture, race, and gender (Atwater, 1994; Haberman, 1991, 1995). Haberman (1995), in describing exemplary or "star" teachers who are successful in teaching children in poverty from diverse cultural backgrounds, stated, "Stars do not use direct instruction as their primary method. . . . Stars use some variation of the project method" (p. 13). He added,

> Star teachers find projects that interest children. They then plan with children how they will set about studying the topics. Usually the projects are built around a set of questions or problems that are of vital interest to the children. (p. 34)

Project-based science has several fundamental features. The fourth scenario reflects some of these features, which we will discuss thoroughly throughout the book. However, to provide you with an initial framework, let's examine some of them briefly now. First, project-based science is relevant to students' lives. Students find solutions to questions that are meaningful and important to them. Second, students engage in inquiry and perform investigations to answer their questions. Third, students, teachers, and members of society collaborate on the question or problem. Fourth, students use learning technologies to investigate, develop artifacts or products, collaborate, and access information. The result is a series of artifacts or products that address the question or problem. As such, project-based science stands apart from these other forms of science teaching in that it situates the learning of science in students doing inquiry to find solutions to questions that they find meaningful. As a result, it shifts the responsibility of learning to the child.

Establishing Relevance to Students' Lives

Science classes should have children explore solutions to questions (NRC, 1996, 2000). Project-based science calls for questions or problems that are meaningful and important to learners (Blumenfeld et al., 1991; Krajcik & Blumenfeld, 2006). We refer to such questions as **driving questions**. An example is "What is the pH of rainwater in our city?" Students need to find the questions they explore meaningful and important. The driving questions are the primary vehicle for establishing relevance to students' lives.

The driving question of a project is also the first step in meeting all of the other key components of project-based science. It sets the stage for planning and carrying out investigations to measure the acidity of rainwater and test the impact acid rain has on living and nonliving things. Once the stage has been set, students might use technology to investigate the question. For example, they might use electronic pH meters and find information on the World Wide Web about acids and bases. As students collaboratively pursue solutions to the driving question, they develop a meaningful understanding of key scientific concepts, such as acids and bases, pH, and concentration. Hence, instruction is anchored in real-world situations that students find meaningful and from which questions emerge, and it leads to students developing deep understandings. Finally, students can develop concrete representations, such as posters, of the results of their investigations of the driving question. Creating these representations solidifies students' learning.

The questions that students can explore can originate either from the students or the teacher. What is critical is that students find the questions important and meaningful to their lives. In the fourth scenario, how did the question about the health needs of pets emerge? As in this scenario, it is possible to set up a learning situation that will lead to natural questions from the students. In one science curriculum, for example, second- and third-grade students set up a series of aquariums with clear plastic shoe boxes, water, sand, fish, and water plants, but no water pumps or filters. As you can imagine, a layer of "black stuff" appears on the bottom of the aquariums in a week or so, and always one or more students asks, "Where does the black stuff on the sand come from?" The teacher can then build with the children a clear set of investigations/experiences to find the answers to this question.

In other situations, teachers may present the driving question to the students, but they do so in a real-world context, one with which the students can identify and about which they can ask **subquestions**. For instance, if a teacher starts a project with the question "Is our

> **CONNECTING TO NATIONAL SCIENCE EDUCATION STANDARDS**
>
> **SCIENCE TEACHING STANDARDS**
>
> Teachers select science content and adapt and design curricula to meet the interests, knowledge, understanding, abilities, and experiences of students (p. 30).

water safe?" students have the potential to ask a number of subquestions such as "Is our water safe to swim in?" "Is our water safe to drink?" and "What can live in our water?" Another example comes from consumer products. For instance, the teacher might set up this question for the class, "Are our products environmentally friendly?" Students' subquestions might include "Is my ink pen environmentally friendly?" "Are my batteries environmentally friendly?" and "Is my chewing gum environmentally friendly?"

The central aspect behind driving questions is that they need to help establish relevance to students' lives. We will discuss establishing relevance to students' lives more thoroughly in Chapter 3.

Students Engage in Inquiry

Sustained **inquiry** based on important and meaningful questions forms one of the hallmarks of science. In project-based science, students investigate a question over an extended period of time, rather than taking part in short-term activities that are out of the context of real life. Questions, such as "What do pets need to stay healthy?" and "Where did the black stuff come from in the bottom of the aquarium?" can provide the basis for long-term inquiry in which students perform a number of **investigations** to find solutions to their questions. These investigations are meaningful to students and, therefore, keep the students' attention for long periods of time—sometimes over the course of the entire school year.

When students engage in inquiry, they pursue solutions to questions of importance, debating ideas; making predictions; designing plans for investigations; measuring, collecting, and analyzing data and/or information; drawing conclusions; making inferences; constructing artifacts; communicating their ideas and findings to others; and asking new questions. Although we discuss how to support students in inquiry more thoroughly in Chapters 4 and 5, we will briefly illustrate students engaging in inquiry by looking at an example of how they might find solutions to the question: "Where does the black stuff on the sand come from?" Students might refine this question by asking additional questions, such as "Did the fish leave the black stuff?" "Is the black stuff alive?" "Does it grow?" "Did the black stuff grow because the aquarium was in the sun near the window?" and "Would we still have the black stuff if we added a filter to the aquarium?" Students can look for and find information about aquarium maintenance in books and magazines in the

CONNECTING TO NATIONAL
SCIENCE EDUCATION STANDARDS

PRINCIPLES AND DEFINITIONS

Inquiry is a multifaceted activity that involves making observations; posing questions; examining books and other sources of information to see what is already known; planning investigations; reviewing what is already known in light of experimental evidence; using tools to gather, analyze, and interpret data; proposing answers, explanations, and predictions; and communicating the results (p. 23).

school library, on the World Wide Web, or in pet stores. One group of students might decide to test the idea of whether the black stuff is caused by placing the aquarium in the sun. They might set up several different aquariums in different locations in the room, collect data about the growth of the black stuff, analyze the data, and make a conclusion. These students might communicate their findings by creating a newsletter that tells owners of aquariums what to do to limit the growth of the black stuff in the bottom of their aquariums.

Collaborating to Find Solutions

As we will discuss in Chapter 2, learning occurs in a social context. Project-based science involves students, teachers, and members of society collaborating to investigate questions. In Chapter 7, we discuss in detail how teachers can develop and implement **collaboration** in the classroom. In this manner, the classroom becomes a community of learners. Students collaborate with others in their classroom and with their teacher to form conclusions, make sense of information, and present findings. The use of World Wide Web, chat rooms, and e-mail allows students access to a wider community in which they can communicate with knowledgeable individuals, take advantage of more extensive resources, communicate with other students in different parts of the world, and share data with other student scientists and professional scientists.

CONNECTING TO NATIONAL SCIENCE EDUCATION STANDARDS

SCIENCE TEACHING STANDARDS

Teachers of science guide and facilitate learning. In doing this, teachers orchestrate discourse among students about scientific ideas (p. 32).

For instance, students investigating the black stuff in the bottom of the aquarium might interview a local pet store owner. Students investigating the needs of pets might collaborate with experts from a local pet store or veterinary hospital to find answers to their problem. In each situation, students can use the World Wide Web to obtain additional information, and they can use electronic communications to describe their projects to others.

Using Technology Tools to Learn Science

Technology tools can help transform the science classroom into an environment in which learners actively construct knowledge (Linn, 1997; Novak & Krajcik, 2005; White & Fredrickson, 1995). We refer to technology tools that can support students in learning as **learning technologies**.

Using technology tools in project-based science makes the environment more authentic and relevant to students. The students can use technology to access *real* data on the Internet, expand interaction and collaboration with others via networks (such as e-mail), use tools to gather data (such as light and motion probes that are plugged into computer ports), employ graphing and visualization tools to analyze data, and produce multimedia artifacts. These are activities that they see and read about professionals doing. In addition, the multimodal and multimedia capabilities of technology make information more accessible, not only physically (by providing easy access to information), but intellectually (by helping students incorporate new information into their understanding) as well (Krajcik, Blumenfeld, Marx, & Soloway, 2000).

For example, students actively "construct" knowledge when they use technology, such as a computer-based pH probe to gather data about the pH of the aquarium with the black stuff at the bottom. Real-time graphing (simultaneous graphing with the pH probe) makes information intellectually accessible because it allows students to make immediate interpretations of their data. Another example involves the study of weather. Students can study how temperature fluctuates during the day and night by taking continuous temperature readings with an electronic temperature probe. They can download satellite weather maps from the World Wide Web and then predict the weather just as meteorologists do. In these ways, technology makes weather information physically accessible. When children use simple draw programs to create pictures to represent their ideas, technology has helped make their ideas intellectually accessible. By sharing data with others, learners make their information physically accessible to others.

Technology should be used as a tool to support science teaching. Chapter 6 explores how learning technologies support students in inquiry. Where appropriate, we have integrated discussion of its importance throughout all of the chapters of this book.

Creation of Artifacts

Because **artifacts** or products show what students have learned, teachers can use artifacts to assess students' understanding of science (Krajcik & Blumenfeld, 2006). Project-based science results in students creating a series of artifacts that address the driving question and show what children have learned. Often, teachers have students share their artifacts with other class members and with teachers, parents, and members of the community. We refer to the products that students construct as artifacts because, like historical artifacts, they serve as objects and records of students engaging in science.

The creation and sharing of artifacts serves several purposes. First, artifacts are real and motivating. For example, making a display of appropriate habitats for classroom pets is more enjoyable and meaningful and, therefore, more motivating than taking a test about animal habitats. The creation and sharing of artifacts also more closely approximates the activities of "real" science. Scientists frequently expose their ideas to public scrutiny through the processes of publishing and presenting their work at conferences. Presenting an artifact to an audience of peers, professionals, and community members provides an outcome for the investigation and gives students an opportunity to talk with others about their work.

Second, artifacts help students develop and represent understanding. Because artifacts (such as physical models, reports, videotapes, and computer programs) are

CONNECTING TO
NATIONAL SCIENCE
EDUCATION STANDARDS

SCIENCE TEACHING
STANDARDS

Teachers of science engage in ongoing assessment of their teaching and of student learning (p. 37).

concrete and explicit, they can be shared and critiqued. Such feedback permits learners to reflect, extend their understanding, and revise their artifacts.

Third, artifacts allow students to show what they have learned throughout an investigation, and they document broad learning—sometimes over an entire school year. Because artifacts show learning over time, they track how student understanding develops. For these reasons, artifacts are excellent forms of assessment.

While studying the question "What kind of insects live on our playground?" students could construct maps of where on the playground they found various insects. Students could then compare their maps with the maps of other students in the class (making the investigation real and motivating). By comparing and contrasting their maps, students might construct new knowledge. They might discover that the playground provides several different types of habitats for insects (sandy area, grassy area, wooded area) and that different insects live in different habitats. Finally, students in the class could study these habitats throughout the school year to document changes in insect populations during different seasons, thereby providing a measure of learning over time. We discuss artifacts more thoroughly in Chapters 9 and 10.

The Nature of Science and Its Relationship to Project-Based Science

What do you think **science** is all about? Don't be surprised if you have difficulty answering this question. It is never easy to describe what science is. However, to teach science to children, it is important to develop some understanding of this question. More thorough descriptions of science can be found in the works of Kuhn (1962) and Phillips (1987), but we will spend some time exploring here the question "What is science?"

CONNECTING TO
NATIONAL SCIENCE
EDUCATION
STANDARDS

PRINCIPLES AND
DEFINITIONS

The goal of science is to understand the natural world (p. 24).

Theories, Models, and Principles

Humans created science to predict, explain, and understand events and phenomena in the natural world. These predictions and explanations are dependent upon the ideas or, more formally, the theories and models that scientists have developed that are consistent with observations. **Theories** represent detailed explanations of how the world works. However, they are not the "real world." Often people confuse the theories, models, and principles of science as the phenomena. **Scientific models,** however, are ideas that are consistent with observations that provide an explanation for phenomena. For instance, the theory of plate tectonics gives us a detailed explanation of the origin of the continents. The fact that this theory has also helped us explain other related phenomena, like earthquakes and volcanic activity, enhances its usefulness and viability as a theory.

Theories provide predictive and explanatory power, but when theories can no longer explain and make predictions, humans create new theories to replace the old ones. The new theory explains everything the old theory does, but also accounts for observations that the old theories could not explain. For instance, chemists once thought that atoms were small indivisible spheres. However, this model of the atom could not explain all observable data related to how atoms behaved; nor could it help us explain how various atoms reacted with different atoms. Chemists replaced the indivisible atom with an atom that had various components. Theory construction and the replacement of old theories with new ones also illustrate the tentative nature and the dynamic, recursive process of science.

Theories influence how we see and interpret (or make sense of) data and the world around us. To understand how our world views influence what we see and how we interpret data, let's examine a theory that influenced interpretations for many years. In 1817, William Bucklund found a giant pointed tooth that resembled the smaller tooth of a modern-day lizard. He built a theory that the tooth came from a giant lizard. Later, other researchers found giant teeth that were flat. In 1841, Richard Owen put together a theory that the pointed and flat teeth came from giant animals that were extinct, and he named these extinct animals *dinosaurs*, which means "terrible lizards." Because theories shape the way we see the world, for many years scientists believed dinosaurs were giant lizards or cold-blooded reptiles. Scientists believed that dinosaurs buried their eggs and left them to hatch, much like many reptiles today lay eggs in the sand and leave them to hatch. Bob Bakker, a paleontologist, developed a theory that dinosaurs, because of their large rib cages and huge chest area, had very large hearts—and most modern animals with large hearts are warm blooded. Further, he suggested that the large hip sockets and thighs were characteristic of fast-moving animals and similar to the hips and thighs of modern chickens or turkeys. Jack Horner, another paleontologist, found evidence that dinosaurs did not lay eggs and leave them to hatch like reptiles. Instead, they tended to their eggs and reared their young in families (Czerniak, 1995a, 1995b). As you can see, as humans gained more evidence about dinosaurs, our theories changed. Our current theories about dinosaurs are also tentative, and it is quite likely that these too will change some day.

Based on questions that scientists ask about real-world problems, they use theories to form **hypotheses** (best guesses). Scientists test these hypotheses by collecting data, analyzing data, making conclusions, and communicating findings because, to be classified as scientific, the observations, measurements, and conclusions made by one group of scientists must be verified by others.

The understanding that results from science is tentative and changes when new observations cannot be explained by current theory, and it is dependent upon the agreement of other scientists. Science allows us to revise ideas, gather more data, or change predictions and test again. In this sense, science is a dynamic, recursive process that results in tentative findings that help explain the way the world works.

Scientific Practices

When scientists engage in inquiry, they perform a series of practices to explore the natural world. These **scientific practices** involve asking and refining questions, finding information, planning and designing, building the apparatus and collecting data, constructing models, analyzing data, constructing explanation, making conclusions, finding solutions, and communicating findings. In some respects, we employ these practices in our daily lives. Trying a variation of a recipe in making bread, for example, is an investigation to help make better bread. We might change the ratio of whole wheat to white flour to experiment with making the bread heavier. We might add more sugar to make it sweeter, or we might add more yeast to make if fluffier.

> **CONNECTING TO NATIONAL SCIENCE EDUCATION STANDARDS**
>
> **PRINCIPLES AND DEFINITIONS**
>
> Inquiry is a multifaceted activity that involves making observations; posing questions; examining books and other sources of information to see what is already known; planning investigations; reviewing what is already known in light of experimental evidence; using tools to gather, analyze, and interpret data; proposing answers, explanations, and predictions; and communicating the results (p. 23).

How many times have we groaned at the results when we realized we changed two or more ingredients (variables) at the same time and couldn't figure out which one (or ones) caused the different results, either bad or good? Inevitably, the confounding results taught us to be more careful and use more scientific ways of testing the recipe so that it could be replicated the next time we made bread.

Why have we introduced these ideas about the nature of science? Like scientists, students in science classes ask questions and try to find solutions that will help them explain their world. Students engage in inquiry similar to what scientists do. Students find solutions to questions by exploring phenomena, asking and refining questions, finding information, planning and designing, building apparatus, collecting data, analyzing data, constructing explanations, making conclusions, and communicating findings. Students create artifacts and share these with members of a learning community; likewise, scientists share their research findings with others. Students and scientists alike generate new ideas and questions as a result of their investigations and communicate their ideas with others. Finally, students develop creativity, open-mindedness, and imagination, qualities essential for successful scientists. As such, project-based science is a teaching approach that develops learning environments that reflect the nature of science.

Reasons Young Learners Should Study Science

As we watch children play, we realize that they are imaginative tool designers and theory builders. For example, children turn cardboard boxes into space ships or sets of kitchen utensils into medical equipment as they play. Just as scientists construct theories, children construct their view of reality. For example, a child watching a tree's leaves blowing in the wind might construct his or her own belief that trees make wind. Although this is not a scientifically accepted explanation of the phenomenon, the process used to generate the theory is not that far from the processes used by scientists.

Helping young children learn science can be one of the most enjoyable experiences in teaching. Watching children develop skills and learn principles and facilitating that development and learning can be empowering experiences for teachers. Students have a natural inquisitiveness that generates many questions, such as "Why do pumpkins decompose?" "What happens to all the garbage?" "How does my electrical train work?" and "Why does Ming Na run faster than Caitlin?" This is what project-based science is all about—helping students explore solutions to questions they find of value.

Most educators agree that science is an essential and crucial subject for all students. The elementary and middle school grades are especially important years for students to construct understanding about the world and develop interest in science so that they will pursue scientific studies as they continue into upper grades where science may become optional (NRC, 1996).

Science teaching also receives a great deal of attention in the national media. It is not uncommon to read about how this generation's science achievement compares with that of a previous generation or with those of other countries. The media continually review and examine state achievement scores and goals for science education. Why has the teaching of science in elementary and middle schools received such substantial attention? Why is it important for children to learn science? Take a few minutes to reflect on the activities you performed today. In Learning Activity 1.2, you will think about why it is important for students to learn science.

Learning Activity 1.2

WHY SHOULD CHILDREN LEARN SCIENCE?

Materials Needed:

- A sheet of paper
- Something to write with
  A. On a sheet of paper, make three columns. Label the first column "Things I Did Today," label the second column "Related Science Concept," and label the third column "Related Questions." Take about 5 minutes to reconstruct what you did since you got up this morning. List as many things as you can remember in the first column.
  B. In the second column, identify as many related science concepts as you can. For example, you may have used a curling iron and hair spray to style your hair. The curling iron is heated with electricity, and the hair spray is a chemical, so the related concepts are electricity and chemicals.
  C. In the third column, write down related questions for each situation, such as "How can I make hair spray work better?" and "Where does electricity come from?"
  D. Reflect upon the list you have constructed. How does it help answer the question "Why should children learn science?" Record your thoughts and your lists.

Science Affects Every Aspect of Our Lives

Science (or technology that results from the efforts of science) affects every aspect of our lives in the workplace, at home, at school, in transportation, and in entertainment. Young children are especially curious and interested in understanding phenomena in their lives. They ask a lot of "Why?" and "What if?" questions. Much of what they are curious about is related to science.

Like you, a child may awake to the sound of an alarm clock. The clock is operated by electricity, and the sound reaches her ears through vibrating air molecules. She showers using water heated by gas and uses soap and shampoo, which are chemicals. She eats cereal or bread that is processed from plants, fortified with vitamins, packaged with technology, and delivered to stores through other technology. Her clothes are made from cotton (a plant) and acrylic (a synthetic fabric). The morning newspaper she reads is made from trees. Headphones bring radio waves to her ears. The music is created by vibrations. Her school bus runs on gasoline burned in an internal combustion engine. The brakes on the school bus cause it to stop through friction. At school, the child uses computers and CD-ROMs; and at home she plays with computer toys, she watches DVDs, and she listens to CDs—all inventions and technology that resulted from basic scientific discoveries about electricity, light, and magnetism. Her day, like yours, contains hundreds of events that are related to science. Each of these situations gives rise to questions that students could ask and investigate: "What chemicals are in our homes?" "What kinds of foods promote better health?" "How does technology improve our lives?" Understanding these questions requires students to develop an understanding of important underlying ideas in science.

Because science clearly affects every aspect of our lives, we need a basic understanding of science in order to understand our lives and explain the world. There are also many other reasons for students to learn science. It helps them acquire knowledge and skills that will be useful throughout their lives; it teaches them to think critically, solve problems, and make decisions that can improve the quality of their lives; it develops attitudes, such as curiosity or sensitivity to environmental concerns, that foster taking responsibility for one's actions; and it guides students in understanding real-life issues and participating in a global society—the hallmark of scientifically literate citizens. Finally, some students will be encouraged by their studies in the elementary and middle level to pursue science studies in high school grades, when the study of science is often optional, and in postsecondary education.

Students Acquire Useful Knowledge, Skills, and Attitudes

By studying science, children acquire knowledge, skills, and attitudes that will be useful to them throughout their lives as they engage in such activities as choosing lifestyle habits related to food and exercise, conducting everyday activities that affect the environment, making informed voting decisions, and solving everyday problems. For example, knowledge about the systems of the human body and nutrition provides the basis for making informed decisions about the food one eats. Skills, such as comparing and contrasting, are involved in selecting one food over another. Attitudes, such as curiosity, lead one to seek more information about nutrition and exercise in journals or on the World Wide Web.

Poor instruction and shallow development of science concepts in elementary and middle school grades are often at fault for persistent, inaccurate beliefs about scientific phenomena. Students come to school with their own ideas about how things in the natural world work, ideas that are not always consistent with the conceptions of experts in the field. For instance, children often believe that a force is needed to keep an object in motion or that mass is lost when substances are burned. Both of these ideas make intuitive sense and are consistent with casual observations and life experiences. For instance, when driving a car, the car doesn't keep moving unless we keep our foot on the gas pedal. Typically, we need to apply a force to keep it moving because of the opposing force of friction. When we burn a log in a fireplace, we see that only ashes remain, and it appears that other mass is gone. Therefore, we find that many adults, in spite of school instruction in science, still hold ideas that differ from those of scientists (Presidents and Fellows of Harvard College, 1995).

Science Teaches Critical Thinking, Problem Solving, and Decision Making

In science, students learn to think critically, solve problems, and make decisions. For example, students investigating the black stuff at the bottom of the aquarium critically analyze and review possible causes of the growth. They learn to solve the problem by investigating different variables, such as the amount of light reaching the aquarium, the pH of the water, the filtration systems, and the types of gravel. Finally, they make decisions regarding the most effective procedures for maintaining the aquarium. This science investigation, of course, is related to real life, because some students also have aquariums in their homes and many students are interested in animals. More important, however, through this study students learn skills they can apply to other situations. For example, the next time they encounter an everyday problem, such as trying to figure out why certain plants are not growing well in the garden or why a toy doesn't work properly, they will have acquired the skills of analysis, review, investigation, and making conclusions.

Science Helps Students Take Responsibility for Their Actions

As students investigate many scientific questions, they acquire knowledge and develop attitudes that encourage them to take responsibility for their own actions. For example, if students learn in an investigation that motor oil kills plants, their future behavior regarding the disposal of motor oil will be affected.

Science Develops Scientifically Literate Citizens

Science guides students in understanding real-life issues and helps them participate in local, national, and world issues—the hallmarks of scientifically literate citizens. In this new millennium, we find that now, more than at any other time in our history, students face crucial decisions about global issues. These include environmental pollution, global warming, AIDS, overpopulation, world hunger, nuclear power, and genetic engineering. Science and engineering have generated solutions to some of these issues: cars with better fuel mileage, more fuel-efficient homes, and great quantities of more nutritious foods. Scientific investigation, invention, and technology have also resulted in discoveries that make life easier for humans, provide cures for diseases, and inspire exploration of new frontiers, such as the ocean or space. We need professional scientists to investigate these matters, but we also need a scientifically literate

> ## CONNECTING TO NATIONAL SCIENCE EDUCATION STANDARDS
> ### INTRODUCTION
> The goals for school science that underlie the *National Science Education Standards* are to educate students who are able to experience the richness and excitement of knowing about and understanding the natural world; use appropriate scientific processes and principles in making personal decisions; engage intelligently in public discourse and debate about matters of scientific and technological concern; and increase their economic productivity through the use of the knowledge, understanding, and skills of scientifically literate people in their careers (p. 13).

citizenry who can participate in making informed decisions about them. Today's students, more than those of any previous generation, will need to understand basic scientific principles.

Students in project-based science classrooms learn to investigate, understand, and interpret information and interface with people in their own communities. As a result, they are better prepared to deal with the scientific, technological, and social issues that will face them in the future.

Science Develops Curiosity and Motivation to Explore the World

Most educators agree that science is an essential and crucial subject for all students. The elementary and middle school grades are especially important years for students to construct an understanding about the world and develop interest and curiosity in science so that they will pursue scientific studies as they continue into upper grades where science may become optional (NRC, 1996).

The elementary and middle grades are also important years for developing curiosity and interest in science. Research by Yager and Yager (1985) has shown that negative attitudes toward science increase by grade level; and students, especially girls, as early as grade 3 exhibit dislike of and anxiety toward science. The Third International Mathematics and Science Study (Schmidt et al., 2001; TIMSS, 1997, 1998) found similar patterns in students' attitudes toward science. Eighty-five percent of 4th graders indicated that they liked science. However, by the time students reached the 12th grade, their favorable attitudes toward science had decreased. The percentage of students who liked science ranged from 49% to 68%, depending upon the discipline, with chemistry on the low end and biology on the high end. The reason students gradually come to dislike science frequently lies in the way science is taught (passively) and the type of curriculum used (not relevant to students' lives). As a teacher, you will be in the position to stimulate student interest in science or extinguish it.

Goals of Science Education

In this section, we examine the **goals** of science education in the United States, and we consider why project-based science enables you to teach in a way that meets those goals. This is an exciting and challenging time in science education. Many parents, educators, scientists, and public officials are concerned that students will be unable to cope with the massive explosion of knowledge and the tremendous changes that science and technology will bring about in our future. More and more, we are concerned about the ability of our students to develop an understanding of science that they can apply to real life and that will allow them to take part in society as productive citizens. We all realize that the understandings students develop in school will impact what they are able to do throughout their lives, regardless of their chosen careers. It is not surprising that one fundamental national goal of science education today is to develop a scientifically literate society. Tying national goals to our own classroom goals can be helpful in guiding our students to become scientifically literate citizens. Learning Activity 1.3 will help you think abut the goals you have for science teaching.

Learning Activity 1.3

WHAT ARE YOUR PERSONAL GOALS FOR SCIENCE EDUCATION?

Materials Needed:

- A sheet of paper
- Something to write with
  A. Before reading the section on the goals of science education, brainstorm in small groups the goals that you think are important for science education. Try to reach group consensus. Share your group's goals with the rest of the class.
  B. How do the goals of science education differ among the groups? What are the similarities?
  C. Think about one national goal already mentioned—the development of scientifically literate citizens. What do you think characterizes a scientifically literate citizen? What does a scientifically literate citizen need to know? What skills does this person need to have? What attitudes should this person possess? How does this goal compare with your group's goals?
  D. Record your thoughts.

Below, we examine and summarize several reform documents that have established the standards for science teaching and learning. Specifically, we summarize the *National Science Education Standards* (National Research Council, 1996), *Project 2061. Science for All Americans* (Rutherford & Ahlgren, 1989), and *Benchmarks for Science Literacy* (AAAS, 1993.)

The National Research Council, with the assistance of the National Academy of Sciences, developed the *National Science Education Standards* (National Research Council, 1996). Subsequent publications (National Research Council, 2000, 2001) explore the national science standards with respect to inquiry and assessment.

> CONNECTING TO
> NATIONAL SCIENCE
> EDUCATION STANDARDS
>
> OVERVIEW
>
> The intent of the *Standards* can be expressed in a single phrase: establish science standards for all students. The phrase embodies both excellence and equity. The *Standards* apply to all students, regardless of age, gender, cultural or ethnic background, disability, aspiration, or interest and motivation in science (p. 2).

The goal of the *National Science Education Standards* is to educate *all* students

> to experience the richness and excitement of knowing about and understanding the natural world; use appropriate scientific processes and principles in making personal decisions; engage intelligently in public discourse and debate about matters of scientific and technological concern; and increase their economic productivity through the use of the knowledge, understanding, and skills of the scientifically literate person in their careers. (p. 13)

To accomplish these goals, the *National Science Education Standards* simultaneously addresses standards for teaching science, the professional development of teachers of science, assessment in science education, K–12 science content, science education programs, and the science education system. These standards represent a consensus of teachers, science educators, scientists, and the public.

The *National Science Education Standards* stresses that science should be inquiry based and adapted to meet the interests, abilities, and experiences of students. It emphasizes that science teachers should use strategies that develop science understanding through a community of learners; use resources outside the school that support inquiry, guide and facilitate learning by promoting collaboration and discourse among students, help students become responsible for their own learning; and work with colleagues within science (biology, chemistry, geology, and physical science) and across disciplines (mathematics, language arts, social studies, art, music, and physical education). Throughout this book, we cite these standards as they apply to topics we are introducing in the chapters.

The American Association for the Advancement of Science (AAAS) published three important documents that provide guidance to science education: *Science for All Americans: Project 2061* (Rutherford & Ahlgren, 1989), *Benchmarks for Science Literacy* (AAAS, 1993), and *Atlas of Science Literacy* (AAAS, 2001). The purpose for these publications is to promote achievment in science literacy nationwide. *Science for All Americans* lays out the important ideas of five panel reports (a) biology and health science; (b) mathematics; (c) technology; (d) physical science, information sciences, engineering; and (e) social and behavioral sciences. The ideas found in this seminal publication represent a consensus document from scientists in different fields of research. *Benchmarks for Science Literacy* shows the various benchmarks[2] that learners should attain at different grade bands: K–2, 3–5, 6–8, and 9–12. It is designed to serve as a curriculum guide for curriculum developers, state departments of education, and school systems developing science curriculum. The benchmarks are not instructions on how to teach; they are statements about what *all* students should know or be able to do in science, mathematics, and technology at grades 2, 5, 8, and 12. A unique aspect of the benchmarks is that they become more complex as the grades progress and show how a concept becomes elaborated throughout the grades. The *Atlas of Science Literacy* (AAAS, 2001) illustrates in graphical form how various benchmarks develop over time and how they are related to each other. The *Atlas* shows how the ideas that students learn in one grade depend on and support what they will learn at another grade level.

The publications from AAAS focus on students' developing meaningful learning of science, rather than on sheer coverage of numerous science topics. In order to ensure scientific literacy, the amount of material covered throughout the academic year needs to be reduced. All children, not only those who have traditionally been successful in science, learn more thoroughly when fewer topics are covered in greater depth. Instead of reading vast amounts of material, students need to think deeply about central concepts, themes, and principles of science. Because the boundaries of science are much grayer than they are portrayed in the science textbooks, the AAAS publications emphasize the integration of and connections among the various science fields, as well as with social science, mathematics, language arts, and technology.

Summary of National Goals

The publications from NRC and AAAS have influenced the character of science education in K–12 schools since their publication and will continue to influence science education for years to come. These efforts have resulted from the work of some of the best thinkers in science, science education, teaching, and learning. What these people have advocated for science education is consistent with how children learn. What follows is a summary of the common themes of these recommendations. We can cluster these themes under three broad categories: learning, curriculum, and teaching. Although this summary is a good start in understanding the standards, we encourage you to read the original publications because of their impact on the field of science teaching and learning.

Learning

With respect to learning, six major themes are apparent:

1. Students should explore broad concepts or "big ideas" instead of isolated facts or skills. Because of the rate of knowledge growth, no individual knows all there is

to know and never will. For this reason, teachers should focus on broad topics. For instance, students should explore ecosystems instead of isolated topics, such as rivers, plants, and animals. In this way students learn the connections between concepts and principles and are able to apply their understanding to as yet unencountered situations.

2. *All* students should learn to think critically, solve problems, and make decisions. Science is for all students, not just those embarking on scientific careers. We live in a world that is scientifically and technologically based. As our children develop into adults, the world will continue to change as a result of scientific and technological decisions. To make informed decisions regarding their own lives and society, children need to have a firm understanding of both the content and process of science so that they may, for example, one day be able to make informed decisions regarding the management of natural resources.

3. Children should construct meaning from experiences with concrete materials, rather than passively. Literature on child development is clear about the fact that children develop understanding by cognitively engaging in the exploration of phenomena. While some children learn about plants simply by reading about them, most children also need to grow plants and observe their development to construct deep understandings.

4. Students should learn how to apply science and technology to everyday life. Science should be relevant to students' lives and not consist solely of learning concepts, but also of their applications. Students who study decomposition of matter, for example, might also be asked to create school and home composting systems.

5. Science should foster the development of students' natural curiosity, creativity, and interest. For example, if students are interested in magnets, but their textbook only covers matter and energy, the teacher should take advantage of the students' interest in magnetism and find ways to link the students' interest to curriculum goals.

6. Science instruction should foster the development of scientific attitudes. Students should learn to seek out knowledge, be skeptical, rely on data, accept ambiguity, be willing to modify explanations, cooperate in answering questions and solving problems, respect reason, and be honest. For instance, students typically believe that what they read in a book, hear on television, or read in the newspaper is true. Science instruction can teach students to question what they read and see. Additionally, elementary and middle grades science should aim to promote positive attitudes about science.

Curriculum

With respect to curriculum, three major themes come to the forefront:

1. Less content should be covered. Students should be allowed to discover and learn in depth a few major concepts and principles that will allow learners to explain a broad spectrum of phenomena. We like to refer to these central concepts and principles of science as the "big ideas" (National Research Council, 2005; Smith, Wiser, Anderson, & Krajcik, 2006). Instead of covering weather, the human body, magnetism, electricity, chemical change, and the solar system in 1 year, students might explore only three or four of these major areas.

2. Science should be portrayed to students as interdisciplinary, connected to other fields of study (Czerniak, 2007). In actual practice, science areas are interwoven. Paul DeHart Hurd, in an article entitled "Why We Must Transform Science" (1991), wrote,

> Science today is characterized by some 25,000 to 30,000 research fields. Findings from these fields are reported in 70,000 journals, 29,000 of which are new since 1978. Traditional disciplines have been hybridized into such new research areas as biochemistry, biophysics, geochemistry, and genetic engineering. . . . These changes in the way modern science is organized have yet to be reflected in science courses. There is little recognition that in recent years the boundaries between the various natural sciences have become more and more blurred and major concepts more unified. (p. 33)

Topics should be connected. For example, in exploring a question such as "What is in our stream?" students study such basic chemical concepts as concentration and pH, as well as perform basic chemical tests for various substances. They explore biology concepts, by examining living organisms found in the water. They study earth science concepts, such as the water cycle and the watershed. All of these areas are studied in an integrated and unified manner, rather than as isolated facts.

3. Students should explore the interrelationships among science, technology, and society. The study of science should highlight the integration of technology with societal issues. For example, in investigating "What is in our lake?" students could explore how our waterways get polluted. They might also explore how our society makes decisions and passes laws that govern many of the actions taken by individuals and industries.

4. Curriculum materials need to be built that help students develop science concepts and principles across the grade levels. As described in the *Atlas of Scientific Literacy* (AAAS, 2001), curriculum documents need to show the "rich fabric of mutually supporting ideas and skills" that students need to develop over time. Too often in science education, students move from one grade level or even within a grade level without ideas systematically building on one another.

Teaching

With respect to teaching, five major themes are apparent:

1. The teacher serves as a guide in the classroom, encouraging student exploration and learning, rather than as an authoritative presenter of knowledge. For instance, instead of lecturing about decomposition and bacteria, a teacher might have students explore why pumpkins decompose by setting up various conditions to investigate decomposition. Students can set up a compost pile at the school, build a worm bin for their classroom, or create a decomposition column. In these situations, the teacher serves as a facilitator, setting up the situations, but the students conduct the actual investigations.

2. The content of science should be taught through the practices of science, involving investigation and answering questions. For example, in the process of exploring why pumpkins decompose, students would ask questions, design plans, collect and analyze data, and make decisions. Assessment should be consistent with these

practices. In other words, it should measure what is taught. Assessment should also typically be embedded into instruction, rather than conducted solely at the end of a chapter or unit.

3. Science instruction needs to be integrated with instruction in the other discipline areas. Teachers should focus on the relationships among science, language arts, social studies, and mathematics. For instance, a class exploring air pollution might write letters to legislators, create posters expressing their opinions, or make graphs.

4. Science instruction should encourage students to challenge conceptions and debate ideas. In this process, they form communities of learners who collaborate together. For example, students might work together to challenge each other's ideas about which factors contribute most to the decomposition of leaves.

5. Science instruction should build upon children's prior experiences and knowledge. For instance, if a child believes that air does not have mass or take up space, instruction should be designed to foster a new understanding. Many ideas that children hold result from their experiences while playing. For instance, a student might develop the idea that gases do not have mass from playing with helium balloons.

National Goals and Project-Based Science

Now that you have been introduced to the major national goals, you might wonder how they match with teaching science in the project-based approach. Project-based science is congruent with current goals as represented in the *National Science Educa- tion Standards* (National Research Council, 1996, 2000, 2001), *Project 2061: Science for All Americans* (Rutherford & Ahlgren, 1989), and *Benchmarks for Science Literacy* (AAAS, 1993). In fact, project-based science is considered one of the most effective approaches for better schools (www.edutopia.org/bigideas).

First, the national goals stress teaching less content at each grade level. Project- based science is consistent with this emphasis, because its focus is on covering less content in greater depth. Students *investigate authentic questions* that encompass central concepts over an *extended period of time*. The process of *asking and modifying questions, performing investigations,* and *building artifacts* to answer questions might take place over weeks or months. For instance, in order for students to discover what bugs they have in their school yard and when the bugs disappear and reappear, they will have to conduct investigations throughout the entire school year. By covering broader, "big ideas" instead of isolated facts, students in project-based classrooms develop richer, fundamental understandings of major themes or issues. Project-based science takes more time to teach than covering many topics superficially; this method helps teachers use the time they have to teach deeply so students know more.

Second, all of the major national goals stress that science should be relevant to students' daily lives and that students should apply their understanding to the real world. Project-based science's driving questions are important, meaningful, and worthwhile. Students apply the solutions they develop to their own life situations. Students' explaining the black stuff in the bottom of the aquarium learn how to set up and maintain aquariums at home. They learn to make informed decisions regarding pollution in an aquarium.

CONNECTING TO NATIONAL SCIENCE EDUCATION STANDARDS

SCIENCE TEACHING STANDARDS

Teachers of science plan an inquiry-based science program for their students (p. 30). Teachers focus inquiry predominantly on real phenomena, in classrooms, outdoors, or in laboratory settings, where students are given investigations or guided toward fashioning investigations that are demanding but within their capabilities (p. 31).

Third, the national goals call for the integration of science throughout the curriculum, within the sciences and across all disciplines. Project-based science promotes this interdisciplinary approach to instruction because, in the investigation of real-world questions, science cannot be separated from other subject areas. When students explore the quality of water, they combine a number of science disciplines, including chemistry (chemicals in the water), biology (effect on plant and animal life), and earth science (polluted water seeping into groundwater supplies below rock layers). Opportunities to connect with other curricular areas also exist. Students might explore public policy governing the pollution of waterways by private industry (social studies). They might create three-dimensional models of their local watersheds (art).

Fourth, the national goals stress that the teacher should serve as a guide to the instruction that takes place in the classroom. In project-based science, the teacher's role is not to impart knowledge to passive learners, but to guide students through the processes of inquiry: modifying driving questions, developing investigations, engaging in explorations, collaborating with others, and creating artifacts. The teacher *facilitates* learning by teaching lessons that focus on big ideas or important scientific practices, helping students find resources, putting students in touch with members of the community, fostering collaboration, and asking questions that lead to new investigations.

Fifth, national recommendations call for classrooms composed of communities of learners. In project-based science classrooms, students are encouraged to collaborate with other students and with members of the community, debate ideas, challenge the thinking of peers, share ideas, and communicate with people around the world.

Sixth, a major focus of the national goals is on the way students learn: Prior experiences should be taken into account and hands-on materials should be used. In project-based science classrooms, teachers identify students' prior knowledge of a concept and use this information to guide lessons and assess learning. Students in project-based science classrooms construct meaning by exploring phenomena. They don't just read about water pollution; they use such materials as Secchi disks, pH meters, and water test kits to investigate water quality.

Seventh, the national goals argue that science should be learned through inquiry. Inquiry is the core of project-based science because students do various investigations to find solutions to their questions. Investigations provide students with the opportunity to ask questions, explore and initiate ideas, plan, seek information, construct designs, collect and interpret data, make inferences, reevaluate their understandings, and construct useful connections among real-life ideas. Students, investigating the cause of the black stuff at the bottom of the aquarium, ask and refine questions about variables related to the growth of the black stuff, design experiments to identify what causes the growth, collect and analyze data about the amount of black stuff, construct explanations, and make decisions about aquarium maintenance.

Eighth, an essential element of the national goals is building interest in science. Meaningful questions are a fundamental part of project-based science as they capitalize upon students' natural curiosity. Students' questions drive the curriculum, leading to investigation of topics that are meaningful and interesting to the students. Because project-based science builds on students' interests, it helps meet the many needs of a diverse classroom.

Ninth, national goals stress the development of scientific attitudes and habits of mind, such as accepting ambiguity, being skeptical, and respecting reason. Project-based science supports the development of habits of mind through the processes of investigation (asking questions, developing experiments, collaborating with others, making conclusions, constructing explanations, and producing artifacts).

Tenth, each of the national reports calls for new methods of assessing student understanding in science that are fair, are reliable, and match instructional goals. Project-based science stresses assessment methods that are embedded in the instructional process. Student-developed artifacts, for example, represent students' understandings of the topic they have investigated. For instance, while investigating what insects live on the playground, students might make drawings of insect life cycles. Such activity is developed during the process of investigation, but it also serves as a tool of assessment because it can tell the teacher whether students understand the concept of the investigation.

You probably have many questions at this point about teaching project-based science and how you implement the national goals in your teaching. Throughout this book, we will explore in greater depth many of the ideas that were introduced in this chapter.

Chapter Summary

Project-based science represents an exciting way to teach science. In project-based science classrooms, students investigate and collaborate with others to find solutions to real-world questions. Using technology, students investigate, develop artifacts, collaborate, and make products to show what they have learned. This method of teaching science motivates young learner to learn and explore, and it meets the national goals for and standards of science education. Because project-based science parallels what scientists do, it represents the essence of inquiry and the nature of science. Throughout the remaining chapters of this book, you will encounter more in-depth information about the fundamental features of project-based science and strategies for implementing the approach. In the process, we hope you see how enjoyable teaching science can be.

Chapter Highlights

- Project-based science situates the learning of science in questions that children find meaningful and relevant to their lives.
- Project-based science is sensitive to the varied needs of diverse students with respect to culture, race, and gender.
- Driving questions establish the relevance of learning science to the children and serve to organize and guide instructional tasks and activities.
- Students engage in investigations to answer their questions.
- Communities of students, teachers, and others collaborate on the questions or problems.
- Students use technology to investigate, develop artifacts, collaborate, access information, and actively construct knowledge.
- A series of artifacts or products document what students have learned.
- Scientists test these hypotheses by collecting data, analyzing data, making conclusions, constructing explanations, and communicating findings.
- To be classified as scientific, the observations, measurements, and conclusions made by one group of scientists must be verified by others.
- The understanding that results from science is tentative and changes with new observations.
- Science study helps students
  - Acquire knowledge and skills that will be useful throughout their lives.
  - Develop positive attitudes that foster responsibility for one's actions.
  - Understand real-life issues.
- The study of science helps develop scientifically literate citizens.
- The elementary and middle grades are important years for developing curiosity and interest in science.
- Science education goals advocate science literacy for *all* students and set forth recommendations about learning, curriculum, and teaching:
  - Students should explore broad concepts and principles or "big ideas" instead of isolated facts or skills.
  - *All* students should learn to think critically, solve problems, and make decisions.
  - Children should construct meaning from experiences with concrete materials.
  - Students should learn how to apply science and technology to everyday life.
  - Science should foster the development of students' natural curiosity, creativity, and interest.
  - Less content should be covered in the curriculum.
  - Science should be portrayed to students as interdisciplinary, connected to other fields of study.
  - The teacher serves as a guide in the classroom, encouraging student exploration and learning, rather than as an authoritative presenter of knowledge.
  - The content of science should be taught as scientific practices, involving investigation and answering questions.
  - Science instruction should encourage students to challenge and debate ideas.
  - Science instruction should build upon children's prior experiences and knowledge.
- Because project-based science parallels what scientists do, it represents the essence of inquiry.

Key Terms

Artifacts	Learning technologies
Collaboration	Project-based science
Driving question	Science
Goals	Scientific models
Hypotheses	Scientific practices
Inquiry	Subquestions
Investigations	Theories

Notes

1 All scenarios and names of people in scenarios throughout this book are fictitious. Any similarity to actual teachers, students, or schools is coincidental.
2 For clarity, we refer standards in the National Science Education Standards and Benchmarks as standards (lower case "s") and the National Science Education Standards as "Standards" (upper case "S").

References

American Association for the Advancement of Science. (1993). *Benchmarks for science literacy*. New York: Oxford University Press.

American Association for the Advancement of Science. (2001). *Atlas of science literacy*. Washington, DC: American Association for the Advancement of Science and National Science Teachers Association Press.

Atwater, M. M. (1994). Research on cultural diversity in the classroom. In D. L. Gabel (Ed.), *Handbook of research on science teaching and learning*. New York: Macmillan.

Blumenfeld, P., Soloway, E., Marx, R., Krajcik, J., Guzdial, M., & Palincsar, A. (1991). Motivating project-based learning: Sustaining the doing, supporting the learning. *Educational Psychologist, 26*, 369–398.

Czerniak, C. M. (1995a). Dinosaur! *School Science and Mathematics, 95*(3), 160–161.

Czerniak, C. M. (1995b). Dinosaurs: Fantastic creatures that ruled the earth. *School Science and Mathematics, 95*(3), 161–162.

Czerniak, C. M. (2007). Interdisciplinary science teaching. In S. K. Abell & N. G. Lederman (Eds.), *Handbook of research on science education*. Mahwah, NJ: Lawrence Erlbaum.

Dewey, J. (1991). *The school and society: The child and the curriculum*. In P. W. Jackson (Ed.), Centennial Publication Series, Chicago: University of Chicago Press.

Haberman, M. (1991). The pedagogy of poverty versus good teaching. *Phi Delta Kappa, 73*(4), 290–294.

Haberman, M. (1995). *Star teachers of children in poverty*. West Lafayette, IN: Kappa Delta Pi.

Hurd, P. D. (1991, October). Why we must transform science education. *Educational Leadership*, 33–35.

Krajcik, J., Blumenfeld, P., Marx, R., & Soloway, E. (2000). Instructional, curricular, and technological supports for inquiry in science classrooms. In J. Minstrell & E. Van Zee (Eds.), *Inquirying into inquiry: Science learning and teaching* (pp. 283–315). Washington, DC: American Association for the Advancement of Science Press.

Krajcik, J. S., & Blumenfeld, P. (2006). Project-based learning. In R. K. Sawyer (Ed.), *The Cambridge handbook of the learning sciences*. New York: Cambridge University Press.

Kuhn, T. S. (1962). *The structure of scientific revolutions*. Chicago: The University of Chicago Press.

Linn, M. C. (1997). Learning and instruction in science education: Taking advantage of technology. In D. Tobin & B. J. Fraser (Eds.), *International handbook of science education* (pp. 265–294). Dordrecht, The Netherlands: Kluwer.

Novak, A., & Krajcik, J. S. (2005). Using learning technologies to support inquiry in middle school science. In L. Flick & N. Lederman (Eds.), *Scientific inquiry and nature of science: Implications for teaching, learning, and teacher education*. Dordrecht, The Netherlands: Kluwer Publishers.

National Research Council. (1996). *National science education standards*. Washington, DC: National Academy Press.

National Research Council. (2000). *Inquiry and the national science education standards*. Washington, DC: National Academy Press.

National Research Council. (2001). *Classroom assessment and the national science education standards*. Washington, DC: National Academy Press.

National Research Council. (2005). Systems for state science assessments. In M. R. Wilson & M. W. Bertenthal (Eds.), *Committee on test design for K–12 science achievement*. Washington, DC: The National Academies Press.

Phillips, D. C. (1987). *Philosophy, science, and social inquiry*. Oxford: Pergamon Press.

Presidents and Fellows of Harvard College. (1995). *The private universe project*. South Burlington, VT: Annenburg/Corporation of Public Broadcasting Mathematics and Science Collection.

Rutherford, J., & Ahlgren, A. (1989). *Science for all Americans: Project 2061*. New York: Oxford University Press.

Schmidt, W. H., McKnight, C. C., Houang, R. T., Wang, H. C., Wiley, D. E., Cogan, L. S., & Wolfe, R. G. (2001). *Why schools matter: A cross-national comparison of curriculum and learning*. San Francisco: Jossey Bass.

Smith, C. L., Wiser, M., Anderson, C. W., Krajcik, J. (2006). Implications of research on children's learning for standards and assessment: A proposed learning progression for matter and the atomic molecular theory. *Measurement: Interdisciplinary Research and Perspectives, 14*(1&2), 1–98.

Stake, R. E., & Easley, J. A. (1978). *Case studies in science education*. Urbana, IL: Center for Instructional Research and Curriculum Evaluation, University of Illinois.

Third International Mathematics and Science Study (TIMSS). (1997).

Third International Mathematics and Science Study (TIMSS). (1998).

White, B. Y., & Fredrickson, J. R. (April 1995). *The Thinker Tools Inquiry Project: Making scientific inquiry accessible to students and teachers*. Causal Models Research Group Report 95-02. Berkeley, CA: School of Education, University of California.

Yager, R. E., & Yager, S. O. (1985). Changes in perceptions of science for third, seventh, and eleventh grade students. *Journal of Research in Science Teaching, 22*(4), 347–358.

Chapter 2

How Children Construct Understanding of Science

Introduction

Children face an increasingly scientific and technological world. Information and access to information are growing at an exponential rate. Thus, science and technology are needed increasingly in daily life. Additionally, many jobs will require students to understand science and technology. Because of the global economy that exists today, many believe that America's economic future is dependent upon a knowledge-based economy where individuals can use and reason from ideas. Thus, we need students who are well educated in science and technology and who can compete in a global marketplace (National Academies Press, 2005; Project Kaleidoscope, 2006). Schools, starting in very young grades, need to prepare our youth to learn and apply scientific knowledge to solve real-world problems. Do students in

Chapter Learning Performances

– *Explain the difference between inert and meaningful knowledge.*
– *Describe the three types of knowledge—content, procedural, and metacognitive.*
– *Critique examples of teaching to determine if they represent receptional or transformational approaches to teaching and learning.*
– *Explain what is meant by social constructivism and describe the various features of a social constructivist model of teaching.*
– *Critique lessons using a social constructivist model of teaching.*
– *Apply the idea of scaffolding to help children accomplish a difficult learning task.*
– *Discuss how authentic tasks help women and minorities participate and stay interested in science.*

our schools develop understandings that help them use science? What is a *usable understanding of science?* How can teachers help students develop usable science understanding?

This chapter examines the educational literature that anchors project-based science in the social constructivist theory of learning. We will examine types of knowledge and ways to help students develop integrated understandings through the construction of knowledge in social situations. Finally, we will explore a number of techniques to help teachers scaffold students' learning of science. First, however, let us develop a working definition of *understanding.*

Student Understanding

What kinds of science understandings do students develop in school? Does school help them develop understandings that are useful for their lives? Unfortunately, a number of research studies (AAAS, 1993; Linn, 1998; Osborne & Freyberg, 1986; Rutherford & Ahlgren, 1989) have indicated that students at the elementary, middle, and high school levels do not develop an understanding of science that is useful for their everyday lives. Most students memorize science terms without understanding, and they memorize how to solve problems (Eylon & Linn, 1988; Osborne & Freyberg, 1986). Students learn bits of factual information and how to solve problems at the end of chapters by using formulas. Scientific facts and algorithmic problem solving make up much of the science curriculum taught in U.S. schools. Most children do not develop rich understanding and cannot apply what they learn to explain scientific phenomena. This kind of knowledge has been defined as **inert knowledge**. The learner with inert knowledge lacks connections and relationships between ideas and cannot retrieve or use knowledge in appropriate situations (Perkins, 1992). For example, a fifth grader might be able to define *atom,* but not know how to use the definition to explain properties of matter. In fact, many children think that a virus is smaller than an atom!

Students' decreased interest in science appears to be a contributing factor to decreased understanding of science. Although most elementary children start school with an interest in the physical world, they soon lose interest in learning about science. As students transition to middle school, their interest in science decreases further, and students lose motivation to continue their learning of science. Yager and Penick (1986) summarized the data from various national studies and concluded that "The more years our students enroll in science courses, the less they like it." This conclusion is also supported by the Third International Mathematics and Science Study (TIMSS, 1998), which indicated that interest in science decreases from elementary grades to high school. One possible reason for this decrease in motivation stems from the predominant use of textbooks to teach science and the emphasis on memorizing science facts. Most students need to experience science as an active process of engagement for interest to be high.

Let's examine three scenarios that illustrate the variety of understandings students have of scientific ideas.

Scenario 1: What Is Alive?

The first scenario comes from a common biology topic: What is alive? Mr. Ramirez asked his second graders if a seed, such as that from an apple, is alive. Most of his students responded, "No." When Mr. Ramirez probed them to explain their answers, many said that a seed does not move and so cannot be alive.

Young children tend to categorize *alive* and *not alive* according to superficial physical characteristics and to the presence or absence of locomotion (Carey, 1985; Driver, Squires, Rushworth, & Wood-Robinson, 1994). Older students and students more knowledgeable of biology retain aspects of this categorization scheme, but seem to broaden their criteria for *life* to include more carefully differentiated ideas of function (autonomous movement is distinguished from simple movement) and ideas about additional functions (growth, metabolism, and reproduction). If we ask older children (seventh or eighth grade) if an apple seed is alive, many will say that it has the "potential" to grow, but that it is not alive. This example shows how students do not develop rich ideas about basic biological phenomena.

Scenario 2: Boiling Water

The second scenario centers on a common physical science topic: the boiling and condensing of water. Ms. Beacher asked two of her fourth-grade students, Kristen and Shawn, to make observations of and explain a laboratory setup in which water was boiling in a beaker and condensing on a cool glass surface held above the beaker. Figure 2.1 shows the experimental setup. Kristen said that the water was boiling. Ms. Beacher asked how she knew, and Kristen answered, "Because of the bubbles in the water." Shawn added that water was condensing on the glass surface. Ms. Beacher probed them further to elicit more in-depth responses. She asked, "What is inside the bubbles?" Kristen responded, "Air," and Shawn agreed. Then Ms. Beacher asked, "Well, if there is air in the bubbles, where does the air in the bubbles come from?" Both Kristen and Shawn looked at Ms. Beacher with blank faces. She then asked them, "Shawn, you mentioned that water was condensing on the glass surface. Where does that water come from?" Shawn responded, "It evaporated from the beaker." Ms. Beacher probed further: "Can you tell me what it means to evaporate from the beaker?" Shawn replied, "You know; it came from the beaker."

This scenario shows that the children were able to use scientific terms, but they could not use the terms to explain some everyday phenomena. Children might use terms such as *evaporate, condense,* and *boil.* However, if we use probing questions to elicit from them more in-depth descriptions and explanation of the phenomena, we find that many students do not have a very rich understanding of the terms they use. As the example illustrates, many students cannot go beyond just using vocabulary

terms (Osborne & Cosgrove, 1983; Smith, Wiser, Anderson, & Krajcik, 2006). Interestingly enough, the responses of eighth graders are not very different from those of these fourth graders. In their explanation of boiling, some eighth-grade students say that the bubbles in the water are made of air or hydrogen and oxygen gas instead of explaining that the bubbles are filled with water molecules in the gaseous phase.

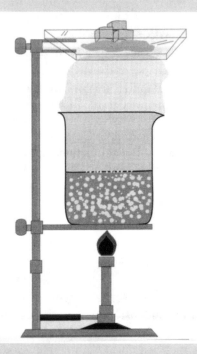

Figure 2.1   Children cannot frequently explain simple phenomena, such as the boiling and condensing of water.

Scenario 3: Does air have mass?

The third scenario centers on students' understanding of the nature of matter—the mass of a volume of air. Ms. Jackson asked her fifth-grade students if air has mass. Most of the students said that air had no mass because it was "weightless" or because "you can walk right through it." Ms. Jackson then asked her students what would happen to the mass of the deflated volleyball if she pumped it up with air. Most of the students said that the reading on the scale would not change because air did not weigh anything. Some students said the volleyball would be lighter because air is so light. Ms. Jackson had several students weigh a deflated volleyball and wrote the mass of the volleyball on the board. She then pumped the volleyball up with air using an air pump. She then asked students to read the mass of the volleyball. Students were surprised that the weight of the volleyball went up! Several students thought that Ms. Jackson tried to fool them and asked that she do it over.

Many children think of air as not having mass (Driver et al., 1994), but like all matter, air does have mass and it can be measured. The responses of older students do not differ substantially from those of these fifth graders. The example illustrates that students lack understanding of the nature of matter and that all matter has mass and occupies space.

These scenarios illustrate that, when we ask students to qualitatively explain scientific phenomena, students typically use the "correct" words, such as *evaporation, condensation,* or *mass and volume,* but they lack understanding of the underlying scientific concepts (Driver et al., 1994; Osborne & Cosgrove, 1983; Smith et al., 2006). These scenarios also illustrate that often students bring ideas related to scientific phenomena that fit their experience and make sense to them, but are often very different from the thinking of scientists (Bransford & Donovan, 2005). These alternative concepts can serve as a hindrance to further learning. As scenario three illustrates, teachers need to elicit students' ideas by making predication (Linn, 2006).

How would you have answered the questions in the three scenarios? Qualitative explanations of scientific phenomena are a challenge, not only for many elementary and middle school students, but also for many high school, undergraduate, and even graduate students. When the three scenarios were replicated in science methods classes and in graduate courses, even students with backgrounds in science had a difficult time giving solid, qualitative explanations of the observed phenomena. What these scenarios show is that the ideas that students bring with them into the classroom can have a powerful and lasting effect on what they will learn. What students learn depends on the prior knowledge they bring with them to the classroom (Donovan & Bransford, 2005).

You can learn more about the scientific understandings that children hold by talking with some children about common physical, chemical, or biological phenomena. Conversing with children of different ages and comparing their responses can help you discover some surprising things about what they know. Learning Activity 2.1 is designed to help you begin talking to children about their science understandings.

Integrated Understandings

Instead of developing inert knowledge, project-based science helps students develop integrated, meaningful understandings. Integrated understanding results from the learner's building relationships and connections among ideas and blending personal experiences with more formal scientific knowledge (Bransford, Brown, & Cocking, 1999; Krajcik, 2001; Linn, 2006). Imagine a child who understands from a prior lesson the relationship between molecules in gaseous, liquid, and solid states. The child learns a new concept about sound vibrations traveling through different substances. The child notices during some activities that sound travels best through solids, next best through liquids, and least well through air. The teacher asks, "Why do you think sound travels best through the solid and least well through the air?" The child answers, "Well, it probably has something to do with what it is made of." The teacher replies, "What did we learn about what all matter is made of?" The child says, "Atoms and molecules!" The teacher prompts the student by saying, "What do you know about the molecules in different states of matter?" The child answers, "Well, the molecules are close together in solids and far apart in gases." The teacher questions, "And what might this tell you about sound travel?" The child says, "Oh, I get it. In solids the molecules are close, so vibrations happen because the molecules can bump into each other. And that's not true in gases; the molecules can't bump into each other as well

Learning Activity 2.1

INVESTIGATE YOUNG LEARNER'S IDEAS

Materials Needed:

- Something to write with
- Baking soda
- A 250 ml beaker
- Vinegar
- Water
- Two students of different ages to talk with

A. Set up a conversation with two students of different ages to learn about how they understand some science concepts. You might want to tape your interview. Make sure to find out the following information:
   - Age of student.
   - Current grade.
   - What science topics they remember studying in school or at home.

NOTE: Probing a child to determine his or her conceptual understanding is challenging; it is a good idea to practice with a classmate or a friend before you conduct these interviews with children. Tape these practice sessions and study the way you asked questions. In general, observe the following guides when interviewing your learners:

1. Avoid using leading questions that suggest responses. For example, don't say, "How do you think this will react?" By saying *react*, you lead the students to think something is going to happen.
2. Avoid praise or negative comments. For example, don't say, "Good." Rather, say, "I see."
3. Use phrases that allow a learner to clarify and expand on his or her ideas. Be sure to probe when scientific or technical terminology is used. For example, if the student says, "It will evaporate," ask, "What do you mean by 'it will evaporate'?" To clarify ideas, say things like, "Could you tell me more about what you mean?" Here are other phrases:

   - "Please describe that further."
   - "What do you mean by...?"
   - "Tell me more about..."
   - "Is this what you mean...?"
   - "Please explain that further."
   - "Hmm. That's interesting. Tell me more about that."
   - "What else would you like to tell me about what you observed?"

4. Listen to the learner—use active listening techniques. For example, look at the child when he or she is speaking. Paraphrase what the student says. For example, if the child says, "I think it will disappear," you might say, "Do you mean you wouldn't be able to see it?" Ask for explanations. Summarize what you think the student said.

5. Use the learner's language to rephrase and further probe the learner's response. For example, if a student says, "I won't be able to see it," say, "Why can't you *see* it?" Don't say, "Do you mean it will dissolve?"
6. Provide the learner with ample time to construct a response. Wait at least three to five seconds after asking a question before you say anything more.
7. Establish a calm and accepting atmosphere. Don't rush the conversation by talking hurriedly or being judgmental about responses.
8. If a learner cannot answer a question, try rephrasing the question to find out what the student does know and what his or her thoughts are. For example, if you first ask, "What do you think will happen?" and the child doesn't respond, rephrase the question to say, "What do you think will happen when I put this baking soda into this water and stir it?" However, be careful; don't add more than what the student said.

B. Complete the following tasks with two students:

Dissolving Change Task. In this task, you will mix a teaspoon of baking soda in 100 ml of water. The baking soda will dissolve in the water.

- Show students a teaspoon of baking soda and a 250 ml beaker with 100 ml water. Ask students to predict what will happen if you mix the baking soda in the water.
- Place the baking soda in the beaker of water and stir. Ask the students to describe what they see. Ask the students to explain what they observe. Find out what the students mean by the words they use. For instance, if they say, "It dissolved," ask them what they mean by *dissolved*.
- Ask the students what they might see if they could magnify the contents of the beaker 100 million times. Ask the students to draw what they think they would see.

Chemical Change Task. In this task, you will mix a teaspoon of baking soda in vinegar. This is a chemical change; the baking soda will react with the vinegar to form new products.

- Show students a teaspoon of baking soda and a 250 ml beaker with 100 ml of vinegar. Ask students to predict what will happen if you mix the baking soda in the vinegar.
- Place the baking soda in the beaker of vinegar. Ask students to describe what they see. Ask students to explain what they observe. Find out what the students mean by the words they use. For instance, if they say, "It reacted," ask them what they meant by *reacted*.
- Ask what they might see if they could magnify the contents of the beaker 100 million times. Ask the students to draw what they think they would see.

C. Process what you found. Combine your data with the data from other students in your class. Summarize your data. What conclusions can you draw? What educational implications do your conclusions indicate?

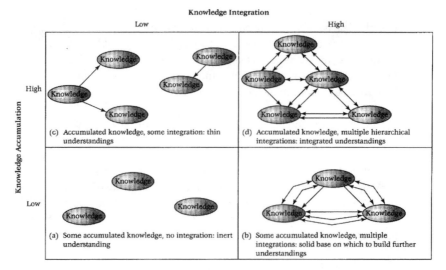

Figure 2.2   Understanding is a function of knowledge accumulation and knowledge integration. Theoretically, it is possible to accumulate a large number of discrete bits of knowledge without creating relationships between those pieces of knowledge or with only a few weak relationships (Cells A and C). It is also possible to have a limited amount of knowledge of which each piece is well integrated or linked to other knowledge, perhaps in multiple ways (Cell B). Rich, integrated understandings are achieved as knowledge becomes both structured and integrated (Cell D; Talsma, 2004).

because they are too far apart." This student has integrated understandings: She used her understanding of molecules in different states of matter to come up with a new explanation of the effect of states of matter on sound travel.

Students who have integrated understandings can use those understandings to solve problems (Linn, 2006). For instance, a child with integrated understanding might use Newton's laws of motion to explain why a person needs to wear a seat belt when riding in a car. Newton's First Law of Motion states that an object will stay at rest or continue in motion unless acted upon by a force. Therefore, we wear seat belts so that our bodies do not continue in motion (hit the windshield or are thrown out of the car) if the car comes to a sudden stop.

Figure 2.2, created by Valerie Talsma, illustrates the difference between inert and integrated understanding. The cell in the lower left corner, Cell A, represents a learner who has very little understanding and whose understandings are not connected to each other. This learner has inert knowledge—disconnected, unusable fragments of facts. The cell in the lower right corner, Cell B, represents a learner with some understanding that is tightly integrated. This learner has the type of understanding we want a student to develop at the third- or fourth-grade level, where their understandings, although limited, should be integrated. The cell in the upper left corner, Cell C, represents a learner who has accumulated knowledge but has not integrated the knowledge. This is the type of understanding that often emerges in schools. The cell in the upper right corner, Cell D, represents a learner who has accumulated knowledge and developed integrated understandings. This is the rich, coherent understanding that learners achieve in project-based science.

Types of Knowledge

One way to think about integrated and meaningful knowledge is to think about three types of knowledge: content, procedural, and metacognitive (Anderson & Krathwohl, 2001; Perkins, 1992). All three types of knowledge are needed for understandings to become integrated and meaningful.

Content knowledge refers to the central concepts, principles, and theories in an area of study. One example of a central science concept is *force*. An example of a principle is Newton's First Law of Motion. A learner who understands why an object stops moving once it is given a push, for example, has a better foundation to understand central concepts and principles of physics, such *as force, acceleration, velocity,* and Newton's three laws of motion.

Content knowledge, however, is only one type of knowledge that students need to develop (Krajcik, 2001; Perkins, 1992). Understanding also depends upon accumulating supporting knowledge that helps the learner know how to learn. This supporting knowledge includes procedural knowledge and metacognitive knowledge. **Procedural knowledge** is essential in helping a student engage in inquiry and in finding solutions to problems. A child uses procedural knowledge to find solutions to questions, design an experiment, and find and evaluate background information related to a question. For instance, knowing how and when to control variables is procedural knowledge. (Chapter 4—"Developing Scientific Investigations"—discusses the control of variables in more detail.) Procedural knowledge also involves knowing when the evidence is sufficient for a person to draw a valid conclusion.

Metacognitive knowledge (Anderson & Krathwohl, 2001) is knowledge about cognition in general, as well as awareness and knowledge about one's own cognition. Anderson and Krathwohl described three types of metacognitive knowledge: strategic knowledge, knowledge of cognitive tasks, and self-knowledge. **Strategic knowledge** refers to knowing different strategies for learning, such as knowing how to use search terms to find information on the World Wide Web. **Knowledge of cognitive tasks** refers to the understanding that cognitive tasks are different, such as understanding that data analysis might include transforming the data into a different representation. **Self-knowledge** refers to knowing one's own strength and weaknesses in learning. For instance, a child might realize that he has difficulty spelling science words, so he might need to spend extra time on them.

A child uses metacognitive knowledge to monitor her progress on a report or to decide when to seek help with a procedure. A student with strong metacognitive strategies will be able to track how well he is completing a task, such as following procedures, that is part of an investigation that extends over time. A child with metacognitive knowledge will know when she has collected enough information to make a conclusion.

We present this summary of the three types of knowledge because all three are essential for students to develop the integrated and meaningful understanding needed for continued learning. Limiting a student to content knowledge robs the learner from learning more. Meaningful understandings will occur only if content knowledge is developed in the process of engaging in inquiry. For example, if a child knows only definitions of molecular states, she will not be able to use that understanding to

TABLE 2.1   Three Types of Knowledge

Type of Knowledge	Definition	Examples
Content	Knowing the central concepts and principles in a domain	In biology, understanding predator and prey relationships; in chemistry, understanding physical and chemical change; in physics, understanding inertia
Procedural	Knowing how to solve problems and design and carry out investigations	Knowing how to design an experiment; knowing how to analyze data; knowing what couts as evidence
Metacognition	Knowing different strategies for learning; knowing the difference between different cognitive tasks; knowing one's own strengths and weaknesses in learning	Knowing different strategies for solving problems; knowing that memorizing definitions of terms is not as difficult as explaining what a term means; knowing that one finds it difficult to write careful observations; knowing strategies for monitoring one's progress

explain a new situation related to sound travel. Table 2.1 summarizes the three types of knowledge.

Models of Teaching

You can think of teaching on a broad continuum from transformational approaches to receptional approaches. **Transformational approaches** to teaching and learning involve teachers supporting students to make sense of material. Project-based science is one example of a transformational approach to teaching and learning. In contrast, **receptional approaches** to teaching and learning involve teachers transmitting information and students receiving it. Receptional approaches to teaching and learning have long dominated our school systems. Armstrong (1994) wrote,

> For most Americans, the word *classroom* conjures up an image of students sitting in neat rows of desks facing the front of the room, where a teacher either sits at a large desk correcting papers or stands near a blackboard lecturing students. (p. 86)

A Receptional Approach to Teaching

Marie, a sixth-grade teacher, believes that children need to be quiet and in their seats to learn the information she is telling them. In a recent lesson in which she taught about sound, she first had the students read from the textbook the chapter on sound. Students read aloud one paragraph at a time. Marie indicated to students what the important concepts were by writing them on the board as they were read. Students wrote definitions of the important vocabulary words—frequency, pitch, amplitude, waves, rarefaction, and compression—in their science journals. Marie attempted to prove to the students that sound is produced by waves by demonstrating the waves that are created by striking a tuning fork and placing it in water. Next, Marie showed the children the different frequencies produced by tuning forks of various pitches. Finally, Marie reinforced the important concepts by showing a videotape covering the same ideas. Students received a homework assignment: They had to answer three questions at the end of the textbook chapter. Marie evaluated the students with a quiz at the end of the week. On this quiz, students were asked to choose the correct definitions for words, such as pitch, amplitude, and frequency.

A Transformational Approach to Teaching

Roberta, another sixth-grade teacher, introduced a similar sound lesson as Marie in a very different manner. Roberta was investigating with her class the driving question "What Makes Music?" She introduced a lesson to help students understand basic concepts of sound. First, she handed each child in the class a balloon, a rubber band, and a ruler. The students were instructed to spend the next 10 minutes exploring ways to make sounds with each of the objects. At the end of the exploration time, students were asked to discuss their findings with each other in small groups and later share with the whole class what they had discovered about sound. Roberta did not evaluate these ideas, but listed all of them on the board. One child said that the balloon was louder than the rubber band. Another said that the ruler changed sound as it was hung off the table at different lengths. Yet another child said that all the objects were moving when the sounds were produced. Roberta wrote the word *vibration* on the board and told the children that that is the term scientists use for such movement.

Then she asked the class to stop and compare each object again. She asked, "Is it true that all the objects need to be moving to produce sound?" The students debated the evidence supporting this conjecture. One student asked how the sounds made by a thick rubber band would differ from those made by a thin one, just as the strings on his guitar made different sounds. Roberta encouraged the students to explore this idea. Through questioning, she guided the students through a comparison of the lengths, thicknesses, and stretched widths of the various objects and a comparison of the pitches. She asked students to think about why the pitch would be different—was there anything different about the vibrations? Using the materials, students explored the question in small groups. By analyzing the evidence they collected, the students explained that the differences in pitch seemed to be caused by variations in how fast the rubber bands were moving.

Roberta introduced the class to the term *frequency* and helped students see how their findings matched the definition of this term. Students were encouraged to think about their own experiences that would support the concept that sound is caused by vibrations and that pitch is caused by variations in the frequency of the vibrations. A student mentioned the digital monitors on her stereo system that visually displayed the frequency of vibrations. Another told about a time that he dropped salt on the top of the TV; the salt "danced" up and down and vibrated while the TV was turned on but was still when the TV was turned off. A third student discussed how she would place her fingers and move her lips to play her clarinet. Yet another talked about watching the pictures hung on her brother's wall vibrate when he turned up his rock music. Roberta consistently asked students to elaborate on their examples and

> CONNECTING TO NATIONAL SCIENCE
> EDUCATION STANDARDS
>
> SCIENCE EDUCATION PROGRAM STANDARDS
>
> Student understanding is actively constructed through individual and social processes. In the same way that scientists develop their knowledge and understanding as they seek answers to questions about the natural world, students develop an understanding of the natural world when they are actively engaged in scientific inquiry—alone and with others (p. 29).

to explain how the examples supported the question asked. Students were frequently asked to confirm or refute comments made by others in the class. Finally, to evaluate what students learned, Roberta asked them to write down in their journals what they had learned in the lesson.

What can be done to help children develop integrated and meaningful understanding of science? How can we help children apply scientific concepts and principles to solve real-world problems that are of interest to them? What can be done to help children develop content, procedural, and metacognitive knowledge? Research findings from the last 2 decades have shown growing support for the notion that integrated and meaningful types of knowledge are best learned when what occurs in schools is less receptional and more transformational. In Chapter 8, we discuss more thoroughly instructional strategies that help make science teaching and learning more transformational.

Social Construction of Knowledge

How do students come to make sense of material rather than only memorize it? Brown, Collins, and Duguid (1989) argued that "Knowledge is ... in part a product of the activity, context, and culture in which it is developed and used" (p. 32). These researchers see knowledge as contextualized. By contextualized, they mean that knowledge cannot easily be separated from the situation in which it is developed. Blumenfeld and colleagues (1991) reiterated this notion: "Knowing and doing are not separated; knowledge is not an abstract phenomenon that readily can be transferred from how it is learned in the classroom for use in other situations." This learning model, which suggests that learning cannot be separated from the social context in which it takes place, is often referred to as **social constructivism**.[1]

The Influence of Jean Piaget

Jean Piaget (1971), arguably the most influential developmental psychologist of the 20th century, believed that a person's reasoning and intelligence developed like other biological systems—through maturation, or innate biological changes, as well as through the process of equilibration. According to Piaget (1970), an individual actively modifies what he/she knows through interactions with the environment and others. Maturation, in addition to environmental and social interactions, influences how an individual's reasoning and intelligence developes. **Equilibration** refers to the process of modifying one's mental structures, or schemas, to fit new experiences. **Schemas** can be thought of as blueprints that guide behavior and reasoning, but that can be modified.

Piaget's ideas about biological maturation and interaction with the environment and others had important consequences for curriculum and instructional activities. If an individual does not reach a particular maturation point, it will be impossible for him to develop understanding; similarly, if he lacks appropriate experience, he will not be able to develop the appropriate understanding. Hence, according to Piaget, curricular materials should fit the cognitive development of the student.

Piaget's ideas about adaptation also play a key role in his theory of reasoning and intellectual development. Adaptation refers to an individual's attempt to create an accurate view of the world so that she can continue the process of development. Two basic principles allow adaptation to occur: assimilation and accommodation. Assimilation refers to fitting new information into existing mental structures or schemes. For instance, a child who has goldfish will think of a fish as an animal that lives in water. When he learns about whales in school, sees a picture of a whale in a book, or sees a video about whales, he classifies the whale as a fish because it lives in water. Accommodation refers to making a permanent change in a mental structure. For instance, as the child gains more knowledge about fish and mammals, he changes his mental structure of a fish as an animal that lives in water. He now has schemas for fish that live in water and mammals that live in water.

Many psychologists refer to Jean Piaget as the "founder of constructivism." Piaget's ideas on how environment and social interactions influence knowledge construction have had a major impact on the constructivist theories at the foundation of new inquiry-oriented curricula of the 1990s and 2000s and on the use of collaborative learning strategies in science. Piaget regarded socialization as significant in the development of reasoning (Piaget, 1959). He argued that, when learners discuss ideas with their peers, they become aware of the different views of a problem. This confrontation requires them to seek new information and formulate new hypotheses and, over time, to structure a new way of thinking.

The Influence of Lev Vygotsky

Although influenced by Piaget, social constructivist theories in education developed primarily from the works of Lev Vygotsky, a Russian psychologist, who concluded that children construct knowledge or understanding as the result of thinking and doing in social contexts (Vygotsky, 1986). Vygotsky believed that development depends on biological factors (such as brain growth and maturation) and on social and cultural forces (such as the influence of others at home, in school, or on the playground). He also believed that learning takes place in social contexts (such as during playtime with peers, in conversations with classmates, or while parents or teachers are speaking) and that children internalize information they gather in those contexts to form understanding. Children gradually become more independent and autonomous through social interactions with others, such as teachers and other adults. Researchers in science education have found evidence to support the social constructivist viewpoint about how children learn science (Brown & Campione, 1994; Driver, 1989; Roth, 1995).

Social constructivism holds that children learn concepts or construct meaning through their interactions with and interpretations of their world, including essential interactions with others. From the moment a child is born, she (supported through interactions with others) is constructing knowledge of her environment. Knowledge is not something that is simply memorized; it is constructed by the learner according to her experiences in the world. For example, a child may learn that leaves fall from trees after they turn colors during the autumn by observing this happen several years in a row. Discussions with family members or peers may influence his understanding. For example, a parent might explain that leaves fall when the weather becomes cold. Knowledge is the result of individual interpretations (the child notices that leaves fall after they turn color) and constructions of reality as it occurs in a social context (a parent says this happens when it gets cold).

Children receive information, interpret it, and relate it to other prior knowledge and experiences. As a result, they come to school already holding concepts related to concepts teachers will be expecting to help them learn. These prior experiences and conceptions will influence any new knowledge they attempt to acquire. For example, a teacher who attempts to teach a class about why trees drop leaves will encounter any number of prior ideas about this phenomenon in the minds of her students. Some might think the leaves drop because they are dead. Some might think the tree runs out of food. Others might think that the color of the leaves causes them to fall. Some might think trees sleep in the winter. Still others might think that the cold winter winds blow the leaves off. These prior beliefs will affect the teaching and learning going on in the classroom. Because prior knowledge and experiences influence the learning of new knowledge, it is important to reflect frequently on prior experiences.

Because children come to school with prior understandings about their world, their concepts and theories are not always the same as those developed over the years by scientists. As a teacher, you might try, for example, to teach elementary students that air is matter and that matter is something that takes up space and has mass. A typical activity in many elementary textbooks demonstrates to children that two deflated balloons will balance on a scale but that one deflated and one inflated balloon will result in the tilt of the inflated balloon toward the ground.[2] This activity provides students with some evidence that air has mass and takes up space—the balloon filled with air has more mass than the one without air.

What if, before showing children the result of this activity, a teacher asked them to predict the event? What do you think young children would predict? Think about some of the prior experiences elementary children may have had with air. Air and gases (helium) in a balloon causes it to float away. Air in a raft or beach ball causes it to float on water. Air cannot be seen. These prior experiences may have convinced students that air is nothing or that air has no mass. They may predict, therefore, that nothing will happen—air is a nonentity, so the inflated and deflated balloons should still balance. Perhaps they will predict that the inflated balloon will float upward and the empty balloon will sink. Young students will probably not be convinced otherwise by a teacher's demonstration.

The teacher needs to create a learning environment that encourages students to revise their own concepts to accept these new formulations. This is not an easy task. Learning is a continuous process that requires many new experiences in which students can construct and reconstruct knowledge by interacting with others and materials. Children need many opportunities to express and explore their ideas. These ideas about social constructivism have a number of implications for teaching science and are the foundations for project-based science.

A Social Constructivist Model of Teaching

Social constructivist theory asserts that children take an active role in constructing meaning; they cannot construct meaning by passively absorbing knowledge transmitted from a teacher. An ancient Chinese proverb captures this idea in three simple lines:

Tell me, and I forget.
Show me, and I remember.
Involve me, and I understand.

Accordingly, teaching based on social constructivist theory focuses on the child as an active builder of knowledge in a community of learners. Social constructivist theory has implications for the way a teacher creates the learning environment, sets up lessons, asks questions, reacts to students' ideas, and carries out lessons. Lorsbach and Tobin (1992) suggested that teaching science using a constructivist approach means that teachers do *not* teach science as "the search for the truth." Instead, they teach science more as scientists really do science—by actively engaging children in the "social process of making sense of experiences." This approach differs greatly from much of what can be seen in "school science" today, where science teaching consists of asking students to memorize terminology and find correct answers.

Table 2.2 lists the main features of this model of learning. We will now explore each of these in depth.

Active Engagement with Phenomena

Children construct understanding in science by actively engaging with phenomena. Active engagement describes several experiences: Students ask and refine questions related to phenomena; they predict and explain phenomena; and they mindfully interact with concrete materials. **Active engagement**, then, is both mental and physical.

TABLE 2.2   Features of the Social Constructive Model of Learning

Active Engagement with Phenomena
- Students ask and refine questions related to phenomena.
- Students predict and explain phenomena.
- Students mindfully interact with concrete materials.

Use and Application of Knowledge
- Teachers and students use prior knowledge.
- Students identify and use multiple resources.
- Students plan and carry out investigations.
- Students apply concepts and skills to new situations.
- Students are given time for reflection.
- Students take action to improve their own world.

Multiple Representation
- Teachers use varied assessment techniques.
- Students create products or artifacts to represent understanding.
- Students revise products and artifacts.

Use of Learning Communities
- Students use language as a tool to express knowledge.
- Students express, debate, and come to a resolution regarding ideas, concepts, and theories.
- Students debate the viability of evidence.
- Learning is situated in a social context.
- Knowledgeable others help students learn new ideas and skills that they could not learn on their own.

Authentic Tasks
- Driving questions focus and sustain activities.
- The topic or question is relevant to the student.
- Learning is connected to students' lives outside school. Technology tools can help connect learning to life outside of the school.
- Science concepts and principles emerge as needed to answer a driving question.

Students Ask and Refine Questions and Predict and Explain Phenomena

To actively engage students intellectually, teachers must create a learning environment in which students can *ask questions* freely, *dialogue* with classmates and more knowledgeable others to *refine questions,* and *predict and explain phenomena.* Such cognitive activities help students make connections and develop in-depth understandings (Brooks & Brooks, 1993). It is through discourse (asking questions, having discussions about important questions, making predictions, and providing explanations) that students come to understand what they know. (The role of questioning and ways to set up project-based learning environments that support questioning are covered in several chapters. Chapter 4 explores the role of predicting and explaining phenomena.)

Imagine a second-grade teacher asks, "How could we find out if the apple seed is alive?" She is stimulating the students to express ideas and ask questions. Students might ask in return, "Could we plant it?" "Can we cut it open to see if something is growing inside?" "Could we ask a farmer?" The teacher has the students work in groups to discuss and refine these questions. They might debate the fact that cutting open the seed could kill it and, therefore, defeat the purpose of their investigation. This thought might lead them to settle on planting the seed to see if it is alive.

If the students planted the apple seed, they could predict what would happen next. This could lead to more refined questions, such as "Why isn't it sprouting yet?" and "Do we need to water it more?" If the apple seed began to sprout, students could explain that the seed was alive. However, a lively debate could still take place about whether the seed was alive before being planted. It is through the process of asking and refining questions, making predictions, making observations, and providing explanations about the apple seed that students begin to understand whether or not the seed is alive.

Students Mindfully Interact With Concrete Materials

One hallmark of social constructivist teaching is that students mindfully interact with concrete materials. Children retain more of what they are taught if they engage in more active, concrete types of learning. In fact, it is estimated that the more active and concrete their learning, the more they retain (Bruner, 1977). Students need to experience phenomena in order for the ideas of science to make sense to them and to link ideas together. Experiencing phenomena serves as important prior experience for students to build and link science ideas. Young learners especially can benefit from firsthand experiences with phenomena. The experiences make the ideas associated with explaining the idea plausible to learners, and learners see the usefulness of the ideas in understanding their world (Kesidou & Roseman, 2002). For students who live in the northern portion of the United States, understanding what snow is is not difficult, but for children who live in the southern portion of the country and never experience snow, understanding what snow is poses difficulties. Similarly, it is very difficult for students to understand chemical reactions if they never experience a variety of chemical reactions.

Use and Application of Knowledge

To develop integrated understandings, students need to use and apply their knowledge. This use and application of knowledge is supported by six strategies:

1. Teachers must elicit and use students' prior knowledge.
2. Activities must encourage students to identify and use multiple resources.
3. Activities must involve students in planning and carrying out investigations.
4. Learned concepts and skills must be applied to new situations.
5. Students should be allowed time for reflection.
6. Teachers must help students take action to improve their own world.

Teachers and Students Elicit and Use Prior Knowledge and Experience

In order for a teacher to help students use and apply their knowledge, she or he must consider what it is that students already know. The following paragraph serves as an illustration of this point:

> It is really actually simple. First, you arrange things into different groups depending upon their makeup. Of course, one pile may be enough depending on how much there is to do. If you have to go somewhere else due to a lack of equipment, that is the next step; otherwise, you are pretty well set. It is important not to overdo any particular part of the job. That is, it is better to do too few things at once than too many. In the short run, this may not seem important, but trouble from doing too many can easily arise. A mistake can be expensive as well. Working the equipment should be self-explanatory, and we need not dwell on it here. Soon, however, it will become just another facet of life. It is difficult to see an end to the necessity for this task in the immediate future, but then one can never tell. (Bransford, 1979)

What is this paragraph about? You probably had difficulty understanding it because you were not given any clue about its topic. If you had known from the start that it was about washing clothes, you would have understood it instantly. Reread the paragraph now that you know the topic. Doesn't it make more sense? In this case, activating your prior knowledge clearly would have enhanced your ability to understand new material.

Constructivist teaching approaches focus on the learner's prior knowledge, because it is the learner who must integrate new ideas into his or her current understandings. Shapiro (1994) wrote,

> The role of the teacher [in a constructivist view] is not to simply present new information, correct students' "misconceptions," and demonstrate skills. It is to guide the learner to consider new ways of thinking about phenomena and events. In order to do so, the teacher must have some understanding of what the learner brings to the learning experience, that is, his or her prior ideas, and thoughts. (p. 8)

To help students integrate their understandings, teachers must know about their prior understandings. Recall the examples earlier in this chapter of the role played by prior knowledge and experience in learning. Children have had years of experience with leaves falling or with air (blowing, in balloons, in balls) before they come to school to receive formal instruction about these concepts. When teachers attempt to teach students that air has mass, prior experiences may conflict with the new ideas as

CONNECTING TO NATIONAL SCIENCE
EDUCATION STANDARDS

SCIENCE EDUCATION PROGRAM STANDARDS

Learning science is something students do, not something that is done to them. In learning science, students describe objects and events, ask questions, acquire knowledge, construct explanations of natural phenomena, test those explanations in many different ways, and communicate their ideas to others (p. 18).

students try to integrate the new understandings into their conceptual framework of "air." With some awareness of students' past experiences, teachers can help students reconcile what seem to be conflicts between those experiences and new learning. Without this awareness, teachers will still be able to ensure that teaching takes place, but not that learning does.

There are many ways to probe for students' prior knowledge. For example, in the washing clothes scenario, a teacher could provide students with a mental organizer (such as telling them they are about to read about washing clothes) and discuss what they know about washing clothes before students read the paragraph. Another strategy is to remind students of an activity or investigation done earlier in the year that has some connection to new learning.

Students Identify and Use Multiple Resources

An important strategy in helping children construct understandings in science is having them identify and use multiple resources. Multiple resources, which can be used in the course of answering a driving question, include books, journals, science equipment, supplies, and computers. They reinforce student understanding through different presentation and through emphasis of different information. While a book might explain via text and photographs how a landfill works, a Web page might present a slightly different explanation through illustrations and through interactive charts of statistics on landfill usage and cleanup efforts. When students analyze and synthesize this different information presented in these different ways, they create more solid, integrated understandings.

Students Plan and Carry Out Investigations

The cycle of asking a question, making observations, designing an investigation, collecting data, analyzing results, and asking new questions, which is a key feature of project-based science, requires children to use and apply their understandings. (The investigation web is further explored in Chapter 4.) Assume a class is exploring the question, "Where does all our garbage go?" To explore this driving question, students might set up a decomposition column (or other organic material), which is a small composter in a 2-liter pop bottle filled with leaves, dirt, banana peels, and worms. Students ask such questions as "Why do worms help the materials in our pop bottles decompose faster?" or "How does moisture affect how fast decomposition occurs?" Students gather information related to their questions from multiple resources, such as books, journals, and the World Wide Web. Students plan investigations to answer these questions, considering the materials they will need, how they will collect and analyze the data, and how they will present their findings to the class. They make predictions about what will happen to the items in the bottle. Then students carry out their investigations. They debate whether two pop bottles can provide ample evidence for whether moisture affects the rate of decomposition. Students create artifacts, such as posters or multimedia products, to represent their understanding of decomposition. Finally, students share and explain ideas about the problem of decomposition as it relates to landfills. This process of asking and refining questions, debating ideas, making predictions, designing experiments, gathering information, collecting and analyzing data, constructing explanations, drawing conclusions, and communicating ideas and findings to others helps students construct a solid, integrated understanding of the topic and related concepts.

Students Apply Concepts and Skills to New Situations

Students develop rich, integrated understandings when they apply their knowledge to new situations. This phenomenon is illustrated as the change in understanding from Cell C to Cell D in Figure 2.2. By applying concepts and skills to new situations, students elaborate on their understandings, form new connections with old ideas, and build connections between new ideas and old ideas. For instance, in the decomposition example, students can apply their knowledge of how oxygen affects decomposition to real-world situations, such as landfills. Through discussions with others, students can make connections between the understanding that oxygen is needed for decomposition and the awareness that materials do not decompose quickly in a landfill.

Students Are Given Time for Reflection

In science, **reflection** involves thinking about alternative questions, considering hypotheses, contemplating a variety of answers, speculating on outcomes, deliberating on steps that can be taken, and meditating on conclusions found. Reflection takes *time*. Teachers cannot rush through topics, lessons, and examples and expect students to learn new material thoroughly. For this reason, constructivist teachers present fewer topics in the curriculum and spend more time on them. Teachers provide students with time in class to discuss ideas with others, write about experiences, and revise ideas and products. Recall the teacher who had students discuss whether the apple seed was alive. During discussions, teachers use the technique of wait-time: They wait 5 to 10 seconds after asking a question before calling on a student. In addition, teachers ask *probing* questions, which are questions designed to elicit more details. (Wait-time and probing questions are discussed in more detail in Chapter 8.) Brooks and Brooks (1993) wrote,

> Classroom environments that require immediate responses prevent these students from thinking through issues and concepts thoroughly, forcing them, in effect, to become spectators as their quicker peers react. They learn over time that there's no point in mentally engaging in teacher-posed questions because the questions will have been answered before they have had the opportunity to develop hypotheses. (p. 115)

Students Take Action to Improve Their Own World

The idea of taking action to improve the world has become popular in science education in the last few decades. The Science Technology Society (STS) movement that was popularized in the 1980s is one example. STS focuses on such topics as health, population, resources, pollution, and environment—topics that people must understand in order to become active citizens who make decisions and take actions related to society or to improving their lives.

Children are particularly interested in studying questions that can be applied to their own lives and improve their own world. Adolescents see themselves as emerging adults and feel a particular urgency about the future. Many adolescents are interested in ecological topics and environmental issues (Barnes, Shaw, & Spector, 1989). Adolescents see exploration of these topics as a way of taking action to improve their world. Middle grade students may be interested in more far-reaching *local* issues, such as the health of a local stream. They may be interested in monitoring the quality of the stream and reporting it to a local governmental agency. Young elementary students have fewer global concerns. They are interested in more personal concerns, such as improving the living conditions and nutrition of their pets at home. Children of all ages care about improving the school environment. They can put up bird feeders, plant wildflowers, and organize a litter pickup day. When learning includes taking action to improve their world, children see the importance of it, and the action solidifies their knowledge.

> ### CONNECTING TO NATIONAL SCIENCE EDUCATION STANDARDS
>
> ### SCIENCE EDUCATION CONTENT STANDARDS K–4
>
> As a result of activities in grades K–4, all students should develop understanding of personal health, characteristics and changes in population, types of resources, changes in environments, and science and technology in local challenges (p. 138).

Multiple Representations

Constructivist theory asserts that learning involves developing **multiple representations** of ideas that integrate understanding. In the decomposition activity described earlier, students constructed a decomposition column, wrote about their investigation, and built a poster to express their ideas. Writing, building a poster, and manipulating materials are examples of different representations. These three activities require the translation of understandings into three different formats—concrete, textual, and graphical representations. When students make connections among these representations, they develop integrated understandings that can be applied to new situations.

> ### CONNECTING TO NATIONAL SCIENCE EDUCATION STANDARDS
>
> ### SCIENCE EDUCATION CONTENT STANDARDS 5–8
>
> As a result of activities in grades 5–8, all students should develop abilities of technological design and understandings about science and technology (p. 161).

Teachers Use Varied Assessment Techniques

There are two reasons to use varied assessment techniques. First, different types of assessment techniques can better assess different types of understanding formed through a variety of intellectual activities. Second, multiple forms of assessment help different learners succeed in demonstrating their understandings.

In project-based science, students develop different types of understandings that each must be assessed, and these understandings may be assessed in a variety of ways. (Assessment and evaluation are covered in detail in Chapters 9 and 10.) For example, how might students demonstrate understanding about water quality in a stream? They might write reports, present artifacts to classmates, take photographs, or create multimedia products that document a stream cleanup effort.

Because different learners will respond to different assessment techniques in different ways, using a variety of assessment techniques ensures that all students are evaluated appropriately. Whereas one child might flourish on a written exam, another might present her ideas more effectively with a poster or three-dimensional model, and yet another might express his ideas better through a written journal. By using a variety of forms of assessment, teachers let students express their understandings in ways consistent with their individual strengths.

Because constructivist theories focus on students' prior experiences to frame teaching and learning, multiple paths (from different students' ideas) to learning emerge. Therefore, assessment also needs to include students' interpretations of their learning. Students' interpretations can be captured with self-evaluation techniques. Teachers can interview students about how they think they have progressed in learning about a topic or skill. Students can also keep journals to track their own progress throughout a project.

Students Create and Revise Products or Artifacts to Represent Understanding

Another approach to helping students create understanding is through the building and revision of artifacts. **Artifacts** are tangible representations of student understanding that answer a driving question. They include models, reports, videos, and computer programs. They can be thought of as external, intellectual products. Papert (1980, 1993) referred to artifacts as "objects-to-think-with" because they are concrete and explicit and serve as tools of learning. Students build understanding by constructing artifacts and explaining to each other the meaning of their artifacts. Artifacts and products can be critiqued by others—students, teachers, parents, and community members. As a result, learners have many opportunities to reflect on and revise their artifacts, and so they have many opportunities to further enrich their knowledge.

Imagine that students decide to make a poster presentation to explain how they will explore the influence of fertilizer on plant growth. In the process of developing this artifact, students need to select plants and explain their selections, and they need to determine the amount of fertilizer. The students need to provide answers to questions, such as "How many plants?" "How much fertilizer?" and "How often?" All of these decisions focus the students' thinking and help develop deeper understanding. Students again have an opportunity for enhanced understanding when their teacher and classmates give them feedback on their poster.

Learning Communities

The social constructivist model holds that students learn in a social context—a **learning community**. Within such learning communities, language is a primary tool for developing understanding. In this section, we explore five aspects of learning communities:

1. students using language as a tool to express knowledge,
2. students using language to debate and come to a resolution about science ideas and theories,
3. students debating the viability of evidence,
4. students learning in a social context, and
5. students learning from knowledgeable others.

We will also discuss ways that knowledgeable others can use scaffolding to help students learn.

Students Use Language as a Tool to Express Knowledge

Children (and adults) learn best when they can talk about and share their ideas with their peers and with concerned adults. Hence, meaningful learning develops as an interplay between the child thinking about ideas (an *intra*personal use of language) and talking with others (an *inter*personal use of language). This interplay helps the child build connections among ideas, integrating understanding. When a child uses the words push or pull to describe force to another child, a learner demonstrates his or her understanding of the concept.

Teachers can help students use language in a variety of ways. First, students can share information in written form. For example, students can create reports to share the results of their investigations with others or exchange journals so that others can read about their investigations. Second, students can learn through oral language. For example, many teachers have students explain their investigations to classmates, using artifacts or products.

Students Express, Debate, and Come to a Resolution Regarding Ideas, Concepts, and Theories

Healthy debates are a critical aspect of developing understanding in science. Constructivists define science as knowledge that has been publicly debated and accepted by scientists (Shapiro, 1994). Scientists continually obtain new information from investigations, and they debate ideas based on this information. Similarly, in science classrooms, teachers want students to debate ideas, concepts, and theories and come to resolutions about them. The student who believed dinosaurs were pets because he collected them debated this idea with classmates, who offered reasons that dinosaurs were not pets. Finally, all students came to the resolution that dinosaurs were not pets.

Students Debate the Viability of Evidence

Scientists constantly debate the viability of evidence. One group of scientists may speculate that mutations being found in frogs around the world are the result of depleted ozone. Another group of scientists argues that the deformations are a result of pollutants in the environment. The viability of evidence is debated until one supposition or another is supported by most in the scientific community. Just as scientists debate the viability of evidence, students need to do this in science classrooms. For example, students might debate whether the height of a plant is a good indicator of growth. Some students might argue that number of leaves and color are more important indicators than is height, because height might only indicate that the plant is stretching to reach sunlight (tropism).

Learning Is Situated in a Social Context

Children do not construct understanding in isolation, but in a social context. Parents, friends, teachers, peers, community members, books, television, movies, and cultural customs all affect the construction of student understanding. For example, students learning about decomposition share with classmates their findings from various experiments performed on decomposition columns. They visit local landfills or recycling plants and talk with the managers. They connect to the World Wide Web to talk with other students around the country about the problem of landfills in their region. They talk with peers to see if any use composters at home. Each of these social interactions helps students build integrated understandings. In isolation, these understandings might not be connected, but through social construction of ideas, students link new understandings with old ones.

Knowledgeable Others Help Students Learn New Ideas and Skills That They Couldn't Learn on Their Own

A basic idea that stems from these ideas about language, culture, and community is that more competent others can assist children in accomplishing a more difficult cognitive task than they otherwise could on their own. The development of understanding occurs as a result of social interaction with more knowledgeable others and with peers. Even second graders are able in learning communities to carry out sophisticated investigations under the guidance of a classroom teacher.

Vygotsky (1978) developed the construct of the zone of proximal development to represent the hypothetical space between assisted and unassisted performance of a learner, or the distance between the actual developmental level as determined by independent problem solving and the level of potential development as determined through problem solving under adult guidance or in collaboration with more capable peers (pp. 85–86). He studied the difference between concepts learned spontaneously, or out of school, with those learned in school, and he concluded that children learned more when the teacher or more knowledgeable adults assisted them. The presence of a more knowledgeable other appeared to foster the discussion and debate of ideas. Vygotsky concluded that the social interaction between students and teachers or other adults is an important aspect of intellectual development, because the zone of proximal development allows learners to take part in more cognitively challenging tasks and problem solving than they could on their own.

Others have expanded Vygotsky's ideas to include collaborative interactions with peers. Researchers have found that students learn more effectively when working in collaborative groups with advanced or knowledgeable peers than they do working alone. Forman (1989) coined the term bidirectional zone of proximal development to describe the expertise levels in collaborative groups and the fluctuations among group members—sometimes members include the teacher and sometimes they are the learner. A collaborative classroom thus becomes an environment that comprises what Vygotsky called multiple zones of proximal development. Students are exposed to overlapping zones as they learn by interacting with many different people—the teacher, peers, and community members. Each person with whom the student interacts can become a support that will enable the student to climb to the next level of learning. This is one reason that collaborative learning is stressed in constructivist classrooms. We discuss collaboration in detail in Chapter 7.

By identifying the learners' zones of proximal development, a teacher can provide the assistance that is needed to move a learner to a higher level of understanding than would be possible without the support. Good teachers have always provided this support by doing such things as modeling ideas for students. The zone of proximal development simply serves to name these types of support so we can talk about them.

Scaffolding. The concept of scaffolding stems from Vygotsky's notion of the zone of proximal development. **Scaffolding** (Bruner, 1977; Wood, Bruner, & Ross, 1976) is a process in which a more knowledgeable individual provides support to another learner to help him or her understand or solve a problem. In scaffolding, the more knowledgeable other directs those aspects of the intellectual task that are initially beyond the capacity of the learner. This allows the learner to take part in intellectual activities that otherwise would be unwelcoming, activities that he or she doesn't completely understand. Imagine a parent helping her child put together a Lego structure. The parent might say, "Let's find all the pieces that are blue." Next, the parent might suggest snapping all the blue pieces together. Although the child might not see how the individual steps fit together to build a castle, because the parent has structured the activity, the child can participate in building the structure.

Scaffolds used by knowledgeable others (parents, teachers, other adults, or peers) include modeling, coaching, sequencing, reducing complexity, and marking critical features. In the Lego example, the parent has provided a very important aspect of instruction that is known as **sequencing**. Sequencing, which is just one of the many types of scaffolds that can be used in a classroom, breaks a difficult task into much smaller, manageable subtasks. What follows is a summary of scaffolding strategies that teachers can use to support learning:

- **Modeling** is the process through which a more knowledgeable person illustrates to the learner how to complete a task. For example, a teacher could demonstrate how to use the concept of *average* to analyze data or how to read a scale. Many science processes can be modeled for students. Some of these processes include *asking questions, planning and designing investigations, constructing explanations,* and *forming conclusions.*

- **Coaching** involves providing suggestions to help a student develop knowledge or skills. For example, a teacher might make suggestions to a student about how to make more precise measurements when reading a spring scale. The teacher might suggest, for instance, that the student make sure the scale is calibrated to start at zero before beginning to use the scale. Other forms of coaching include asking thought-provoking questions (such as "How do your data support your conclusion?"), giving students sentence stems (such as "My data support my conclusion because ..."), and supplying intellectual or cognitive prompts (such as asking students to write down predictions, give reasons, and elaborate answers).

- **Sequencing** is breaking down a larger task into subtasks so a child can focus on completing just one subtask at a time, rather than worry about the entire task at once. A teacher might break down the process of investigations into various components, not allowing the learner to proceed to the next step until completing a step. For example, the teacher could require the learner to complete a rough blueprint plan before moving on to building an apparatus.

- **Reducing complexity** involves withholding complex understandings or tasks until the learner has mastered simpler understandings or subtasks. The classical example is helping a child learn to ride a bicycle by using training wheels. In science classrooms, a teacher might use an analogy to reduce the complexity of a concept. For instance, a teacher might compare how cells use DNA to the instructions for building a model airplane.

- **Marking critical features** is highlighting the essential elements of a concept or task. For instance, a teacher might point out to young students that animals called *mammals* all have hair. In teaching a student how to focus a microscope, a teacher might point out that it is important to always start with the lowest power lens first.

CONNECTING TO
NATIONAL SCIENCE
EDUCATION STANDARDS

SCIENCE EDUCATION
PROGRAM STANDARDS

Teaching Standard B—at all stages of inquiry, teachers guide, focus, challenge, and encourage student learning (p. 33).

Oftentimes, it is the teacher who provides the scaffolds during instruction. Other vehicles for scaffolding are peers (same age or older), community members, parent volunteers, and technology. Many teachers find that, at first, they (or knowledgeable others) must provide much support and structure in the classroom. However, gradually, the learner can take more and more responsibility for structuring his or her own learning. For example, at the beginning of the year, a teacher might model for students how to ask good questions and demonstrate why they are appropriate questions. However, as the year progresses, the teacher expects students to ask their own questions and give justifications for why they are good questions. When students have reached this point, it is still critical that the teacher provide feedback.

For scaffolding to be beneficial to the learner, the following conditions must be present.

- Support must be relevant to the student and the task. To be relevant, the support must be related to a task a student needs to complete. For instance, the teacher can reduce the complexity of data analysis by providing a chart for organizing data.
- Support must correspond to the level of help needed by the student. If the support is geared too high for students, it will not match their understanding. If it is too low, it will not be useful. The analogy that DNA is like a set of instructions for putting together a model airplane might be appropriate for a fifth grader who has put together an airplane. Comparing DNA to the instructions in a computer program, however, would be beyond most fifth graders' experience.
- The support must be given in close proximity in time to the student's request for help. Delaying support might mean that an educational opportunity is lost and that support is not given until a student no longer needs it.
- The student must take action on the opportunity (or apply what he or she has learned). For example, a student needs to apply her understanding of finding averages to her own data.
- Scaffolds need to be **faded**—This means that the support must decrease over time. A child first learning to ride a bike uses training wheels. The training wheels are gradually raised. Next, the training wheels are removed, and the child tries to ride the bike with an adult running alongside the bike. Finally, the child rides the bike on his or her own. In science class, a teacher might start by showing students how to create a bar graph. Next, the class and the teacher coconstruct bar charts. After that, students working in groups construct bar graphs, with the teacher merely giving them feedback.

Authentic Tasks

Social constructivism also points to the need for students to learn by addressing problems that they see as authentic (Brown et al., 1989; Newman, Griffin, & Cole, 1989; Resnick, 1987). To develop integrated understandings, students must develop knowledge related to real situations. Because deep learning occurs only when a task is anchored in meaningful contexts, tasks must take on meaning beyond what occurs in the school. **Authentic tasks** are those that have meaning for the child beyond the classroom. We will explore four aspects of authentic tasks: the driving question, the relevance of a question or topic to students, the connection of learning to students' lives outside of school, and the emergence of science concepts and principles when they are needed.

Driving Questions Focus and Sustain Activities

Driving questions are a vehicle for bringing authentic tasks into classrooms. (Chapter 3 explores the driving question in greater detail.) In project-based science, the driving question contextualizes learning in the lives of learners, organizes concepts, and drives activities. As students pursue solutions to a driving question, they develop integrated understandings of key scientific concepts. For instance, the question "What is the pH of rainwater in our city?" allows students to explore concepts, such as acids

> CONNECTING TO
> NATIONAL SCIENCE
> EDUCATION STANDARDS
>
> SCIENCE EDUCATION
> PROGRAM STANDARDS
>
> Teaching Standard A—Inquiry into authentic questions generated from student experiences is the central strategy for teaching science (p. 30).

and bases, pH, and concentration. The question also organizes the activities involved in planning and carrying out investigations of the acidity of rainwater and of the impact acid rain has on living and nonliving things. A question such as "How do you light a structure?" allows students to explore principles, such as parallel and series circuits, voltage, resistance, current, and power. Activities might be organized around students designing and building structures that they can light.

The Topic or Question Is Relevant to the Student

A guiding principle of constructivism is that problems, questions, and topics must be relevant or pertinent to children. Greenberg (1990) suggested that relevant questions are testable by children, involve the use of equipment in testing ideas, are complex enough to elicit problem-solving approaches, and can be solved through group efforts. Haberman (1995) argued that many students' misbehaviors in schools are caused by not being involved with meaningful learning activities. He stressed that project-based methods can make learning relevant and meaningful. Similarly, Blumenfeld and colleagues (1998), in a summary of the literature, argued that student interest is enhanced when (a) tasks are varied and include novel elements, (b) the problem is authentic and has value, (c) the problem is challenging, (d) there is closure through the creation of an artifact or product, (e) there is choice about what and/or how work is done, and (f) there are opportunities to work with others.

Some people criticize the idea of relevance, arguing that a certain curriculum, whether pertinent to students or not, must be covered in schools. Brooks and Brooks (1993) wrote that, "Relevance does not have to be preexisting for the student...Relevance can emerge through teacher mediation" (p. 35). (For this reason, Chapters 3, 4, and 12 discuss ways teachers can stimulate curiosity and develop relevant investigations.) Few children will come to school interested in the topics of force, momentum, and acceleration. However, a teacher can help make these topics relevant to students by turning the topics into driving questions, such as "How do we stay on a skateboard?" or "Why do I have to wear a helmet, knee pads, and wrist protectors when I'm rollerblading?"

Learning Is Connected to Students' Lives Outside School

Children (and adults) usually exert greater effort to study questions that relate to their own lives outside of school. If a lesson being presented to students is not connected to their lives, it is not uncommon for an upper elementary or middle school student to ask, "Why do we need to know this?" Unfortunately, teachers sometimes reply, "Because it is in the book" or "You will need it when you get older." Few children have the patience to study something simply because it is in the book, and most children do not have the cognitive capacity to accept or care that they will need something far in their future. Research on women's and minorities' involvement in science also points out the need to connect science to children's lives outside of school (Barton, 1998; Haberman, 1995). The lack of relevance to life outside of school is believed to be a strong factor in leading girls and minorities away from science.

Most topics in elementary and middle grades curriculum can be situated in students' lives, For example, teachers can connect the concept of insulation to students' lives by emphasizing the role of insulation in coats and gloves, thermos bottles that keep drinks hot or cold, and coolers that keep food from spoiling.

Science Concepts and Principles Emerge When Needed to Answer a Driving Question

In constructivist classrooms, the concepts and skills presented are used to answer driving questions, rather than presented for their own sake. When concepts and principles emerge on an as-needed basis, students can more easily integrate them into their understandings. For example, students learn to read a thermometer so they can measure the temperature in their decomposition column. They learn about oxygen so they can understand why worms help the decomposition process. In each of these examples, students are not learning something because it is in the textbook; they are learning it so they can solve a problem or apply it to a particular situation. Before continuing, complete Learning Activity 2.2.

Using Technology Tools to Extend Learning

Technology tools extend learning by helping students perform cognitive tasks. Humans have always used cognitive tools to help them learn, engage in intellectual tasks, and make tasks easier (Salomon & Perkins, 1991). For example, humans used the abacus as early as 500 B.C. for counting and performing calculations. Adding

---

Learning Activity 2.2

CRITIQUING A CLASSROOM LESSON

Materials Needed:

- A video of an elementary or middle grade science lesson or a classroom that you can observe in person
- Video equipment and video
- Table 2.2: Features of the Social Constructivist Model of Learning

A. Obtain a video of an elementary or middle grades science lesson or observe a teacher teaching a science lesson.
B. Use Table 2.2 to evaluate the teaching. For each idea in the table, rate the lesson on a scale of often to seldom.
C. What would you do to make the lesson more consistent with the social constructivist view? Record your ideas.

machines and calculators are more modern technological tools for helping us calculate. **Concept maps**, or diagrams to visualize relationships among ideas or concepts, are used widely in schools to clarify concepts. New technology tools also serve as cognitive tools. Electronic spreadsheets allow accountants to collect and analyze numbers. Graphs allow them to visualize and interpret data.

In science classes today, a wide variety of technology tools is available to help students learn. For example, computers can be used to access data on the World Wide Web, calculate data on spreadsheets, visualize concepts on a diagram or picture, and express ideas as written words. Students can use technology tools to acquire, process, organize, and visually represent information. Temperature, light, and motion probes can help students conduct investigations and learn important concepts that are more difficult to understand without the use of technology. Because technology tools have assumed such an important role in science teaching, we devote Chapter 6 to a more thorough review of technology tools.

Chapter Summary

In this chapter, we discussed student understanding and examined several scenarios of student understanding. Students need to construct content, procedural, and metacognitive knowledge to develop understanding. Project-based science strives to create integrated understandings.

The chapter reviewed the historical roots of constructivism and outlined a social constructivist model of teaching. The first feature of this model is active engagement of students with phenomena. In a constructivist classroom, students ask and refine questions, predict and explain phenomena, and engage with concrete materials. The second feature is student use and application of knowledge. In constructivist teaching, teacher and students consciously address prior knowledge; and students identify and use multiple resources, plan and carry out investigations, apply concepts and skills to new situations, devote time to reflection, and take action to improve their own world. The third characteristic is the use of multiple representations of understandings. Multiple representations include varied assessment techniques and student products or artifacts. The fourth characteristic is the use of learning communities. Learning communities are created when students use language to express knowledge, debate, and come to a resolution regarding ideas, concepts, and theories; debate the viability of evidence; learn in a social context; and obtain help from knowledgeable others in learning new ideas and skills that they couldn't learn on their own. The fifth characteristic is the focus on authentic tasks. What makes learning authentic are driving questions that focus and sustain activities, topics or questions that are relevant to the student, learning that is connected to students' lives outside school, and science concepts and principles that emerge as needed to answer a driving question. These five characteristics characterize a transformational, as opposed to a receptional, approach to teaching.

We discussed the important role teachers play in scaffolding student learning. Scaffolds allow learners to take part in cognitive activities just beyond their reach of cognitive development. Teachers can scaffold learners by modeling, coaching, sequencing, reducing complexity, and highlighting critical features.

Tasks in school need to be authentic for children to find them meaningful. Four aspects of authentic tasks are driving questions, the relevance of questions or topics to students, the connection of learning to students' lives outside of school, and the emergence of science concepts and principles when they are needed. Technology tools play an important role in making connections to students' lives in science classrooms today.

Chapter Highlights

- Integrated understanding results when learners have built relationships among ideas, can explain these relationships, and can use their ideas to explain and predict phenomena.
- School science often results in students developing inert knowledge—disconnected, unusable fragments of ideas.
- Project-based science can help students develop integrated understandings.
- To develop integrated understanding, students need to develop content, procedural, and metacognitive knowledge.
- Children construct meaning through their interactions with and interpretations of their world, including essential interactions with others.
- The features of social constructivist teaching include
  - active engagement in phenomena,
  - use and application of knowledge,
  - multiple representations,
  - development of learning communities, and
  - authentic tasks.
- Language serves as a tool to develop understanding.
- Teachers scaffold students so they can engage in tasks just out of their cognitive reach.
- Teachers can scaffold learners by
  - modeling,
  - coaching,
  - sequencing,
  - reducing complexity, and
  - marking critical features.
- Tasks in school need to be authentic to have meaning.
- Four aspects of authentic tasks are
  - the driving question,
  - the relevance of a question or topic to students,
  - the connection of learning to students' lives outside of school, and
  - the emergence of science concepts and principles when they are needed.
- Technology tools extend learning in science classrooms.

Key Terms

Active engagement	Knowledge of cognitive tasks	Reflection
Artifacts	Learning community	Scaffolding
Authentic tasks	Marking critical features	Schemas
Coaching	Metacognitive knowledge	Self-knowledge
Concept maps	Modeling	Sequencing
Content knowledge	Multiple representations	Social constructivism
Equilibration	Procedural knowledge	Strategic knowledge
Faded	Receptional approaches	Transformational approaches
Inert knowledge	Reducing complexity	Zone of proximal development

Notes

1 This book adopts a social constructivist perspective. Other forms of constructivism are radical constructivism and contextual constructivism. For more information on other variants of constructivism, see Tobin, K. (Ed.) (1993). *The practice of constructivism in science education*, Washington, DC: American Association for the Advancement of Science. Also consult Shapiro, B. (1994). *What children bring to light: A constructivist perspective on children's learning in science*. New York: Teachers College Press.

2 When doing this demonstration, the balloon must be filled with a source of dry air, so do not blow up the balloon using your mouth. Your lungs and mouth contain much moisture and will add water vapor to the balloon, as well as other gases. Instead, use a bicycle pump to blow up the balloon.

References

American Association for the Advancement of Science. (1993). *Benchmarks for science literacy*. New York: Oxford University Press.

Anderson, L. W., & D. R. Krathwohl (Eds.). (2001). *A taxonomy for learning, teaching, and assessing: A revision of Bloom's taxonomy of educational objectives*. New York: Longman.

Armstrong, T. (1994). *Multiple intelligence in the classroom*. Alexandria, VA: Association for Supervision and Curriculum Development.

Barnes, M. B., Shaw, T. J., & Spector, B. S. (1989). *How science is learned by adolescents and young adults*. Dubuque, IA: Kendall-Hunt.

Barton, A. C. (1998). *Feminist science education*. New York: Teachers College Press.

Blumenfeld, P., Soloway, E., Marx, R., Krajcik, J. S., Guzdial, M., & Palincsar, A. (1991). Motivating project-based learning: Sustaining the doing, supporting the learning. *Education Psychologist, 26*(3 & 4), 369–398.

Blumenfeld, P. C., Marx, R. W., Patrick, H., Krajcik, J. S., & Soloway, E. (1998). Teaching for understanding. In B. J. Biddle, T. L. Goode, & I. F. Goodson (Eds.), *International handbook of teachers and teaching* (pp. 819–878). Dordrecht, The Netherlands: Kluwer.

Bransford, J. D. (1979). *Human cognition: Learning, understanding, and remembering*. Belmont, CA.: Wadsworth.

Bransford, J. D., Brown, A. L., & Cocking, R. R. (Eds.). (1999). *How people learn: Brain, mind, experience, and school*. Washington, DC: National Academy Press.

Bransford, J. D., & Donovan, M. S. (2005). *Scientific inquiry and how people learn. How students learn: History, mathematics, and science in the classroom.* Washington, DC: National Academic Press.

Brooks, J. G., & Brooks, M. B. (1993). *In search of understanding: The case for constructivist classrooms.* Alexandria, VA: Association for Supervision and Curriculum Development.

Brown, A. L., & Campione, J. C. (1994). Guided discovery in a community of learners. In K. McGilly (Ed.), *Classroom lessons: Integrating cognitive theory and classroom practice* (pp. 229–270). Cambridge, MA: MIT Press.

Brown, J. S., Collins, A., & Duguid, P. (1989). Situated cognition of learning. *Educational Researcher, 18,* 32–42.

Bruner, J. (1977). *The process of education.* Cambridge, MA: Harvard University Press.

Carey, S. (1985). *Conceptual change in childhood.* Cambridge, MA: Harvard University Press.

Driver, R. (1989). The construction of scientific knowledge in school classrooms. In R. Miller (Ed.), *Doing science: Images of science in science education* (pp. 83–106). Lewes, East Sussex, UK: Falmer Press.

Driver, R., Squires, A., Rushworth, P., & Wood-Robinson, V. (1994). *Making sense of secondary science.* New York: Routledge.

Eylon, B., & Linn, M. C. (1988). Learning and instruction: An examination of four research perspectives in science education. *Review of Educational Research, 58*(3), 251–302.

Forman, E. A. (1989). The role of peer interaction in the social construction of mathematical knowledge. *International Journal of Educational Research, 13,* 55–70.

Greenberg, J. (1990). *Problem-solving situations. Volume I.* Corvallis, OR: Grapevine Publications.

Haberman, M. (1995). *Star teachers of children in poverty.* West Lafayette, IN: Kappa Delta Pi.

Kesidou, S., & Roseman, J. E. (2002). How well do middle school science programs measure up? Findings from Project 2061's curriculum review. *Journal of Research in Science Teaching, 39*(6), 522–549.

Krajcik, J. S. (2001). Supporting science learning in context: Project-based learning. In R. Tinker & J. S. Krajcik (Eds.), *Portable technologies: Science learning in context* (pp. 7–29). New York: Kluwer Academic/Plenum Publishers.

Linn, M. C. (1998). The impact of technology on science instruction: Historical trends and current opportunities. In B. J. Fraser & D. Tobin (Eds.), *International handbook of science education* (pp. 265–294). Dordrecht, The Netherlands: Kluwer.

Linn, M. C. (2006). The knowledge integration perspective on learning and instruction. In R. K. Sawyer (Ed.), *The Cambridge handbook of the learning sciences* (pp. 243–264). New York: Cambridge University Press.

Lorsbach, A., & Tobin, K. (1992, September). *Research matters to the science teacher: Constructivism as a referent for science teaching.* Columbus, OH: National Association for Research in Science Teaching.

National Academies Press. (2005). *Rising above the gathering storm: Energizing and employing America for a brighter economic future.* From http://newton.nap.edu/catalog/11463.html

Newman, D., Griffin, P., & Cole, M. (1989). *The construction zone: Working for cognitive change in school.* Cambridge, UK: Cambridge University Press.

Osborne, R., & Freyberg, P. (1986). *Learning in science: The implications of children's science.* London: Heinemann.

Osborne, R. J., & Cosgrove, M. M. (1983). Children's conceptions of the changes of states of water. *Journal of Research in Science Teaching, 20*(9), 825–838.

Papert, S. (1980). *Mindstorms: Children, computers, and powerful ideas.* New York: Basic Books.

Papert, S. (1993). *The children's machine: Rethinking school in the age of the computer.* New York: Basic Books.

Perkins, D. (1992). *Smart schools: Better thinking and learning for every child.* New York: The Free Press.

Piaget, J. (1959). *The language and thoughts of a child* (3rd ed.). London: Routledge and Kegan Paul.

Piaget, J. (1971). Advances in child and adolescent psychology. In *Science of education and the psychology of the child* (pp. 25–41). New York: Viking Press.

Project Kaleidoscope. (2006). *Transforming America's scientific and technological instruction: Recommendations for urgent action.* Washington, DC: Author.

Resnick, L. B. (1987). Learning in school and out. *Educational Researcher, 16*, 13–20.

Roth, W. M. (1995). *Authentic school science.* Dordrecht, The Netherlands: Kluwer.

Rutherford, J., & Ahlgren, A. (1989). *Science for all Americans: Project 2061.* New York: Oxford University Press.

Salomon, G., Perkins, D. N., & Globerson, T. (1991). Partners in cognition: Extending human intelligence with intelligent technologies. *Educational Researcher, 20*(3), 2–9.

Shapiro, B. (1994). *What children bring to light: A constructivist perspective on children's learning in science.* New York: Teachers College Press.

Smith, C. L., Wiser, M., Anderson, C. W., & Krajcik, J. (2006). Implications of research on children's learning for standards and assessment: A proposed learning progression for matter and the atomic molecular theory. *Measurement: Interdisciplinary Research and Perspectives, 14*(1&2), 1–98.

Talsma, V. (2002). Student scientific understandings in a ninth-grade project-based science classroom: A river runs through it. Unpublished doctoral dissertation, Ann Arbor: University of Michigan.

Third International Mathematics and Science Study (TIMSS). (1998).

Vygotsky, L. (1986). *Thought and language* (A. Kozulin, Trans.). Cambridge, MA: MIT Press. (Original English translation published 1962)

Vygotsky, L. S. (1978). *Mind in society: The development of higher psychological processes.* Cambridge, MA: Harvard University Press.

Wood, D., Bruner, J. S., & Ross, G. (1976). The role of tutoring in problem solving. *Journal of Child Psychology and Psychiatry, 17*, 89–100.

Yager, R. E., & Penick, J. E. (1986). Perceptions of four age groups toward science classes, teachers, and the value of science. *Science Education, 70*(4), 355–364.

Chapter 3

Establishing Relevance to Students' Lives

Introduction

Project-based science has several features, one of which is establishing **relevance** to students' lives (Krajcik, Blumenfeld, Marx, & Soloway, 2000). Science instruction should have children exploring solutions to questions (NRC, 1996, 2000). Project-based science calls for a question or problem that children will find relevant to their lives. By relevant, we suggest meaningful and important to learners (Blumenfeld et al., 1991; Krajcik & Blumenfeld, 2006). One of the primary ways in which we establish relevance is by the use of a **driving question** (Krajcik & Mamlok-Naaman, 2006). Driving questions meet the learning and motivational needs of *all* students (Atwater, 1994; Haberman, 1991, 1995) and, therefore, are critical for creating learning environments for both boys and girls, as well as for children from different cultures and races. Another way to establish relevance is to show how what students are learning is connected to real-world issues.

The driving question also serves to organize and drive the diverse activities of a project. Examples of some driving questions are "Will it rain tomorrow?" "Why do I look the way I do?" "What care do our classroom pets need?" "What types of trees

Chapter Learning Performances

- *Describe a driving question.*
- *Explain the features of a driving question.*
- *Create driving questions and defend why they are good driving questions.*
- *Describe how to help children generate driving questions.*
- *Distinguish between driving questions and topic-based questions.*
- *Explain the value of using driving questions to teach science.*

grow in our neighborhood?" "How healthy is our stream?" and "Does my community have acid rain?" In the fall of the school year, a good question for many early elementary students might be "Do pumpkins all have the same number of seeds?" This question would give students many opportunities to explore characteristics of pumpkins and the growth of plants.

This chapter will explore important characteristics of driving question. The chapter also explores how teachers and students can develop driving questions. The chapter ends with a discussion of the value of using a driving question throughout a project. Before we begin to explore how to make science instruction relevant to students, let's examine three scenarios that describe ways in which teachers might organize science instruction. These scenarios will illustrate the importance of using driving questions to establish relevance.

Scenario 1: Reading About Force and Motion

Think about your own elementary and middle school science experiences. Perhaps your science class was like this: Mr. Simmons, your fifth-grade science teacher, started class by saying, "Today we're going to find out about force and motion. Open your books and read pages 121 to 123, which focus on motion and force. Pay attention to the photographs and diagrams." You scanned the pages and read carefully the captions under the graphs and photographs. After about 15 minutes, Mr. Simmons directed students to work in groups at their tables to answer several questions at the end of the chapter. After another 15 minutes, Mr. Simmons called on several students to give their responses to the questions.

You followed most of the reading, but occasionally you drifted off, thinking about the conversation you had with Carmella on the phone last night. You weren't sure why you were learning about force and motion. It was fun working with the students at your table to answer the questions. You were pleased that Mr. Simmons said, "Very good," when you responded, "Because of a force," to the question, "Why does an object stop moving?" You had picked up the answer from your reading. It was confusing to you because it seemed from your everyday experiences that force must be applied to *keep* something moving. You kept thinking about pedaling your bike and how you needed to keep pedaling to keep moving.[1]

Sound familiar? In Chapter 1, we referred to this method of science instruction as *reading about science.* **Although students were introduced to new ideas, they experienced science largely as a reading activity. The concepts of science were not linked to the children's everyday lives. Although reading is an important part of science learning, when reading about science is not tied to a larger picture that has relevance for students, it can become a routine and uninteresting way of learning.**

Scenario 2: Learning About Force and Motion Via an Activity

Maybe your class was like this: Mrs. Wilson, your fifth-grade teacher, always had you doing activities in science. Mrs. Wilson directed the class to work in groups to test if the mass of a block at the bottom of a ramp influenced how far it would move when hit by a moving toy car. Previously, your class had concluded that the number of washers in a toy car influenced how far it could push a wood block at the bottom of an inclined plane (a ramp supported by two books). Mrs. Wilson wrote on the board, "Does mass influence how far the block will move?" She handed out a sheet with directions on how you should set up and carry out the activity. She also handed out a sheet of paper with two columns on it. The columns were labeled "Number of Blocks (Mass)" and "Distance the Blocks Moved."

You were pretty sure that the more blocks you had, the shorter the distance they would move, so you were not sure why you were doing the activity. It was pretty easy, except for measuring the distance the blocks moved. Sometimes you and your partners disagreed about how far the blocks moved. Overall, the time went by pretty fast. You always looked forward to doing the activities, although you weren't always sure why you were doing them.

Once the class completed the testing, Mrs. Wilson had different groups summarize what they found. Overall, the class results supported what you had expected and your own group's findings. Mrs. Wilson wrote on the board, "The more blocks, the less distance they moved when hit by the toy car." You were pretty pleased because this seemed to support your ideas. It made sense. In the class discussion, however, Mrs. Wilson also said and then wrote on the board, "The more mass an object has, the more force that is needed to change its motion." This was confusing. You still didn't really understand "force." You wrote down both of the statements Mrs. Wilson had written. You knew that any sentence Mrs. Wilson wrote on the board would probably be on a test.

As described in Chapter 1, this kind of science teaching is called *process science teaching*. The students performing this activity did learn some new ideas, and they enjoyed themselves; but the students' observations were not connected to other ideas they were learning. Although Mrs. Wilson used the activity to answer a question— "Does mass influence how far the block will move?"—the question was tied directly to learning a topic (motion and force). Students did not see how the question was tied to their lives, except that it might appear on a test.

Scenario 3: Learning About Motion and Force Through a Driving Question

Perhaps your class was more like this: Your class was exploring the question "Why do I have to wear a helmet and knee and elbow pads when I'm roller blading?" As part of this exploration, your class performed a number of activities. Mrs. Vasquez started each of the activities by saying, "If we are going to explore the question 'Why do I have to wear a helmet and knee and elbow pads when I'm roller blading?' then we should know something about the laws of motion and about what causes injuries" Mrs. Vasquez selected the driving question because she knew that a number of students in her class liked to roller blade and would find the question relevant to their lives. She also knew that it would let the class explore a number of ideas related to force and motion, as well as some ideas in biology related to human anatomy. She also knew that it would allow her students to do a number of related activities and also plan some investigations.

Here is how Mrs. Vasquez used the driving question. Mrs. Vasquez directed the class to work in groups to test whether mass influences how far an object will move when pushed by another object. Mrs. Vasquez handed out a sheet of directions on how to set up and carry out the activity. She directed you to set up an inclined plane (a ramp raised on one end by two books). Mrs. Vasquez had the class relate this activity to the previous day's activity, which consisted of defining *force*. She asked the class why it was important to have the ramp at the same height each time. On the board in big letters she had written, "A Force Is a Push." Because of the previous day's work, your class was able to come up with the idea that it is important to have the same push on the blocks each time. She also handed out a sheet of paper with two columns on it. The columns were labeled "Number of Blocks (Mass)" and "Distance the Blocks Moved."

Once the class finished its testing, Mrs. Vasquez had various groups compare their findings and come to an agreement. All the groups posted their results on the board. Then Mrs. Vasquez asked, "How should we analyze the data?" The class had a debate on whether it was more useful to just look at it or create bar graphs. In the end, the class developed bar graphs. Mrs. Vasquez directed each group to write a conclusion based on the graphs. Your group wrote, "The more blocks (mass), the less the blocks moved when pushed by the same thing." Another group wrote, "The fewer the blocks, the farther they moved when hit by the car." Mrs. Vasquez recorded these various conclusions on the board. From the various conclusions, the class developed a consensus conclusion: "The greater the mass (more blocks), the less it will move when the same force (push) hits it; and the smaller the mass (fewer blocks), the farther it will move when the same force (push) hits it." At the end of class, Mrs. Vasquez asked, "How is this activity related to the driving question of the project: 'Why do I have to wear a helmet and knee and elbow pads when I'm roller blading?'" Mrs. Vasquez first had each student write his or her own response. You wrote in your notebook, "I wear knee pads when I roller blade because, when I fall, my body isn't heavy enough to move the sidewalk, and my knees will take the full blow. The knee pads soften the hit." Next you shared and compared responses with students at your table. The science lesson ended with various groups sharing their responses.

This third scenario illustrates what is called project-based science. The main activity is very similar what occurred in the second scenario. However, this lesson goes far beyond that one. Rather than teaching an isolated lesson, the teacher tied all lessons to the driving question of the project. The students learned an important idea related to the motion of objects, but this important idea was connected to their lives through the driving question. The scenario illustrates how the driving question can be used to link activities in a project. The activity of testing whether mass influences how far an object will move was not done in isolation but rather was an essential component in the process of answering the driving question.

What Is a Driving Question?

The driving question is the primary vehicle used in project-based science to establish relevance. What is a driving question? A **driving question** is a well-designed question that is elaborated, explored, and answered by students and the teacher. As depicted in Scenario 3, the driving question is the central organizing feature of project-based science. In fact, the driving question sets the stage for all activities and investigations of project-based science. Project-based science *requires* a question that is meaningful and important to children and serves to organize and drive activities (Blumenfeld et al., 1991; Krajcik et al., 2000). As students collaboratively pursue answers to the driving question, they develop understanding of key scientific concepts associated with the project.

To learn more about driving questions, let's explore further the example driving question presented in Scenario 3 "Why do I have to wear a helmet and knee and elbow pads when I'm roller blading?" How is this question useful? First, the question serves the purpose of organizing and driving activities that take place in a science class. Second, the teacher can use the question to introduce the students to a number of important science concepts and principles—motion, force, distance, and Newton's laws of motion—that are needed to answer it. Third, the question can be used to link these various concepts together. Fourth, the question can inspire applied use of technology—students might want to use motion probes[2] in their investigations, for example, or the question might prompt exploration of the World Wide Web as students try to find out about the different types of materials used to make helmets and pads. Fifth, the question is motivating because it is meaningful to the children in the class; many children enjoy roller blading. Sixth, the question can lead to the creation of products, such as posters or videos, that show what students have learned about why they need to wear helmets and pads and that can, thus, be used as assessment. Hence, the driving question "Why do I have to wear a helmet and knee and elbow pads when I'm roller blading?" ties together activities, skills, concepts, and educational objectives.

The driving question can also be thought of as creating a context that is problematic and needs resolution. A situation that is problematized leads students to encounter and grapple with important ideas or processes (Reiser, 2004), creating a need to know for learners. Reiser discussed three conditions necessary for **problematizing**. First, problematizing involves focusing students' on a situation that needs resolution. Second, problematizing engages students in reasoning about an aspect of a problem that may involve creating a sense of dissonance or curiosity. Finally, it creates interest in some aspect of a problem or getting students to care about understanding or resolving the problem.

Features of Driving Questions

Driving questions have several key features. Table 3.1 summarizes the key features.

Feasibility

A key feature of a good driving question is **feasibility**. A driving question is feasible if (1) students can design and perform investigations to answer it, (2) resources and

TABLE 3.1   Key Features of Driving Questions

Feasibility
- Students can design and perform investigations to answer the question.
- Materials for the investigations are readily available.
- The question is developmentally appropriate for the students.
- The question leads to further questions.

Worth
- The question is related to what scientists really do.
- The question is rich in science content/concepts.
- The question helps students link science concepts.
- The question meets district, state, or national curriculum standards.

Contextualization
- The question is anchored in real-world issues.
- The question has real-world consequences.

Meaning
- The question is interesting and important to learners.
- The question intersects with learners' lives, reality, and culture.
- The phenomena covered by the question are of interest to students.

Ethical
- The practices used to answer the question do not harm living organisms or the environment.

Sustainability
- The question allows students to pursue solutions over time.
- Students can pursue answers to the question in great detail.

materials are available for teacher and students to perform the investigations necessary
to answer it, (3) it is developmentally appropriate for students, and (4) it can be broken
down into smaller questions that students can
ask and answer. We will now explore each of
these components.

Good questions allow students to design
and perform investigations to help find solutions to them. For instance, in the project
with the driving question "What birds live
in our neighborhood?" students can conduct
bird-watching inquiries. They can track bird
types throughout the seasons. They might try
to attract different birds to their playground

> **CONNECTING TO
> NATIONAL SCIENCE
> EDUCATION STANDARDS**
>
> **SCIENCE EDUCATION
> PROGRAM STANDARDS**
>
> Science content must be
> embedded in a variety of
> curriculum patterns that are
> developmentally appropriate,
> interesting, and relevant to
> students' lives (p. 212).

or nature area by erecting birdhouses and hanging bird feeders offering different
types of bird seeds. A question, such as "How can you cure a sick animal?" although
meaningful for children, doesn't lend itself to students doing investigations because
students don't have the knowledge or materials needed to cure a sick animal. Such
a project also has the potential to be harmful to animals, and any project involving
sick animals is potentially harmful to students (see Chapter 11 for more information
about safety in the classroom).

Another component of feasibility is the availability of materials. Teacher and students should be able to find resources and materials to perform a variety of investigations related to finding solutions to the driving question. For instance, in the project
with the driving question "Does our community have acid rain?" students can readily collect and measure the pH of rainwater using a variety of techniques, including
pH paper, a pH meter, and a pH probe attached to a computer. Because pH paper is
inexpensive, all students in the class can measure the pH of rainwater several times
during the school year. Students can also take pH paper home to measure the pH
of rainwater near their homes. Some questions, however, involve materials that are
much more difficult to obtain. For instance, the question "How clean is my air?" does
not readily allow students to design investigations. Although students can measure
particulate matter easily, testing for pollutants often requires sophisticated pumps
and filters that are too expensive and advanced for most school systems.

Good driving questions are also developmentally appropriate for students. For
instance, the project with the driving question "What birds live in our neighborhood?" is developmentally appropriate for children of various ages because, while
investigations might differ among age groups, children of many ages can investigate
what birds live in their neighborhood. However, a question, such as "What is the
quality of my air?" although potentially very meaningful for most children, is probably not developmentally feasible for students in grades three and lower. The content
and the investigations for this question, which involve understanding the particle
nature of matter, are more suitable for students in middle and high school.

A feasible driving question will also allow students to ask and answer their own
subquestions. When a driving question is broad enough and is meaningful to

students, students should be able to think of many related questions they would like to pursue. For instance, in the project with the driving question "How healthy is our stream or lake?" students might ask, "Does our stream quality change with the seasons?" or "What are some biological indicators of our stream's health?" In the project with the driving question "What happens to all our garbage?" students might ask, "What materials decompose the fastest?" and "Does the presence of worms speed up the process of decomposition?"

Other questions don't lend themselves to students generating their own questions and pursuing solutions to those questions. For instance, it would be difficult for students to ask questions related to the driving question "Can we travel to Mars?" It's not that the curriculum content contained in the question is not important. It's that the question will not allow students to explore through investigations the questions they ask. Allowing students to ask their own questions has several benefits. First, student motivation increases with student ownership of a project. Second, by designing their own experiments, investigations, or presentations, students learn design and process skills and increase their knowledge of a topic. Finally, asking questions and pursuing solutions to these questions are essential to experiencing and "doing science."

A feasible driving question helps students realize that they can learn about the natural world by setting up and conducting their own investigations. Even the most interesting driving question will fail to generate a successful learning experience if appropriate investigations cannot be done, there is a lack of resources, or students are not developmentally ready for it.

Worth

Another key feature of the driving question is its **worth**. A question that is worthwhile contains rich science content that students can explore and that helps meet district, state, or national standards.

Perhaps the most important feature of a worthwhile driving question is the quality of science content and process that it can encompass. The most worthwhile driving questions cannot be answered without students gaining an understanding of the intended science content. Students should be able to see how the science content relates to the driving question, and the driving question should be used to help place all the science content into a real-world setting. For instance, in Scenario 3, students explore a number of important science concepts related to force, motion, and the human body. Students cannot answer the question "Why do I have to wear a helmet and knee and elbow pads when I'm roller blading?" without understanding these science concepts.

Another way of looking at this component of worth is to consider that the driving question must subsume important science content that helps teachers meet district, state, or national curriculum standards. If a question cannot help a teacher meet curriculum requirements, then it lacks worth. A question, such as "Does our community have acid rain?" provides students with the opportunity to explore numerous concepts related to chemistry, ecology, and geology. In exploring the question, students are exposed to much science content: They learn about acids and bases, they find out about weather, they study the effect of acid rain on the environment, and they learn about how experiments are designed. However, if these outcomes are not included in the objectives of a school's curriculum for that year, the question's worth is diminished.

Contextualization

Contextualization is a key feature of good driving questions. A contextualized question anchors the project in an important real-world situation and has important consequences. A contextualized driving question is critical in students' being able to see the relevance of the project. However, students may not see immediately how a question relates to the real world or perceive its consequence. A good driving question presents the opportunity to draw students in and *teaches* them to see how it is related to their lives. A number of instructional strategies can enhance contextualization of driving questions. Teachers can use investigations, observations, readings, and discussions to help students appreciate how a driving question relates to their lives. Many upper elementary and middle school students will not immediately perceive the importance of the question "What happens to all of our garbage?" Most children are not interested in what happens to their trash. Imagine that a teacher asks students to bring to class a number of items that typically get thrown out at home. Children bring in paper cups, plastic cups, newspaper, and orange peels, among other things. The teacher asks the students to put these items in some "mess stockings" (nets to hold the items), bury them in a fenced-off area of the playground, and then make predictions about what will happen to the various items they buried. As students pursue this investigation, they come to realize that some materials, such as Styrofoam, do not decompose at all and that some materials decompose more slowly than others do. From this activity, students can make the leap to a larger context and begin to think about large landfills and the cumulative effect of garbage on the environment. At this point, the driving question, thanks to the contextualizing activity, assumes importance for the students.

In contrast, think about a question that is not contextualized, such as "What are the names of rocks in my environment?" Although identification of elements of the natural world is important scientific content, unless the identification of rocks is tied to something that the children value, the question lacks the potential to motivate students and generate a successful, meaningful learning experience. Another common activity in schools is for children to make objects, such as terrariums, solar cookers, or oceans in a bottle. Although these activities could easily be contextualized, they often are not; and after children make the objects, they don't do anything with them.

Meaning

Driving questions should have **meaning** in students' lives. Usually, meaningful questions are those that students see as important and interesting to them. Meaningful questions intersect with their lives, reality, and culture. For instance, students might see the question "Why do I have to wear a helmet and knee and elbow pads when I'm roller blading?" as meaningful because they (or others they know) roller blade. The question is directly related to their lives and what they do.

It is important to point out, however, that learners might not immediately perceive the real-world issue involved in this question, which is the prevention of serious head injuries. In addition, students might not initially see all the ways that a driving question is related to their lives. For instance, many students might not initially care about the quality of their stream. However, through finding out about water quality and its effects on humans, students come to care about the driving question "What is the quality of our stream?"

Driving questions can also be meaningful just because they refer to phenomena that are inherently interesting to students. For instance, many early elementary students are fascinated with magnets. A good driving question for early elementary school children could be "How are magnets used around my home?" Although many upper elementary students would not find this question of interest, most early elementary students would be captivated by it. A number of phenomena have this potential to attract students' interests. Young learners are frequently interested in sinking and floating things in water, how toys work, rainbows, and animals. Such phenomena can be turned into driving questions, such as "Why does a rainbow form?" that early learners can pursue. Many children also love to find out how to light a bulb with one battery, one bulb, and one wire (Elementary School Science Program, 1970). Once they learn this, they are on to new challenges, such as lighting a model house or finding out how their battery-operated toys work.

Ethical

Good driving questions are **ethical**. As students investigate their questions, they should not harm living organisms or the environment. Further, as students explore and find solutions to questions, the procedures they use should be ethical. There should be no leeway in this position.

A driving question is ethical if it upholds the safety, health, and welfare of living things and the environment. Students sometimes have questions that are of interest to them but, if investigated, would harm living things or the environment. For example, although students might be interested in polluting an environment to observe what happens to various organisms in it, such experiences cannot be supported in the classroom. They are unacceptable in schools and in scientific practices. As is true in professional science organizations, many school districts have written guidelines about the use of vertebrate animals and other living things in scientific research. Such guidelines provide strict rules about ethical treatment of living things. It is a teacher's responsibility to help children select driving questions that are ethical. Rather than purposely polluting an environment to observe the results, students can easily develop a driving question that investigates existing streams and ponds for levels of pollutants and observe the already existing effect on plant and animal life.

Sustainability

A final feature of a good driving question is **sustainability**—its ability to sustain student engagement over time. Teachers and students can work on a good question for weeks or even months. For instance, to develop a full understanding of the driving question "How healthy is my stream?" students not only must learn content in a variety of disciplines but also explore the content several times during the school year as the changing seasons contribute to variations in temperature, water levels, and animal life in the stream.

In addition to holding student interest over time, a sustainable driving question encourages students to study information in great depth. For instance, in pursuing solutions to the question "What birds come to my bird feeder?" students can identify, count, and keep track of various birds. The bird-feeder question also encourages students to learn about the behavior of the different birds, their eating and nesting habits, and their migration patterns.

"How much water can different paper towels hold?" and "Do plants grow better in the shade or light?" are examples of questions that teach important scientific concepts and skills, but are not sustainable. In one class period, students can answer the question about paper towels, and in as little as 2 weeks students can answer the question about the growth of plants. Both of these investigations could make worthwhile components of a project if they were subsumed under a larger driving question. For instance, the question about whether plants grow better in the shade or light could be subsumed under the broader driving question "What do plants in my environment need to grow?" In this case, students could plan and carry out a number of other experiments to explore the influence of fertilizer, temperature, moisture, light, soil type, and plant type that would keep the project going over time.

Examples of Driving Questions

In this section, we examine the key features of three driving questions. For each example driving question, we summarize the key features.

Example 1: What Kind of Insects Live on Our Playground?

Imagine that the district you are working in has a curriculum objective related to students learning about insects. Rather than covering insects as a topic, you decide to ask the driving question "What kinds of insects live on our playground?" Does this question meet the features of a good driving question? Let's explore.

First, the question is feasible because it gives children opportunities to investigate and ask new questions of their own. They might ask, "When do the various insects first appear on the playground?" "When do the insects disappear?" "What kind of food do they eat?" "Where do they live?" "What type of environment do the insects live in?" and "How do they eat?" Students can design and perform investigations to answer these questions. For example, they can keep journals about when insects appear and disappear. They can collect insects and investigate what they eat or how they react to light. They can take nature walks to investigate where insects live and how they move. In addition, there isn't much sophisticated equipment—besides a hand lens, jars, and maybe a net for collecting—that is needed for this project.

Second, the question is worthwhile because it enables students to explore rich science content, including insect classification, life cycles, and behaviors. Also, the question enables students to relate to what scientists really do (ask many questions and pursue answers to these questions). This question matches up with a number of benchmarks for elementary students cited in *Benchmarks for Science Literacy* (AAAS, 1993). For instance, one of the benchmarks for grades 3–5 is "Insects and various other organisms depend on dead plant and animal material for food." By exploring the subquestion "What do insects eat?" students can attain this benchmark. The question also helps meet a number of other benchmarks (AAAS, 1993) and standards (National Research Council, 1996) for inquiry and the nature of science. For example, as students observe insect behavior and take notes, they are observing and recording information, two important benchmarks.

Third, the question is contextualized and meaningful. The type of insects found can give an indication of the quality of soil and the local environment. By observing the insects, finding out information about them, asking subquestions, and conducting investigations to answer the subquestions, students can come to understand how this is a real-world question with meaning for them: Students are not reading about insects or exploring them through classroom activities; they are exploring and learning about insects in their environment, in the backyard, on the playground, and/or in a local nature area.

Fourth, this question is ethical because it does not lead students to harm insects. By taking nature walks, for example, they are able to observe the insects without collecting or killing them. The question does not promote practices that would endanger the environment in which the insects live.

Fifth, this question is sustainable: It can take students an entire year to study all the insects in their environment, given that insects appear and disappear in different seasons. Moreover, the question might propel students to learn other information about insects.

Example 2: What Kind of Land Features Are in My Environment?

Often, school districts have curriculum objectives related to earth science. Imagine that your class is studying the question "What kind of land features are in my environment?"

The question is feasible because it allows students to ask a number of subquestions such as "What type of soil do we have?" "What kind of rocks do we have?" "How was the shape of the river formed?" and "How were the hills and valleys formed?" However, these subquestions will not come easily to upper elementary students, and the teacher might need to lead walks to help focus students on various geological features. Some of these questions allow students to perform investigations. For instance, students can collect a variety of rocks from around the school or at home and then try to identify what they are. Although students can't *explore* how a river was formed, they can do various explorations on a stream table (a long tray set up with sand and water to simulate erosion) to duplicate the formation of streams and rivers. They can also search the library or the World Wide Web for historical pictures of the river to see if the shape of the river has changed. Other questions, such as "How were the hills and valleys formed?" cannot be explored easily through investigations. However, students can read and discuss how such phenomena occur, and they can manipulate modeling clay to simulate how a hill is formed as pressure below the earth presses up on it.

Next let us explore whether the question is worthwhile. The driving question and subquestions provide learners with opportunities to learn a number of important earth science concepts. For instance, the students can explore classifications of rocks and soil, compositions of rock and various soils, types of land formations, and reasons for various land features. Hence, by pursuing an answer to the driving question, students have the opportunity to explore important science content that meets a number of benchmarks cited in *Benchmarks for Science Literacy* (AAAS, 1993). For instance, one of the benchmarks for grades 3–5 is as follows:

> Rock is composed of different combinations of minerals. Smaller rocks come from the breaking and weathering of bedrock and larger rocks. Soil is made partly from weathered rock, partly from plant remains—and also contains many living organisms. (AAAS, 1993, p. 72)

Taking students on walks to explore various local phenomena helps link the driving question and subquestions to the real world. Although some children will not immediately see the question as meaningful to them, allowing them to bring in different soil samples and various rocks from their backyards as part of their investigations will help them gain ownership of the question. For instance, students can compare various sample cores (drillings into the soil with a tube to show the soil from different layers) from their backyards. They can explore such questions as "How are the samples different?" and "If they are different, why are they different?"

This exploration of land features is ethical because it does not involve students endangering the environment or polluting it. In fact, this particular driving question may also help students observe whether other human practices in their region, such as strip mining or blasting for stone in quarries, have destroyed land features. A study of land features could also initiate some interesting discussions of ethics among the students in the classroom.

Finally, as students begin to see the value of the driving question, the question becomes sustainable because there is much rich content associated with it. Many of the subquestions take time to explore and investigate. For example, "What type of soil do we have?" is a question that can be explored over time as students investigate what plants can grow in the soil type, whether it can be used for making things (such as bricks), and if it drains well (such as sand) or holds water (such as clay).

Example 3: How Can I Care for the Various Animals in Our Classroom?

Let's look at a third example that focuses on biological objectives. Many children will be interested in the animals that are in the classroom. Most students, as well as adults, love to watch various animals eat and play. The driving question "How can

I care for the various animals in our classroom?" appears to have much potential to engage students in learning about animals.

The question appears to be very feasible. Students can ask a number of related questions and pursue answers to many of these subquestions through investigations. For instance, students could ask, "How do the eating habits of our animals differ?" "How do the various behaviors of the animals differ?" and "What kind of habitat do the various animals prefer?" All of these questions can be explored through investigations. Students can observe the eating habits of the different animals and find different feeding patterns. Note that some subquestions, such as "What should I do if one of our animals gets sick?" should not be explored via hands-on investigations. Students can explore this type of question by finding literature on the topic or talking to experts, such as veterinarians. As mentioned earlier, the care of animals should always be left to the expert.

In pursuing an answer to the driving question and the various subquestions, students will need to explore a variety of different science content areas. For instance, they will study such content as nutrition and environmental needs, habitat, body systems, body function and structure, and behavior. These content areas match *Benchmarks for Science Literacy* (AAAS, 1993) and *National Science Education Standards* (NRC, 1996). Hence, with respect to students exploring important content, the driving question appears to be worthwhile.

The question also appears to be very motivating. Observing and caring for classroom animals is meaningful. Teachers can enhance interest and ownership by assigning each student to care for a particular pet. Because the pets are in the classroom and the learners must care for them, the question is contextualized for the learners.

This question is clearly an ethical driving question as it focuses on the welfare of animals. Occasionally, students will be tempted to investigate whether animals like some types of food more than others, prefer a warmer or colder temperature, or respond to direct light. Although children might learn from such investigations, they may not be ethical. For example, if an investigation of food preferences provided an animal with food that was not nutritious, it could endanger the animal's health or shorten its lifespan. Because the driving question posed here focuses on caring for the animal, it directs children away from problematic practices, such as investigating unhealthy foods or temperature.

Finally, the scope and the nature of this driving question and its subquestions appear to hold students' attention for a long period of time. Caring for and learning about the animals can be done throughout the entire school year.

How Does Using a Driving Question Differ from More Traditional Instruction?

Traditional science classes are organized by topic, rather than by a driving question. Topics might include the solar system, phases of the moon, weather, nutrition, plants, the human body, force, and so on. Non-project-based science classes might cover the same science content as project-based science classes, but the material is not connected in the same way. By using driving questions, students can see relevance in what they learn. When they learn a new skill or concept, they can immediately apply it to help answer their questions.

Typically a textbook chapter will mention reasons for learning the topic. For instance, the study of genetics has led to the development of different hybrids of corn and other vegetables, and so is worthy of study. Although knowing such reasons might increase interest, the reasons are not the driving forces behind learning; and teachers frequently find it difficult to hold students' attention in such cases.

Contrast the three driving questions just examined with questions that focus on such topics as "What are the six simple machines?" "What are the nine planets of the solar system?" and "What is light?" Although some students might be interested in the structure of the solar system, most children don't have a reason to be. Although these topics might encompass worthwhile content, they are not likely to engage learners because they are not directly related to children's lives. These questions lack the ability to help students develop connections between their studies and their own lives.

In addition, traditional science instruction and textbook activities lead to the development of isolated skills and understandings, without context and meaning. Typical activities that students pursue in school science are designed to demonstrate or verify concepts. For example, a traditional science textbook chapter on plants will tell students that plants need sunlight to grow. Then, the textbook activity will likely verify this information by having students grow one plant in the sun and another in a dark closet.

Take time now to complete Learning Activity 3.1. It will help you see the difference between a traditional question and a driving question.

How Is a Driving Question Developed?

Coming up with driving questions with the features discussed earlier in this chapter is challenging. Teachers might make a number of failed attempts before settling on a question for a project. Slowly, however, teachers can develop a repertoire of driving questions that helps show the relevance of projects.

Teacher-Generated Driving Questions

Driving questions that originate from the teacher or the curriculum have a number of advantages. Perhaps the most important is that such questions can match school district outcomes and standards with a classroom project. Another is that orchestrating a class driven by a teacher's questions is much easier than orchestrating one driven by students' questions. However, teacher-developed questions may not be meaningful to the student or provide opportunities for students to generate subquestions. The teacher needs to ensure that students see the relevance of the question.

There are many ways that a teacher can develop a good driving question. Some teachers like to begin with ideas they come across in their reading. For example, a teacher might use a question found in this book. Others like to begin with ideas

Learning Activity 3.1

EVALUATING VARIOUS QUESTIONS

Materials Needed:

- Paper and pencil or a computer

A. Listed here are several questions. Evaluate whether the questions can serve as driving questions. Are the questions feasible, worthwhile, meaningful, ethical, contextualized, and sustainable? Be sure to state your reasons.

B. Contrast your answers with those of another person in your class. Try to come to a consensus.

Question	Can It Serve as a Driving Question?	Reason
What is gravity?	_____	_____
Why do I need to wear a bicycle helmet?	_____	_____
Why does my water taste bad?	_____	_____
What is matter?	_____	_____
What foods are good for me to eat?	_____	_____
Do all apples have the same number of seeds?	_____	_____
How can I run faster?	_____	_____
How good is the soil in my playground or yard for growing plants?	_____	_____
Why do only the weeds grow in the cracks of the sidewalk?	_____	_____
What kind of leaves are these?	_____	_____
Why is too much junk food bad for me?	_____	_____
Why is it colder in the winter and warmer in the summer?	_____	_____

they have heard about or experienced in methods classes, at teacher workshops, or in working with students. Some teachers like to develop questions from the school's curriculum. Some great driving questions come from listening to students' questions and ideas. A teacher's hobbies or personal interests can provide the catalyst for driving questions. So can the local newspaper, a news event, or a television program. Finally, other teachers are great sources of ideas for new driving questions and projects. We will now discuss each of these sources for developing driving questions. We will also discuss concerns that must be addressed when using teacher-developed questions.

Developing Driving Questions From Personal Experiences

It is possible to glean many driving questions from your reading. This book has presented several questions that can be used to drive projects:

- What care do our classroom pets need?
- What type of trees grow in our neighborhood?
- How healthy is our stream?
- Does my community have acid rain?
- Do apples all have the same number of seeds?
- Why do I have to wear a helmet and knee and elbow pads when I roller blade?
- What kind of land features are in my environment?
- What kind of insects live on our playground?

These questions can be used at a variety of grade levels. For instance, the question "Why do I have to wear a helmet and knee and elbow pads when I roller blade?" could be used to drive a fourth-grade unit on motion, or it could be used to drive a seventh-grade class on the same topic. We have used the question "How healthy is my stream?" successfully at both the fifth-grade level and at the eighth-grade level. Other questions are probably best used only for lower or upper grade levels. For instance, the question "Do apples all have the same number of seeds?" is most appropriately used at the early elementary level (grades K–2).

Other personal experiences besides reading that are good sources of driving questions are methods classes, teacher workshops, and interactions with students during teaching. A typical science methods class exposes teachers to a variety of physical, life, and earth/space science topics that can be turned into driving questions. For instance, a teacher might have learned about electricity in a methods class by lighting a bulb using a wire and dry cell, constructing series and parallel circuits, and connecting small motors and switches to circuit paths. This experience could be used to develop the driving question "How is electricity used in my environment?" Teaching experience may tell a teacher that students are interested in classroom animals. This knowledge would help the teacher develop the driving question "What care do our classroom pets need?"

There are several advantages to using driving questions with which you have some experience. First, although you cannot possibly anticipate all of the rough spots in a project, knowing what students will be studying helps smooth out the rough spots. Second, with familiar questions, you can map out clearly what curriculum standards a project will meet. Third, teacher-designed familiar questions are often the most effective initial driving questions, because they allow you to start from a familiar basis instead of treading into unknown territory. Finally, with a familiar question, you can plan and prepare resources, such as equipment, readings, and guest speakers.

Brainstorming Driving Questions Associated with the Curriculum

One of the authors of this book likes to start to develop a driving question by reviewing the curriculum outcomes or standards that need to be met by a project. Next, he likes to brainstorm the investigations that students might conduct. He then considers ways to contextualize the project for students. At the same time, he jots down related questions that might be meaningful to students. Figure 3.1 illustrates this process of brainstorming a driving question to meet curriculum outcomes.

One middle school teacher rejected a number of questions related to her curriculum on forces and motion before she settled on the driving question "How do I stay on a skateboard?" Science content was the teacher's first consideration in selecting a driving question for the skateboard project. Even before she began thinking about what kind of question would be meaningful and interesting to students, she clearly

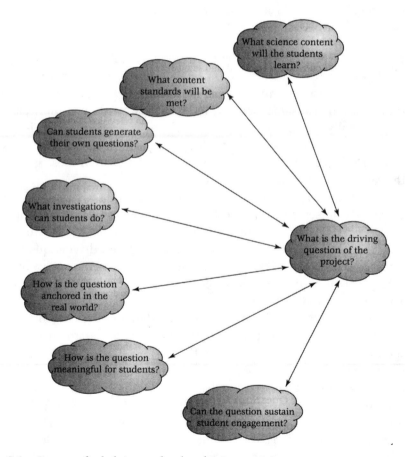

Figure 3.1   Prompts for helping to develop driving questions.

delineated what science content (force and motion) would need to be covered in order to meet district curriculum standards. Student motivation and interest were also a primary concern for her in the skateboard project. By using "How do I stay on a skateboard?" the teacher helped her students see tangible and interesting connections between abstract physical concepts and their own world.

The strongest advantage to brainstorming driving questions from curriculum is that the teacher can thus ensure that important curriculum outcomes are met. This is critical in most school districts, because the curricula at different grade levels are interdependent. In addition, in some areas, achievement tests or graduation requirements are based on particular content.

Developing Driving Questions by Listening to Students

Another good way to develop a driving question is to listen to questions that students ask. Some teachers create "test banks" of student ideas. A test bank is a notebook in which teachers record student questions and ideas. For example, a student might ask, "What kind of bird is that pretty red one?" This question might later become the driving question "What birds visit our feeder?"

Another source of student questions is the dialogue journal. This is a journal of correspondence between student and teacher. Science notebooks also provide question possibilities. A teacher might notice that a student writes in her notebook," My mom was nagging me again last night to wear that stupid helmet when I was roller blading!" This idea might prompt the teacher to develop the driving question "Why do I have to wear a helmet and knee and elbow pads when I'm roller blading?"

Driving questions developed from listening to students and using their experiences are potentially more meaningful to students (MacKenzie, 2001). This is a powerful reason to use these types of questions. However, because a driving question must meet district curriculum standards and other criteria, use Figure 3.1 as a guide in evaluating the questions.

Developing Driving Questions From Hobbies and Personal Interests

Teachers can come up with ideas for driving questions from their hobbies or personal interests. Rock collecting, fishing, cooking, outdoor activities, traveling, amateur astronomy, and jewelry making can lead to excellent driving questions.

For example, if a teacher's hobby is outdoor activities, she can take the class on various field trips to set the stage for asking questions related to environmental issues and nature. Driving questions that might arise from this are "What kind of trees and plants are in my neighborhood?" "What kind of animals are in my neighborhood?" and "What kind of land forms are in my neighborhood?"

There are several reasons for developing driving questions from hobbies and personal interests. First, teachers often show a great deal of enthusiasm for questions related to personal interests, and this enthusiasm can be contagious, making the questions meaningful to students. Second, questions related to hobbies and personal interests are frequently sustainable because the hobbies or interests themselves are sustainable. Finally, teachers know a great deal about their personal interests, enabling them to easily make connections with curriculum outcomes and standards.

Developing Driving Questions From the Media

Driving questions can also result from reading the newspaper or magazines, listening to the news, or watching documentaries and other shows on television. Newspapers and news shows present ideas that are current, usually less than a day old. Magazines and television documentaries provide information that is only a month or two old. Other ideas presented in magazines or on television shows may be interesting enough to become driving questions.

For instance, imagine that a gasoline tanker overturned a block away from a school in a large urban district. The area had to be evacuated, the sewers were pumped, the air smelled like gasoline, children and adults acquired headaches, and the expressway was closed for hours. A teacher in the district could quickly develop this driving question: "What hazards exist in my environment?" This question might lead to the study of the hazards and how chemicals become airborne, why the people got headaches, and what breathing problems exist for adults and children.

Questions that result from media sources can be very appealing to students, especially if they are familiar. Familiarity with news items helps contextualize the question for students, making the question interesting and meaningful. However, a major problem with using driving questions that come from media sources is that they are hard to plan for ahead of time. This lack of planning often prevents teachers from reaching the curriculum standards. However, a teacher who is familiar with curriculum standards ahead of time might be able to quickly capitalize on a news story. For this reason, it is a good idea to become very familiar with district, state, or national standards before the planning stage.

Developing Driving Questions by Listening to Other Teachers

The personal experiences of others are good sources of driving questions. Your colleagues have probably attended interesting classes and had varied experiences with students. They may participate in novel hobbies or watch different television shows or read different magazines than you.

In addition to providing fresh ideas for driving questions, colleagues can provide invaluable insight into dealing with the challenges of doing a project. They may have ideas for solving a problem, meeting curriculum standards, or classroom investigations. They may also have connections to community agencies, guest speakers, or resources.

Using Driving Questions That Come From Published Curriculum

A number of projects and driving questions can result from using published curriculum. Published curriculum includes textbook series, computer programs, CD-ROM applications, and kit-based materials. Textbook series and kit-based programs are frequently adopted by school districts to meet their science outcomes and standards. Computer and CD-ROM applications commonly augment the adopted curriculum. Often these published curriculum materials are a good place to start when doing project-based science for the first time, because they provide a structure for selecting driving questions that meet curriculum standards. In addition, they usually contain some ideas for investigations or activities, resources (such as science supplies, hands-on materials, and student handouts), and background information about the science content for the teacher. Many of these materials provide the structure for lessons designed to teach basic knowledge and skills necessary to investigate a driving question.

For example, most elementary science textbook series include a chapter on such a topic as land and water. Each of the topics can be turned into a number of driving questions. For instance, in a chapter on land and water, students can explore the quality of their school's water supply and investigate a number of subquestions, including "What is the nitrate and chloride concentration of our school's water supply?" and "What are their sources, and how do they enter the water supply?" In answering these questions, students would deal with important science content. Students would learn about watersheds, examine where their water comes from, investigate how local topography influences their watershed, and see how sources of pollution in one component of a watershed can influence water quality in other areas. They could study pollution in terms of the context and concentration of various substances, focusing especially on the influence of nitrates in fertilizers on water pollution. Students could also engage in teacher-directed investigations. Students could, for example, measure the nitrate and chloride concentrations of their school's water supply and examine water treatment by investigating the influence of chlorine on yeast growth. Class data about nitrate levels could be shared with a research team (a team made up of representatives from other schools) via an electronic network. Students would consider how local environment, commerce, agriculture, and industry might account for the differences reported by various research teams. Students could also create a number of products that reflect their understanding. For instance, they might debate water-use policy, design models of sources of pollution, and create posters and videos of their debates and panel discussions. Similarly, textbook chapters on weather, geology, or pollution may also lack driving questions; but, again, it is very easy to develop some: "What makes our weather?" "What affects the landforms in my region?" or "Does our community have acid rain?"

Text-based curriculum materials, such as *The Pillbug Project: A Guide to Investigation,* by Robin Burnett (1999), available through the National Science Teachers Association, also serve as sources of possible driving questions. Although there is no explicit driving question associated with *The Pillbug Project,* one easily comes to mind: "What are these critters anyway?" Teachers can show students some pillbugs, and the project work can begin. The *Great Explorations in Math and Science (GEMS)* materials, available from the Lawrence Hall of Science, can be used as a base for developing many driving questions and projects. There are more than fifty GEMS books, including *Ant Homes, Crime Lab Chemistry, Bubble-ology, Acid Rain, Global Warming and the Greenhouse Effect, Mystery Festival,* and *Fingerprinting.* For example, the book *Ant Homes* can be used to develop the driving question "What kinds of insects live on our playground?"

Although topically organized, kit-based curricula, such as the *Full-Option Science System (FOSS),* published by Delta (Lawrence Hall of Science, 1993, 2000), and *Science and Technology for Children (STC),* published by Carolina Biological (National Science Resource Center, 1991, 1997), are excellent project starters because they offer a wealth of resources and provide the structure for many lessons. For example, FOSS's *Trees* kit can be used to develop the driving question "What kinds of trees are in my neighborhood?" The kit entitled *Insects* can be used to study insects that live on the playground. *Earth Materials, Landforms,* or *Pebbles, Sand, and Silt* can be used to study the driving question "What kinds of land features are in my environment?" STC's kit, *Food Chemistry,* can be used for lessons to study the driving question "How do we stay healthy?"

An extensive list of science curriculum materials is available from the National Science Teachers Association Web site (www.nsta.org). Another great resource for examining published curriculum materials is the Eisenhower National Clearinghouse for Mathematics and Science Education, available on the World Wide Web (http://www.goenc.com).

Using Technology Tools To Develop A Driving Question

Teachers also can use the World Wide Web as a source of driving questions and project ideas. A number of research groups are developing projects, and many of these projects can be found on the Web. The Web-based Inquiry Science Environment (WISE) and the Center for Highly Interactive Classrooms, Curriculum, and Computer in Education (hi-ce) are two worthwhile sites to visit (http://wise.berkeley.edu/ or http://www.hice.org/).

Making Sure Teacher-Generated Driving Questions Work for Students

No matter what the source of a driving question, students need to find the driving question relevant to their lives. Driving questions that provide opportunities for students to ask their own questions and explore solutions to those questions are more likely to be motivation to students. Driving questions that are teacher generated should not be so highly constrained that the outcomes are predetermined, leaving students with little room to develop their own approaches to them. In Learning Activity 3.2, you will practice generating driving questions and turning curriculum materials into good driving questions.

Learning Activity 3.2

GENERATING DRIVING QUESTIONS FROM TOPICS

Materials Needed:

- Paper and pencil or a computer
- A variety of curriculum materials (such as from GEM, AIMS, FOSS, STC, and *National Geographic*)

A. Use the following topics to create teacher-developed driving questions. Write an explanation of why you think they make good driving questions.

Grades K–4 topics from the National Science Education Standards (NRC, 1996):
- Changes in the earth and sky
- Light
- Electricity
- Magnetism
- Life cycles
- Organisms and environments
- Properties of earth materials
- Objects in the sky

Grades 5–8 topics from the *National Science Education Standards* (NRC 1996):
- Changes and properties of matter
- Motions and forces
- Transfer of energy
- Heredity
- Reproduction
- Structure of the earth
- Solar system
- Environmental hazards
- Personal health

B. Examine a variety of commercially published curriculum materials. Match available materials to the topics listed from the *National Science Education Standards* (NRC, 1996). Write about how the commercially published materials could support the driving question.

C. How could the driving questions you developed provide opportunities for students to ask their own questions and explore solutions to them?

Student-Generated Driving Questions

Initially, teachers may feel more comfortable creating questions and activities themselves. However, driving questions can also arise from students' personal interests or surface as subquestions to the main driving question of a project. With practice, students can take responsibility for creating both the questions and the investigations. One of the biggest benefits of student-generated questions is that the meaningfulness criterion is practically ensured. However, teachers must still work carefully to determine whether questions are feasible, worthwhile, ethical, and contextualized. Sometimes students select topics that have high interest, but that cannot be answered through student investigations. Dinosaurs are a common topic of interest among young children, for example, but there aren't many investigations that students can perform related to this topic. For this reason, although dinosaurs might be a good topic to read about, they don't lead to feasible projects. Teachers must also worry about whether content meets district curriculum standards and whether students' questions are real world questions. For example, students might be interested in spaceships, but spaceships are not a part of students' daily lives and probably don't match any of the district's curriculum standards.

To develop driving questions from students' interests, a teacher must provide students time to develop questions and establish a setting in which questions can emerge. Teachers must develop ways to support students in their asking of questions, select from among the many possible driving questions that are worth pursuing, and sensitively handle those questions that are not selected. In this section, we explore techniques for helping students ask questions, selecting students' questions, and developing students' ideas into driving questions.

Student-generated questions require a little more work on the part of the teacher, because supports need to be used to get students to develop them. The teacher needs to make sure that students work through the process of determining whether all of the key criteria of driving questions are met. However, most teachers find that, with experience, children will begin to ask, refine, and develop questions that they can explore. Finally, teachers who are familiar with their school district's curriculum standards are usually able to adapt students' questions to school outcomes. Table 3.2 summarizes ways that students' questions can be developed into driving questions.

CONNECTING TO NATIONAL SCIENCE EDUCATION STANDARDS

K–4 CONTENT STANDARD: SCIENCE AS INQUIRY

Scientific investigations involve asking and answering a question and comparing the answer with what scientists already know about the world (p. 123).

5–8 CONTENT STANDARD: SCIENCE AS INQUIRY

Students should develop the ability to refine and refocus broad and ill-defined questions (p. 145).

TABLE 3.2   Ways to Help Students Generate Driving Questions

- Provide or create an environment in which students can make observations.
- Have students brainstorm what they know.
- Focus on students' hobbies and personal interests.
- Help students modify topics into questions of interest.
- Model the process of evaluating questions using the criteria in Figure 3.3.
- Have students share, evaluate, and refine questions in small groups.
- Have groups share questions and rationales with the whole class.

Supporting Students' Asking Questions

Teachers can use various ways to encourage students to ask questions that might be turned into a driving question for a project or that might be used as subquestions in a project. One of the best ways to accomplish this is to have students make observations of their surroundings. To do this, teachers must build a classroom environment that fosters students' asking questions and that exposes students to situations that they can observe. Teachers can help students recommend subquestions to the main driving question of the project. Finally, teachers also can have students brainstorm ideas they know about and help them identify hobbies and personal interests.

An Encouraging Classroom Environment. There are a number of ways that a teacher can build an environment that encourages students to ask questions. One technique involves setting up situations in the classroom that allow students to make observations and ask questions. For instance, a teacher might set up an aquarium in the classroom or plant a variety of seeds in a classroom terrarium. Imagine a series of aquariums that are missing pumps and filters, but that contain sand, goldfish, snails, and duckweed. Students make careful observations, recording them on a daily basis. After several days, students notice a buildup of "black stuff" on the sand at the bottom of the aquariums. Students will likely ask questions, such as "What is the black stuff on the bottom of the aquarium?" This situation forms a great learning experience for early elementary students in which they can test theories about where the black stuff comes from. Students might, for example, infer that the black stuff comes from "the snails and the fish going to the bathroom" or "the fish vomiting" or the sand itself being dirty and the dirt rising to the top of the sand as it settles over time.

A second technique involves teachers exposing students to situations in which they can explore their environment. Teachers can take students on walking excursions or on field trips or bring in guest speakers. For instance, teachers can support students asking questions about the quality of the water in a stream near the school by taking the students on a stream walk. Or they can have students explore the types of trees in the neighborhood by walking around the block. On the walk around the block, students can make careful observations of the types and locations of trees. By looking at the shape of the leaves, the texture of bark, and the color variations in the leaves and the bark, students can begin to ask such questions as "Why do some trees have rough bark and others have smooth bark?" This question can naturally lead to the driving question "What trees are in my neighborhood?"

Helping Students Generate Subquestions. When teachers are working with a driving question, perhaps one that comes from district- or state-mandated curriculum, student-generated subquestions can increase students' interest with and ownership of the question. For example, a driving question derived from the *Science and Technology for Children* (STC) kit entitled "Food Chemistry" might be "What affects the nutrition of the food I eat?" While this driving question may not interest students at first, various techniques can help students form subquestions of their own. One tool that can help students generate their own subquestions is the brainstorming wheel pictured in Figure 3.2. Students add "spokes," or questions related to the driving question, to the wheel. Students might add questions, such as "Which snack foods have less sugar in them?" "What foods would make my hair shiny?" or "What is in potato chips that makes them high in calories?" These subquestions become projects for investigation. The question "Which snack foods have less sugar in them?"

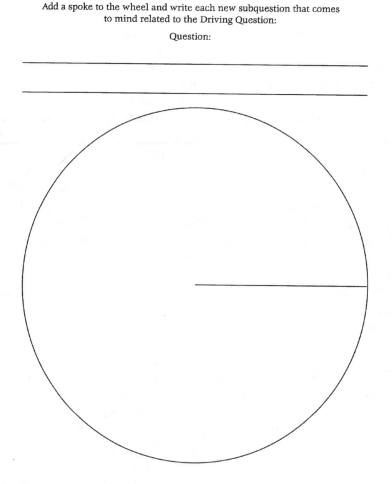

Figure 3.2 Brainstorming wheel.

becomes the impetus to investigate glucose levels, a topic in the STC kit on "Food Chemistry."

What Students Already Know. Another strategy that can be used to help students generate their own driving questions is to have them brainstorm about what they know. Students will come up with many ideas about animals, dinosaurs, cars, and toys. Frequently, students will come up with ideas based on interests and hobbies, such as their pets, television shows, baseball card collections, doll collections, and musical instruments. Many teachers list these ideas on a chart with three columns labeled "What we know," "What we want to know," and "What we learned." The ideas that students generate can be listed under the "What we know" column. This method, called the *KWL* (for *Know, Want to Know,* and *Learned*) *method* and developed by Debra Ogle (1986), can be used in a variety of contexts.

Next, teachers help students turn these ideas into questions. For example, in the first column students might list "Animals live outdoors," "Some animals live indoors," "Animals need food to grow," "Animals are big and little," and "Animals can be dangerous." These ideas might lead to an entry in the middle column "What animals live in our neighborhood?" One reason this is a good driving question is that it can be investigated by students. Future chapters will explore the KWL method in greater detail.

Besides having students brainstorm topics and related questions in a large group, teachers can have the class generate topics and questions in small groups. This method allows students to critique the questions as well since small-group review is much less threatening than having questions exposed to the whole class for review. Some teachers use dialogue journals to accomplish this small-group generation of topics and sharing. Dialogue journals are journals that students keep and share with peers or the teacher and that include responses from readers. Now is a good time to begin Learning Activity 3.3.

Selection of Students' Questions. After students have generated possible driving questions, teachers must help them select the questions that will be used for projects. This is not an easy task. Many students want their question selected, and their feelings can be hurt and self-esteem damaged if their questions are not selected. Several techniques can be used to sensitively select the driving questions from the pool of questions students generate.

One technique is to have students evaluate and refine their questions. In the process of evaluating and refining questions, they often discard many of them. Teachers can use the features of a driving question (see Table 3.1) and the questions in Figure 3.3 to support evaluation and refinement of questions. Most students will not know how to evaluate questions, so a critical first step is to model the process of evaluating. For example, imagine the class has generated the question "How do robots work?" The teacher can respond to this question by saying,

> My driving question is "How do robots work?" I want to know more about how to make a robot. Let's see . . . I will need to know about electricity and machines. I will also need to know about computers. I could break this question into smaller ones: How is a robot made? How expensive is a robot? Could a robot do my homework? What investigations can I do? Hmmm. I don't know. I don't have money to buy a robot, so we can't experiment on it. I have

Learning Activity 3.3

DEVELOPING DRIVING QUESTIONS FROM WHAT STUDENTS KNOW

Materials Needed:

- A classroom to visit or an elementary or middle school student to interview
- Paper and pencil or a computer

A. Obtain permission from appropriate school personnel to interview an elementary or middle school student (or interview a student you know, such as a son or daughter, neighbor, cousin, or student you baby sit). Ask the student about his or her hobbies and personal interests. Find out what topics the child thinks he or she knows much about. Make a list of these ideas.
B. Analyze the list of ideas and decide how these could be turned into driving questions.
C. What challenges might you have meeting the key features of driving questions?
D. Record your responses.

a toy robot. I could take that apart, but once I took it apart, I'd be done in a few hours. This is interesting, but I guess it's not a good driving question. I can't investigate this very much!

From the teacher's modeling of the process, students can see how the teacher eliminated the robot question because it wasn't feasible or sustainable. Once the teacher models the process, students in the class are usually ready to evaluate some of their own questions using the criteria in Figure 3.3.

Another strategy for selecting a driving question from a list of students' questions is to have two groups of students share and critique each other's driving questions. When using this technique, teachers should make sure that the groups who are having their questions critiqued also have the right to defend their questions. It also is critical that students give reasons for their comments. Just claiming that they don't like a question is not a sufficient reason. The criteria in Figure 3.3 can be used to focus and support student evaluation of the questions posed by others. While students are evaluating and refining their questions in small groups, it is vital that the teacher give students feedback on their evaluations. This helps students stay focused. Active monitoring of group work and encouragement are critical if teachers want students to complete the task with a high level of investment.

Teachers should also help students realize that good driving questions, like most products that have value, don't emerge complete from the first round of revision. To arrive at a good product, people often must perform several revisions. Therefore, modifying questions is a critical step in developing good driving questions. To accomplish this, some teachers like to bring the separate groups back into a large group and have the groups present their questions and give justifications for selecting them. Then, the members of other groups can add their comments to the discussion of a driving question candidate. This technique helps reinforce the qualities of a good question.

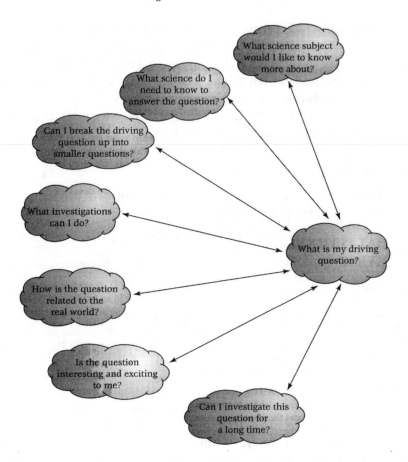

Figure 3.3 Students' prompts for developing and evaluating their own driving questions.

What Is the Value of the Driving Question?

Often in schools students see what they are learning as separate from their own interests and lives. How many times have students been told, "You will need this for life" or "You will need this for the next class"? Unfortunately, these reasons are not motivating for students of any age. Projects, however, help make learning meaningful (Crawford, Marx, & Krajcik, 1999). Students find the questions and, thus, the learning relevant to their lives and their world or culture, and they find the related projects exciting and interesting.

What typically happens in schools is that ideas are taught in isolation from each other. There often is no organizer that can pull together the overall ideas. For example, a common science curriculum might teach students how to classify animals, about habitats needed by animals, and about behaviors of animals. These topics are usually taught as separate ideas—perhaps even through different chapters of a book covered at different times during the year. Driving questions help teachers link the various concepts taught, which helps students develop integrated understandings. The questions become anchors for all the new information that students are learning, which results in the information making more sense.

Students frequently lack motivation to expend a great deal of energy over an extended period of time when there is no driving question to focus that energy and give them a reason for their efforts. A driving question ties the learning of concepts to the project activities and ties both to a meaningful purpose. With this combination, students are engaged throughout the duration of a project.

A well-written driving question gives coherence and continuity to project-based science, and it links content with practice. Deborah Brown, a teacher from Detroit Public Schools, illustrated this in the case report she wrote after doing the "What's in Our Water?" project (Krajcik et al., 1996):

> One of the biggest advantages to using a driving question as the focus was that it gave students a sense of purpose. As we worked together through this first unit, my students began to realize that there was a reason for everything that we did. No more questions like, "Why do we have to learn this?" No matter how different the activities seemed, the end results were the same. We were trying to find out about our tap water.
>
> This sense of purpose did not come about quickly or easily. Many times as I was teaching the first unit ("What's in Our Water?"), I would forget to emphasize our main question. The students may have understood the main concepts of that lesson and still had no idea how it was connected to the unit.
>
> When I did remember to ask the students about our main question, I would many times get blank stares. They were not used to making these kinds of connections. So, when I first asked them to go back to research they had done3 weeks ago and apply it to activities they were doing today, they thought I was out of my mind. [But later] students became more adept at making connections. Students were able to draw on previous data to explain new concepts.

This quote also illustrates that a sense of purpose comes about only after the teacher does a lot of active teaching. The teacher must continually help students see how the content and practices are linked to the driving question. The teacher is central to the success of project-based science, because the teacher, as an active and supportive leader and guide, orchestrates the educational environment. If the teacher does not play this role, students are not able to focus on the demanding intellectual activities of projects, and they cannot see the purposes of the activities of the project.

A common problem in science classrooms is that students learn only fragmented knowledge without ever relating school science content to the world in which they live. But when science class is focused an interesting, real-world driving question, students can more easily make connections with the content, and they can see first-hand how the study of science relates to issues and problems in their own lives.

Often students don't have experiences in school that help them attach much of what they learn in one subject area to what they learn in other subject areas. For example, students might learn how to use a ruler in mathematics class to solve math problems, but not about applications to science. In science class, they might learn about various animal habitats, but fail to learn about the connections to history and politics. They may learn how to write a business letter in language arts, but they never learn about the use of writing skills in math and science. As a result, students develop fragmented knowledge that they can't apply in other areas of study.

Through the driving question, project-based science explicitly links different subject areas. For example, through the driving question "How do we care for the pets in our room?" students might learn to use a ruler in mathematics class so they can measure the amount a pet has grown. They might learn about animal habitats in order to build the best home for their classroom pet. They might learn to write a business letter in language arts so they can write to companies that test products on animals. When learning is integrated in this way, students develop cognitive structures for processing and linking new information. In this way, learning occurs at a deeper, more permanent level. (See Chapter 12, where we discuss integration across the curriculum.)

How Can a Driving Question Be Used Throughout a Project?

Once a teacher has chosen a driving question and begun a project, he or she must keep referring to the driving question. The driving question gives a focus to class activities and to what the students are learning. By keeping the driving question in focus as teachers lead students through different activities, the driving question contextualizes what students are learning and helps maintain continuity throughout the project. The driving question can be an excellent vehicle to maintain focus and support integration, but only if the teacher takes an active role in using the driving question throughout a project.

In a project-based science classroom, investigations and activities should play a role in answering the driving question. However, making students aware of these connections can be a challenge. These connections are not always obvious to students until the teacher takes the time to help students understand the reasons behind activities. There are several techniques for applying the driving question throughout a project.

One excellent and simple technique is to refer frequently to the driving question. At the end of every class and activity, before beginning each new activity, and periodically throughout the day, the teacher can ask, "Why are we doing all this?" Soon enough, students will be able to join the teacher in responding, "To learn how to care for the pets in our room" or "To find out what birds visit our feeder." It won't be long before students begin to think about the driving question as they work, even without the teacher's prompts.

Another technique is to develop ways to link activities and investigations in class to the driving question. Teachers need to consider how to introduce a project's driving question so that the connections between the question and classroom activity start out firmly planted in students' minds. Throughout the project, teachers can reinforce this connection by asking periodically. "Why is this investigation important to our project?" and "How does it help us answer our driving question?" Students will come to expect teachers to ask these questions and will develop good responses.

Another way of linking the content and the class activities to the driving question is to have the driving question visible at all times. The teacher can point to the question every time she asks, "Why are we doing this?" Students will soon learn to look at the wall or bulletin board where the question is posted to help them stay on track. One teacher hung a large banner with the driving question on it on the wall outside her classroom. From the banner, she hung various student products. The banner with the driving question and products formed a time line of project activities. The banner also generated excitement among other students in the school.

CONNECTING TO NATIONAL
SCIENCE EDUCATION STANDARDS

SCIENCE TEACHING STANDARDS

Student understanding is actively constructed through individual and social
processes. In the same way the scientists
develop their knowledge and understanding as they seek answers to questions about the natural world, students
develop an understanding of the natural
world when they are actively engaged in
scientific inquiry—alone and with others
(p. 29).

Another technique is to have
students periodically write in their
journals about how the driving
question relates to what they are
doing. Some teachers use this technique in conjunction with class discussion that encourages students to
tell how what they do in class on a
given day is related to their driving
question. Because these projects are
long term, the potential for losing
focus is high, so continual reminders of the driving question can prevent fragmented learning and keep
students from getting lost.

Chapter Summary

One of the key features of project-based science is establishing relevance to students'
lives. The driving question is the main vehicle for establishing relevance. A driving
question helps to organize content and drive activities that take place during the
project. The driving question is the first step in meeting all of the other key features
of project-based science. The question sets the stage for planning and carrying out
investigations. This chapter elaborated on the features of good driving questions:
feasibility, worth, contextualization, meaning, ethics, and sustainability. It also discussed ways that driving questions can be teacher generated from the curriculum,
colleagues, personal interests, and the media and how they can be student generated.
Techniques for encouraging and selecting student questions or subquestions were
explored. The chapter ended with an exploration of the value of a driving question:
The question is meaningful for students; it helps teachers link lessons, practices, and
content, and different other subject areas; and it allows students to engage in an intellectual problem over time. Developing driving questions is clearly worth the time
and effort because they allow students to learn in a more educationally sound manner, and they make teaching more satisfying and rewarding.

Chapter Highlights

- Relevance is established through the driving question.
- Driving questions organize and drive sustained inquiry.
- Driving questions have six features:
  - Feasible.
  - Worth.
  - Contextualization.

- Meaning.
- Ethical.
- Sustainability.
- Teachers can develop driving questions from
  - Personal experience.
  - School curriculum.
  - Listening to students.
  - Hobbies and personal interests.
  - Media, including the World Wide Web.
  - Listening to other teachers.
  - Published curriculum materials.
- Teachers can help students develop driving questions by
  - Creating rich classroom environments.
  - Helping students generate subquestions to the main driving question.
  - Linking to students' prior experiences.
- Driving questions are valuable because they
  - Are meaningful to students.
  - Allow teachers and students to link various parts of the project.
  - Engage students in intellectual problems over time.
  - Link content and practices.
  - Help students connect school to their lives.
  - Connect subject areas.
- Throughout projects teachers continually make connections to the driving question.

Key Terms

Relevance	Meaning
Contextualization	Problematizing
Driving question	Subquestions
Ethical	Sustainability
Feasibility	Worth

Notes

1 This is a classic misconception, The appropriate scientific concept is Newton's First Law of Motion that states that every object continues at rest or in motion, unless acted upon by a force.

2 Motion probes are sonic range detectors (like those found in Polaroid cameras) that are electronic instruments attached to a computer's USB port through an interfacing box. The probe sends out sound waves that cannot be heard by humans. The sound waves bounce off objects in front of them and are deflected back to the probe. The probe can then calculate the distance the object is from the probe. Accompanying software plots the data on a computer monitor, allowing learners to view real-time, distance-time, or velocity-time graphs.

References

American Association for the Advancement of Science. (1993). *Benchmarks for science literacy.* New York: Oxford University Press.

Atwater, M. M. (1994). Research on cultural diversity in the classroom. In D. L. Gabel (Ed.), *Handbook of research on science teaching and learning.* New York: Macmillan.

Blumenfeld, P., E. Soloway, R. Marx, J. Krajcik, M. Guzdial, & A. Palincsar. (1991). Motivating project-based learning: Sustaining the doing, supporting the learning. *Educational Psychologist, 26*(3&4), 369–398.

Burnett, R. (1999). *The pillbug project: A guide to investigation.* Washington, DC: National Science Teachers Association.

Crawford, B., Marx, R., & Krajcik, J. (1999). Developing collaboration in a middle school project-based science classroom. *Science Education, 83*(6), 701–723.

Elementary Science Study. (1970). *The ESS reader.* Newton, MA: Education Development Center.

Haberman, M. (1991). The pedagogy of poverty versus good teaching. *Phi Delta Kappan, 73*(4), 290–294.

Haberman, M. (1995). *Star teachers of children in poverty.* West Lafayette, IN: Kappa Delta Pi.

Krajcik, J., Blumenfeld, B., Marx, R., & Soloway, E. (2000). Instructional, curricular, and technological supports for inquiry in science classrooms. In J. Minstrell & E. Van Zee (Eds.), *Inquiring into inquiry: Science learning and teaching* (pp. 283–315). Washington, DC: American Association for the Advancement of Science Press.

Krajcik J., Soloway, E., Blumenfeld, P. C., Marx, R. W., Ladewski, B. L., Bos, N. D., & Hayes, P. J. (1996). The casebook of project practices—An example of an interactive multimedia system for professional development. *Journal of Computers in Mathematics and Science Teaching, 15*(1&2), 119–135.

Krajcik, J. S., & Blumenfeld, P. (2006). Project-based learning. In R. K. Sawyer (Ed.), *The Cambridge handbook of the learning sciences.* New York: Cambridge University Press.

Krajcik, J. S., & Mamlok-Naaman, R. (2006). Using driving questions to motivate and sustain student interest in learning science. In K. Tobin (Ed.), *Teaching and learning science: An encyclopedia.* Westport, CT: Greenwood Publishing Group.

Lawrence Hall of Science. (1993). *Full option science system,* Nashua, NH.: Delta Education.

Lawrence Hall of Science. (2000). *Full option science system.* Nashua, NH.: Delta Education.

MacKenzie, A. H. (2001). The role of teacher stance when infusing inquiry questioning into middle school science classrooms. *School Science and Mathematics, 101*(3), l43–53.

National Research Council. (1996). *National science education standards.* Washington, DC: National Academy Press.

National Research Council. (2000). *Inquiry and the national science educational standards: A guide for teaching and learning.* Washington, DC: National Academy Press.

National Science Resource Center. (1991). *Science and technology for children.* Washington, DC: Smithsonian Institute, National Academy of Sciences.

National Science Resource Center. (1997). *Science and technology for children.* Washington, DC: Smithsonian Institute, National Academy of Sciences.

Ogle, D. (1986). A teaching model that develops active reading of expository text. *The Reading Teacher, 39*(2), 564–570.

Reiser, B. J. (2004). Scaffolding complex learning: The mechanisms of structuring and problematizing student work. *Journal of the Learning Sciences, 13*(3), 273–304.

Chapter 4

Developing Scientific Investigations

Introduction

Investigations form the essence of inquiry science. What is an investigation? How are investigations developed? How do I help students plan and design experiments? How can I help students be systematic in analyzing data? How do I help students make conclusions from data? In this chapter, we examine asking and refining questions, planning and designing experiments, and assembling and carrying out procedures. In the next chapter, we discuss analyzing data, constructing explanations, drawing conclusions, and sharing information with others. In the two chapters, we consider the various components as a web of related activities. Throughout both chapters, we examine the various instructional supports a teacher can provide to help learners through the various components of investigations; and we explore answers to numerous questions that might arise about investigations.

First, we consider several types of school science activities, some of which you may have experienced yourself as a student. As you read each scenario, focus on various

Chapter Learning Performances

- *Explain the value of elementary and middle grade students engaging in investigations.*
- *Clarify the components of the investigative web.*
- *Design and carry out investigations with students.*
- *Critique and give feedback to elementary and middle grade students on the various components of an investigation.*

features of the instructional setting. What are the students doing and thinking? What is the role of the teacher? What instructional supports does the teacher provide to develop investigations?

Scenario 1: A Step-by-Step Activity

You may have observed a situation similar to this. Ms. Pilarski, a fifth-grade teacher, presents information on the physical and chemical properties of materials. Ms. Pilarski shows the students some white powders and tells the class they will observe what happens when they mix a small amount of each powder with some liquids the teacher calls *solutions*. Ms. Pilarski hands out a sheet of paper with step-by-step directions about the amounts of powder and liquid to use and about how to mix the materials together. Ms. Pilarski also passes out a sheet of paper with a carefully labeled table on which the students are to record their observations. The powders are labeled across the top of the table, and the solutions are labeled down the side. Ms. Pilarski directs the class to describe and illustrate their observations in the table.

Ms. Pilarski allows the students to work in groups of four. Each group has 30 minutes in which to complete the activity. Ms Pilarski directs each group to assign one member to gather the required materials—the powders and solutions—from the table at the front of the class. Students from various groups volunteer to gather the materials; but in one group, none of the students wants to go get the materials. After the students return to their tables with the materials, one student, Frank, grabs some of the powders and begins mixing them with one of the liquids. In some groups, students begin by reading Ms. Pilarski's directions to decide what they need to do first. In one group, all the students begin working independently of each other. Some students become involved by reading off the directions and handing materials to the other students. Some students use magnifying glasses to make observations of the powders, and they draw pictures of what they see. The classroom is loud, but everyone appears to be on task. Students make comments like, "Hey, that was cool." A number of the groups really like it when one of the powders bubbles and fizzes when mixed with one of the solutions.

In this first scenario, the teacher selected and gave the students directions about how to carry out the activity. Although the students manipulated materials and worked in small groups to make and record observations, they spent little, if any, time planning what they were to do and little, if any, time discussing what the observations meant. Moreover, the activity the students performed was not tied to a question that the students asked or that they found meaningful.

Scenario 2: A Trial-and-Error Activity

You may have observed a situation like this. Mr. Warren, a fifth-grade teacher, passes out one light bulb, a wire, and a battery. He challenges the students to find as many ways as possible to light the bulb, using just the wire and the battery. Mr. Warren directs the class to work in groups of two and to sketch "all the ways you can get the bulb to light, using the battery and wire." He hands out plastic bags that contain the various materials.

Almost all of the teams jump into the task with little or no discussion about what to do. A number of the teams try connecting the bulb, wire, and battery in a straight-line configuration, but this method fails to light the bulb. Some students comment that they have a "bad bulb" or a "battery that doesn't work." After several failed attempts, some of the teams discuss ways to get the bulb to light. As some of the groups get their light bulb to light, comments like "cool" or "wow" are heard. Some of the teams that fail to light their bulbs become frustrated and stop trying, but most of the groups persist, possibly because trying to light the bulb is more fun than answering the questions at the end of the chapter. The excitement in the room grows as various teams light the bulb. After about 15 minutes, most teams have lit their bulbs. Mr. Warren walks around the room, asking each team to demonstrate its strategies.

Mr. Warren challenges the teams to find out how to light the bulb in different ways. Some teams that are having difficulty finding a way to light their bulbs ask other teams for help; the successful teams seem happy to share their techniques. Mr. Warren encourages the groups that are having difficulty to think of different ways to light the bulb. Mr. Warren also challenges the teams that had success lighting the bulb in more than one way to "make two light bulbs light so that, if you disconnect one light bulb, the other goes out too." Some teams talk about how lighting the bulb is related to switching on the light in the classroom.

In Scenario 2, as in Scenario 1, the teacher selected the activity for the students. Once again, the activity was not tied to the driving question or to a question that the students asked. Unlike Scenario 1, in this case the teacher did not tell the students how to do the activity, so students conceivably could plan how to light the bulb; but most just used trial-and-error methods. The activity was fun, but most of the students did not understand why the bulb would or would not light. Little, if any, discussion occurred regarding why the bulb lit in one configuration and not in others. Moreover, the class did not follow through with this activity to further explore issues related to electrical phenomena. Although some of the students were able to make a connection between lighting the bulb and lighting their school, most students did not understand how lighting the bulb related to how the lights work in a building.

Scenario 3: An Investigation

Perhaps you experienced a situation like this. In a fifth-grade science classroom, one group of four students is sitting in a circle talking with other students writing in notebooks. Another group is standing behind a table, putting soil in a pot, and another is gathered around a computer at the back of class. Ms. Tamika Brown, the teacher, is talking with a fourth group of students.

The group sitting in a circle is making a list of possible questions to explore: Do worms help decomposition? Will new worms be born in the decomposition column? What will decompose first? One student says that the group can't design an experiment to study whether worms will be born in the decomposition column because it would be hard to count all the worms. Another student responds that it doesn't matter if they count *all* the worms; they just need to see if there are *more*.

The group putting soil in a pot has already made a plan for investigating the effects of compost on the growth of bean plants. The students are planting eight bean plants. They plan to plant two plants with a 75/25 mixture of compost and soil, two plants with a 50/50 mixture, another two with a 25/75 mixture, and two with just soil. Some of the students are still discussing whether they need to plant some plants with 100% compost.

The group at the computer is composing a letter to another fifth-grade class. The students are explaining to the other class, which is located in another state, their plan for investigating the influence of light on decomposition. They will e-mail their letter to this class when they have finished writing it.

The fourth group is discussing with Ms. Brown how to investigate the effects of air flow on decomposition. This group's plan is to build a decomposition column with numerous holes. Ms. Brown, through a series of questions, is trying to help the group understand the need for a control and a larger sample. Ms. Brown asks, "If you use only one decomposition column, how will that help you understand how air flow influences decomposition?" The students in the group are debating this question. One student says, "You really won't know if you use only one column because there will be nothing to compare it to." Another student argues that they could investigate "how fast the various stuff decomposes," using just one column. Another student responds that "how fast" won't help them answer their question. The group continues to debate. Ms. Brown moves on to work with another group.

This scenario shows students engaging in different aspects of the process of investigation: asking and refining questions, planning and designing experiments, assembling materials and apparatus, carrying out procedures, and sharing information with others. The scenario also shows the important role that teachers play in helping students work through the various stages of investigations. Unlike the first and second scenarios, this scenario does not represent an activity that the teacher selected: What the teacher created was an educational environment in which students could learn about decomposition. Then the teacher provided instructional support by working with each of the groups, asking probing questions, and providing materials that students could use to investigate their questions.

There are times in the curriculum when step-by-step and trial-and-error activities, like those illustrated in the first two scenarios, are important. Such activities, however, are not investigations, because they are determined and set up by the teacher.

The third scenario shows students involved in many aspects of science inquiry. Students pursue solutions to questions of importance by asking and refining their own questions, debating ideas, making predictions, designing plans and/or experiments, measuring, collecting and analyzing data and/or information, drawing conclusions, making inferences, communicating their ideas and findings to others, and asking new questions. Research supports the notion that young children are capable of performing all these aspects of investigations (Krajcik et al., 1998; Metz, 1995; National Research Council, 2000). Metz (1995) presented a strong argument for the investigative ability of young children and for their ability to learn from investigations, although their investigations are not as sophisticated as those of adolescents and adults because of their limited prior knowledge. The teacher plays the critical role of orchestrating the investigative process and selecting curricular areas that allow students to ask questions that result in developing understanding of important concepts and principles.

Research also suggests that students should be involved throughout their school experience in the process of scientific investigation and that children should be involved in the business of thinking scientifically:

> From their very first day in school, students should be actively engaged in learning to view the world scientifically. That means encouraging them to ask questions about nature and seek answers, collect things, count and measure things, make qualitative observations, organize collections and observations, discuss findings, etc. Getting into the spirit of science and liking science are what count the most. (AAAS, *Benchmarks for Science Literacy*, 1993, p. 6)

CONNECTING TO NATIONAL SCIENCE EDUCATION STANDARDS

SCIENCE EDUCATION PROGRAM STANDARDS

Standard A: The program must emphasize student understanding through inquiry (p. 214).

Moreover, the National Research Council in the *National Science Education Standards* (1996) argued that students need to experience **inquiry-based learning** that focuses on developing a deep understanding by exploring the world. Students of all ages need to ask their own questions that are important to them, design their own experimental procedures, make sense of their data, analyze their data, construct explanations, share their plans and findings with others for feedback and criticism, and generate new ideas. Asking questions, designing experiments, analyzing data, and sharing what they know should be a routine part of students' learning science. However, students seldom have opportunities to ask their own questions and then define experiments to gather data related to the questions. Science textbooks seldom engage students in the process of investigation. As in Scenario 1, children are often given the purpose of activities, step-by-step directions, and procedures for how to collect data. Activities of this sort do not engage students in scientific thinking. For activities of this sort, students often are given the outcomes and not challenged to analyze the purpose or predict outcomes. Often, students know the expected outcome to the activity before the activity is conducted.

The Investigation Web

Engaging in the process of investigation is akin to seeking the answer to a question, for which humans seem to have a deep need. In this section, we examine the investigation web, a process of carrying out an investigation that includes "messing about," asking and refining questions, finding information, planning and designing, building the apparatus and collecting data, analyzing data, constructing explanations, drawing conclusions, and communicating findings.

Figure 4.1 presents one visualization of the investigation process. The investigation web represents the nonlinear quality of inquiry and illustrates the way that students revisit various components of an investigation. Science is truly a nonlinear endeavor. Each component of an investigation provides feedback for another part. For example, finding information about a topic might lead students to refine their questions that took them to the information in the first place.

In Scenario 3, students might have uncovered information about building a compost column that caused them to refine their question regarding the impact of air flow on decomposition. Preliminary data analysis could have suggested ways for students to modify their procedures to collect more reliable data. Students may well have noticed that their decomposition column felt warm, and this observation might have encouraged them to collect data on temperature changes. Completing an investigation

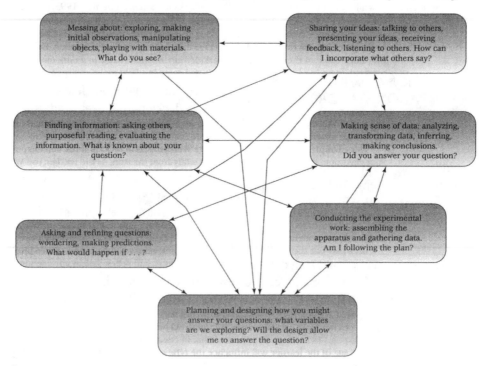

Figure 4.1   The investigation web.

should lead to other questions. For example, once students found out that oxygen is needed to help materials decompose, they could have posed new questions regarding decomposition of materials buried in a landfill. Some investigations may well have begun with the gathering of data or with the search for a pattern. Students might have designed an investigation of decomposition only after "messing around" with different decomposing materials and noticing that air affects the rate of decomposition.

Science is a messy process. Although scientists often have good hunches about important questions based on theories and their own prior knowledge, the questions

> **CONNECTING TO NATIONAL SCIENCE EDUCATION STANDARDS**
>
> **SCIENCE EDUCATION TEACHING STANDARDS**
>
> Teachers structure the time available so that students are able to engage in extended investigations. Teachers create a setting for student work that is flexible and supportive of science inquiry (p. 43).

for investigation and the processes they ultimately employ often differ greatly from their original conceptions. Some great examples of scientists' work that illustrate the investigative web in practice come from the works of Barbara McClintock, who was a famous cytogeneticist, and Eugenie Clark, a world-renowned icthyologist and authority on sharks.

Let's look at an example of the investigation web. Imagine that a fourth-grade class is making observations of various crustaceans in a nature area. One student group, which is turning over logs and rocks, notices that sowbugs can be found under the logs (messing about). Based on their observations, the students wonder if sowbugs like dark areas (wondering). In their classroom, the teacher has students share their observations and questions with classmates (sharing ideas). The group that has observed the sowbugs asks the question, "Do sowbugs like the dark?" (asking questions). (See *The pillbug project: A guide to investigation* by Burnett, 1999) Based on feedback from the class, the students refine their question to "Will sowbugs choose a dark area over a lighted area?" (receiving feedback, refining questions). The students design an experimental procedure to answer their questions (designing). They then share their plans with the class (sharing ideas). Students in the class wonder if they are using enough sowbugs (wondering). The students incorporate the class feedback into their design (incorporating what others say). Next, the students look for information on sowbug habitats (finding information). They use the World Wide Web to obtain more information about sowbugs and pillbugs. Then they gather sowbugs from the nature area, build a habitat so the sowbugs can live, and construct their experimental chamber for testing light and dark habitats (assembling the apparatus). Next, they perform their experiment, carefully recording their data (gathering data). To make sense of their data, they create a bar chart of the number of sowbugs found in dark and light places (transforming data). They draw conclusions based on their findings and decide that sowbugs like dark areas (inferring). After some debate in the group regarding the number of trials needed, they decide to collect more data by carrying out two additional trials to see if their conclusion holds up (gathering data). They analyze their new data by comparing the old and new results and decide that they support their original conclusion (analyzing data). They then present their findings to the class (sharing ideas).

Some classmates suggest that they should have ruled out moisture to determine if the sowbugs like dark areas (receiving feedback). The students argue that moisture did not make a difference; however, after some debate, they realize that their classmates might have a point (talking to others). The students then redesign their investigation: This time, they keep some of the sowbugs in a dark dry environment, some in a dark moist environment, some in a light dry environment, and some in a light moist environment (planning and designing). The students once again analyze the data and draw a conclusion (making sense of data). The group then presents its findings and reasons to the class (sharing ideas). As students work, they ask a number of new questions about temperature and time. When the students complete their investigations, they return the sowbugs and the habitat materials to the nature area.

This example illustrates the investigative web in its ideal form; don't expect your classroom investigations to run so smoothly from the start. As the teacher, you will need to provide instructional supports throughout an investigation. Being good at science, like learning to play an instrument or a sport, takes practice, knowledge and experience. You would not expect to be a good tennis player the first time you pick up a tennis racket, and you shouldn't expect a student to be good at carrying out scientific investigations at first either. Excellent tennis players don't develop even over the course of a season, and so you cannot expect students to develop into good investigators during a single school year; but over the course of a season, if a tennis player practices, listens to feedback, and studies the game, they will improve — and students will improve on various aspects of scientific inquiry as well.

Let's play out the tennis analogy a little further: Tennis can be broken down into a variety of activities, such as forehand, backhand, serve, and volley. Skill in any one tennis activity doesn't make a good tennis player; neither does skill in all of the activities of tennis without putting them together. Practicing all the components together helps make a person a good tennis player. Similarly, conducting investigations has a number of components, such as planning, sharing, and gathering data. Skill in planning or gathering data alone does not make a good investigator. Students need to practice all of the activities of investigating, not in isolated applications, but in concert.

The age of the learner will influence the type of investigations the learner can perform. Children who are 5 to 7 years old should be observing, measuring, and identifying properties of materials. For example, children in first or second grade could observe the different types of insects or birds found on their playground. Six- to nine-year-old children should be looking for evidence and identifying patterns. For example, fourth-grade students could chart throughout the day the length of a shadow cast from a stake placed in the ground to notice the pattern of the sun's "movement" across the sky. As children gain experiences and become more developmentally ready, they can perform more complex investigations. Eight- to 12-year-old children should begin to design cause and effect experiments and develop their observations, measurements, and evidence-gathering skills; and 11- to 13-year-olds should design and conduct experiments that involve manipulating variables (AAAS,

1993). For example, middle school students could explore the influence of temperature on seed germination.

Messing About

How do you start students in the process of engaging in an investigation? How do you help children ask questions related to a content area? The first step in the investigative web is messing about (Hawkins, 1965). Messing about includes exploring, manipulating materials, making initial observations, reading about phenomena, and taking things apart.

Messing about implies exploring aspects of the world you find of interest. Children can learn much by examining what is in a pond, taking apart an old flashlight, watching how various gears work in unison, or playing with a stream table. Messing about with natural phenomena creates situations that encourage children to wonder. Messing about also allows children to have essential prior experiences in exploring phenomena (Krajcik, Blumenfeld, Marx, & Soloway, 2000). Children cannot form ideas or ask questions about the behavior of sowbugs unless they first observe their behavior. Such experience allows students to build the prior experiences necessary to develop conceptual understanding of science concepts (Bransford, Brown, & Cocking, 1999).

Despite the implications of its name, this first step in investigation is not unstructured time. In fact, messing about needs to be carefully orchestrated by the classroom teacher. Although it may look very unplanned, the teacher selects situations that will lead children to ask questions that have important curriculum outcomes.

Students often have inadequate background knowledge to ask questions about some science content areas. For example, students may not know enough about physical science concepts, such as friction and momentum, to ask questions about simple machines. With careful structuring, however, worthwhile investigations of even difficult and unfamiliar content can occur. For example, a teacher can structure a messing-about session to help students ask good questions about skateboards (which are a type of simple machine involving wheels and axles) and provide the motivation to seek out information about simple machines. Teachers can use a number of techniques to create learning environments in which children can mess about and ask questions. We will consider two types: initial observations and manipulation of materials.

Initial Observations

One way to structure science lessons so that students can mess about productively and ask worthwhile questions is to encourage initial observations. These **initial observations** allow students to observe phenomena that might acquaint them with concepts, pique their curiosity, and motivate them to ask questions. Almost any type of physical object is a great catalyst for initial observations in elementary and middle grade classrooms. You might consider outdoor nature areas, playgrounds, simple machines around the school (such as pencil sharpeners), terrariums, aquariums, and classroom pets.

Imagine that you want to involve students in a biology investigation on insects. To focus early elementary students' observations, have the students map out 8-inch by 8-inch areas of the playground or of a nature area. Some of the areas should be shady, some sunny, some moist, and some dry. Then have the students divide each area into a 2-inch by 2-inch grid. Ask each to select one area, or microbiome, for observation. If possible, supply students with magnifying lenses. While in the field, students might draw what they see onto 8-inch by 8-inch sheets of paper that are also marked off in 2-inch by 2-inch grids. Once in the classroom again, have students compare their grids with one another. Ask them to write down and share questions that emerge from their observations. Students might ask, for example, "Do each of the areas have the same insects?" "What is different about them?" "Why do you suppose they are different?" Stream or pond walks are also excellent catalysts for initial observations about ecosystems. Learning Activity 4.1 will help you experience an initial observation activity; it places you in the role of the learner on a stream or pond walk.

Learning Activity 4.1

GOING ON A STREAM WALK

Materials Needed:

• Writing materials
• Walking shoes

A. Where do questions come from? Questions often arise from exploring our environment. Take a 30-minute walk around a stream or pond. As you walk, examine the bank. Look closely at the water. What do you see? You might dip a bucket into the water and draw a water sample so that you can make more thorough observations.

B. As you walk and observe your stream or pond, take notes of your observations. What do you find interesting? What catches your eye?

C. Also note any questions that come to mind. What would you like to find out more about?

D. When you have completed your walk, record your observations and questions on a chart like the following:

Observations	Questions
_____	_____
_____	_____
_____	_____
_____	_____

E. What other questions come to mind now? You might decide to order your questions from most to least interesting.

F. Compare your observations with those of other students in your class.

If going to a nature area or a stream is not possible for your students, you can set up some exciting observation environments for students to explore in your classroom. For example, you and your students can create chameleon environments so that you can observe the behaviors of chameleons. Craig Berg (1994) developed some excellent techniques for using 2-liter plastic soda bottles as chameleon houses. Building such habitats will allow students numerous opportunities to observe the feeding habitats of chameleons or the life cycles of crickets. *Bottle Biology* (Ingram, 1993) offers a number of ideas for using 2-liter pop bottles. You can also use 2-liter plastic soda bottles to create decomposition environments or terrariums. What is wonderful about these environments is that you can easily alter conditions to make noticeable changes. For instance, you can manipulate moisture in the environment to see how it influences plant growth.

Initial observations need not be limited to biological phenomena; students can just as easily focus on physical phenomena. For example, you might have students observe the weather over an extended period. Students might keep weather journals over the course of a school year, which can lead them to ask numerous questions about how weather changes.

Manipulation of Materials

Another way to help students become involved in investigations is through the manipulation of materials. **Manipulation of materials** may include building apparatus, taking things apart, or handling or playing with objects. For instance, you might have students set up an aquarium, take apart a flashlight, look at different objects through a magnifying glass, or plant some seeds. Many objects are good items for students to manipulate: seeds, toy cars, balls, magnets, and scientific equipment, such as balances, tape measures, magnifying glasses, and measuring cups.

Imagine that you want students to learn about the relationship between streams and streambeds and banks. You can help your students set up stream tables in your classroom. These long, flat pans filled with sand can be used to demonstrate the flow of a river or stream. Once the students have set up the tables, have them manipulate the conditions of the stream. For example, they might place clay balls, marbles, and metal objects within the stream of the water and on the banks. Ask students to make observations about what happens with each

CONNECTING TO NATIONAL SCIENCE
EDUCATION STANDARDS

SCIENCE EDUCATION CONTENT STANDARDS

K–4—Standard A: As a result of activities in grades K–4, all students should develop abilities necessary to do scientific inquiry and build understanding about scientific inquiry (p. 121).

5–8—Standard A: As a result of activities in grades 5–8, all students should develop abilities necessary to do scientific inquiry and build understanding about scientific inquiry (p. 143).

manipulation. You might want to tie this activity to a stream walk so students can make connections between what they observed in the stream table and what happens in an actual stream. *River Cutters* (1989), developed by the Lawrence Hall of Science, describes how you can make some inexpensive stream tables for your classroom.

Manipulating materials and making initial observations should not be isolated activities. Manipulation ought to be accompanied by observation of the effects of manipulation. Likewise, observations help students make decisions about what variables to manipulate. For example, once students have built something like a stream table or aquarium, they make observations of this environment to determine what to manipulate.

The Role of the Teacher

Making initial observations and manipulating materials can motivate students to ask questions, but it is likely that student observations at first will be scattered and unfocused. It is also probable that children's descriptions will not be detailed. For example, in describing a white powder, a child might say that it "looks like salt," but we want students to go beyond this simple labeling of the material. As the classroom teacher, your role is to provide instructional support to help students focus their observations, write clear descriptions, and become more thorough and complete. With your help, a student might eventually be able to say about the powder, "The stuff is white. Each piece looks like a cube when I look at it with the magnifying glass. Some of the cubes look like they are scratched or chipped."

At the beginning of the school year, you may find it useful to implement lessons that teach a necessary concept or skill to help students improve their observation skills. Throughout the school year, you can continue to focus students' observations to help them become skilled.

CONNECTING TO NATIONAL SCIENCE EDUCATION STANDARDS

SCIENCE EDUCATION CONTENT STANDARDS

Grades K–4—Standard A: Students should be able to ask a question about objects, organisms, and events in the environment (p. 122).

SCIENCE EDUCATION CONTENT STANDARDS

Grades 5–8—Standard A: Students identify questions that can be answered through scientific investigations (p. 145).

Asking and Refining Questions

Once students have "messed about" and made initial observations, they need to ask meaningful and worthwhile questions based on those observations. Worthwhile questions are related to the learning goals you hope to meet, and meaningful questions are those that students find of value. The students who are observing insects in a nature area might begin to ask questions, such as "Where do different insects live?" "When do the insects first appear?" and "When do we stop seeing various insects?" Making observations and asking questions are not necessarily separate activities. Children often begin to ask questions while they are observing. Frequently, students refine these questions as they make more observations and as they find and synthesize information. Figure 4.1 illustrates how making observations and asking questions is a back-and-forth process.

Experiences may also prompt student questions. Reading the newspaper, talking to parents or other family members, or going on a family trip can lead to students' asking questions. As students experience the world around them, questions about the environment can and will arise. For instance, if a child's family composts lawn clippings, the child might be interested in why the compost pile doesn't smell. It is the role of the teacher to support students as they move from making observations to asking questions.

The Role of the Teacher

The teacher plays a critical role in helping students develop meaningful and worthwhile questions that will help them learn important curricular content. The teacher is the curriculum leader who guides or selects the driving questions for the project and who supports the selection of the phenomena that students will explore and observe. One technique for helping students generate questions is having them set up their experimental notebooks with two columns: *Initial Observations* and *Questions*. Table 4.1 shows how a student might set up his or her notebook.

Another technique to help students generate questions is to provide students with question stems. Question stems serve as cognitive supports and focus student questions. The following stems are useful in helping students generate questions: "I wonder what would happen if...?" "What if...?" "How does...?" and "What does...?"

It is also likely that students will ask more questions than they can answer; you will need to help students determine which questions are worthwhile, feasible, and ethical. Students may ask questions that are not related to the curricular goals of the project; you need to help them understand, without discouraging, that they need to select new, worthwhile questions. For example, let's say your class is exploring the driving question "What lives in our nature area?" One group of students asks the question "How do things move?" Although this might be an interesting question to the children, it is not directly related to the current curriculum objectives. As their teacher, you will need to redirect their question to perhaps "How do sowbugs move?" Questions also need to be ethical. For example, it is important that students

TABLE 4.1   Observations and Questions

Observations	Questions
Sowbugs under logs	Do sowbugs like dark places?
Green slimy stuff on top of pond	What causes the green slimy stuff on top of the pond?
Some leaves with different shapes	I wonder how many different kinds of leaves I can find in my neighborhood?

not torture the insects to get them to move. Other students might ask questions that are beyond their experience or developmental level or beyond the resources of your classroom; while emphasizing the value of the questions, you need to direct the students to ask more feasible questions. For example, in a project on the environment, students might be interested in exploring the technology of how plastics are recycled. Although students can find information about this topic, elementary and middle school are not equipped with the appropriate laboratory equipment and ventilation system for students to conduct an investigation on this topic. Moreover, such synthesis (related to chemistry topics beyond their developmental level) is most likely beyond the experience of students.

Types of Questions

There are three types of questions that students might ask as part of an investigation. Descriptive questions permit students to find out about observable characteristics of phenomena. Relational questions allow students to find out about associations among the characteristics of different phenomena. Cause-and-effect questions provide opportunities for students to make inferences about how one variable affects another variable. Table 4.2 gives examples of the different question types.

Students in early elementary grades will ask many descriptive questions that will lead them to find out information about a phenomenon through making systematic observations. "What kind of foods do mealworms eat?" "How bright can I get a flashlight bulb to shine?" and "What kinds of leaves can a caterpillar eat?" are descriptive questions. Students can answer these questions by making systematic observations either in natural settings or in controlled situations. For example, students can answer the question "What kind of foods do mealworms eat?" by placing different kinds of food in a box that contains mealworms and making observations on which foods the mealworms select.

Students in upper elementary grades should make a transition from descriptive questions to relational questions. "Which dissolves faster in water—salt or sugar?" "What's a better insulator—paper, Styrofoam, or aluminum foil?" and "What conducts sound better—water, metal, air, or wood?" are relational questions. To answer these types of questions, students need to set up experimental procedures with which

TABLE 4.2   Types of Questions

Descriptive questions	Relational questions	Cause and effect questions
What materials dissolve in water?	Does salt dissolve faster than sugar?	Does the temperature of water affect the rate that salt and water dissolve?
What macroinvertebrates are found in a stream?	Are different macroinvertebrates found in different areas of a stream?	Does water quality affect the types of macroinvertebrates found in a stream?
How fast does my heart beat?	Who has a higher heart rate—boys or girls?	If people hear a loud sound, do their heart rates go up?

they can compare and contrast one or more characteristics. Such experimental situations provide opportunities for students to collect and analyze data and draw conclusions. To answer the question "What's a better insulator—paper, Styrofoam, or aluminum foil?" for example, students would need to compare and contrast the temperature of a substance when wrapped in the different materials. For instance, they could wrap soda cans in the different materials such as aluminum foil, plastic, or newspaper and collect data on how the temperature of the soda changes. To conclude which material is the best insulator, students would need to analyze the data from the different conditions.

Upper elementary and middle grade students need to make the transition to cause-and-effect questions that provide opportunities to explore the influence of one variable on the outcome of another. "Does water quality influence the type of macroinvertebrates found in a stream or pond?" "How does fertilizer affect the height and size of plants?" and "How does the surface of the ground affect how fast I can move on roller blades?" are cause and effect questions. To answer these questions, students need to design experiments in which they manipulate one variable (independent variable) to observe the effect on another variable (dependent variable). Such experimental situations allow students to collect data that they then can analyze to determine the reason for an outcome or result. One way to answer the roller-blade question is for students to design simulated environments in which they modify the surface of a 1-foot by 8-foot board (by adding sandpaper, rubber, oil, carpet, and vinyl flooring) and move wheels on axles across the surface. Students might choose to measure the different frictional forces with a spring scale or force probe.

Although early elementary students are more likely to ask descriptive questions and middle school students begin to make a transition to more cause-and-effect questions, it is not the case that middle school students should never ask descriptive questions or that elementary students do not ask cause-and-effect questions. Some projects might generate a natural progression through all three types of questions. For example, in a project exploring "What insects live in our playground?" students might first ask a very descriptive question: "What insects can I find?" Investigating this question may lead to a relational question: "Do I find different insects at different times of the day?" After conducting investigations, with guidance from the teacher, the students might be ready to ask a cause-and-effect question: "Do certain insects prefer dark or light environments?"

Although young children are unlikely to design an experiment to answer a cause-and-effect questions on their own, you can set up an investigation in which the whole class explores the influence of one variable on another. For example, to explore if the amount of sunlight influences how green a plant is, you could set up an experiment in which one plant is kept in the sun and one in the shade. The class could make periodic comparisons between the two plants.

You will need to help students move from asking descriptive to relational to cause-and-effect questions. The types of stems that you give students to help generate questions can have an effect on the types of questions they generate. "How does...?" and "What does...?" question stems are more likely to lead to descriptive questions. The "What if...?" stem tends to encourage more cause-and-effect questions. Another way to facilitate the transition is to list students' questions on the board or overhead, have students specify what each type of question is, and then transpose each question into another type of question. Table 4.2 shows how descriptive questions, relational questions, and cause-and-effect questions are related and how one type of question can be modified to another type of question. Reading across each of the rows, we see how the various questions are related to each other.

You will also find students asking basic information questions (Scardamalia & Bereiter, 1991). Students can find answers to basic information questions in reference source books or on the World Wide Web, but they cannot answer them through the design of an investigation. "What is the diameter of the earth?" and "How hot is the sun?" are basic information questions. Although these questions have value and might be intellectually interesting to a student, project-based science stresses students' asking questions that they can find answers to by setting up their own investigations. There is, however, a place in project-based science for basic information questions. Students might need to ask and answer basic information questions to complete the background information for an investigation. For an investigation of the influence of the amount of fertilizer on plant growth, for example, a student might need to find out what fertilizer is made of and if various brands of fertilizer differ in their composition. You can design lessons that answer basic information questions when a whole class needs the information.

Questions for Investigation and the Driving Question

If similar questions arise for a number of students, you might pick one to explore as a class. In doing so, this selection could lead to a driving question for a project. This is particularly valuable when students are first learning how to design and conduct investigations. However, as students gain more experience in doing inquiry, students should have opportunities to explore their own questions. Working on their own questions leads students to feel ownership of a project and to be more engaged.

Conducting investigations to answer student questions should contribute to answering the driving question of the class, however. For example, students might complete a number of decomposition investigations, such as exploring what causes and promotes decomposition, to answer their own questions: "Do worms cause materials to decompose in soil?" and "Does airflow speed up decomposition?" Even though such investigations would be designed to answer the student questions, they would be related to the driving question of the project: "Where does all our garbage go?"[1] Maintaining a consistent focus ensures that what happens in the classroom is purposeful and meaningful.

Hypothesizing

Hypotheses are questions stated in testable form that relate how the independent variable affects the dependent variable. A student can design an experiment to collect data that will either give support for a question or refute a question; such a question becomes a hypothesis. "I wonder if sowbugs like cool places?" is a fine question for

Learning Activity 4.2

IDENTIFYING TESTABLE QUESTIONS

Materials Needed:

- Chart
- Writing materials

A. Analyze which statements in the first column of the following chart are testable and which are not.

B. Fill in the missing sections of the chart.

WHICH QUESTIONS ARE TESTABLE?

Question	*Testable?*	*Testable revision of question*
Do different types of apples have different numbers of seeds?	Yes	_____
Why does my bike need to be painted?	No	What happens to metal if it is exposed to air?
How does fertilizer affect plant growth?	_____	_____
What types of objects fly?	_____	_____
Do worms like to eat garbage?	_____	_____

fourth-grade students, but it is not a testable question. It is not possible to measure if pillbugs *like* cool places. However, the question "Will pillbugs choose cool places over warm places?" can be transformed into the hypothesis "Pillbugs will move to cool places rather than warm places." Students can design an experiment to collect data that will either support or refute this hypothesis. They could count the number of pillbugs that go to the lower temperature section of the apparatus, for example.

How might you phrase the question "Will wrapping my soda can in newspaper keep it cool?" in terms of a hypothesis? What is missing from this question is a comparison. A testable hypothesis for this question might be "A soda can wrapped in newspaper will stay cool longer than will a unwrapped soda can." A thermometer could be used to measure the temperatures of the different cans of soda. Learning Activity 4.2 helps you learn to identify testable questions.

Helping your students make hypotheses from their questions should not destroy the excitement students feel about the world around them. Making hypotheses is an intellectual skill that can be developed over time. From a developmental standpoint, you might be more concerned with helping early elementary students ask *observable* questions. You might want to have upper elementary and middle school students practice formulating *hypotheses* from their questions. What is most important is that students ask questions that are testable.

One way to help your students change questions into hypotheses is to use a table, such as Table 4.3, to show students the transformation. Display the table on the board or on an overhead. First, have students list their questions. Next, have them write

TABLE 4.3   Refining Hypotheses

Student questions	Tentative hypotheses	How might you test your hypotheses?	Revised hypotheses

tentative hypotheses and decide if these are testable by coming up with ways to test them. Finally, based on classroom feedback, including your own, have students write revised hypotheses.

Remember that the investigation web is not a linear process. As students continue in their investigative work, you should allow them to modify their questions and hypotheses. The questions and hypotheses that are published in research reports are seldom, if ever, the questions and hypotheses with which the researchers started.

Making Predictions

Predictions are what students think might happen. Students make predictions based on previous experiences, knowledge, and observations of the phenomena they are exploring. Having students make initial predictions on questions, such as "What insects will we see on the playground?" or "How many different types of birds will we see?" is important because predictions focus students' thinking. Also, students' commitment to work increases when they make predictions. Finally, making predictions helps students become cognitively engaged in the work because it uses their prior knowledge. Making predictions also helps students synthesize their prior knowledge with the new understandings they gain from exploring their topic.

One technique you can use to help children make their own predictions and take ownership for their work is to have each child in a group make his or her own prediction. Then have group members share their predictions. Next, give group members an opportunity to comment on their predictions, giving reasons for agreeing or disagreeing with the other predictions. Once all group members have shared their predictions, ask groups to generate group predictions, while protecting each student's right to dissent from the group prediction. One technique for generating student predictions is to expand Table 4.3 to include a column for predictions.

Finding Information

Seeking information is a vital component of the investigative web. It is through the information-seeking process that students learn the background information so essential to a successful investigation. The term *information* can mean what is known about a topic, and it can mean data that others have *collected*. Students can seek out two types of data: current data and archival data. **Current data** refers to data scientists collected in the recent past; typically, it is anywhere from a few hours old to a month old. Data on today's weather is an example of current data. **Archival data** refers to data that scientists collected in the past and then stored. Data on upper atmospheric ozone readings over the past several years is an example of archival data.

Numerous resources can help students find information. Elementary and middle grade students might consult trade books, magazines, encyclopedias, CD-ROMs, and the World Wide Web. *My Big Back Yard* is an excellent resource magazine for early elementary students, and *Ranger Rick, Science World, National Geographic,* and *Scholastic* are excellent magazines for upper elementary and middle school students. To find current information, such as a new discovery or a recent news story, students need to use a different set of resources. These include the newspaper, telephone interviews, person-to-person interviews, e-mail, and the World Wide Web. Television specials, such as Public Broadcast Station programs and "NOVA" programs, can provide both archival and current information. The use of Google, Google Scholar, and Wikipedia[2] — a free online encyclopedia that is carefully vetted — and the other increasing number of resources available on the Web make it a vital source of both background and current information for students of all ages. The National Science Teachers Association (NSTA) has a section at its website called SciLinks. SciLinks is a partnership between many textbook publishers and NSTA, which provides science resources that are linked to particular text programs.

Imagine that students asked the question "Does my community need to worry about lead poisoning from cars?" Students might look for basic information on lead poisoning to help them understand the problem. They might contact the Environmental Protection Agency to find out if others have collected current data on lead from cars, or they might seek archival data on lead poisoning in their community. Although current data and archival data are very different types of information, both can provide important information for a project.

The Role of the Teacher

A critical component of any investigation is for students to explore relevant background information. You will need to help your students develop skills in searching out information related to their questions. Young children will need several types of assistance.

First, many students will need to learn how to use a variety of resources. You may need to teach lessons on how to find information using computerized library search systems and the World Wide Web.

Second, you will need to help your students learn how to determine which sources to consult for different types of information. Most children will rely on the encyclopedia for much of their information, but encyclopedias do not provide the depth of information many investigations will require. You can use a checklist like the one in Table 4.4 to guide students in their search.

Third, students need assistance in sifting through, abstracting, and outlining the information they find. Imagine a fifth-grade science class is studying acid rain. The class could seek basic information related to acid rain or archival data on acid rain to compare to the data they have collected. To help students determine what information is relevant, the teacher could use a chart like the one in Table 4.5.

TABLE 4.4   Checklist for Finding Information

Data on car emissions

Sources of new information (less than several months old)	Sources of old information (more than several months old)
☐ Magazine	☐ Trade book
☐ Newspaper	☐ Encyclopedia
☐ Telephone interview with an expert	☐ Compact disk
☐ Person-to-person interview	☐ Laserdisc
☐ E-mail	☐ Book
☐ World Wide Web	☐ World Wide Web
☐ Television documentary	☐ Television documentary

Fourth, with the explosion of available information today and the ease of its distribution due to electronic networks, you also need to help learners assess the validity of a source of information; that is, you need to help students decide if it is propaganda or scientific information. The World Wide Web contains a great deal of information presented by businesses or special interest groups that are trying to advocate certain positions. Such information is often biased or invalid. Students can be taught to think critically about the validity of information by answering questions like those in Table 4.6.

You might want to use tabloid articles to show extreme examples of "information" that is invalid because it cannot be supported, proven, or substantiated by others. Information about a three-headed goat found in the Himalayas by one person who has no photograph of the goat or other evidence is a good example of information that is invalid. Another good source to teach skepticism about such stories can be found on the World Wide Web. For example, there are many Web sites now that talk about *intelligent design*, which proposes alternative explanations for evolution without the scientific data to support it. Eventually, you can have your students assess the validity of more reputable information from the World Wide Web, local newspapers, and magazines.

Finally, it is important that the teacher help students synthesize information. For elementary and middle grade students, this often requires having them summarize the key ideas and related concepts. A graphic organizer like the Central Idea Dia-

TABLE 4.5   Evaluating Information

Information found	Does it add to our understanding base? How?	Is the information new or different from what we have already found?	Is the information more recent than what we have already found?	Does it help answer our question? How?
Acid rain is thought to be a cause of declining frog populations.	Yes. It provides biological information related to acid rain.	Yes.	Yes. It is a recent discovery.	No. We are looking for information about how acid rain erodes surfaces of buildings.

TABLE 4.6   Assessing the Validity of Information

Information found	What are the facts to support the claims?	Can a claim be proven? Substantiated by others? If so, how?	Is the source trying to sell something, advocate an opinion, or persuade you to believe something?	How reputable is the source?
What was said?	_____	_____	_____	_____
Who said this?	_____	_____	_____	_____
When did they say it?	_____	_____	_____	_____
Why did they say it?	_____	_____	_____	_____
Where did this occur?	_____	_____	_____	_____
How did it occur?	_____	_____	_____	_____

gram in Figure 4.2 helps students focus on synthesizing the information from an investigation by identifying the main topic and related concepts.

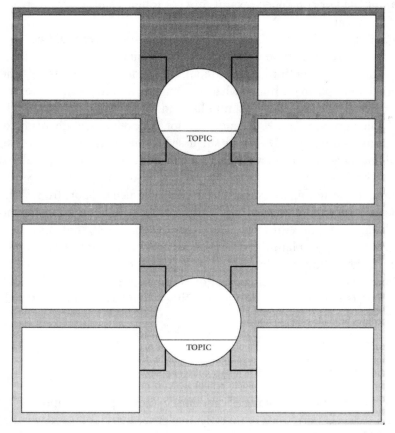

Figure 4.2   Central idea diagram. Source: S. Park & H. Black. (1992). *Organizing thinking: Graphic organizers.* Pacific Grove, CA: Critical Thinking Press and Software.

Planning and Designing

Planning and designing are essential components of an investigation. In planning and designing, students think through and define their variables, consider how long to carry out their investigation, and write out the procedures they will follow. Planning refers to students' thinking about and working out how their investigation will take place and answering questions, such as "Who will measure the data?" and "Who will get the equipment?"

Designing refers to the structure of the experiment and answers the questions, "What data do I need to answer my question?" and "How will I obtain the data?" Using a plant example, planning includes determining who will take measurements of the height of plants and which days the measurements will be taken. Designing, includes determining what observations will be made (for example, observing the germination of bean plants), how often the data will be collected, what types of observations will be made (drawing or taking photographs of the seeds), and what variables need to be manipulated or controlled (examining the influence of temperature on germination of bean plants).

Many children find planning and designing difficult. This should not be surprising as these tasks require students to engage in difficult cognitive work. Planning and designing make students think ahead, require them to specify materials they need, force them to develop an outline of what they will do, and make them create or modify materials. Many elementary and middle grade students do not have such experience, and some primary grade students may not be developmentally ready to plan or create designs on their own. Do not be surprised if some students' initial designs and plans will not allow them to answer questions adequately. Frequently, students' initial designs will be flawed, including confounded and ill-defined variables. Students will need to practice designing investigations, receiving much support from you.

Although planning and designing are difficult cognitive activities for most children (and adults too), the effort produces many benefits. First, students become more reflective. Focusing on planning helps students think through issues and become thorough in their thinking. Second, when students make a plan and design, they develop more ownership of an investigation. Because they make the plan and design, they also understand them. This understanding also helps them understand the science involved.

Designing Experimental Procedures

Designing an investigation requires students to develop four important capabilities: (a) write clear descriptions of investigative procedures, (b) identify variables, (c) define variables operationally, and (d) control variables.

Learning Activity 4.3

WRITING PROCEDURES

Materials Needed:

- Lego™ building blocks
- Writing materials

A. Have your students work in pairs. Give each pair ten types of Lego building blocks in different colors. Ask each pair to construct a structure using all the Lego blocks you gave them. Then ask each pair to write a detailed description of how to make the structure so that someone else, following their instructions, could make the same structure. Instruct the pairs not to look at one another's structures.

B. Have the pairs exchange their written directions. Now ask the pairs to follow the directions they have been given in building a new structure.

C. Ask the pairs to inspect the structures built from their directions. How could the directions be improved?

D. You might also have your students try to explain how to tie a shoe, make a peanut butter sandwich, or make a cube out of paper.

Writing Clear Procedures

The first thing students need to learn to do when designing an investigation is to clearly communicate the procedures they will follow. Learning Activity 4.3 demonstrates just how difficult it is for most people to write clear descriptions that others can follow. Because writing clear procedures that others can understand is difficult, you might encourage your students to read their plans to one another and receive feedback. Do other children understand what is intended? How can the procedures be made easier to understand?

Depending on the age and experiences of the student, designs of investigations will range from descriptions of what observations will be made to descriptions of what variables will be manipulated. Elementary students might design an investigation to observe the life cycle of mealworms. Their data collecting could include drawings of how the mealworms look over time. Middle school students may well design an investigation to examine the effects of temperature and moisture on seed germination. Their experimental designs will show more sophistication and describe ways of collecting data systematically.

Identifying Variables

Students need to identify variables as part of their design effort. **Variables** are the characteristics that will change in an investigation. For example, if students are studying the influence of the amount of fertilizer on the height of grass seedlings, the amount of fertilizer and the height of the seedlings are variables. The amount of fertilizer is the **independent variable** because students can change it purposefully. A good way to help students remember which is the independent variable is to call it the *"I changed it"* variable. Since the phrase starts with the word "I," it helps students associate it with the word "independent," since it also starts with the letter "i." The height of the plants is the **dependent variable** because it will change as a result of the conditions of the investigation (Cothron, Giese, & Rezba, 1993). A good way to remember the dependent variable is to call it the *"it changed"* variable. This experiment also has a number of other variables involved in it, such as sunlight and temperature.

Defining Variables Operationally

A critical component of an investigation design is clearly defining variables so that others can understand what is being explored. This clarity helps others replicate the experiment to see if they get the same results. For example, if students were exploring how fast students in their class could run, they would need to give an operational definition of *fast*. They could operationally define *fast* as "the time it takes for a student to run a particular distance" or as "how far a student can run in 10 seconds." These definitions are **operational definitions** because they define the operation or process used to measure a variable. Defining variables operationally is a difficult cognitive activity. You will need to work with young learners to help them express variables operationally. One technique is to have students share their operational definitions with their classmates and receive feedback from classmates and from you.

Controlling Variables

Controlling variables is holding all variables constant except for the one that is manipulated. In the fertilizer investigation, there were variables other than the dependent variable (the height of the seedlings) and the independent variable (the amount of fertilizer). These other variables, which include amount of sunlight and temperature, will have to be controlled, or held the same, for all the plants; or students will not be able to determine how the amount of fertilizer influenced the height of the seedlings.

Controlling variables is another challenging cognitive activity for students. You should expect children to have difficulty understanding how to control variables. With younger students, you might use the term *fair* to

CONNECTING TO
NATIONAL SCIENCE
EDUCATION STANDARDS

SCIENCE EDUCATION
CONTENT STANDARDS

K–4—Standard A: In the earliest years, investigations are largely based on systematic observations. As students develop, they may design and conduct simple experiments to answer questions. The idea of a fair test is possible for students to consider by fourth grade (p. 122).

introduce the idea of controlling variables. Asking, "Is this a fair test?" may help students understand. Students have little trouble understanding that card games in which not all the players have the same number of cards are not fair. It is not a huge leap from that understanding to thinking about making sure that all the plants in an investigation get a "fair" chance.

The Role of the Teacher

Students will find planning and designing an investigation challenging intellectual work. However, if you support students in their designing, your students will show considerable improvement throughout the school year (Krajcik et al., 1998). Designs will become more complex as students include more variables, define variables with greater detail, consider more ways of measuring the variables, and use multiple samples. Students' procedures will also become more clearly defined; they will think of more ways in which to represent their data, identify necessary materials, and divide up responsibilities for conducting the investigation.

Table 4.7 lists questions to prompt and support students in creating their plans and designs. It should be modified to match the age and experiences of the children you are teaching. For example, in first grade, learners are not developmentally ready for controlled experiments and should not be concerned with independent and dependent variables. However, early elementary students are capable of thinking about the observations they will make, how they will record the data, and how data relate to their question.

TABLE 4.7   Planning and Design Chart

Our question:

Our design:
- What variables and terms do we need to define?
- What data need to be gathered to answer our question?
- What observations should we make?
- What will I need to measure?
- What variables need to be controlled to answer the question?
- What is the independent variable?
- What is the dependent variable?
- What is my control?
- How often should I take measurements?
- How often do I need to make observations?

Our plan:

Equipment needed:
- What do we need?
- How much do we need?
- How can we obtain the material?

Procedure:
- What is it?
- How often do we do it?
- How much of each material do we need?
- Who will do it?

TABLE 4.8   Expanded Goal Sheet

What do we need to do?	Who will do it?	How will it be done?	When will it be done?
Find information about acid rain	Bill	Go to the library	By Friday, May 18th
Contact the EPA	Alicia	Go to the school office and call	Tomorrow during class
Make a design for collecting rainwater and measuring the pH of rainwater	Precious	Look for information on collecting rainwater. Ask the teacher and share plans with the class	Today By Friday, May 18th

There are a number of other ways to support students in learning how to plan and design. These include using goal sheets, critiquing plans and designs, modeling planning and designing, and creating class plans and designs.

You can help your students learn to be more thoughtful planners and follow through on decision making by using goal sheets. Students can use goal sheets to track plans, determine the division of labor, develop time lines and schedules, and make decisions about resources. A goal sheet is simply a grid in which students record their steps in one column and their plans about who will take responsibility for the steps in a second column. Additional columns can specify how it will be completed and in what time frame. Table 4.8 is an example of an expanded goal sheet.

Another technique that is valuable in helping students plan is to use a checklist to provoke students' thinking and expand their ideas about possible resources. In an information-based world, making decisions about what information to use is no easy task, and students will need your help. Table 4.9 shows a checklist that can help students select resources. You might even decide to work with students to develop their own checklist.

Critiquing Plans and Designs

One way to help students learn the initial components of planning and designing is to have them critique some good and poor example plans and designs. Learning Activity 4.4 presents the design of an experiment you might have students critique. Students can use the questions in Table 4.7 to guide the critique.

Although critiques of made-up plans and designs can help students learn some basic ideas of planning and designing, it is important to remember that such critiques are artificial activities akin to practicing a serve in tennis without learning to use a serve in a real game. Within a project-based science classroom, there certainly is a place for these types of lessons because they *introduce* students to important ideas; however, the real thinking occurs when students develop their own investigations and give and receive feedback on those plans and designs.

A very useful spin-off of critiquing made-up plans and designs is to have students critique a plan and design for an investigation they will conduct in class. Research conducted at the University of Michigan (Krajcik et al., 1998) indicated that giving

TABLE 4.9   Resources Checklist

INFORMATION

What?	Can we use it?	How?
Encyclopedia	Yes	Find out more about acid rain.
Magazines/journals	Yes	Go to library and look for information on acid rain.
CD-ROMs	Maybe	The school doesn't have any on acid rain. Check the public library.
Software	Maybe	The school doesn't have any on acid rain. Check the public library.
Guest speaker	Yes	Get from the EPA.
Telephone call	Yes	Call the EPA.
E-mail		
Interview	Yes	Talk with someone at the EPA.
World Wide Web	Yes	Do a search on acid rain.
Letter		

SUPPLIES AND MATERIALS

What?	Can we use it?	How?
Writing materials		
Computer/probes	Yes	
Drawing materials	Yes	
Science equipment	Yes	Check with the teacher.
PH paper	Yes	
Calculator	Yes	Check with the teacher.

students a plan and design and critiquing the important features of the plan with the students is an important first step in helping students create and modify their own plans and designs. Have them ask critical questions, such as "Will the data collected help answer the question?" "How and why?" and "What are the independent and dependent variables?" Use other questions found in Table 4.7 as a guide for discussion.

An essential step in helping students learn how to plan and design is to model the process. Model for the students how *you* might go about planning and designing an investigation. This might include thinking aloud about what materials you would need and what procedures you would follow. For example, you could think aloud about how you might explore how shifting gears on your bike influences how fast you can go: "I wonder if the larger gear will make the bike go faster or slower." Write down the steps on the board or the overhead projector as you say them: "Shift the gear to the largest size and determine how fast I can go." Allow students the opportunity to comment on your design and to make their own suggestions. The questions in Table 4.7 can help you think aloud about the various steps you would take.

Another important step in supporting students' planning and design work is to have students work on a plan and design together (Krajcik et al., 1998). This is especially helpful for early elementary students or students new to making their own

Learning Activity 4.4

SPROUTING BEAN PLANTS

Materials Needed:

• Scenario below

A. What follows is a scenario that describes an experiment that a group of students performed to study the influence of temperature on the sprouting of seeds. Read the scenario carefully.

Heather, Michael, and Jackie wonder how temperature influences the sprouting of bean plants. They decide to modify a technique they used earlier in the school year. First, they will soak some kidney beans overnight. The next day, they will place moistened paper towels into two resealable plastic bags. Then they will place five soaked kidney beans and a thermometer in each of the plastic bags. They will set one of the bags on top of a heating pad on the window sill and the other bag just on the window sill. Each day they will observe the beans in each bag, record the temperature of each bag, and moisten the paper toweling.

B. What are the independent and dependent variables in this experiment? What variables are being controlled?
C. Is the procedure complete? What is missing from the procedure? What else, if anything, might influence the sprouting of the bean seeds?
D. What are other pluses and negatives of Heather, Michael, and Jackie's plan and design?
E. How might you modify the design to better determine the influence of temperature on bean plants?

plans and designs. Again, questions in Table 4.7 can be used to guide discussion of the plans.

Perhaps the best way for students to learn how to plan and design an investigation is to create, in small groups, a number of designs and plans. Start off the school year with some quick opportunities for planning and designing. For example, you could give students the challenge to plan and design a bridge, using plastic straws and pins, that could hold the most weight. The *Elementary School Studies'* curriculum materials (1970) and Annenberg's *Video Case Studies* (1997) contain a number of ideas for planning and design practice.

While students are creating their plans and designs, provide feedback. Students should also share their plans and designs with the class to receive feedback. Feedback should point out the positive features of plans and designs, as well as the features that students need to think about and, perhaps, revise. You should make sure that students record the comments made regarding their work. Students will learn about the planning and design process not only by listening to what others have to say about their work, but also by listening to others' plans and designs and by contributing comments to others. It is critical for students to have time to revise their plans and designs based on the feedback they receive. After students have revised their designs, it is important that you give them one last round of feedback.

Carrying Out the Procedures

Conducting the procedures of an investigation includes a wide range of activities. It includes students' gathering the equipment, assembling the apparatus, following through on procedures, and making observations. As children carry out their procedures, they oftentimes will think of new ideas for modifying their work. As discussed in Chapter 6, technology tools can play a major role in assisting children in collecting data. A **quicktrial** is an experimental technique that involves trying out experimental procedures to see if they work. For example, students might have designs to make a rain collector in order to measure the amount of precipitation. However, students might first perform a quicktrial and build a model of the apparatus to see if they left out an important design feature, such as a suitable way to anchor the rain collector during high winds.

Gathering the Equipment

To perform an investigation, students need to have the necessary equipment. Equipment can be everything from a hand lens for observing insects to a thermometer for measuring the temperature of a classroom habitat or a computer-based probe for gathering data over time. Equipment is essential in extending students' abilities to make observations. Much of this equipment should be standard in the elementary and middle school science classroom, but it is often a challenge to obtain the necessary materials and supplies. School budgets sometimes stand in the way. Students can find many common materials at home, but others will need to be purchased. By enlisting the participation of members of the community, who might donate needed items, you can often obtain the support you need.

Special attention should be given to obtaining equipment that will enable all students, including those who are visually impaired or have physical handicaps, to engage in investigations. *Science Activities for the Visually Impaired/Science Enrichment for Learners With Physical Handicaps* (SAVI/SELPH), available through the Lawrence Hall of Science, is a program designed for this purpose. SAVI/SELPH includes Braille, for example, on the measuring cups, test tubes, and other science equipment in its program.

Assembling the Apparatus

Many investigations involve building an apparatus. If students are completing a project on acid rain, they might need to build a rain collector to collect their rainwater. If students are exploring the effect of a surface on the speed of an object, they might need to build ramps with different surfaces to investigate how different types of surfaces influence the speed at which an object moves.

Another challenge you and your students will face is learning how to use the tools needed to build the apparatus. This will be the first time many of your students have ever built something from basic materials. Many students will not know how to hammer, screw, glue, or solder. Often you will need to model these techniques for students, or you will need to solicit support from volunteers in the community who can come into your classroom and supervise construction.

> ### CONNECTING TO NATIONAL SCIENCE EDUCATION STANDARDS
>
> #### SCIENCE EDUCATION CONTENT STANDARDS
>
> Grades K–4—Standard A: Students employ simple equipment and tools to gather data and extend their senses (p. 122).
>
> #### SCIENCE EDUCATION CONTENT STANDARDS
>
> Grades 5–8—Standard A: Students use appropriate tools and techniques to gather, analyze, and interpret data (p. 145).

Following Through on Procedures

Do not be surprised if students forget a number of important experimental procedures, like watering plants, adjusting temperature, or recording observations. In some circumstances, students might "overdo" their procedures—they might water plants too much or provide too much sunlight. Another common occurrence is "changing procedures on the fly," or modifying experimental procedures without recording the modification or telling anyone about it. For instance, a student might decide to add twice as much fertilizer to the plants.

To prevent these common experimental errors, you should monitor investigations carefully. It is likely that you will need to remind students to make observations (record temperatures, draw diagrams, measure volumes) and to carry out procedures (add fertilizer, water plants, adjust temperature, or feed fish). It is wise to watch students as they carry out procedures to make sure they are doing so correctly, without overdoing anything and without making unplanned modifications.

One useful technique to use to help students learn to follow through is to pair students. Each member of a pair can help keep the other one on track. Another technique is to have students record their observations on charts, tables, or checklists. These record-keeping systems organize data. They also serve to remind students about what data need to be collected and what procedures need to be completed. Even though students create a table for recording data, they may not remember to use it. Take the time to discuss the importance of recording data appropriately.

Making Observations

At first, students will not be very thorough and precise in making either qualitative or quantitative observations. Qualitative observations are detailed descriptions of what is seen. Recording qualitative observations might include making diagrams and labeling them, writing a paragraph about observations, or taking a photograph of observable results. It is helpful to point out to students that real scientists keep careful notes and drawings. Some of the observations and notes of naturalists, such as James Audubon, or scientists, such as Charles Darwin, are useful examples for students. Quantitative observations are those that can be counted or recorded in a numerical format. Students can use instruments, such as thermometers, rulers, and electronic pH probes, to make quantitative measurements. Your students will need support collecting both qualitative and quantitative data. At first, students' descriptions will be very incomplete and measurements might not be precise.

One way to support students in becoming more precise and detailed in their observations is to have all students in the class observe the same phenomenon. Then have the students in the class compare the observations they made. Typically, there will be a great deal of variation among observations. For example, have all of the students in the class make observations of the sprouting of grass seeds. This is a particularly nice activity for a number of reasons. First, it can be used at a variety of grade levels. Second, both qualitative and quantitative measurements can be collected. Third, grass is easy to sprout and grows rapidly. Each group should grow its own grass, and each student should make his or her own observations. Students may make qualitative observations about the color and texture of the sprouting grass. They may make quantitative observations of the temperature and pH of the soil, as well as of the height of grass seedlings. Have each member of a group share each of his or her observations with others in the group. Students should discuss how their observations are similar or different. You need to provide feedback regarding the level of detail and precision of the observations. Whole-class feedback is also very valuable. Ask each group to report its observations to the class. Then, have class members comment on the thoroughness and precision of the observations. Once the class has given its comments, you can give some feedback as well.

Keeping Track of Ideas

Often, as students are conducting their investigation, they think of new ideas to modify their work. Encourage students to keep track of these ideas and to record them in their notebooks. They can use these ideas to modify a current investigation or in future experiments. For instance, a group of second-grade students was observing the germination of bean seeds, and one student wondered if all seeds germinate in the same amount of time. This idea lead to a future investigation that examined the time it takes for seeds to germinate.

One of the difficult decisions you will face as a teacher is when and how to give students the time to incorporate new ideas into their current work. Although this incorporation of new ideas is time consuming, it is a valuable learning experience, and it reflects the reality of scientific work. The revision process is an essential component of the investigation web. Remember, good science is not a linear process.

The Role of the Teacher

Teachers play a critical role in helping students carry out procedures. They need to help students track down and organize equipment. Teachers can save time in investigations and help students build routines if essential equipment, such as scissors, magnifying glasses, rulers, and test tubes, is accessible and always placed in the same location in the classroom. Teachers also need to help students assemble apparatus. Young students may not have the strength or manual dexterity to do the necessary cutting and fitting. Although the teacher should give students opportunities to try doing things themselves, a teacher or an adult helper might need to give assistance. This is particularly true if it looks like a student might injure herself in the process of trying.

Teachers need to support students in their observations. One lesson that many teachers have used successfully is the mystery bag activity. In this activity, a number of different kinds of materials, such as a piece of fur, a piece of a tennis ball, and odd-shaped pieces of wood, are placed into opaque bags. Students stick their hands into the bags to feel the objects. Without looking, they describe what they feel. Students then can engage in interesting discussions about what they felt, comparing their observations, asking others to explain further what they mean, or disagreeing on what they felt. After discussion, give students a second chance to make observations; this way they may reach a consensus about what is in the bag.

There are a number of spin-offs from this activity. For example, you can build mystery boxes: Glue odd- or regular-shaped objects inside shoe boxes. Then seal the boxes so that students cannot open them. Have students explore what is in the boxes by inserting long wood or metal probes, such as 10-inch wood skewers, through small holes drilled into the box. Students then draw to scale the objects that they examine in this way to predict what they think is inside the mystery box.

It is most fruitful to create lessons that focus on observations specific to a project. For example, you can begin a project on the behavior of reptiles by having students describe what a chameleon looks like and how it behaves. Record the observations on the board. Then the class can discuss whether the observations contain sufficient detail. After the discussion, the class can create a common description of the chameleon.

Many students have a tendency to make premature inferences, a conclusion based on evidence and reasoning, rather than observations. You need to help your students distinguish between observations and premature inferences. Imagine that a group of students is recording observations of the crustaceans they find in a nature area. Some students record, "Sowbugs eat wood," and others record, "I found sowbugs under a fallen tree branch." "Sowbugs eat wood" is a premature inference or a speculation; "I found sowbugs under a fallen tree branch" is an observation. One way to help students make this distinction is to ask, "Did you see it?"

Teachers can encourage students to use tools to help them make observations. Elementary students can work with a magnifying glass, a ruler, and a scale. Older students can use a microscope, voltmeter, or light probe. By collaborating with community members, students obtain access to more sophisticated equipment, such as an electron microscope or a telescope.

Students need guidance to become more systematic in recording their observations. Encourage students to keep their observations in a notebook. Their observations should be specific and detailed enough so they can be understood weeks or months later. Imagine that your students are exploring when various insects disappear and reappear. The student who writes, "Insects disappeared when it got cold," and months later writes, "Insects reappeared when it got warm," does not know what temperature was related to the insects' disappearance or appearance. Terms such as *cold* and *warm* are relative and tell the student very little.

One technique that you can use to help students improve their recording of observation notes is to have them read their notes aloud in class to receive feedback from other students. You can also give them written comments. Gentle reminders can also serve as strong cognitive support in helping students gain clarity and become more systematic.

Chapter Summary

In this chapter, we discussed why investigations should be an integral part of elementary and middle school science instruction. We examined the seven important components of the investigation web: (a) Messing about so students can initially explore ideas, (b) asking and refining questions that can be investigated by students, (c) hypothesizing, (d) making predictions about the results, (e) finding information that will provide direction for the investigation, (f) planning and designing a procedure, and (g) carrying out the procedures and sometimes refining them. We discussed a number of ways to help students implement these components of the investigation web. We considered numerous strategies for getting started, asking and refining questions, changing questions to hypotheses, using predictions, finding information, planning and designing, and conducting the experimental work. We also examined ways that teachers can support students by critiquing and modeling these components of the investigation web.

Chapter Highlights

- Investigations are an important part of elementary and middle school science instruction and meet important aspects of the National Science Education Standards.
- One way of thinking about the process of investigation is to think of a nonlinear investigation web.
- The investigation web can be thought of as consisting of several iterative processes:
  - Messing about so students can initially explore ideas,
  - Asking and refining questions that can be investigated by students,
  - Hypothesizing,
  - Making predictions about the results,
  - Finding information that will provide direction for the investigation,
  - Planning and designing a procedure,
  - Carrying out the procedures and sometimes refining them,
  - Making sense of data by transforming it and analyzing it,

- • Drawing conclusions and making judgments about the conclusions,
- • Sharing with others, which might lead to changes and modifications, and
- • Moving into the next round of investigation based on things learned.
- Research supports the idea that children of all ages are capable of designing and performing investigations.
- Messing about helps students build essential knowledge and experiences related to the investigation.
- Researchable questions can stem from children's initial observations.
- Descriptive questions provide opportunities for students to ask about observable characteristics of phenomena.
- Relational questions allow them to find out about associations among the characteristics of different phenomena.
- Cause and effect questions enable students to make inferences about how one variable affects another variable.
- Hypotheses, expressed in testable form, allow students to investigate questions.
- Predictions based upon previous experiences, knowledge, and observation allow students to forecast what they think might happen.
- For students to be successful with investigation, they need to explore relevant background information.
- In designing a procedure, students need to clearly identify, define, and control variables.
- Teachers play critical roles in helping students by critiquing, modeling, and creating plans and designs.
- Teachers play a critical role in helping students following through on plans and making and recording accurate observations.

Key Terms

Archival data	Manipulation of materials
Cause and effect questions	Messing about
Controlling variables	Operational definitions
Current data	Planning
Dependent variable	Predictions
Descriptive questions	Qualitative observations
Designing	Quantitative observations
Hypotheses	Quicktrial
Independent variable	Relational questions
Initial observations	Seeking information
Inquiry-based learning	Variables
Investigation web	

Notes

1 Thanks to Ann Novak and Chris Gleason from Greenhills School for this project idea.
2 The URL for Wikipedia is http://en.wikipedia.org/wiki/Main_Page.

References

American Association for the Advancement of Science. (1993). *Benchmarks for science literacy.* New York: Oxford University Press.

Annenberg/CPB. (1997). *Case studies in science education.* Washington, DC: Author.

Berg, C. (1994). *Chameleon condos: Critters and critical thinking.* Shorewood, WI: Chameleon Publishing.

Bransford, J. D., Brown, A. L. , & Cocking, R. R. (1999). *How people learn: Brain, mind, experience, and school.* Washington, DC: National Academy Press.

Burnett, R. (1999). *The pillbug project: A guide to investigation.* Washington, DC: National Science Teachers Association.

Cothron, J. H., Giese, R. N., & Rezba, R. J. (1993). *Students and research: Practical strategies for science classrooms and competitions* (2nd ed.). Dubuque, IA: Kendall-Hunt.

Elementary Science Study. (1970). *The ESS reader.* Newton, MA: Education Development Center.

Hawkins, D. (1965). Messing about in science. *Science and Children, 2*(5), 5–9.

Ingram, M. (1993). *Bottle biology.* Dubuque, IA: Kendall/Hunt.

Krajcik, J. S., Blumenfeld, B., Marx, R. W., Bass, K. M., Fredricks, J., & Soloway, E. (1998). Middle school students' initial attempts at inquiry in project-based science classrooms. *Journal of the Learning Sciences, 7*(3&4), 313–350.

Krajcik, J., Blumenfeld, B., Marx, R., & Soloway, E. (2000). Instructional, curricular, and technological supports for inquiry in science classrooms. In J. Minstell & E. Van Zee (Eds.), *Inquiring into inquiry: Science learning and teaching* (pp. 283–315). Washington, DC: American Association for the Advancement of Science Press.

Lawrence Hall of Science. (1989). *River cutters.* Berkeley, CA: Author.

Metz, K. E. (1995). Reassessment of developmental constraints on children's science instruction. *Review of Educational Research, 65,* 93–128.

National Research Council. (1996). *National science education standards.* Washington, DC: National Academy Press.

National Research Council. (2000). *Inquiry and the National Science Education Standards: A guide for teaching and learning.* Washington DC: National Academy Press.

Parks, S., & Black, H. (1992). *Organizing thinking: Graphic organizers.* Pacific Grove, CA: Critical Thinking Press and Software.

Scardamalia, M., & C. Bereiter. (1991). Higher levels of agency or children in knowledge building: A challenge for the design of new knowledge media. *Journal of the Learning Sciences, 1,* 37–68.

Chapter 5

Making Sense of Data and Sharing Findings

Introduction

A hallmark of science is that it rests upon explanations and models that are consistent with empirical data. Yet learners, especially young learners, often see science as mysterious, with laws and principles that lay beyond what they can explore. Project-based science shows students that they can learn about the world they live in by gathering and analyzing data and by drawing appropriate conclusions from those analyses. In this chapter, we discuss how teachers can help young learners realize that they can learn about the world by examining and making sense of the data they collect in investigations. We discuss three important components of the investigation web (see figure 4.1 in Chapter 4): how to make sense of data, drawing conclusions, and sharing information. Because science is about explaining phenomena, we also discuss how to construct scientific explanations. Making sense of data is a critical component of scientific explanation. This aspect of engaging in science is perhaps the most challenging for students. Another critical aspect of scientific investigation is sharing ideas with others. Often this important aspect of science is left out of classroom practice. We conclude this chapter by describing a method that teachers can use to support students in all parts of the investigation web.

Chapter Learning Performances

- *Illustrate how to support students in making sense of data.*
- *Describe how to draw conclusion.*
- *Define the meaning of a scientific explanation.*
- *Describe how to support students in making explanations.*

As we discussed previously, research supports the notion that young children are capable of performing all aspects of investigations, including analyzing data and drawing valid conclusions from it (Krajcik et al., 1998; Metz, 1995; National Research Council, 2000). Young learners might not use as sophisticated analysis procedures as older students or scientists, but they can make claims and draw conclusions using evidence. Constructing explanations and making conclusions that are data driven are a challenging aspect of science for all students. However, when students grapple with data to support claims, their perceptions of science and their understanding of science content change (Bell & Linn, 2000; Driver, Newton, & Osborne, 2000). The teacher plays the critical role in supporting students in analyzing data and drawing conclusions.

Scenario 1: A Step-by-Step Process for Analyzing Data

You may have observed a situation similar to this. Ms. Welch, a fifth-grade teacher, has her students collect temperature data three times a day—morning, afternoon and evening—for a week. Her goal was to show how the temperature rises and then decreases each day. Ms. Welch distributed a data table with four columns on it: day, morning temperature, afternoon temperature, and evening temperature. She wanted to help the students out, so she filled in the column marked day with the appropriate information to guide the students to record data three times during the day. The table looked like this:

	Morning Temperature	Afternoon Temperature	Evening Temperature
Monday			
Tuesday			
Wednesday			
Thursday			
Friday			

Once all the data was collected by the students, Ms. Welch walked the students through a step-by-step process of creating a bar graph that would show the temperature in the morning, afternoon, and evening temperature by date. She then told the students to color the morning temperature blue, the afternoon temperature red, and the evening temperature green. The students then colored in the various bars. Once all the students had the coloring completed, Ms. Welch instructed the students to notice how the morning temperatures, the columns colored blue, were all lower than the afternoon temperatures. Timmy really did not understand it because the afternoon temperature on Friday was actually lower than the morning temperature of Monday.

In this first scenario, the teacher allowed students to collect data, gave them a way to organize data, and showed them how to analyze the data. Knowing how to create and read bar graphs is an important skill to learn. However, Ms. Welch did not provide opportunities for the students to make sense of the data on their own. For instance, Ms. Welch could demonstrate making the bar graphs for the first day and then ask students to explain what they would do for the next day. She might then ask them to compare across Monday and Tuesday. However, because Ms. Welsch showed students a process of collecting and analyzing the data, the students had little opportunity to make sense of the data on their own.

Scenario 2: Keeping Track of Data

Mr. Henderson, a fourth-grade teacher, designed an activity in which his students mixed various liquids and solutions with four powders. He wanted his students to realize that the liquids would interact differently with different solutions. He made test tubes containing the liquids (solutions) and jars holding the various powders so that each group could have a complete set. He spent the evening before writing instructions and creating a grid to record data. The introductions directed the students to place a fourth of a teaspoon of each powder in one test tube. Then, they were to add four drops of the liquid and observe what happened. Mr. Henderson's grid looked like the following:

	Liquid 1	Liquid 2	Liquid 3	Liquid 4
Powder 1				
Powder 2				
Powder 3				
Powder 4				

In class the next day, Mr. Henderson wrote a big warning on the chalkboard that all students needed to wear their goggles during the activity. The students seemed to understand what Mr. Henderson wanted them to do. They all had fun mixing the liquids and powders together. During the class, you could hear various students saying, "Wow" and "How cool."

However, the next day, when Mr. Henderson asked students to describe which liquid reacted with the most powders, Mr. Henderson heard several different responses. When he asked students for their evidence, only a few students could point to data that they collected. Many students just tried to recall what they had seen. Tim was adamant that Liquid 3 was the most reactive because he remembered it that way. Tonya, however, insisted that the most reactive was Liquid 2 and pointed to evidence from her data table.

In this scenario, although Mr. Henderson was very careful in describing to students how to set up the activity and students had fun in class doing the activity, Mr. Henderson did not prepare them to collect and record the data systematically. Therefore, students could not use the data for evidence in making claims about the various materials. Accurate record keeping is a critical step in helping students make claims supported by evidence.

Scenario 3: Making Sense of Data

You may have observed a situation similar to this. Ms. Nowakee, a seventh-grade science teacher, had her students collect several measurements of water quality data from a stream behind her school. She started the unit by taking her students on a stream walk so they could make initial observations. She asked her students, "What do you think is the quality of our stream?" Before going out to collect the data, the class studied various measures of water quality. Ms. Nowakee assigned a different part of the stream to different groups of students. The class and the teacher coconstructed a data table. Ms. Nowakee also shared standards for water quality and discussed what they meant with the class. She introduced the class to constructing evidence-based scientific explanations by sharing explanations written in previous years. She then asked the class to use the water quality standards and their data to construct a scientific explanation. After the class shared their results with each other, Ms. Nowakee invited others from the community so the students could present their results to them.

This scenario shows students engaging in different aspects investigation web: analyzing data, constructing explanations, and sharing information with others. The scenario also shows the important role that teachers play in orchestrating these challenging components of the investigation web. The teacher shared and worked with students to develop understanding of water quality standards, helped students understand how to construct scientific explanations, and arranged for community members to come into the classroom.

Unlike the first and second scenarios, this scenario supported students in analyzing data and constructing explanations, but did not give the students step-by-step directions on how to analyze the data. The teacher created an educational environment in which students collected real data and used that data to construct an explanation that could be presented to community members. There are times in the curriculum when the teachers will need to support students in analyzing and drawing conclusions from data. As indicated in scenario two, students drew faulty conclusions because they did not analyze and record their data correctly.

Making Sense of Data

Students make sense of their data by analyzing and interpreting them. The capability to analyze data is an essential aspect of scientific literacy and of scientific investigation. Unfortunately, not many children get the opportunity to take part in analyzing data and making sense of what data mean. Oftentimes, students simply respond to questions and, as they get older, complete formulas. Frequently, what is expected of children is to answer a series of questions regarding the data they collected following a cookbook-type investigation that spells out every step or to use the data in some formulas. The thinking aspect, as well as the interesting aspects of science, is destroyed in such cases. Even many adults do not have good experiences with this feature of investigative work. Observations in the classroom have demonstrated that this is perhaps the most difficult component of the investigative web (Krajcik et al., 1998). Commonly, students think that an investigation is complete when they are finished collecting the data, yet it is only through analysis that students can learn about the phenomena they are exploring. Making sense of the data is an intellectual challenge (NRC, 2006). How can teachers support students in analyzing data? How can they find trends and patterns in their work?

A few techniques can be used to support students in making sense of their data. One technique is to transform the data into a different form. Seeing data in new ways gives students more opportunities to understand the data. Children can transform their data by creating tables, graphs, diagrams, or other visualizations. As students become older, they might use statistics that are more descriptive in the process, such as means or averages. Another technique is to create models. This helps students combine the disparate aspects of their work. In general, helping students think of different ways to represent their data helps them perceive patterns and trends, which is an important element of analysis.

> ### CONNECTING TO NATIONAL SCIENCE EDUCATION STANDARDS
>
> #### SCIENCE EDUCATION CONTENT STANDARDS
>
> Grades K–4—Standard A: Students use data to construct a reasonable explanation (p. 122).
>
> #### SCIENCE EDUCATION CONTENT STANDARDS
>
> Grades 5–8—Standard A: Students will develop descriptions, explanations, predictions, and models using evidence. Students think critically and logically to make the relationships between evidence and explanations (p. 145).

Transforming Data

Perhaps one of the best ways to help students make sense of their data is to have them **transform**, or change them, into some other form or representation. Scientists use tables, graphs, diagrams, and maps to transform their data into other forms that might help them understand and see patterns. We will discuss a few of these forms that are particularly helpful for elementary and middle grades students.

Using Tables

In early elementary grades, students can create tables from tallies or counts they have made. Imagine that students made a record of the different types of insects they observed on the playground. They could then transform this data into a large table on newsprint. The table might include a column with the name of each insect, a second

TABLE 5.1 A Data Table

Insect	Drawing	Number
Ant		20
Bee		3

column with a drawing of each insect, and a third column with the number of sightings. Table 5.1 is an example of a table early elementary students might create.

Students will need support and practice making tables. One good way to help students learn how to make tables is to model the process for them. As you model how to develop a table, allow students the opportunity to contribute their own ideas to the process. Once you have modeled creating tables, you can also help students develop a sense of how to make tables by letting them design their own tables and by giving them feedback on their tables. As the year progresses, students can construct more elaborate tables (Wu & Krajcik, 2006). Another useful technique is to have students share their tables with classmates. Allow classmates to give comments on the tables and discuss how the tables will help them make sense of the data.

Tables can include both qualitative and quantitative data. When recording qualitative data, the cells in the table can include descriptions of observations. Remember that students need to write thorough and complete descriptions of their observations. Table 5.1 shows how a table can include both qualitative and quantitative data. In scenario two, students could use the table that the teacher constructed to guide their procedure, but they can also use the table for recording their data.

Once students have constructed their tables, they should write summaries describing what their tables show. Summaries can be written for both qualitative and quantitative data. Such summaries are just another way to transform the data. For example, the young children who created Table 5.1 could write a simple paragraph describing going outside to look at insects, counting them, and drawing the pictures of the types observed. However, not all tables will be applicable to such summary statements. Depending on what the question is and how the table is organized, students could write summaries by column or by row. For instance, using the data table from scenario one, students could summarize across a row and describe how the temperature changes throughout each day, or they could summarize down a column and describe how the temperature changes in the morning, afternoon, and evening across the days. Once again, feedback from you and classmates will help improve what is written. Often descriptive statistics can be used to write summarizes of quantitative data (see section later in this chapter).

The tables students create will vary according to grade level. Imagine that a third-grade class is carrying out a weather project. The teacher has students use a number of pictures to represent the weather. Figure 5.1 illustrates how students might do this.

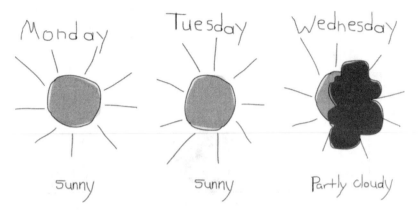

Figure 5.1   Chart created by early elementary students.

Older students should record more quantitative measurements. For example, they might observe and record such measurements as the temperature, barometric pressure, wind speed, and precipitation. Table 5.2 illustrates a table a sixth-grade student might create to record various weather-related measurements.

Using Graphs

Graphs communicate visually the data that are collected in an investigation. Graphs show how dependent variables change with the independent variable. By convention, the dependent variables are depicted along the y-axis (the vertical axis of a graph), and the independent variables are depicted along the x-axis (the horizontal axis of the graph). Transforming data into graphs will help students see trends in quantitative data. It is also critical that students know how to read, interpret, and see trends in graphs. Newspapers, magazines, reports, and Web sites, as well as news reports on the television, show graphs all the time. People need to know how to read graphs to be a scientific literate citizen, since understanding how to interpret graphs is a skill that people use throughout their lives.

In elementary and middle school, students can transform their data into pie charts, bar graphs, histograms, and line graphs. Software programs, such as word processing software and spreadsheets, can support older children in this effort, but it is best to use a graphing program for most elementary and middle grades children. ***Graphical Analysis 3*** is a software program available from *Vernier* software that can support

TABLE 5.2   Weather Data Chart

		Monday	Tuesday	Wednesday	Thursday	Friday	Saturday	Sunday
Week 1	Temp	___	___	___	___	___	___	___
	Pressure	___	___	___	___	___	___	___
	Precipitation	___	___	___	___	___	___	___
Week 2	Temp	___	___	___	___	___	___	___
	Pressure	___	___	___	___	___	___	___
	Precipitation	___	___	___	___	___	___	___

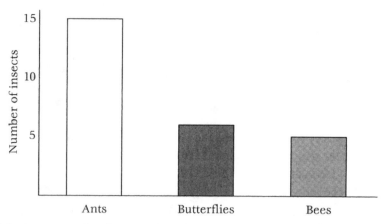

Figure 5.2   Insects on the playground.

middle school students in building a variety of graphs (http://www.vernier.com/soft/ga.html). *Graph Club 2* is graphing program distributed by *Tom Synder* (http://www.tomsnyder.com/) that elementary students can use. *Tinkerplots*, by Key Curriculum Press (http://www.keypress.com/x5715.xml), helps students make sense of data and recognize patterns as they unfold in their investigations.

Rather than giving a simple count, graphs offer a different visual representation of the data that support making interpretations. Figure 5.2 shows a bar graph created by elementary students who explored the number of insects they found on their playground one afternoon. The graph provides a visual representation of the number and types of insects found on the playground. Figure 5.3 is a line graph created by seventh graders to show the differences in height over time of grass seedlings growing in different soil types.

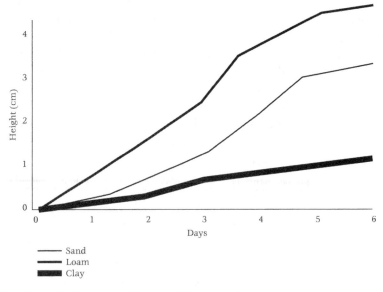

Figure 5.3   Growth of grass seedlings in different soil types.

It is not always easy to determine when to use bar graphs and when to use line graphs. Bar graphs are typically used when the data are discrete; that is, when they can be divided into categories and when the intervals between the data have no meaning. Types of insects, types of trees, and color of flowers are examples of **discrete data**. Line graphs are best used when the data are continuous; that is, when the intervals among the data have meaning. The concentration of fertilizer, the heights of plants, and the speed of a ball rolling down a hill are all **continuous data**. You should give your students practice in using the various kinds of graphs.

Once students have collected data, classroom teachers need to support elementary and middle grades students in making and interpreting the graph. Making graphs and interpreting them are cognitively challenging tasks for children, as well as adults. Knowing how to interpret the meaning of various graphs will help students understand various graphs found in daily life in newspapers, on the Web, and on television.

A teacher can support students by modeling, sharing graphs, and giving feedback. One way to support students in interpreting graphs is to have them come up with statements that describe what the graph means. Have them look for a pattern or a trend in the shape of the graph. Once students have written their own interpretation, members of the groups can compare the statements. Next, have the group construct a group response and share the group statement with the class to receive additional feedback. For instance, a student could write the following interpretation of Figure 5.3, Growth of Grass Seedlings: "The grass in the loam grew the best." Another student might write: "The grass grew worse in the clay." After sharing and discussing their interpretations with each other and receiving feedback from the teacher, the group might compose the following group response: "The grass grew best in the loam and worse in the clay. The grass in the sand grew a little better than in the clay but not by much."

Young children who do not have an understanding of number concepts should be encouraged to make concrete, physical graphs. For example, if early elementary students are counting the number of seeds in different fruits and vegetables, they can line up on a table the number of seeds from a pumpkin, a watermelon, and a cantaloupe. Then you might ask questions about which fruit has more seeds. Young learners first need to see the real seeds lined up in a row to perceive a concrete graph. Watch for conservation problems. Because the cantaloupe seeds are smaller than pumpkin seeds, for example, young children may say the pumpkin has more seeds. It is critical, therefore, to line seeds up with one another, as if they were intervals on a graph.

To help young children make the transfer to using numbers in graphs, teachers must first work with nonnumerical ideas, such as "Which is longer?" "Which is

> CONNECTING TO NATIONAL SCIENCE EDUCATION STANDARDS
>
> BENCHMARKS FOR SCIENTIFIC LITERACY
>
> The Mathematical World—Symbolic Relationships: By the end of fifth grade, students should know that tables and graphs could show how values of one quantity are related to values of another (p. 218).

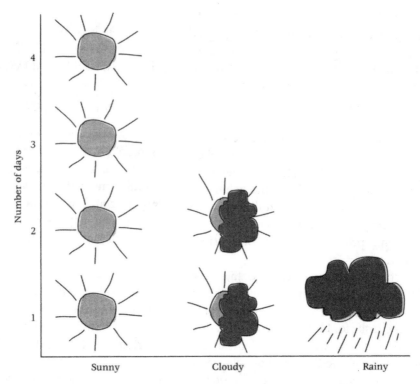

Figure 5.4　Weather pictograph.

shorter?" "Which is bigger?" "Which is smaller?" "Which is closer?" and "Which is farther?" To make the transition to "units," use nonstandard units such as blocks, fingers, or shoes. For example, you might ask, "How many blocks high is the bean plant?" "How many fingers wide did the toy car move?" or "How many shoes long did the turtle walk?"

To help young learners make the transfer from concrete graphs to more abstract paper-and-pencil graphs, use pictures of the objects in the paper-and-pencil graphs. For example, to graph the number of seeds in a pumpkin, a watermelon, and a cantaloupe, the student draws the seeds on graph paper to correspond to the real seeds. Figure 5.4 shows an example of this pictograph technique using weather data. Eventually, your students will be ready to make the leap to using numbers when you match up the objects or events on the graph with numbers.

Using Visual Techniques

Other nontraditional visualization techniques can be used by students to transform data. For instance, students could combine the use of quantitative data, qualitative data, and images. Imagine that some students are exploring how different soil types influence the growth of grass. The students decide to measure the height of the grass over time as an indication of growth. One excellent way to represent this data would be to transform it into a graph of the height of the different plants over time. Students could also represent this data by taking photographs of the plants over time (with a ruler in the background, behind the plants, to record height) and then link the visual pictures to the quantitative data. Photographs of the soil type could also be matched to the quantitative data. If students do not have a digital camera, they could draw a picture or a diagram of their plants. In this case, they need to specify the scale of their drawings. Students could also study weather patterns and create a weather map (or download weather maps from the World Wide Web on the day the plant heights were measured) corresponding to plant photographs.

Combining graphing techniques with other visual techniques should be encouraged strongly at all age levels. When students represent data and ideas in different forms, they are developing deep understanding of the content. Combining techniques is also valuable, because it helps meet different learning styles of students. In Chapter 12, we discuss multiple intelligence theories, which are supported by using a variety of techniques.

Using Descriptive Statistics

Descriptive statistics are another way to transform data. Students can use **descriptive statistics** to describe the basic features of the data in a study—such as the average and the spread of scores. Descriptive statistics provide simple summaries about the sample. For instance, students could calculate the average temperature for a week, but also describe the range of temperature from the lowest to the highest. Along with graphs, descriptive statistics give students a good sense of the trends in data. Upper elementary students and middle school students should be encouraged to use descriptive statistics, such as differences, averages, and ranges, to make sense of their data. Such basic descriptive statistics give a measure of central tendency and variation in their data. For example, a way to transform data about the influence of soil on the height of grass plants is to calculate the average heights of the grasses in each soil type. Students can then create graphs of the average heights of the grass over time. You can introduce young children to descriptive statistics by using such terms as *more, less, big, little, taller,* and *shorter.* See Cothron, Giese, and Rezba (2004) for further discussion on descriptive statistics.

Modeling

A model is a simplified representation of a phenomenon that suggests how the phenomenon works. There are three types of models: physical, conceptual, and mathematical. Models help scientists describe, explain, and predict a host of related phenomena. Models allow scientists and learners to explore phenomena and processes that are too small or too big to observe and examine directly, those that might occur too fast or too slow to examine, or might be too dangerous to investigate directly (AAAS, 1990). For instance, how the flow of rivers changes the land over time occurs too slowly to examine directly; however, stream tables allow scientists and learners to examine the process of stream flow and how stream flow changes land forms.

Physical models are actual devices or processes that behave like phenomena. A model airplane and a stream table are examples. Physical models are most obvious to

CONNECTING TO
NATIONAL SCIENCE
EDUCATION STANDARDS

BENCHMARKS FOR
SCIENTIFIC LITERACY

Common Themes—Models: By
the end of the fifth grade, students should know that seeing
how a model works after changes
are made to it may suggest how
the real thing would work if the
same were done to it (p. 268).

young learners. For instance, a stream table
allows students to visualize how the process of the flow of water makes a river and
the land around the river change over time.
One challenge that teachers have is that
many young learners think that something
that looks more like the real object is a better model, rather than realizing the model
needs to be able to describe and explain
some aspect of the object or phenomena.
For instance, a stream table decorated with
toy cars and people does not make a better
model of stream behavior.

Conceptual models link the unfamiliar to the familiar through metaphors and
analogies. Conceptual models depend upon the ability of students to imagine that
something they do not understand is similar to something they do understand. Ping-
Pong balls bouncing around (an idea that students understand) is a conceptual model
for visualizing a gas at the molecular level (an idea that students do not understand).
Because conceptual models are an abstraction, the age and experience of students
play a major role in determining their applicability. Direct, hands-on experiences
will help students become adept at using conceptual models. By the end of the fifth
grade, students should know that seeing how a model works after changes are made
to it may suggest how the real thing would work if the same were done to it.

Mathematical models specify a relationship among the variables described and
the behavior of phenomena. A mathematical relationship that describes how much
work is necessary to move an object is "work = force × distance." The relationship
specifies that the amount of work it takes to move an object is equal to the force
applied to the object times the distance the object is moved. For instance, a person
does work when a box is lifted vertically from the floor to a shelf. If you lift the box
twice as far, you do twice the work. Mathematical models are more abstract than
are physical and conceptual models. However, new computer tools allow students
to build dynamic models with a natural language interface. Netlogo is one example
of a computer modeling tool middle school students could use to build models of
predator–prey relationships (Netlogo can be downloaded at http://ccl.northwestern.
edu/netlogo/). Such modeling programs enable learners to develop an understanding
of scientific concepts, despite the learners' lack of mathematical sophistication.

Models, however, have their limitations. Models can be too simplified or too
complex to be of value, and care must be taken to generate truly useful models. For
example, round ball models of atoms have their usefulness in illustrating the particle
nature of matter, but they give no insight into the structure of atoms or into how
atoms combine to form molecules. Learners must come to realize that even simple
models can help to explain certain phenomena and that models can change to explain
a greater range of phenomena.

The Role of the Teacher

Students will need your support in making sense of data (NRC, 2006). A variety of instructional strategies can provide this support. As a teacher, you should model, share, and give feedback frequently. You might start by having the class construct either a class table or a graph. Next, you might have each student write a summary of what the table or graph means. You might have students share and compare their summaries. In each step, you need to provide students with feedback. You should try to give clear and explicit suggestions regarding ways students might improve. By just saying "incorrect" or "improve," your students will not know what aspects made their answers incorrect or what to do to improve. A teacher might point out strengths and weaknesses of students' responses in an effort to give guidance for improving their answers. If you just point out the strengths, students cannot improve; and if you just point out the weaknesses, students might lose motivation to improve.

In addition to these supports that are designed to help students with the cognitive task of making sense of data, students sometimes need support getting along with others in their group. As students try to make sense of their data, it is likely that they will disagree on their analyses. Such disagreement is healthy because it shows that students are taking responsibility and ownership for their learning. As long as students focus their disagreement on the analyses and do not begin to demean or degrade each other, these disagreements should be encouraged. Scientists frequently examine and question the data and interpretations of other scientists. This open-communication process is what helps move science forward and what gives scientific explanations their validity. You can help students focus their disagreements by having them use evidence to support and critique their statements (Davis & Kirkpatrick, 2002). (Chapter 7 explores in detail how to manage collaborative discussions among students.)

Constructing Scientific Explanation

Scientists continuously seek explanation for new phenomena. Think of a scientific explanation as a discussion of how or why a phenomenon occurs that is supported by evidence. The science education standards stress engaging students in constructing explanations. In *Inquiry and the National Science Education Standards: A Guide For Teaching and Learning* (NRC, 2001), which is a complementary monograph to the National Science Education Standards (NRC, 1996), the National Research Council lists five essential features of inquiry. They state that learners need to: (a) engage in scientifically oriented questions, (b) give priority to evidence in responding to questions, (c) formulate explanation, from evidence (d) connect explanations to scientific knowledge, and (e) communicate and justify explanations. What is striking about this list of inquiry features is that four of these five essential features involve some aspect of scientific explanations. One prominent scientific inquiry practice stressed in these standards documents is the construction, analysis, and communication of scientific explanations. Being able to judge the claims made in an explanation is an important aspect of what it means to be a scientifically literate citizen. Citizens need to be able to identify whether scientists or others make claims that are supported by evidence and scientific reasoning. For example, many scientists have presented evidence of global warming, but some interest-based groups counter these claims with conflicting information aimed at downplaying global warming. Citizens, for example, who make decisions regarding automobile purchases, need to be able to discern the various viewpoints and make informed decisions in real life.

CONNECTING TO
NATIONAL SCIENCE
EDUCATION STANDARDS

BENCHMARKS FOR
SCIENTIFIC LITERACY

Habits of Mind—Values and Attitudes: By the end of fifth grade, students should offer reasons for their findings and consider reasons suggested by others (p. 286).

Research in science education shows that, if students construct scientific explanation as a component of their science learning, they develop a better understanding of what science is all about and change their image of science (Bell & Linn, 2000). Rather than seeing science as a body of facts, constructing explanations helps students see science as a way of understanding their world. Constructing evidence-based explanations can also help students develop better understanding of science content (Driver et al., 2000). When students construct scientific explanations, they require and, hence, demonstrate deep understanding of science content (Barron et al., 1998).

What does constructing evidence-based explanations look like in the classroom? How can teachers support students in such a process? Can early elementary students construct explanations? Although challenging, even students in early grades take part in various aspects of inquiry, including constructing explanations.

How to Support Students in Constructing Scientific Explanation

Central to writing a scientific explanation is the use of data. When scientists explain phenomena, they provide evidence and reasons to justify those claims they make about the phenomenon. The evidence is essential to convince other scientists and individuals of the validity their statements. How can you support students in writing such a statement? A useful framework for helping students construct scientific explanation contains four components: claim, evidence, reasoning, and considering alternative explanations (McNeil & Krajcik, in press; Sutherland, McNeill, Krajcik, & Colson, 2006).

The **claim** is a testable statement or conclusion that responds to the original question. For example, the claim is the simplest part of an explanation and often the part students find the easiest. For instance, a second grader might make the following claim about the type of insects found on their playground, "We have mostly ants on our playground." However, it is not a scientific explanation until students include their evidence and reasoning. Our goal is to move beyond students' making just claims to supporting their claims with evidence and reasoning.

Evidence is **scientific data** that supports the claim a student made. For example, often the data come from a student's investigation, but students can also use others' sources for data, including archived data or other sources of information. As discussed in Chapter 4, this data can be qualitative or quantitative. Of course, students will need to analyze the data and make interpretations of it. Even at very early grades, we encourage teachers and other students to ask for evidence. Following through with the example above, a student could present the following evidence, "We found 20 ants and only 3 bees."

Reasoning is the most challenging aspects of constructing a scientific explanation (McNeill & Krajcik, in press). **Reasoning** involves a justification that shows why the data count as evidence to support the claim. Moreover, in creating their reasoning students need to use scientific ideas and principles to link the data to the claim. The scientific ideas and principles show how the data are evidence for the claim. However, it is acceptable for young students to link the data to the claim. Continuing with our insect example, at the early elementary level, the reason might sound something like this, "Because I counted more ants than bees on the playground." As students enter middle school, we encourage teachers to focus students on including reasoning into their explanations.

As students develop their abilities to construct scientific explanations, they need to consider what **alternative explanation** might account for the claims they are making. Often in science, data can be interpreted in more that one way. Taking into consideration alternative ways of explaining data is complex, and we suggest that this aspect be added when students know how to construct a scientific explanation. However, do not expect students to do this until the upper elementary and middle grades.

Here is the entire explanation the students may have said in class:

> We have mostly ants on our playground (claim), because I counted more ants than bees on the playground (reasoning). We found 20 ants and only 3 bees (evidence).

Notice that the claim, evidence, and reasoning do not need to follow in that order. As a teacher, you want to listen for and give feedback on the parts.

Also, notice that, although we break down explanations into these four components for students, our goal is to help students to create an interconnected explanation in which all four components are linked together into a concise statement. However, when intro-

CONNECTING TO NATIONAL SCIENCE EDUCATION STANDARDS

BENCHMARKS FOR SCIENTIFIC LITERACY

Habits of Mind—Critical Reasoning: By the end of eighth grade, students should question claims based on vague attributions (such as "Leading doctors say...") or on statements made by celebrities or others outside the area of their particular expertise (p. 299).

ducing students to explanation, breaking explanations down into the four components provides a good framework for helping students develop capabilities in constructing sound explanations. Let us look at two other examples that combine these four components of constructing scientific explanation. We will look at an upper elementary and middle school explanation.

A fifth-grade class is exploring the temperature at which ice melts. The class takes a cup of crushed ice and places a thermometer in it. They collect data every few minutes. The class collects the following data:

Time	Temperature	Observation
Start	25°F	All ice
2 minutes	27°F	All ice
4 minutes	30°F	All ice
6 minutes	32°F	Almost all ice and a little water
8 minutes	32°F	Ice and water
10 minutes	32°F	Ice and water
12 minutes	32°F	Almost all water
14 minutes	33°F	All water
16 minutes	35°F	All water

The teacher asks the students to construct a scientific explanation. One group constructs the following explanation:

> Ice melts at 32°F. From the data we collected, the ice chucks were all solid below 32°F. At 32°F, the ice chunks started to change from a solid to a liquid. The temperature stayed at 32°F until all of the ice chunks were gone. When the temperature was above 32°F, there were no chunks of ices. Everything was liquid. Melting is when a solid turns to a liquid. Because the ice changed to a liquid at 32°F, 32°F is the melting point of ice.

In the above example, "Ice melts at 32°F" is the claim. The evidence is, "From the data we collected, the ice chucks were all solid below 32°F. At 32°F, the ice started to change from a solid to a liquid. The temperature stayed at 32°F until all of the ice chunks were gone. When the temperature was above 32°F, there were no chunks of ice. Everything was liquid." Notice that, in reporting evidence, the student also contained some reasoning. This is expected and encouraged because we want students to write coherent statements. The reasoning was, "Melting is when a solid turn to a liquid. Because the ice changed to a liquid at 32°F." This is an idealized explanation for a fifth grader, and we would not expect many students at this age level to compose such a complex statement of reasoning. However, if the student left out the reasoning aspect, it would be appropriate for the teacher to ask the student for their reasoning.

The next example is from a seventh- or eighth-grade class that is exploring properties of materials. The students in the class are given the following task (Krajcik & McNeal, in press):

Substance and Property Explanation

Examine the following data table. Write a scientific explanation that states whether any of the liquids are the same substance.

	Density	Color	Mass	Melting point
Liquid 1	0.93 g/cm³	no color	38 g	–98°C
Liquid 2	0.79 g/cm³	no color	38 g	26°C
Liquid 3	13.6 g/cm³	silver	21 g	–39°C
Liquid 4	0.93 g/cm³	no color	16 g	–98°C

Here is one student's response:

> Liquid 1 and Liquid 4 are the same substance. They have the same density, 0.93 g/cm³. They are both colorless. They both have the same melting point of -98°C. The only thing different about them is their mass, but mass is not a property because it varies with the sample size. This evidence shows that Liquid 1 and Liquid 4 are the same substance because they have the same properties.

Can you identify the various parts of the explanation? In this example, the claim is "Liquid 1 and 4 are the same substance." The evidence is "They have the same density, 0.93 g/cm³. They are both colorless. They both have the same melting point of –98°C." The students' reasoning is "This evidence shows that Liquid 1 and Liquid 4 are the same substance because they have the same properties." The student ruled out an alternative possibility with the following comment, "The only thing different about them is their mass, but mass is not a property because it varies with the sample size." This is a sophisticated response for middle school student (McNeill & Krajcik, in press) thus, we would have given this student full credit for his response. Notice that, at this level, we are not expecting students to rule out other possible explanations. Again, the order in which the student made his response is unimportant. Our goal is to have the student construct an interconnected explanation in which all four components are linked together into a concise statement.

In Learning Activity 5.1, you will practice writing a scientific explanation that includes a claim, evidence, and reasoning. You will also need to consider what alternative explanations could account for the observations.

Learning Activity 5.1
CONSTRUCTING A SCIENTIFIC EXPLANATION

Materials Needed:

- Writing materials or a computer

A. Use the information and data in Figures 5.5, 5.6, and 5.7 (on pages 154–155) to construct a scientific explanation. Remember to include a claim, evidence, and reasoning in your explanation. What alternative explanations could account for the observations?

Role of the Teacher

Constructing evidence-based scientific explanations should be a central aspect of teaching science. You can support students in several ways. Students need support in terms of when, how, and why to use the claim/evidence/reasoning/alternative explanation framework (NRC, 2006). McNeill and Krajcik (in press) made a number of suggestions to help students with this new inquiry practice. We summarize these supports below.

1. **Make the framework explicit.** It is important to help your students understand the four components of explanations. Clearly identify and describe what each part of the explanation is.
2. **Discuss the rationale behind the scientific explanation.** Students need to understand why an explanation is an important part of science. Students should understand that science is about providing explanations.
3. **Model the construction of explanations.** After introducing the components of explanations, teachers need to model constructing explanations through their own talking and writing. Providing students with examples of explanations and having them critique them will also be helpful. Teachers can provide an explanation and then ask students to identify where the claim, evidence, and reasoning are in the example.
4. **Provide everyday examples of explanations.** Some teachers find it easier to introduce explanation to students by asking students to write explanation about common, everyday events. For example, students can create an explanation for the tallest student in the school or the best basketball player on the team.
5. **Encourage students to use explanations in their responses.** During class discussions, if a student makes a claim without providing the evidence, teachers can ask the student to provide evidence for the claim. Also, when appropriate, teachers can have students use reasoning to support their claims.
6. **Have students critique explanations.** When students construct explanations in class, teachers can have them trade their explanations with another student to critique each other's explanations. One way to support this process it to show an overhead of a student's explanation from the previous year and, as a class, critique the explanation. Make sure that students critique all four components of an explanation. If you can find example explanations newspapers, magazines, or Web sites, you could have students critique the explanation in terms of the claim, evidence, reasoning, and alternative explanations.
7. **Provide students with feedback.** By providing explicit and clear feedback on their explanations students can improve on this skill. When students construct explanations, comment on their explanation as a whole, as well as the quality of the individual components. Most students have difficulty with providing the reasoning for the claim. Providing feedback is one of the most important strategies for helping students improve, but teachers need to provide explicit and clear feedback that will help students understand how to improve.

Drawing Conclusions

One of the hallmarks of science is drawing conclusions from experimental data. Unfortunately, learners of many ages have difficulty doing this aspect of scientific inquiry. One of the authors of this book and his colleagues (Krajcik et al., 1998) reported that, although many students in a project-based classroom clearly attempted to analyze data and determine its meaning, they tended to present data and state conclusions without explicitly linking the two. Drawing and justifying conclusions requires sophisticated thinking (NRC, 2006). In contrast to their extensive experience following procedures, most students have had limited experience drawing conclusions and justifying what they have concluded from the data. Of course, these kinds of problems are not limited to elementary and middle grades students; they are typical of beginning researchers in all fields, even at the graduate level.

As in the inquiry process, there is no step-by-step procedure that a student can follow to draw a conclusion. However, there are ways to guide the process.

First, when writing conclusions, encourage students to restate their purpose. For example, imagine that a group of third graders is investigating whether objects move faster down an inclined plane that is lubricated or one that is wet. This investigation is related to their driving question, "How does snow affect our daily lives?" These students should restate that the purpose of their investigation was to explore the effect of ice on students using wheelchairs to get up a school ramp when it is icy in the winter.

Second, students need to describe their findings and show how they answered the initial questions. For example, if students found that water caused less friction on the inclined plane, they would first restate that the purpose of the investigation was to explore the effect of wet ice on the wheelchair ramps at school during the winter. They would then explain that they found that ice would reduce the friction of the wheels against the ramp, which is a problem for those trying to go up the ramp. However, while findings need to be linked to the data students analyzed, the findings should not be simple restatements of the data. Rather, findings should state the *trends* or *relationships* in the data. For example, the students would state that all lubricants, including water and ice, reduced friction on the inclined plane.

Third, students need to write a general statement that relates their findings to their purpose. For example, students might write, "We found that lubricants (including oil, ice, and water) reduced friction on an inclined plane. This leads us to conclude that ice would cause problems for those using wheelchairs in the winter, because they would have a difficult time getting into the school building."

Fourth, students need to explain their findings. Often this is accomplished by making connections to background information students have found and to their data. It also includes comments on how reliable the data might be. For instance, if students performed an experiment on how ice affects friction on an inclined plane, but they forgot to keep the inclined plane frozen during the entire investigation, they would need to mention this problem at this stage.

When students write conclusions, watch for several potential difficulties. Imagine that students are investigating how the amount of water influences how tall grass grows.

1. **Students only restate their data.** Students might say, "The grass that we watered a lot was 5 inches tall. The grass that we watered only a little bit was only 1 inch tall." Here students are restating their data and not making a conclusion. An appropriate conclusion might be, "If you water grass more, it will grow taller than grass that is watered less."

2. **Students make conclusions that are not linked to their findings.** Students might conclude that the amount of water does not make a difference in the height of grass. This finding would not be supported by the data because grass watered more was 5 inches tall and grass watered less was only 1 inch tall.

3. **Students make conclusions that do not relate to the purpose of the investigation.** Students might conclude that the amount of sun determines how tall a plant will grow, when the focus of the investigation was on how *water* affects the height of the plants.

4. **Students might overgeneralize.** Students might say, "All plants grow taller with more water." If their investigation used only one type of plant (grass), it would be a mistake to assume that all plants would grow taller with more water. In fact, some plants, such as cacti, would die with too much watering.

> CONNECTING TO NATIONAL SCIENCE EDUCATION STANDARDS
>
> BENCHMARKS FOR SCIENTIFIC LITERACY
>
> Habits of Mind—Critical Reasoning: By the end of fifth grade, students should seek better reasons for believing something than "Everybody knows that . . ." or "I just know" and discount such reasons when given by others (p. 299).

5. **Students make inferences rather than conclusions.** Students might say, "You have to mow the grass more frequently when it rains." Although this idea is related to the topic of investigation, it goes beyond the data the student collected and is mere supposition.

The Role of the Teacher

As teachers, we need to carefully support students in the process of writing conclusions. One way to support students is to ask guiding questions that help them think about their conclusions as they relate to their data and the purpose of the investigation. Table 5.3 provides a list of questions that can be used to help students make good conclusions and justify them.

Another way to give support is to provide ample feedback in this process. First, share with students a conclusion that you wrote that includes all the important features of a conclusion. You can use the questions in Table 5.3 as a guide. Of course, you should not write your conclusion in a way that makes the children feel incapable of approximating it. Next, share with students a conclusion that they can critique. Include in this conclusion salient points that the children will notice. For instance, you might want the students to notice that the findings are not related to the initial purpose of the study. Next, have the students critique the conclusions of their classmates. Make sure that the students use the questions in Table 5.3 to guide their critiques. After the series of critiques, give students opportunities to revise their own work. Finally, give students thorough feedback on their written work. Once again, use the guiding questions from Table 5.3.

TABLE 5.3   Guiding Questions to Support Students in Writing Conclusions

- Did you restate the purpose?
- Did you present your major findings?
- What conclusion do your findings indicate?
- Do your findings and conclusions link to the purpose?
- Are your findings linked to your data?
- How do your findings compare to information you found on the topic?
- Do the conclusions support the information you found, or are they inconsistent?
- How might you explain your findings?

Sharing with Others

Sharing is an essential component of the investigation web (NRC, 2006). Sharing includes communicating to others your plans and progress throughout the investigation, your findings, and your final products. It also includes receiving feedback on what you have communicated. Sharing helps students gain ownership of their investigation, and students learn new science information when others share with them.

Sharing During the Investigation

Sharing should occur throughout various components of the investigation web. Sharing during the project is critical because it gives learners valuable feedback on their work. Often, this is a place where the teacher can teach students to be skeptical of findings and conclusions. This feedback often leads to revisions and new questions. Sharing is also important because students learn new science concepts when they listen to and communicate with others. At the early elementary level, sharing aspects of a project or an investigation might be part of show-and-tell. For older students, sharing should be a central aspect of science class.

Communicating the Findings

In communicating their findings, students describe in writing the trends they see in their data; they create tables, graphs, or other visualizations; and they make conclusions about how the data answer the question of the investigation.

Students can communicate their findings in a variety of formats: reports, posters, newspaper articles, multimedia documents, videos, and e-mail documents. Multimedia allows students to create links, either to other documents or to other parts of the same document. Multimedia documents can contain various types of media (text, graphics, images, and videos). The World Wide Web is an excellent example of multimedia. Students can now use commercially available tools, such as *PowerPoint*, to create multimedia documents easily, and Wiki pages makes creating Web-based documents even easier for students. Figures 5.5, 5.6, and 5.7 show PowerPoint pages created by two students. No matter what form a final report takes, it should include

Procedure

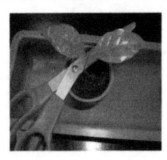

- Remove all but two leaves from both plants.
- Coat one of the leaves on one of the plants with Vaseline.
- Leave all the other leaves alone.
- Give both plants equal soil, water, and sun for an equal period of time.
- Boil the chlorophyll out of all of the leaves.
- Estimate percentage of starch in leaves.

Figure 5.5   PowerPoint document—communicating procedure.

the following typical components: title, question, background information, experimental design and procedures, data, results, conclusions, and future questions.

The Role of the Teacher

The teacher plays a critical role in helping students learn to share information. Elementary and middle grades students do not automatically know how to share their plans, findings, and products; they will need to learn communication skills, speaking skills, and presentation skills to be able to accomplish this task. We discuss ways to teach communication and speaking skills in Chapter 7. These skills, which are taught in language arts classes, create opportunities for the teacher to integrate science with

Data

	Day 1	Day 2	Day 3	Day 4	Day 5
Color of Vaseline plant	Healthy green	Brown line down stem but otherwise healthy	Brown along base of stem otherwise healthy	Healthy green but longer brown stem along base	No change
Color of untouched leaf	Healthy green	Healthy green	Healthy green	Healthy green	Healthy green
Color of untouched plant	Healthy green	Healthy green	Healthy green	Healthy green	Healthy green

Figure 5.6   PowerPoint document—communicating data.

Findings

- Leaves 1 are from the control plant. They had 90-95% starch.
- Leaf 2 is the leaf with Vaseline. It had 10% starch.
- Leaf 3 is the other leaf on the plant with Vaseline. It had 75% starch.

Figure 5.7   PowerPoint document—communicating findings.

language arts. We discuss integration of science across subject areas in more detail in Chapter 12.

In addition to teaching students how to share, teachers need to provide some children with support in transferring hands-on experiences to spoken and written words. Some teachers use graphic organizers to help students make this transition. For example, the KWL technique discussed in Chapter 3 is especially helpful. A teacher can lead a discussion about what was learned (the L portion of the KWL). Once a list of the concepts or skills that were learned is generated, the students can write a summary of this to communicate to others in the class. With young children, the teacher may need to write down students' findings as they dictate them. For very young students, this process itself is a form of sharing. For older students, the teacher may have to review writing skills that focus on supporting main ideas so students can communicate clearly with others.

> CONNECTING TO NATIONAL SCIENCE EDUCATION STANDARDS
>
> SCIENCE EDUCATION CONTENT STANDARDS
>
> Grades K–4—Standard A: Students communicate investigations and explanations (p. 122).
> Grades 5–8—Standard A: Students communicate scientific procedures and explanations (p. 148).

Supporting Students' Implementation of Investigations

Students will find it challenging to implement all the components of the investigation web. They will find it difficult to develop their own unique experimental designs and know how to analyze the data and make appropriate claims. In fact, most scientific investigations proceed from modifications of previous designs. We can support students in doing their own independent investigations by using a similar technique in our science teaching. Teachers first model the process, seek student involvement, and give feedback as students design and conduct their own investigations.

Ann Novak and Chris Gleason from Greenhills School in Ann Arbor, Michigan, use this technique to engage their middle school students in exploring decomposition. In a project with the driving question, "Where does all of our garbage go?" Novak and Gleason's goal is for students to do their own independent investigations, but they recognize that students need support in the process.

They first demonstrate the process by asking the question, "What types of materials will decompose?" In class, they discuss why this makes a good question by asking, "Is it worthwhile?" "Is it feasible?" "Is it important?" Next, they demonstrate how they could investigate this question. They show students how to create decomposition environments using 2-liter pop bottles with holes to allow for the flow of oxygen. They place soil and a variety of materials into the bottles. Some of the materials, like orange peelings, will decompose, and others, like pieces of Styrofoam, will not. During the process, they seek students' input. For instance, students might suggest what types of materials to place in the decomposition environment, justifying each choice. The teachers suggest a possible table format for recording data. They also model how to record the data, using both qualitative descriptions (such as the smell of the decomposition column) and quantitative techniques (such as the mass or temperature). Students then create their own decomposition environments following the teachers' model.

Because it takes several days to observe the decomposition, in the meantime, teachers prompt the students to ask their own subquestions of the driving question. Working in groups, students develop questions, such as "Do worms affect decomposition?" "Does the amount of light influence decomposition?" and "Does the amount of water affect decomposition?" As students discuss their questions, the teachers offer feedback. Both in small groups and in large-class discussion, the questions are critiqued for their feasibility, worth, and importance.

The teachers then ask their students to design investigations by modifying the techniques they used in setting up their first decomposition environments. As students develop their designs, the teachers give the students feedback. Some of the designs are also critiqued in whole-class discussions. Based on the feedback, students can redesign their investigations and then carry them out to explore factors associated with decomposition. Using his technique of first modeling and then allowing students guided by feedback to modify the procedure, the Greenhills teachers have been successful in teaching students to do their own investigations (Krajcik et al., 1998).

Novak and Gleason have used modifications of this technique to engage their students in a number of inquiry-related projects. Professors Hug and Krajcik (2002) further explored this strategy. In a project with the driving question, "Can good friends make me sick?" teachers engage students in exploring the growth of bacteria. The teacher begins by asking the question, "Do I have bacteria on my hands?" The teacher then teaches a lesson, demonstrating how to culture bacteria using agar plates and an incubator. The next day, the teacher demonstrates how to count the bacteria colonies

Learning Activity 5.2

SUPPORTING IMPLEMENTATION OF INVESTIGATIONS

Materials Needed:

- Writing materials or a computer
- Materials to complete an investigation

A. Work with a small group of students or peers to model an investigation of your choice, such as making decomposition columns.
B. After you model the investigation, ask the students to generate some of their own questions for the same question. Record their questions.
C. Have the students design an investigation of their own using the techniques you modeled.
D. What did you find you needed to do to support the students in carrying out their own investigation?

(a noncontaminated plate was used as a control). Students draw conclusions from the data regarding whether or not bacteria are on our hands. Students then ask their own questions. Students have asked such questions as "Does washing my hands make a difference?" "Does a different type of soap make a difference?" and "Is there bacteria on my desk?" This cycle of modeling-investigation-feedback is one technique that teachers can use to support children in the inquiry process. You will try this technique yourself in Learning Activity 5.2.

Criteria for Assessing the Value of an Investigation

Too often, teachers and textbooks perform science for students by providing answers or limiting inquiry. Students do not learn much when most of the investigative work is done for them. It is important that teachers not confuse giving students support with doing the work for students. Table 5.4 lists criteria for determining if the teacher or curriculum materials are performing the cognitive work for students.

Not all of the investigations your students will perform will receive high marks on the criteria provided in Table 5.4. At the being of the year, you might want to constrain or limit the problems your students work on; but, as the year progresses, you would want to provide more options. If, throughout the year, all your investigations are turning out low scores, you need to ask yourself if you are helping students become responsible for their learning and helping them learn to do difficult cognitive tasks on their own. Your goal should be to move students toward *doing* science. Throughout the school year, look for increasingly higher scores on the table. Learning Activity 5.3 provides practice in using Table 5.4.

TABLE 5.4   Reflecting on the Structure of an Investigation

- To what extent did the students define the investigation?

1	2	3	4	5	6
Closely defined					Not defined

- To what extent did the investigation allow students to ask their own questions?

1	2	3	4	5	6
No opportunities					Numerous

- To what extent was the teacher expecting the right answer?

1	2	3	4	5	6
Predefined					Not defined

- To what extent were students allowed to design their own investigation?

1	2	3	4	5	6
No opportunities					Numerous

- To what extent were students allowed to collect data?

1	2	3	4	5	6
No opportunities					Numerous

- To what extent was the investigation meaningful to students?

1	2	3	4	5	6
Not meaningful					Very meaningful

- To what extent was the investigation feasible for students to carry out?

1	2	3	4	5	6
Not feasible					Very feasible

- To what extent was the investigation nontrivial to students?

1	2	3	4	5	6
Trivial					Nontrivial

- To what extent was the investigation based in the real world?

1	2	3	4	5	6
Not real world					Real world

- To what extent did the investigation allow students to analyze data they collected?

1	2	3	4	5	6
No opportunities					Numerous

- To what extent were students allowed to present their findings?

1	2	3	4	5	6
No opportunities					Numerous

- To what extent were students allowed to revise their work?

1	2	3	4	5	6
No opportunities					Numerous

Moving into the Next Round of Investigation

The completion of an investigation should lead to the next round of investigation, either as a result of students coming up with new questions or as a result of students thinking of modifications to their experimental design. Students should be provided opportunities to make predictions that are based on the results of their previous investigation, formulate hypotheses based on results of their previous investigation, and apply experimental techniques to new problems. As students conduct their investigation, they should keep a record of new questions that come to mind and of questions that remain unanswered. You need to capitalize on situations in which

Learning Activity 5.3

COMPARING TWO INVESTIGATIONS

Materials Needed:

- A partner
- Two thermometers
- Two Styrofoam cups
- Hot water
- Stirring rods

A. Read the following investigations.

Investigation 1: Cooling it by Stirring

- Purpose: You will observe how the temperature of a hot liquid changes with stirring.
- Materials: Two thermometers, two Styrofoam cups, hot water, and stirring rods
- Procedures for collecting the data:
    1. Label one Styrofoam cup "Stirring" and one "Control."
    2. Fill the two cups with 150 ml of hot water each. Caution: Hot water can burn you. Make sure both cups are secure.
    3. Place a thermometer in each cup.
    4. Gently stir one of the cups of hot water until the end of the experiment.
    5. Record the temperature of each cup every 3 minutes for 24 minutes.
- Data analysis: For each cup, make a graph of temperature versus time. Place temperature on the y-axis and time on the x-axis. Place each graph on the same grid.
- Questions:
    1. What was the starting temperature of each of the cups?
    2. What was the final temperature of each of the cups?
    3. Which cup changed temperature first? Why do you think this one changed first?
    4. How long did it take before the temperature of the other cup began to change?
    5. How does stirring affect the cooling of a hot liquid?
- Conclusions: Write a conclusion to support your data.

Investigation 2: Stirring Things Up

- Prediction: If one cup of hot liquid is allowed to cool undisturbed while another is constantly stirred for a period of time, will one cool faster than the other? Draw a graph to illustrate your prediction.
- Purpose: In this investigation, you will determine if people who constantly stir their coffee are really affecting the cooling process or just exhibiting a case of nerves.
- Challenge: Design an experimental procedure to find an answer to the question proposed.

- Questions to think about:
  1. Did your results agree with your original prediction? Why or why not?
  2. Will you continue to stir your coffee? Explain.
- Additional investigation: What other variables might have an effect on the cooling process? Set up investigations to explore these other factors.
  B. Use Table 4.11 to discuss the two investigations. How are they alike? Different? Which is most like project-based science? Why?

new questions arise naturally—these are the questions students will be eager to try to answer. Learning Activity 5.4 gives you the opportunity to connect all that you have read and to perform your own investigation with a partner. Remember to keep track of new questions that arise.

Learning Activity 5.4

PERFORMING YOUR OWN INVESTIGATION

Materials Needed:

- Writing materials or a computer
- Mealworms

A. Work with at least one partner to observe mealworms (available at most pet stores). Record your observations.
B. What questions come to mind? Are your questions testable? Can you phrase your questions in terms of a hypothesis?
C. What background information do you need? Obtain that information.
D. Design an experiment to answer your question.
E. Carry out the experimental work.
F. How can you analyze your data? Complete the analysis.
G. What scientific explanation can you construct?
H. What conclusion can you draw?
I. Share your experimental work and data with the members of your class.
J. What other questions come to mind? What would you do differently if you did your investigation over again?

Chapter Summary

In this chapter, we discussed how to make sense of data. Constructing and interpreting graphs and tables is an important part of scientific literacy. Making sense of data is a critical aspect of scientific investigation because it helps students build understanding.

We also discussed how to support students in writing evidence-based explanations. A useful framework for students constructing explanations is for students to write a claim, provide evidence, support it by reason, and rule out alternative explanations. Being able to judge if explanations are supported by evidence and reasoning is an important scientific literacy skill.

We also explore writing conclusions. Students need to make sure that their conclusions link to their findings. We discussed the importance of sharing ideas, particularly explanations and conclusions with others. Sharing and critiquing the work of others is an important component of the investigation web.

Teachers need to support students in constructing and interpreting graphs and tables, in constructing explanations, and in drawing conclusions. One way to do this is to model the construction of an explanation or a table, have students construct their own explanations or table, and provide feedback.

Teachers can support students in doing independent investigation by first modeling the process. Students can then perform their own investigations as teachers give feedback. Students should "do" science themselves, rather than having the curriculum or the teacher perform it for them.

Chapter Highlights

- Students make sense of data by transforming it and analyzing it.
- Students should construct explanations that include a claim, evidence, and reasoning.
- Students should draw conclusions and making judgments about the conclusions.
- Students need to share their findings, explanations, and conclusions with others, which might lead to changes and modifications.
- Interpreting graphs and explanations is an important aspect of scientific literacy.
- Teachers need to support students in constructing and interpreting graphs and tables, and in constructing explanations, and in drawing conclusions. One way to do this is to model the construction of an explanation or a table, have students construct their own explanations or table, and provide feedback.
- Teachers can assess the value of an investigation to determine if students are doing the cognitive work, rather than having science performed for them.

Key Terms

Alternative explanation

Claim

Conceptual models

Continuous data

Descriptive statistics

Discrete data

Evidence

Mathematical models

Physical models

Reasoning

Scientific explanation

Sharing

Transform

References

American Association for the Advancement of Science. (1990). *Science for all Americans: Project 2061*. New York: Oxford University Press.

Barron, B. J. S., Schwartz, D., Vye, N., Moore, A., Petrosino, A., Zech, L., & Bransford, J. D. (1998). Doing with understanding: Lessons from research on problem- and project-based learning. *The Journal of the Learning Sciences, 7*(3&4), 271–312.

Bell, P., & Linn, M. (2000). Scientific arguments as learning artifacts: Designing for learning from the web with KIE. *International Journal of Science Education, 22*(8), 797–817.

Cothron, J. H., Giese, R. N., & Rezba, R. J. (2004). *Science experiments by the hundreds*. Dubuque, IA: Kendall/Hunt.

Davis, E. A., & D. Kirkpatrick. (2002). It's all about the news: Critiquing evidence and claims. *Science Scope, 25*(5), 32–37.

Driver, R., Newton, P., & Osborne, J. (2000). Establishing the norms of scientific argumentation in classrooms. *Science Education, 84*(3), 287–312.

Hug, B., & Krajcik, J. (2002). Students' scientific practices using a scaffolded inquiry sequence. In P. Bell, R. Stevens, & T. Satwicz (Eds.), *Keeping learning complex: The proceedings of the fifth international conference for learning sciences* (ICLS). Mahwah, NJ: Erlbaum.

Krajcik, J. S., P. Blumenfeld, R. W. Marx, K. M. Bass, J. Fredricks, & E. Soloway. (1998). Middle school students' initial attempts at inquiry in project-based science classrooms. *Journal of the Learning Sciences, 7*(3&4), 313–350.

Metz, K. E. (1995). Reassessment of developmental constraints on children's science instruction. *Review of Educational Research, 65*, 93–128.

McNeill, K. L., & Krajcik, J. (in press). Assessing middle school students' content knowledge and reasoning through written scientific explanations. In J. Coffey, R. Douglas, & W. Binder (Eds.), *Science assessment: Research and practical approaches*. Arlington, VA: National Science Teachers Association Press.

National Research Council. (1996). *National science education standards*. Washington, DC: National Academy Press.

National Research Council. (2000). *Inquiry and the national science education standards: A guide for teaching and learning*. Washington, DC: National Academy Press.

National Research Council. (2006). *Taking science to school: Learning and teaching science in grades K–8*. Washington, DC: National Academy Press. Prepublication copy: Uncorrected Proofs.

Sutherland, L., M. McNeill, K. L., Krajcik, J., & Colson, K. (2006). Supporting students in developing scientific explanations. In R. Douglas, K. Worth, & M. Klentschy (Eds.), *Linking science and literacy in the K–8 classroom*. Washington, DC: National Science Teachers Association.

Wu, H., & Krajcik, J. S. (2006). Inscriptional practices in two inquiry-based classrooms: A case study of seventh graders' use of data tables and graphs. *Journal of Research in Science Teaching, 43*(1), 63–96.

Chapter 6

Using Learning Technologies to Support Students in Inquiry

Introduction
Role of Technology in Constructing Science Understandings
Role of the Teacher
Integrating Technology into Instruction

Introduction

Humans have always used tools to help them perform difficult tasks, both physical and intellectual. We use inclined planes to help us move objects from a lower position to a higher position, levers to move large objects, and wheels and axles to overcome friction. Humans have used graphs to see patterns in data and visual diagrams to summarize information. Teachers use concept maps as a tool to help students see and make connections between concepts. New computer-based technology tools can also serve as cognitive tools. The World Wide Web enables scientists to find and share information. Imagery tools enable scientists to peer at the structure of materials. Three-dimensional graphs help scientists visualize and interpret data in new ways.

Computer-based technology tools can also change what we do in classrooms. Computer-based technology tools can extend learning by helping students perform cognitive tasks that otherwise might be too complex or difficult (Salomon, Globerson, & Perkins, 1991). In science classes, a wide variety of technology tools can support

Chapter Learning Performances

- *Describe how students can use learning technologies to support them in carrying out investigations.*
- *Compare and contrast the various ways in which children can use technology to develop meaningful understandings.*
- *Use technology tools to explore scientific ideas.*
- *Explain what to take into consideration when designing instruction using learning technologies.*
- *Explain how learning technologies can extend the boundaries of the classroom.*
- *Discuss some of the pitfalls of using technology in the classroom.*

students in learning and in performing inquiry. For instance, students can use computers to access, find and share data on the World Wide Web, probes to gather data during investigations that otherwise might be too difficult or time intensive, graphing packages to visualize data in various ways, and multimedia development tools to create linked, multiple representations to express their ideas.

At home, in school, or around businesses, children frequently use and see technologies, such as scanners, infrared devices, computers, home entertainment centers, and interactive video games. When children visit the doctor's office, they see professionals use technology to set appointments and collect information about their wellness. New technologies monitor and measure all sorts of physiological functions. Gone are the days of mercury thermometers; nurses and doctors now use digital thermometers. Grocery stores use technologies to check out groceries. Technology is used to measure the speed of pitches during baseball games and to tabulate Olympic skaters' scores. It detects moisture in dryers and records seismic events. It is even used to monitor some playgrounds. Children play with and use new technology gadgets on a daily basis. Children play with video games, download music off the Web, make CDs of their favorite music, find sports information, find pictures of music stars, and chat with friends on e-mail or with instant messenger programs using technology tools.

Similarly, children should use technology in schools to help them accomplish tasks. Technology that helps students collect real data, gives students access to numerous sources of current information, and expands interaction and communication with others makes tasks more authentic. Using learning technologies in the classroom is also important in helping to overcome the **digital divide**. While many middle-class and upper middle-class students have access to technology at home, many minority children and those who come from lower income families may not. This lack of technology access increases the separation between students who typically succeed in school and those who do not. To help reduce this disparity, teachers, particularly those who work in schools with disadvantaged students, need to use technology on a regular basis.

Let us examine three scenarios that illustrate the use of technologies to teach science in elementary and middle grade classrooms.

CONNECTING TO NATIONAL SCIENCE EDUCATION STANDARDS

THE NATURE OF SCIENCE— THE SCIENTIFIC ENTERPRISE

By the end of the eighth grade, students should know that computers have become invaluable in science because they speed up and extend people's ability to collect, store, compile, and analyze data; prepare research reports; and share data and ideas with investigators all over the world (p. 18).

Scenario 1: Technology as a Mindless Activity

You may have experienced a situation similar to this: Students in Ms. Williams' second-grade class are studying a unit on animals. As part of the unit, Ms. Williams has students go to the computer lab to use a computer program that enables them to click on parts of different animals to find out the names of the parts. The program also has a fill-in-the-blank component in which students type in the name of animal structures. If they are correct, colorful graphics appear on the screen to "reward" the students. The program also allows students to get background information on the animals and the various parts. Ms. Williams has students work in pairs on the program. Although students enjoy seeing the various pictures of animals and enjoy seeing the colorful graphics appear when they correctly complete the fill-in-the-blank section, they soon get bored with the program, spend time talking about other ideas, or click randomly around in the program. Although a few students select the "get more information" feature in the program, few students spend time reading the information. A number of students who become bored find the "build-an-animal" game in which they can make animals using various parts from different animals. Most of the students have fun using this part of the program, and they try to build the most ridiculous animal possible.

In this first scenario, Ms. Williams hoped students would learn from the software program concepts that reinforced basic ideas she had taught about animals. Although the students used computers, saw interesting graphics, and had fun, they did not experience the potential that technology can bring to learning. As the scenario illustrated, many students can quickly become bored looking at colorful graphics and filling in blanks. Students spent little time using the technology to solve a problem or learn new ideas; rather, they used the computers simply to reinforce previously introduced ideas. This technology does not extend what students can do without the technology and, in fact, might not be as effective as other resources. For example, using technology in this manner is no different from students using a picture book. In fact, *Ranger Rick* and *National Geographic* often provide better pictures and articles than does this type of software. Also, the use of computers in this way is not tied to a question that students find meaningful.

Scenario 2: Going to the Computer Lab

Mr. Daniels, a fifth-grade teacher, tries to schedule the computer room at least once a month or as frequently as he can. He often tries to think of engaging activities for his students to do in the computer room. To help his students get the most out of their day in the computer room, the day before he demonstrates how to use the program they will be using. Sometimes he has students use software tools that allow them to collect and visualize data. For instance, as part of a heat and temperature lesson, he has students use temperature probes to explore who has the warmest hands in the class and to see if the wooden table-tops are the same temperature as the metal legs. The software is pretty easy to use, so most students complete the activities in the 60 minutes for which they have the computer room reserved. Some students get confused and think the program and probe aren't working correctly, because the wood tabletop and metal leg have similar temperatures, yet the leg feels much cooler to the touch. Mr. Daniels doesn't notice this confusion. On another occasion, Mr. Daniels has his students use the World Wide Web to find information on active volcanoes. Most students find some new information that they report to the class the next day. Although Mr. Daniels works hard to try to monitor as many students as possible during the computer lab, some students quickly write down information about volcanoes and then spend the rest of their time looking for information on their favorite band. A few students either misspell *volcanoes* in the search window or type in the wrong URL, and so they never find any information or appropriate sites.

In this second scenario, Mr. Daniels worked hard to ensure his students used computers for beneficial activities. He also tried to ensure that the activities the students did in the computer room were tied to the science unit they where studying in class. Finally, he attempted to monitor students' activities in the computer lab. However, because of his limited access to computers, students never really developed the idea that computers are tools for learning. Unfortunately, gaining access to computer labs on a need-to-use basis continues to challenge teachers in some schools. As a result, students see using technology as an occasional activity done in school, rather than a necessary tool to support learning in the context of daily lessons. Finally, although the use of the computers was tied to the units of study, it was not tied to questions that students found meaningful.

Scenario 3: Using Technology to Support Students' Doing Investigations

Ms. Wu has students in her sixth-grade class use new technology to help them find information, gather and visualize data, and create products to share what they have learned. As part of a project on "Where does all our garbage go?" students design experiments to explore decomposition. They build decomposition environments to study the influence of a variety of variables on decomposition such as temperature, moisture, type of bedding, and material type. They use digital cameras to take pictures over time of their decomposition environments. They use probes to measure and monitor the pH and temperature of the decomposition environments. Some students use the World Wide Web to find information about other factors that influence decomposition. Other students use the Web to write scientists to ask questions regarding landfills and rate of decomposition. Students also create multimedia documents to share what they learn with each other and with members of the school and community. With help from the school's media specialist, students convert their multimedia documents to Web pages that they post on the school's server.

In this third scenario, Ms. Wu planned for students to use technology as a tool throughout the project. Her idea was that students perceive technology the way they perceive a pair of scissors—as a tool used to accomplish a task. Ms. Wu was fortunate because she had six computers in her classroom that were connected to the Internet. As a member of the computer planning committee, Ms. Wu advocated that computers be in the classrooms so students could have daily use of them. She showed her students how to take digital pictures and use probes to collect data in the context of a lesson. When her own background was limited, she sought the help of others to assist her. Most important, students used technology as needed to help them investigate and answer questions that they asked during a project.

Learning Activity 6.1 will help you examine the ways you have used technology in your own learning.

Role of Technology in Constructing Science Understanding

One of the major global changes that has occurred since the early 1980s is the explosion in technology use. Although the capability of technology has increased since then, its cost has decreased. These two factors have made it

CONNECTING TO NATIONAL SCIENCE EDUCATION STANDARDS

SCIENCE TEACHING STANDARD

Effective science teaching depends on the availability and organization of materials, equipment media, and technology (p. 44).

more feasible for schools to purchase technology and to use it for instructional purposes. We refer to new technologies—in particular the use of computers, software, and various peripherals—that support students' learning as learning technologies (Krajcik, Blumenfeld, Marx, & Soloway, 2000). Learning technologies can support teachers and students in project-based science, because they can help students and teachers communicate, collaborate, carry out investigations, and develop products (Novak & Krajcik, 2005). Technology can play a powerful role in enhancing student and teacher motivation by actively engaging students in the learning process.

Learning Activity 6.1

WAYS IN WHICH YOU HAVE USED TECHNOLOGIES FOR LEARNING

Materials Needed:

- Something to write with

A. Think of the different ways you have used technology to learn.
B. How do your experiences compare to the three scenarios?
C. Share your experiences with others in your class. How do your experiences compare to those of the others?

CONNECTING TO
NATIONAL SCIENCE
EDUCATION STANDARDS

CONTENT STANDARDS

A variety of technologies, such as hand tools, measuring instruments, and calculators, should be an integral component of scientific investigations. The use of computers for collection, analysis, and display of data is also part of this standard (p. 175).

Moreover, the use of technology can help students develop understanding of complex, abstract ideas. For instance, Marcia Linn and Dr. Hsi (2000) found that students using software that enables them to make predictions and explanations were able to understand difficult science concepts, such as temperature, and heat energy.

Dwyer (1994) found that technology transformed the way teachers taught. With technology, the classroom changed from one in which the teacher was the center of attention and mostly lectured to one in which children became the center of learning and children interacted with each other, the teacher, and the computer. The teacher's role changed from "expert" to "collaborator." These changes in a classroom can contribute to student learning. The teacher becomes not a provider of information, but a guide who helps students develop understanding through active engagement with phenomena by asking "What if?" questions.

Technology enables students to explore phenomena otherwise inaccessible to them. Creating multimedia artifacts and products helps students share information and develop meaningful understandings. Throughout this chapter, we will show ways in which students can use technology tools to develop understandings. Table 6.1 summarizes the potential of learning technologies in science classrooms.

Learning Technologies to Support Active Engagement with Phenomena

A number of learning technologies can support students in actively engaging with phenomena. These applications enable students to use technologies much as scientists do—as tools to explore phenomena. This active engagement can help students develop rich ideas from the science content and become proficient in scientific practices, such as collecting and analyzing scientific data.

TABLE 6.1   Potential of Learning Technologies

- Help teachers and students communicate, carry out investigations, and develop products.
- Change the nature of classroom instruction so students can construct understanding.
- Actively engage students in the learning process.
- Allow for exploration not possible in the science classroom.
- Provide dynamic visuals to represent abstract concepts.
- Provide opportunities to ask "What if?" questions.
- Provide opportunities to plan, synthesize, question, predict, and apply.
- Facilitate student creation of multimedia artifacts and products.
- Support students in communicating and collaborating with other students and community members.
- Help students create models of complex systems.

Electronic probes or sensors and accompanying software allow students to use computers as laboratory tools to collect, record, and graph data just as scientists do. Students can interface electronic probes with a computer to detect temperature, voltage, light intensity, sound, distance, dissolved oxygen, or pH while the computer digitally records and graphs the data. Probes can be connected to desktop, portable, or various handheld computers. A temperature probe, for example, interfaced to the computer or handheld device enables students to collect and visualize data related to temperature as it occurs, rather than at intervals after temperature has changed. Thus, students can observe graphs being produced while an experiment is being conducted and obtain immediate graphical results of their data.

Probes used as laboratory tools may support students' development of science concepts (Mokros & Tinker, 1987), as well as their procedural knowledge in a fundamentally new way. For instance, temperature probes allow students to explore the concept of evaporative cooling. Many of us feel cool after we come out

> CONNECTING TO
> BENCHMARKS FOR
> SCIENTIFIC LITERACY
>
> THE DESIGN WORLD
>
> Grades 6–8: Students should use simple electronic devices for sensing, making logical decisions, counting, and storing information (p. 202).

of a swimming pool, and we might be able to explain that this phenomenon is an example of evaporative cooling; but few of us have explored this concept. If you place a probe in water, withdraw it, and allow it to dry in the air, you can observe a graph of temperature versus time. What you observe is that the temperature of the probe dipped in water drops below room temperature! This type of experience helps students develop deeper understanding of science concepts.

With probes, students can collect and visualize data much more quickly than before, giving them the opportunity to ask "What if?" questions. Moreover, probes enable students to perform investigations that were difficult to impossible to perform before (Krajcik & Layman, 1992). In addition, this use of probes may strengthen students' graphing, science process, and problem-solving skills (Linn, Layman, & Nachmias, 1988). The advantage is that probes allow students to do explorations not typically possible in the science classroom. For instance, with the motion probe, students can explore if mass influences the speed at which objects fall. Most students, because of everyday experiences, believe that heavier object fall faster than lighter objects. Using the motion detectors, students can explore and visualize what happens. Such explorations were difficult, if not impossible, in classrooms before the use of probes. Learning Activity 6.2 will help you think more about the use of probes in the classroom.

A variety of studies have illustrated the value of students' using probes to collect and visualize data (Linn, 1998). The computer's capability of providing **real-time graphs** of temperature versus time as students carry out experiments may be one of the key elements in helping students develop science concepts and graphing skills. One possible explanation is that, with probes, students have more time to spend interpreting and evaluating the data because they spend less time gathering and recording it. Probes are also more reliable instruments. Brasell (1987) showed that

Learning Activity 6.2

THE VALUE OF USING PROBES IN THE CLASSROOM

Materials Needed:

- Paper and something to write with

A. Describe the value of using electronic probes in the classroom.
B. Observe a classroom of students using electronic probes.
C. From these observations, how would you modify your original thoughts? Are there features you would add, delete, or modify? Explain.

the simultaneous linking of real-time data collection with the production of graphs results in substantial learning gains. Immediate graphical results help students understand concepts, such as temperature, see trends in their data that allow them to focus on the concepts they are exploring, and ask new questions related to an experiment. Because they can display the results both graphically and numerically, children can more easily interpret the results. Research (Brasell, 1987; Morkos & Tinker, 1987) supports the importance of simultaneous data gathering and graphing to help students form understanding. Although the exact mechanism is not known, the simultaneous creating of a graph and seeing it supports students in learning.

A variety of different types of probes exist, including temperature, motion, pH, dissolved oxygen, conductivity, force, and pressure. One type of probe, the motion detector, works by sending out a sound signal that reflects off an object back toward the detector. The motion detector contains a device that can detect the reflected sound sources. Because the speed of sound is known in air, accompanying software can calculate the distance an object is from the sensor. Figure 6.1 shows a graph of a student walking away from a motion detector at various speeds. Probes and accompanying software can be purchased from a

CONNECTING TO BENCHMARKS FOR SCIENTIFIC LITERACY

THE DESIGN WORLD

Grades 3–5: Students are now beginning to encounter challenging information-processing problems in their school work. These problems have one or more procedures (software) for processing data, and these procedures often can be performed more efficiently with the aid of technology (hardware). Students should be encouraged to identify the data presented in the problem, develop a procedure for processing the data, implement the procedure with the aid of technology, and evaluate the reasonableness of their results. As students encounter more sophisticated problems with more complicated data sets, the procedures and tools that they use should also become more sophisticated. Eventually, students should be gathering data, processing information, and presenting results of their data-analysis activities (p. 201).

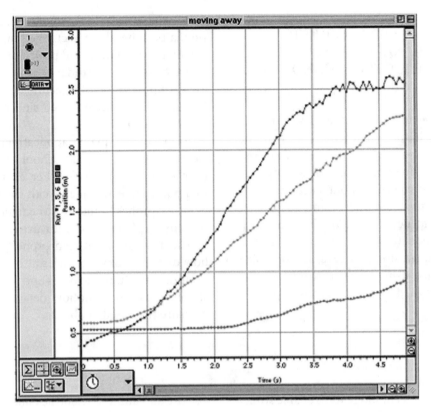

Figure 6.1   A graph showing three different motion-time graphs.

number of manufacturers, including PASCO (http://www.pasco.com) and Vernier (http://www.vernier.com/).

Probes are now available for various hand-held devices, such as the ImagiProbe for the Palm available through PASCO, and the PASPORT System, also available through PASCO (http://www.pasco.com). With probes, these hand-held devices become portable laboratories for collecting data outside of the classroom. For instance, students can use temperature probes to explore how the temperature of the soil changes during the spring. Novak and Gleason (2001) presented a wonderful description of how to use portable probes in middle school classrooms to study water quality. They have students explore the quality of water in a stream behind their school by collecting pH, temperature, dissolved oxygen, and conductivity with hand-held technologies during the fall, winter, and spring. They ask students to use their data to explain why the quality of their stream might change during the different seasons.

Some teachers have students use Texas Instruments' graphing calculators to investigate phenomena in real world contexts (Christmann, 2002). This learning technology lets students explore science and mathematics by connecting Texas Instruments' CBL kit and Vernier probes (for temperature, voltage, light, and so on) to the graphing calculator. With Texas Instruments' CBR, which can be used alone or connected to the CBL, students can collect and analyze distance, velocity, and acceleration. Connectivity software for these technologies enables teachers and students to download data and graphs to the computer. In addition, a peripheral connects the graphing calculator to an overhead projector so an entire classroom of students can view it. To learn more about this technology, visit the Texas Instruments Web site (http://www.ti.com/).

Probes have many applications for elementary and middle school science (Novak & Krajcik, 2005). For example, in a pond project, students can use temperature probes to monitor the temperature, use a pH probe to measure the pH, or use a dissolved oxygen probe to test how dissolved oxygen varies at different locations. In studying forces and motion, students can use motion detectors to measure the speed of objects.

Visualization Tools

Various technology tools enable students to explore and analyze scientific data. **Visualization tools** are software programs that enable students to "see" phenomena and abstract concepts they are trying to understand (Edelson, Gordin, & Pea, 1999). The BioKids project (www.biokids.umich.edu) enables upper elementary and middle school students to use the Web to search and find a variety of information, including text, sounds, and images related to biodiversity of species in local, regional, and global environments. BioKids also provides interactive tools to help students make sense of data. Figure 6.2 shows a screen shot from the BioKids program. The spreadsheet contains data that students collected. The interactive features of the spreadsheed allow students to easily manipulate the columns and rows in the spreadsheet to visualize and find patterns. The BioKids/Deep Think research group (Songer, 2006, 2007; Songer, Kelcey, & Gotwals, 2007) have demonstrated the positive learning gains achieved when students use various aspects of the BioKids program.

With Project FeederWatch (http://www.birds.cornell.edu/pfw/index.html), students can enter data on birds they are tracking and visualize the distribution of different bird populations throughout North America. For instance, students can use the Project FeederWatch to track the migration of different bird species. Visualization software creates opportunities for students to use tools similar to those used by scientists to explore and manipulate data and to perceive trends. Such exploration and manipulation of data help make science learning more like doing real science and help students to develop meaningful understandings (Cohen, 1997).

The Web-based Inquiry Science Environment (WISE) is a free online science learning environment for students in grades 4–12 (http://wise.berkeley.edu) that contains numerous visualization programs. In WISE, students work on exciting inquiry proj-

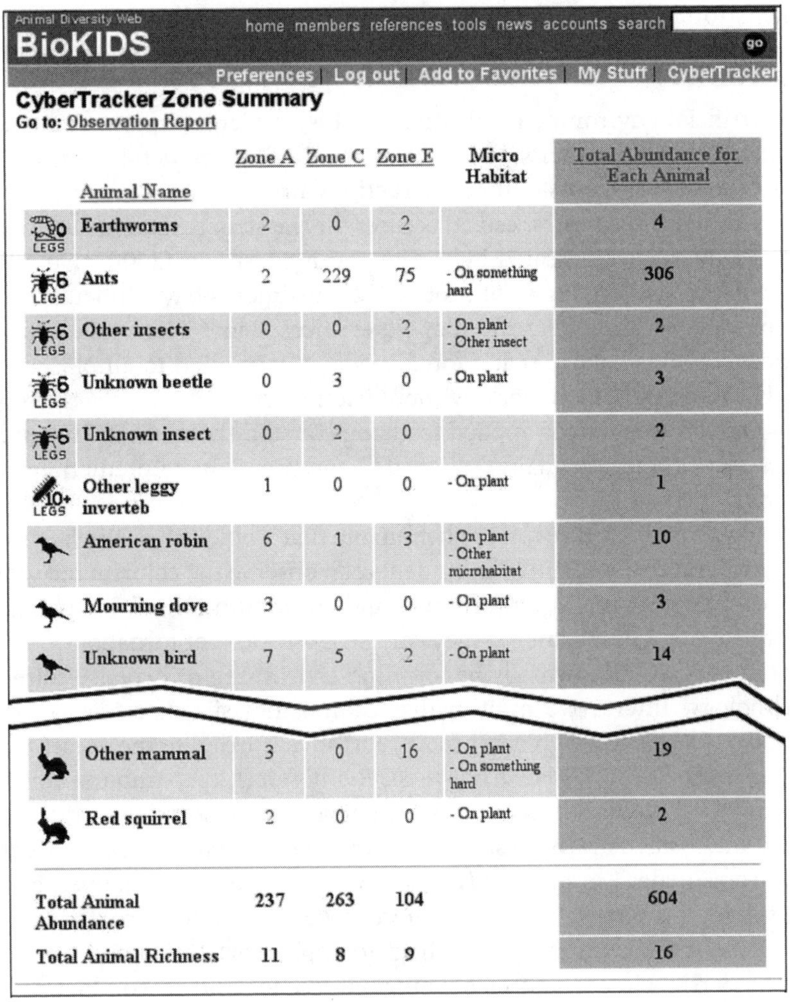

CyberTracker Zone Summary
Go to: Observation Report

Animal Name	Zone A	Zone C	Zone E	Micro Habitat	Total Abundance for Each Animal
Earthworms (LEGS)	2	0	2		4
Ants (6 LEGS)	2	229	75	- On something hard	306
Other insects (6 LEGS)	0	0	2	- On plant - Other insect	2
Unknown beetle (6 LEGS)	0	3	0	- On plant	3
Unknown insect (6 LEGS)	0	2	0		2
Other leggy inverteb (10+ LEGS)	1	0	0	- On plant	1
American robin	6	1	3	- On plant - Other microhabitat	10
Mourning dove	3	0	0	- On plant	3
Unknown bird	7	5	2	- On plant	14
Other mammal	3	0	16	- On plant - On something hard	19
Red squirrel	2	0	0	- On plant	2
Total Animal Abundance	237	263	104		604
Total Animal Richness	11	8	9		16

Figure 6.2   CyberTracker spreadsheet from the BioKids program.

ects in life, earth, and physical science focused on such topics as genetically modified foods, earthquake prediction, HIV, water quality, wolves, and the deformed-frogs mystery. Working in pairs, students learn about and respond to contemporary scientific controversies through designing, debating, collaborating, and critiquing solutions. Most WISE activities are completed on a computer, using a Web browser, such as Microsoft's Internet Explorer or Apple's Safari. Sponsored by the National Science Foundation, WISE software is free for teachers to use.

Microworlds, Interactive Multimedia, and Simulations

Microworlds, interactive multimedia, and simulations combine video, pictures, computer graphics, text, and interactivity to present to students phenomena that otherwise would be inaccessible, too hazardous, too time-consuming, or too expensive for students to observe. Some of this technology enables students to explore difficult questions through simulation that they would otherwise be unable to explore ethically.

In the artificial environments of **microworlds**, students explore and manipulate environments that are otherwise inaccessible. Extraneous details are minimized, making it easier for students to note interactions among variables. For example, students can explore the galaxies, cell structures, or the structure of the atom. Another example is a microworld created by White and Frederiksen (2000) that allows students to explore Newton's Laws of Motion. Most students have difficulty with these physics concepts because their everyday experiences seem to contradict them. When asked, most students believe that a force needs to be applied to an object to keep it moving. However, in a microworld without friction, students can see that objects do keep moving unless a force is applied to change or stop the motion. The microworld challenges students to ask many "What if?" questions, make predictions, and test predictions.

Technology lets students explore phenomena that would otherwise be too hazardous. Using **interactive multimedia**, students can observe the colorful but violent and noxious reaction between liquid bromine and aluminum foil, they can explore the inside of the eye of a hurricane, or they can get close looks at tornadoes.

Activities that would normally be too time-consuming or expensive are possible with technology. Interactive multimedia technologies simulate trips through the human body, voyages through outer space, and adventures into the ocean. Sunburst's *A Field Trip to the Sea* and *A Field Trip to the Rainforest* (http://sunburst.com) ask students to explore ecological issues, participate in a whale research expedition, explore ocean life, and visit the rainforest. The *Great Ocean Rescue* and *Great Solar System Rescue* by Tom Snyder Productions (http://www.tomsnyder.com) actively engage students in solving problems. In the *Great Ocean Rescue,* students participate in challenging rescue missions that require them to apply what they know about marine ecosystems, earth science, and environmental science and to conduct research in these areas to learn more. We have observed numerous teachers implementing these programs, and each time students engaged in lively conversation and debate. All of these programs take students on an electronic adventure, complete with full color pictures and authentic sounds, to places they may not visit in person.

Using **simulations**, students explore what it might be like to manipulate variables that would otherwise be just too unethical, difficult, or impossible to do. For instance, Cooties™ is an interactive simulation program for the Palm Pilot (http://www.goknow.com) that students use to explore the spread of disease. Students meet and possibly "infect" each other by beaming between Palm devices. Typically, one or two of the Cooties programs is set by the teacher to be the initial carrier. The Cooties program tracks which students have met and will tell students if their program became sick or not. After students have finished meeting each other, they work together to determine the initial carriers of the disease and trace the transmission path of the disease. The Cooties software can be set to have high or low resistance and incubation time of the disease. As such, Cooties allows students to ask "What if?" questions and do exploration that would otherwise not be feasible in the classroom.

Although we encourage direct, purposeful experiences whenever possible because they provide the best opportunity for students to develop understanding, micro-worlds, interactive multimedia, and simulations are the *next best thing* to direct, purposeful experiences with phenomena. These technologies, as a type of contrived experience, also provide beneficial experience for students. They allow students to ask "What if?" questions, make predictions, and test out their ideas. Many of these technologies ask students to record their observations or explain what they see. Linking concrete phenomena with writing about the phenomena can help students develop deep understandings. Simulation programs that combine interactive multimedia technologies also give learners opportunities to actively explore phenomena that would be otherwise too small or too big to explore. Such active student engagement can help students develop meaningful science concepts.

Learning Technologies to Support Using and Applying Knowledge

Some learning technologies help students apply their knowledge by supporting them in planning, explaining, and reflecting. Creating models also supports students in the developing integrated understandings (Novak & Krajcik, 2005). In simulations, students manipulate phenomena according to an underlying model; and, in developing models, students must begin by applying the concepts, principles, and rules that govern the phenomena. Modeling is an essential component of what scientist do. Now, with the help of various learning technologies, students can also take part in modeling.

Model-It (Metcalf-Jackson, Krajcik, & Soloway, 2000) is a program that helps students plan models, supports them in building dynamic models using qualitative relationships, and helps them test and evaluate their models. This program supports students in an essential component of learning science—making models—that previously was extremely difficult to accomplish in most middle school audiences. With Model-It, students represent, interpret, and refine their thinking so that they develop more robust understanding. By building models, students also make their thinking visible.

Figure 6.3 shows a screen shot from the Model-It testing window. In this example, students built a model of water quality to explore how the number of people can impact the number of fish in a stream. The left side of the screen shows the tools available for students in the testing window. The screen also shows that the students have tested their model. The graphs across the bottom indicate that, as the number of people increases, the number of fish decreases; and as the number of people decreases, the number of fish increases.

Marcia Linn's Knowledge Integration Environment also helps students explain and reflect on a range of scientific phenomena, including light and temperature. Linn and her colleagues (Linn, 1992; Linn, Songer, Lewis, & Stern, 1993) have also used built-in questions in a probeware environment to encourage students to make predictions and then to compare their results to their predictions. The Web-based Inquiry Science Environment (WISE) is a free online science learning environment

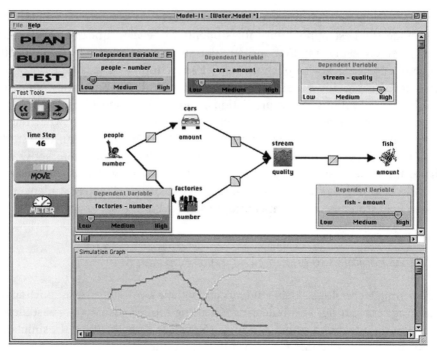

Figure 6.3   Screen shot from Model-It test window.

for students in grades 4–12 (http://wise.berkeley.edu) that contains a causal modeling program that has features similar to Model-It.

Other learning technologies help students use and apply metacognitive knowledge. As discussed more thoroughly in Chapter 2, metacognitive knowledge refers to knowledge about cognition in general, as well as awareness and knowledge about one's own cognition (Anderson & Krathwohl, 2001). Computer programs, such as Inspiration (Helfgott & Westhaver, 2000), Kidspiration (Inspiration Software, 2000; http://www.inspiration.com), and PiCo Map (GoKnow, http://www.goknow. com) help students create concept maps. Concept maps help students with content knowledge, but they are also useful tools to help students examine their own thinking about concepts. Groups of students can discuss their understandings of a topic. Such conversations help students become aware of their own thinking and their own knowledge.

Learning Technologies as a Source of Information

Technology expansion, especially of the World Wide Web, has made information so accessible that students can easily find out about key ideas, concepts, and subject matter topics that arise as they explore solutions to questions. Many of the sources of information on the Web are primary data sources, meaning they often contain the same data sets that scientists use. The Web provides access to instruments, such as telescopes, modeling tools, and remote sensing devices, that can help students carry out their investigations. Finally, the Web makes it possible for students to publish

TABLE 6.2   Advantages of the World Wide Web for Supporting Inquiry

- *Broad range of information:* Learners need access to information to help them refine their questions and find solutions. Online information will bring learners the most current information available.
- *Primary data sources:* Learners need access to demographic data, remote sensing data, and other types of data to support their inquiry. In many circumstances, students will use the same data and information sources as scientists do.
- *Instruments:* Online instruments, such as telescopes and remote sensing devices, serve as sources of direct data as learners carry out inquiry.
- *Computational instruments:* Modeling, simulations, and visualization tools help students analyze and examine online data in new ways.
- *Publishing tools:* Learners need tools that will help them translate information and publish their information for others. This translation is a critical step in helping learners construct their own understanding.
- *Collaboration tools:* Inquiry will be supported if learners have access to synchronous (live) and asynchronous (not simultaneous or live) communication tools that support interaction between students and scientific mentors.

their works—on their own Web sites or on other sites. Many such Web sites enable students to collaborate with scientists and other students about science topics. Table 6.2 summarizes the advantages of using the World Wide Web for supporting inquiry in the classroom.

Wallace and her colleagues (Wallace, Kupperman, Krajcik, & Soloway, 2000) warned that, although the World Wide Web offers many advantages, it also presents difficulties, particularly for young learners:

- Learners may have difficulty locating and taking advantage of information. Often when students do a search, they get back hundreds of hits. Most middle school and high school students (and adult learners) do not have the skills to sift through this many hits. Moreover, most of these hits have little to do with the learner's intended search.
- Learners often become lost and disoriented while navigating.
- Learners require a substantial amount of support to frame their activities.
- Learners often do not evaluate the resources they find.
- Learners often do only one-word searches, rather than taking advantage of multiple word searches to refine their queries.
- Learners unintentionally could end up at nonappropriate Web sites by making simple errors in Web addresses.

Thus, simply providing access to information does not guarantee that it will be useful to learners. Unlike a library, the Web contains many documents that have not been screened for content and validity of information.

> **CONNECTING TO BENCHMARKS FOR SCIENTIFIC LITERACY**
>
> **THE DESIGN WORLD**
>
> Grades 3–5: Children should have the opportunity to use and investigate a range of information-handling devices, such as electronic mail, audio and video recorders, and reference books. They should gather, organize, and present information in several ways, using reference books, paper files, and computers (p. 201).

However, sites, such as Yahooligans (http://yahooligans.yahoo.com) and the Exploratorium (http://www.exploratorium.edu/), overcome many of these problems by screening documents available on the Web for content, validity, and developmental level of the learner. Yahooligans is a Web site designed just for students. It contains links to Ask A Scientist, as well as links to numerous preselected science sites. National Geographic.Com Kids (http://www.nationalgeographic.com/kids/) is a site maintained by the National Geographic Society that has information tailored just to students. The site includes science games and experiments.

SciLinks, developed by the National Science Teachers Association (http://www.nsta.org), provides access to preselected and preapproved online materials. NSTA and textbook publishers have provided an alternative to random searches of the Web. NSTA's teacher-Web watchers have preselected and preapproved Internet science sites for a variety of textbooks. Using SciLinks, students can access Web sites that can provide richer and deeper representations of what they are studying, access the latest news in science, and ask scientists their questions.

Wikipedia is a free online encyclopedia that is collaboratively written by its readers (http://en.wikipedia.org/wiki/). Wikipedia contains much useful science information. Wikipedia uses a special type of Web site, called a wiki, which makes it easy for various people, including children, to create, add, and edit information. People are continually improving Wikipedia, making daily changes. The changes are recorded on wiki pages that keep track of the information. Because information is carefully monitored by a large community that supports the distribution of free and reliable information, you typically will not find inappropriate or misleading information. The wiki community removes inappropriate changes quickly, and repeat offenders are blocked from editing and adding new information.

The Web also contains other new exciting search engines, such as Google Earth (http://earth.google.com). Google Earth combines satellite imagery, maps, and the search power of Google to allow learners to explore, zoom in on, and see various places on the earth. The imagery is unbelievable!

Cameras attached to computers connected to the Internet, **web cams**, allow students to view real-time science phenomena, collect data, and interact with others. For example, the San Diego Zoo's Panda Cam lets children watch the pandas live at: http://www.sandiegozoo.org/zoo/ex_panda_station.html. One of the best sites to direct students to child-safe web cams is EarthCam for Kids (http://www.earthcamforkids.com/). This Web site connects students to hundreds of web cams around the world that feature animals, zoos, aquariums, outer space, and environments.

Learning Activity 6.3 asks that you explore the Web for various sites you can use in your teaching.

Learning Technologies to Support the Creation of Multiple Representations

There has been an explosive growth in the use of media (sound, graphics, color pictures, and video). The variety of media allows for the representation of concepts in multiple, simultaneous ways. For example, when studying tornadoes, students can use media to observe tornadoes, hear sounds associated with a real tornado, exam-

Learning Activity 6.3

USING THE WORLD WIDE WEB TO FIND INFORMATION

Materials Needed:

- Computer with high-speed Internet connection
- A World Wide Web browser such as Microsoft's Internet Explorer or Apple's Safari

A. Use a browser program and various sites in this chapter, such as Wikipedia, national geographic.com kids, or http://www.earthcamforkids.com/, to find information that might be useful for teaching an important science learning goal to children.
B. Explore the various parts of the site.
C. How might you make use of the site in your teaching? Share your ideas with your classmates.
D. Record your thoughts.

ine graphs of wind speed, read text explaining how a tornado works, and interpret graphical illustrations of tornadoes. Much has been written about the value of helping students develop a deep understanding of concepts through the use of **multiple representations** (Spitulnik, Stratford, Krajcik, & Soloway, 1997). The incorporation of multiple representations of concepts within documents is especially important in helping students develop conceptual understanding of scientific concepts (Kozma, 1991). By building thoughtful connections and constructing meaningful relationships among representations, students form integrated understandings (Linn, 1998). For instance, Harel and Papert (1990) have noted substantial gains in mathematics learning among students who devised various graphical and textual representations of fractions in Logo programs.

Technology allows students to manipulate and construct their own representations in several media. Software tools, such as Microsoft Office's PowerPoint and Word, or Web tools, such as wiki spaces, allow students to create multimedia documents of their investigations. For instance, students can take pictures of plants germinating and growing and import them into their documents. They can also record their observations around their pictures and include graphical information, such as how the heights of their plants changed over time.

Figure 6.4 depicts one page of a PowerPoint presentation from an eighth-grade project on photosynthesis. The class explored the question "How do plants get their energy?" The students asked their own subquestion "How does the color of light affect the amount of starch produced?" The figure shows a page from the multimedia document describing their procedure.

A wiki space is a special a Web site that makes it easy for people, including children, to create, add, and edit information. What is wonderful about a wiki space is that others can add information to it, but the original content also remains intact. As an owner, you can always get rid of some modification that you did not want on the site. There are numerous uses of wiki spaces for science teaching. A wiki page allows

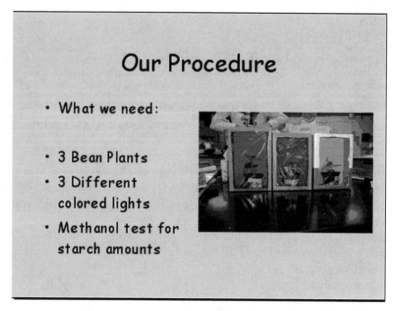

Figure 6.4   A slide from a student's PowerPoint presentation.

students to present conclusions to an investigation and then have other students comment on it. You can use a wiki page from year to year, with students adding more data or information to the wiki space each successive year. For instance, perhaps one of your classes planted a tree. Several times throughout each year you could record the growth of the tree and include a digital photo of the tree. You could also include a graph that plots the height of the tree over time. It is easy to start a wiki space by going to http://www.wikispaces.com.

Learning technologies and application software minimize the physical work that students need to do to create products, freeing them to learn more and develop deeper understandings. Application software, such as word processors, spreadsheets, desktop publishing programs, and drawing programs, can help students create a variety of products to represent what they know in a variety of forms. These application software tools incorporate video, text, and graphics to develop multimedia products and presentations. Using these tools, upper elementary students can easily incorporate diagrams, drawings, graphs, and video into computer-generated products. Moreover, students can link products together in unique ways to represent their understandings.

Using technology tools to create products also allows students the opportunity to more readily revise their work and develop understanding over time (Spitulnik, Zembal, & Krajcik, 1998). Because of the digital nature of multimedia documents, students can revisit their productions over time by continuing to construct new versions as they gather more information and develop their ideas further. For instance, a student could continue to add digital images of plant growth over time to a hypermedia document. The student could also link these images to observations of plant growth and to background information.

Finally, with technology tools that support the creation of multiple representations, students can increase the quantity and quality of their writing. Word processing and multimedia programs help students expand and elaborate on what they have already written. They also let students make corrections or clarifications.

Teachers need to be cognizant of students who want to create "beautiful" multimedia documents that look, at times, professionally done, but that lack substance. Some students will put irrelevant pictures and/or sounds into their documents, not because it helps to present the information, but because the students enjoy doing so. This phenomenon not only happens with multimedia documents, but also when teachers ask students to develop a report or a poster. We certainly want to encourage students' creative tendencies to make nice-looking documents, but not at the expense of the time needed to conduct research, investigate, or understand the topic in depth.

One way to prevent students from spending all their time on beautifying their documents at the expense of the purpose of the lesson is to give them clear criteria about how you will evaluate their product. Point out that presenting information in a clear manner is the most important criterion. Stress to the students that you will look for depth and accuracy of information. Stress also that the images, videos, and sounds that are used in the document should help viewers understand the presentation and not serve as extraneous special effects. You can certainly include appearance and ease of use in your evaluation, but these should not be the major criteria in a science class. See Chapter 10 for further information on how to use rubrics.

To help students learn how to make presentations look good without sacrificing content, you might also enlist the help of a peer teacher. The technology teacher, for example, could teach about Web design, model how to make professional-looking pages, and grade students on the design of their projects. The art teacher could teach about color and aesthetics.

Digital Cameras

Students can use **digital cameras** as data collection and recording tools. For instance, if students are investigating the growth of bean plants, they can take digital images, in addition to taking written descriptions and measurements of their plants. Students can take their digital cameras on stream walks to make visual records of what they see. One of the advantages of using digital photography instead of conventional cameras is the great ease of integrating the images into multimedia documents (Rivet & Schneider, 2001). Students can take a picture of various phenomena, embed the image in a multimedia program, and then annotate their image.

Video

Students can also use video cameras as data collection tools. For instance, students might use a **video camera** to collect information about the quality of their stream. If students use *digital* video cameras, they can easily incorporate a short segment of video into a multimedia document. Apple Computers' iMovie tools have made it simple for even very young learners to edit movies and incorporate them into multimedia documents. iMovie also lets students convert video images that can be placed on the Web for others to download to their computer or play back on a mobile device, such as Apple's iPod.

Learning Technologies to Support the Creation of Learning Communities

Electronic communications, including e-mail, threaded discussions, chat rooms, and blogs, give students access to a wider community across the world. Students can use **e-mail** to communicate with other schools, community members, and even members of the scientific community. E-mail allows users to communicate through mail messages, and users can attach documents to e-mails to share with others. However, electronic communications do pose some threats to children, so adults need to oversee such activities. Adults need to warn children of child predators on the Web and advise students to be careful of giving out too much personal information on the Web. Just like we warn children not to open the door to strangers, the same warning needs to be given for strangers on the Web.

Threaded discussions allow students to follow up on questions and sets of responses that build on each other. One student might submit a question to which several students or scientists might reply, sometimes with new questions. One advantage of this technology is the rapid rate of response that the medium encourages. Another advantage is the breadth of response that can occur. The National Geographic.Com Kids (http://www.nationalgeographic.com/kids/) Web site has a link to Ask A Scientist.

Chat rooms allow a number of individuals from a variety of locations to communicate in real time. Through chat-room technology, live conferences can be set up involving students from one school, scientists, and students from other schools. These conferences encourage and support students in finding solutions to questions of interest and in debating ideas and evidence.

GLOBE, Global Learning and Observations to Benefit the Environment (Rock, Blackwell, Miller, & Hardison, 1997; http://www.globe.gov/globe_flash.html), provides opportunities for children to communicate and collaborate with scientists and other GLOBE students around the world. Students can take part in threaded discussions with scientists on a variety of topics related to various investigations. GLOBE also supports students in the collection and analysis of data. In GLOBE, students take scientifically valid measurements in the fields of atmosphere, hydrology, soils, and land cover. After collecting data, they report their data through the Internet. Students compare their data to archived data collected in previous years. Over one million primary and secondary students in more than 10,000 schools have taken part in the program.

Students can also create blogs related to their projects or investigations. A **blog** is a Web site where individuals write down various messages, ideas, and commentary on just about everything and anything. You can think of blogs as diaries, journals, a site to get feedback on your ideas, a site to place questions and ideas, or anything else you can think of doing in written form. Blogs, however, can be more than just text. Students can also insert graphs and photos into blogs. Blogs can be a multimedia document, allowing students to share data and information with other students. Young adults and middle school children love blogs. Blogs are an innovative, interactive, and easy-to-use tool for creating online learning communities.

Individuals can control who has access to their blog. Teachers, as owners of a blog, can allow their class to have access to the blog or can open it up to others as well. Users also allow people to make comments on their blog. Allowing others to add the comments is an exciting aspect of blogs because it makes it interactive. For instance, teachers could use blogs for students to write down project ideas and receive feedback on them from the class or from others in the community. The following blog, http://hahnsslog.blogspot.com, is from a sixth-grade science teacher. She adds science entries to her blog, and her students respond. You could also use blogs to have your students communicate with other students from different schools from different parts of the country or world. By developing blogs with other students at other locations, your students will communicate with them by writing. Thus, blogging is a wonderful way to get your students into writing and in creating online learning communities. If you are interested in starting your own blog, go to http://www.blogger.com.

Role of the Teacher

The use of the computer as a laboratory tool to support data collection and visualization changes the science classroom into a science laboratory. Achieving this goal takes much support from the teacher. Teachers must encourage students and provide opportunities for them to use the various tools to ask new questions. Many students will not feel comfortable asking their own "What if?" questions. For instance, let's suppose that students are exploring whether stirring helps to cool a cup of hot chocolate. In the process, students might also wonder if blowing over the surface of the cup or the type of cup makes a difference. As a teacher, you need to encourage these types of further explorations using technology tools. One way to do this is to ask the class to keep track of additional questions that come to mind during their initial exploration. As a whole-class activity, students can share the variety of questions that they developed and discuss which might be explored using technology tools. Another way to promote the exploring of additional questions is to have students share the questions with you in their small groups. Because technology tools have been designed to promote easy explorations, as well as the acquisition and visualization of data, give students permission to carry out these additional explorations using the appropriate technology tools.

Although technology tools have features that allow students to explore and take part in a variety of scientific practices, too often technology is used in a way that is as unimaginative and boring as a dry lecture. If the technology tools are used as fill-in worksheets, children will do just that. They will fill in the worksheet, rather than cognitively engage in exploring the science phenomena. As a teacher, you need to create an environment that will encourage students to explore. As you have done in other aspects of inquiry, you will need to support children in the process. (See the section "Supporting Students' Implementation of Investigations" in Chapter 5 for some ideas.) You can do this by modeling how to use the technology to engage in inquiry-based investigations. For example, you could show students how motion detectors can be used to calculate the speed of a moving object.

Teachers also need to make sure that students understand the mechanics and concepts behind the technology. For example, a student should not use a temperature probe unless he understands how to use the probe and what he is measuring. You should first teach students how to use the probe accurately and safely. To teach students what they are measuring, you might first have students use the probe to explore who has the coldest hands in the class. This can be done without the use of technology by having students determine a ranking based on their sense of touch. Next, contrast their sense-of-touch rankings with the recordings from the temperature probe.

Just as you will need to support students in interpreting graphs that result from their investigations, you will need to support students in interpreting the visualizations created by technology visualization programs. Project FeederWatch, mentioned earlier, allows students to see a distribution of a bird population on a color-coded chart of North America. Although this type of visualization is becoming commonplace in newspapers, many students might not know how to interpret it. One way to help students learn how to interpret distribution maps is to allow them to create a distribution map of a bird population across North America using their own bodies. First, download a data chart of a population of birds from the Project FeederWatch site. Select a bird, such as the American Robin, with which all of the children are familiar. Next, using chalk or masking tape, trace a map of North America or a map of a state on the classroom floor. Let's imagine that you trace the State of Illinois. For every five robins found in a particular location of the state, have one student stand on the map in that location. If there were 60 robins seen in southeastern Illinois in November, you would have 12 students stand in southeastern Illinois. Another 50 birds were spotted in southwestern Illinois, so you would have 10 students stand there. However, only 10 birds were spotted in northeastern and five in northwestern Illinois in November. As a result, two students would stand in the northeastern part of the state and only one would stand in the northwestern part of the state. Next, students can look around to observe the pattern. You could ask students where more robins were spotted in Illinois in November of last year. Using such graphic techniques, you can help your students learn how to interpret distribution maps.

Integrating Technology into Instruction

Frequently, teachers, administrators, and parents mistakenly believe that, just because technology is used, students will learn from it. Our work and that of others show that this statement does not come close to representing what needs to happen in classrooms to promote learning. Using technology effectively in the classroom generally requires that teachers spend substantial time creating a learning environment to support it. We suggest that technology can be successfully integrated into instruction if three things are addressed:

1. Teachers have considered carefully the understandings that students might gain from working with the technology tool and how those understandings fit in with the rest of the curriculum.
2. The technology tool is the best instructional strategy to teach the learning objectives.
3. Teachers have planned and created a learning environment that helps students develop skills in using and understanding the technology tool.

Technology Tools Help Students Learn Important Learning Goals in the Curriculum

Technology is rarely an effective standalone teaching tool and, therefore, it should not be used just for the sake of using it. In science classes, the curricular goal is not using the technology. but to learn science. For example, the goal is not to have students surf the Internet. Rather, students may be using the Internet to search for information or communicate with others. Thus, technology tools should only be used if they help the teacher meet learning goals. We discussed more thoroughly earlier in this chapter the role of technology tools in science teaching. These roles help teachers think about uses for technology in science classes to meet learning goals.

Technology Is the Most Effective Instructional Strategy to Meet the Learning Goal

As experienced teachers, we always look for powerful instructional activities to use with our students. However, it is the role of a professional teacher to determine the simplest, most effective instructional strategy to use to teach a concept. Like other instructional strategies (discussed more thoroughly in Chapter 8), technology-based activities need to engage children in the learning process and help them create their own understandings. Teachers oftentimes ask themselves, "Does the tool allow students to be active participants in creating their own understandings, rather than passive recipients of factual material?" The learning technologies that we presented in this chapter have the potential to do this. If the technology presents information in a manner that students simply receive information passively, then you should probably not use the technology. If books or other media can present the information or teach a concept more effectively, use them.

For example, technology can help students carry out investigations. Computer probes let students collect real-time data, whereas other instructional strategies, such as a video only let students watch someone else conduct an investigation. The World Wide Web is a useful instructional strategy for finding up-to-date information related to the investigation, whereas an encyclopedia may be very dated. Thus, technology (probes and the Web) may be the best instructional strategies to help students conduct an investigation and search for current information.

*Teachers Plan and Create a Learning Environment
That Helps Students Use the Technology*

As depicted in Scenario 2, technology cannot be used on an occasional basis if it is going to benefit students. Just as it is too time consuming and complex to learn how to use a table saw to cut just one board, learning to use new technology tools is too time consuming and complex just to use for one day. To benefit from the time and effort it takes to learn to use the tool, students need repeated opportunities to use new technology tools throughout the school year.

Teachers also need to work extensively with technology tools themselves so they fully understand its capabilities and limitations. After gaining these understandings, teachers can help students use the technology tools effectively. For example, teachers of young children will want to be comfortable using a digital camera before handing it to students to use. Furthermore, students need to be taught how to properly use the camera so that time is not wasted, students don't become frustrated with the technology, or students don't accidentally break the camera, which is fairly expensive. There may be certain aspects of the tool that are especially difficult for students to understand and/or use, and teachers need to anticipate this. Initial explorations with the camera can be structured to anticipate the areas where students will have difficulty. For example, students may be able to quickly learn how to focus and shoot a picture, but it may be more difficult for them to learn to download the photos to the computer to incorporate the images into a document.

Finally, teachers need to plan and create learning environments that help students understand how to use the technology to assist in learning. Students can have superficial procedural-level knowledge of pushing buttons or typing computer commands and not understand why they are using the technology. For example, although it is important for students to know how to properly use a digital camera, knowing how to use it doesn't ensure that important learning will take place. Teachers need to help students understand that the camera can be used, for example, to document a finding, communicate with others, or create a log of events over time.

Issues to Consider When Incorporating Technological Tools Into Your Curriculum

Although technology can be a powerful learning tool, there are also some possible pitfalls when using technology with children.

- Do you expect gender or cultural differences in how interested students will be in the tool? If so, what support materials or activities can you design to address this issue?
- Are there potential misunderstandings that the tool might encourage? If so, what support materials or activities can you design to address these potential misunderstandings?
- Are there potential safety issues to consider?

Learning Activity 6.4

USING LEARNING TECHNOLOGIES TO HELP STUDENTS LEARN SCIENCE

Materials Needed:

- Various computer tools discussed in the chapter
- Paper and a pencil

A. Select a content area that students have difficulty learning. What technology tools might be appropriate for helping students learning this idea? Explain.
B. As a teacher, what problems might you encounter? What difficulties would students have using the tool?
C. Record your ideas.

Many studies have documented that boys will dominate in the use of computers over girls and other at-risk students who traditionally do not succeed at science. For example, the American Association of University Women (AAUW) Web site reports research that indicates that girls are not well represented in the area of technology. This research reveals that it isn't that girls aren't interested in technology, but that girls find many computer programming classes dull and tedious and computer games redundant and violent. Equity issues come into play when boys in classes are given more attention by teachers during technology use or allowed to dominate the computer and other technology tools. The AAUW research suggested that teachers need to emphasize the computer as a productivity tool. Further, teachers need to watch for unequal use and ensure that students have equal access to using the computer and keyboard. One easy technique is to tell students to change who is using the keyboard or the equipment every 15 minutes. You could even use a timer. We also recommend that you have no more than two students share a computer or other learning technologies. We have observed, as have others, that once there are more than two students at a computer, one of the students tends to daydream and not be engaged in the activities.

Some programs might unintentionally give students misunderstandings about the world. Although simulation programs are valuable in helping students take part in experiences that they otherwise might not have, some of the generalizations that occur in these programs might give an overly simple view of how a process works. For instance, a program that represents the microscopic world of atoms as small spheres could lead students to think that atoms actually look like spheres. Stress to students that they are only seeing a representation of atoms.

Several issues related to safety come to mind when using technology tools. As mentioned earlier in this chapter, teachers need to be careful in using the World Wide Web to ensure that students do not disclose personal information that could be used inappropriately by others, and students need to be monitored so they do not accidentally stumble upon inappropriate Web sites or talk to people with whom they should not interact. Other safety issues are related to potential electrical hazards when using probes and other technologies around water in science classes.

Learning Activity 6.4 will help you explore the use of learning technologies in your classroom.

Chapter Summary

In this chapter, we discussed the use of learning technologies to support students in developing meaningful understandings of science concepts and in doing investigations. We defined *learning technologies* as the computers, software, and various peripherals that support students' learning. We discussed how learning technologies help students actively engage with phenomena, use and apply knowledge, find new sources of information, create multiple representations of their ideas, and create learning communities. We explored how technology tools, such as electronic probes, are particularly powerful because they allow students to collect, visualize, and analyze data. The simultaneous linking of real-time data collection with graphs that help students visualize data seems to support student learning. Microworlds, interactive multimedia, and simulations allow students to explore phenomena when it is unsafe, too expensive, impractical, or perhaps even unethical for students to have firsthand experiences. Such environments allow students to ask "What if?" questions, make predictions, and test their ideas.

The World Wide Web brings many benefits to the classroom. Through the Web, students have unprecedented access to the newest information and data. In many cases, students can use the same data as scientists. However, using the Web also brings challenges, particularly the high number of sites that are returned when students conduct a search. Many of these sites are irrelevant, contain inaccurate information, or expose students to inappropriate information. We recommend that, because of these issues and the limited time available in classrooms for research, students use Web sites that are preselected and preapproved. As students become more proficient in using the Web, they can learn to navigate on their own and evaluate the sites they find.

Learning technologies also allow students to create their own multimedia documents. Students can thus document their understandings using different types of media. Learning technologies also open up the classroom to sharing and debating with other students and scientists to create an extended learning community. Wiki spaces and blogs offer new opportunities for students to create and share information.

Although we strongly believe that learning technologies present many advantages for instruction in science, we do not support using technology just for the sake of using it. We encourage teachers to make use of technology when it can help students meet important instructional learning outcomes or develop meaningful understandings. It is the responsibility of the teacher to make decisions regarding when and how to best use learning technologies in the science classroom.

Chapter Highlights

- Learning technologies can support teachers and students in project-based science because they can help students and teachers communicate, carry out investigations, and develop products.
- Using learning technologies in the classroom is important in helping to overcome the digital divide.
- Technology can help students develop understandings of complex, abstract ideas.
- Technology transforms the way teachers teach.
- When students become actively engaged with phenomena using technology, they are able to explore phenomena they would otherwise not be able to access.
- The creation of multimedia artifacts helps students share information and develop meaningful understandings.
- Technology supports students in communicating and collaborating with other students and community members.
- Electronic probes, also referred to as sensors, allow students to use computers as laboratory tools to collect and visualize data.
- Visualization tools allow students to explore and manipulate data to see trends they would otherwise be unable to see.
- Microworlds, interactive multimedia, and simulations combine
  - Video.
  - Pictures.
  - Computer graphics.
  - Text.
  - Interactivity.
- Microworlds, interactive multimedia, and simulations present to students phenomena that otherwise would be inaccessible, too hazardous, too time consuming, or too expensive.
- The expansion of technology, especially of the World Wide Web, has enabled students easily to obtain information about key ideas, concepts, and subject matter topics that arise as they explore solutions to questions.
- There are difficulties with the Web, particularly for young learners.
- Multiple representations can enhance students' understanding of concepts.
- Application software, such as word processors, spreadsheets, desktop publishing programs, and drawing programs, helps students create a variety of products to represent what they know in a variety of forms.
- Students can use digital video cameras as data collection tools.
- Electronic communication, including e-mail, blogs, and chat rooms, allows students access to a wider community.
- Technology tools should only be used if they help the teacher meet important learning goals.
- Using technology in the classroom takes much support from the teacher.
- Issues of gender/culture, scientific misunderstandings, and safety need to be addressed.

Key Terms

Blog	Models
Chat rooms	Multiple representations
Digital cameras	Primary data sources
Digital divide	Real-time graphs
Electronic communications	Simulations
Electronic probes	Threaded discussions
E-mail	Video cameras
Interactive multimedia	Visualization tools
Learning technologies	Web cams
Microworlds	Wiki

References

Anderson, L. W., & Krathwohl, D. R. (Eds.). (2001). *A taxonomy for learning, teaching, and assessing: A revision of Bloom's taxonomy of educational objectives.* New York: Longman.

Brasell, H. (1987). The effect of real-time laboratory graphing on learning graphic representation of distance and velocity. *Journal of Research in Science Teaching, 24*(4), 385–395.

Christmann, E. P. (2002). Graphing calculators. *Science Scope, 25*(5), 46–48.

Cohen, K. C. (Ed.). (1997). *Internet links for science education: Student-scientist partnerships.* New York: Plenum.

Dwyer, D. C. (1994). Apple classrooms of tomorrow: What we've learned. *Educational Leadership, 51,* 4–10.

Edelson, D. C., Gordin, D. N., & Pea, R. D. (1999). Addressing the challenges of inquiry-based learning through technology and curriculum design. *Journal of the Learning Sciences, 8*(3&4), 391–450.

Harel, I., & Papert, S. (1990). Software design as a learning environment. *Interactive Learning Environments, 1,* 1–32.

Helfgott, D., & Westhaver, M. (2000). Inspiration (Version 6.0) [Computer software]. Portland, OR: Inspiration Software.

Inspiration Software. (2000). Kidspiration. Portland, OR: Author.

Kozma, R. (1991). Learning with media. *Review of Educational Research, 61*(2), 179–211.

Krajcik, J., Blumenfeld, B., Marx, R., & Soloway, E. (2000). Instructional, curricular, and technological supports for inquiry in science classrooms. In J. Minstrell & E. Van Zee (Eds.), *Inquiring into inquiry: Science learning and teaching* (pp. 283–315). Washington, DC: American Association for the Advancement of Science Press.

Krajcik, J. S., & Layman, J. W. (1992). Microcomputer-based laboratories in the science classroom. In. F. Lawrenz, K. Cochran, J. Krajcik, & P. Simpson (Eds.), *Research matters to the science teacher.* Manhattan, KS: National Association of Research in Science Teaching.

Linn, M. C. (1992). The computer as a learning partner: Can computer tools teach science? In K. Sheingold, L. G. Roberts, & S. M. Malcolm (Eds.), *This year in school science 1991: Technology for teaching and learning* (pp. 31–69). Washington, DC: American Association for the Advancement of Science.

Linn, M. C. (1998). The impact of technology on science instruction: Historical trends and current opportunities. In M. C. Linn (Ed.), *International handbook of science education.* Dordrecht, The Netherlands: Kluwer Publishers.

Linn, M. C., & Hsi, S. (2000). *Computers, teachers, peers: Science learning partners.* Mahwah, NJ: Lawrence Erlbaum.

Linn, M. C., Layman, J. W., & Nachmias, R. (1988). Cognitive consequences of microcomputer-based laboratories: Graphing skills development. *Contemporary Educational Psychology, 12,* 244–253.

Linn, M. C., Songer, N. B., Lewis, E. L., & Stern, J. (1993). Using technology to teach thermodynamics: Achieving integrated understandings. In D. L. Ferguson (Ed.), *Advanced educational technology for mathematics and science* (Vol. 107). Berlin, Germany: Springer-Verlag.

Metcalf-Jackson, S., Krajcik, J. S., & Soloway, E. (2000). Model-It: A design retrospective. In M. Jacobson & R. B. Kozma (Eds.), *Innovations in science and mathematics education: Advanced designs for technologies and learning* (pp. 77–116). New York: Lawrence Erlbaum.

Morkos, J. R., & Tinker, R. F. (1987). The impact of microcomputer-based labs on children's ability to interpret graphs. *Journal of Research in Science Teaching, 24*(4), 369-83.

Novak, A., & Krajcik, J. S. (2005). Using learning technologies to support inquiry in middle school science. In L. Flick & N. Lederman (Eds.), *Scientific inquiry and nature of science: Implications for teaching, learning, and teacher education* (pp. 75–102). Dordrecht, The Netherlands: Kluwer Publishers.

Novak, A., & Gleason, C. (2001). Incorporating portable technology to enhance an inquiry: Project-based middle school science classroom. In R. Tinker & J. S. Krajcik (Eds.), *Portable technologies: Science learning in context* (pp. 29–62). Dordrecht, The Netherlands: Kluwer Publishers.

Rivet, A., & Schneider, R. (2001). Ubiquitous images: Digital cameras to support student inquiry. Paper presented at the annual meeting of the National Association for Research in Science Teaching, St. Louis, MO.

Rock, B. N., Blackwell, T. R, Miller, D., & Hardison, A. (1997). The GLOBE Program: A model for international environmental education. In K. C. Cohen (Ed.), *Internet links for science education: Student–scientist partnerships..* New York: Plenum.

Salomon, G., Globerson, T., & Perkins, D. (1991). Partners in cognition: Extending human intelligence with intelligent technologies. *Educational Researcher, 20,* 2–9.

Songer, N. B. (2006). BioKIDS: An animated conversation on the development of curricular activity structures for inquiry science. In R. K. Sawyer (Ed.), *Cambridge Handbook of the Learning Sciences* (pp. 355–369). New York: Cambridge University Press.

Songer, N. B. (2007). Digital resources versus cognitive tools: A discussion of learning science with technology. In S. Abell & N. Lederman (Eds.), *Handbook of Research on Science Education* (pp. 471–491). Mahwah, NJ: Erlbaum.

Songer, N. B., Kelcey, B., & Gotwals, A. (2007). When and how does complex reasoning occur? Analysis of a learning progression focused on complex reasoning in science. Paper presented at the annual meeting of the American Education Research Association (AERA), Chicago.

Spitulnik, M. W., Stratford, S., Krajcik, J., & Soloway, E. (1997). Using technology to support students' artifact construction in science. In K. Tobin (Ed.), *International handbook of science education.* Dordrecht, The Netherlands: Kluwer Publishers.

Spitulnik, M. W., Zembal, C., & Krajcik, J. (1998). Using hypermedia to represent student understanding: Science learners and preservice teachers. In G. D. Phye (Ed.), *Teaching science for understanding: A human constructivist view* (pp. 363–382). San Diego, CA: Academic Press.

Wallace, R., Kupperman, J., Krajcik, J., & Soloway, E. (2000). Science on the Web: Students on-line in a sixth grade classroom. *Journal of Learning Sciences, 9*(1), 75–104.

White, B. Y., & Frederiksen, J. R. (2000). Technology tools and instructional approaches for making scientific inquiry accessible to all. In M. Jacobson & R. B. Kozma (Eds.), *Innovations in science and mathematics education: Advanced designs for technologies and learning* (pp. 321–360). New York: Lawrence Erlbaum.

Chapter 7

Collaboration in the Science Classroom

Introduction
The Nature of Collaboration
Types of Collaborative Learning
Creating a Collaborative Environment
Challenges That Arise When Students Collaborate in Small Groups
Why Collaboration Almost Always Works Better Than Individual Learning

Introduction

This chapter explores the nature of collaboration and why it almost always works better than individual learning. The topic of collaboration raises a number of questions: *How do I get students to work together on intellectually challenging tasks? How do I get members of the community to work with students to find answers to their questions? What are the benefits of such a classroom environment? What is the role of the teacher in a classroom characterized by collaboration?* This chapter will address these and other questions. Scenarios will illustrate the various types of collaboration among students, teachers, and members of the community. Because collaborative learning doesn't just happen in a classroom, the chapter focuses on creating a collaborative environment over a long period of time. You will learn about the social skills

Chapter Learning Performances

- *Describe various types of collaboration.*
- *Explain the role a teacher can play in supporting collaboration.*
- *Clarify challenges that a teacher might need to overcome to implement collaborative learning groups.*
- *Describe how to help children learn skills needed to work in collaborative groups.*
- *Critique learning environments to determine their ability to foster collaborative learning.*
- *Design and implement different types of collaborative learning environments.*
- *Explain why collaborative learning is almost always better than individual learning, and explain how it particularly helps female and minority students pursue science study.*

students need to work in groups with others and how to hold students accountable during collaboration. The chapter concludes with a discussion of ways to overcome challenges that might arise in the implementing of collaborative groups. We'll begin with scenarios of several types of group situations that you may have experienced in elementary or middle school. As you read each scenario, focus on what the teacher is doing or thinking, what the students are doing or thinking, and how students are working together to complete an assignment.

Scenario 1: Group Work With Little Guidance

Your fourth-grade teacher, Mrs. Weinstein, has covered the topic of dinosaurs for several weeks. Now she explains to your class that you are going to work in teams to complete a project on dinosaurs. Mrs. Weinstein divides your class into teams of four students. She tells you that you and your teammates are to write a report on a certain type of dinosaur, draw a picture of the dinosaur, and give a class presentation about the dinosaur. Shelly, who is really bossy, tells everyone else in the group what to do, and she dominates the entire project. She decides which dinosaur you will research and how the report will be written. She instructs Marcus to search through some encyclopedias from the library, but he doesn't read them as Shelly has asked. Frustrated, Shelly takes the books home, reads them, and writes the report for everyone else. Sarah, who has good handwriting, recopies the report Shelly wrote. You can draw well, so you are left to draw the picture of the dinosaur for the cover. Although you and the others find Shelly's bossy behavior frustrating, you all know you will receive a good grade because she is smart, so you happily let her do all the work. Shelly makes the presentation to the class while you and the others stand behind her. One classmate asks a question; you don't know the answer because you didn't learn much about dinosaurs anyway, but, luckily, Shelly is able to answer it.

In this first scenario, the topic was selected by the teacher. The teacher placed the students in groups but gave very little guidance or assistance that would ensure that students worked together in a joint intellectual effort. Although the students did cooperate with each other enough to complete the project, they did not have the skills to work together effectively. As a result, students were not motivated. Although all the students received the same grade, not all learned about dinosaurs.

Scenario 2: Individual Jobs in Group Work

As a whole class, you read the textbook section on cells. Then the teacher, Ms. Potter, places you into groups of five to learn about cells. Your teacher instructs your group to learn all you can from reference books. Another group learns how to use a microscope to view cheek (animal) and onion (plant) cells. A third group learns how to use a software program on cells. A fourth group watches a video that compares and contrasts plant and animal cells. Ms. Potter assigns each student in each group a job: a runner to get the materials, a recorder of the findings, a timer to keep the group on time, the manager to keep everyone on task, and a cheerleader to keep everyone involved. You are assigned to be the runner. You retrieve the encyclopedias from the shelf for your group and then proceed to watch others in the class. You wish you had been in the computer group because you find reading the encyclopedia boring. When it comes time to share your group's work, you wonder where the information came from, because you hadn't really been paying attention. The other groups share their information about microscopes, the computer, and the video. Their projects seem really interesting, and you learn a lot from them.

In this scenario, the teacher assigned each student a "job" so that the work was shared by members of the group. The teacher provided each group some guidance to complete the activity. However, because the teacher had selected the problems and learning approaches for each group, the technique did not motivate all students to learn. Some team projects seemed more fun than others, and some students' attention wandered. They didn't learn the intended material. In addition, not every "job" was equally important, and not every job was critical to completing the task.

Scenario 3: Using Expert Groups

Mr. Jackson tells you that you are going to investigate the topic of sound and that the goal of this lesson is for you to become an expert on one subtopic or concept and teach it to others. Mr. Jackson divides the class into teams of four and assigns each team member a job. One person serves as a runner who gathers the materials, one is a recorder who writes down the results, one acts as a manager who gets to do the experiment, and one is the timer who makes sure the group completes the project in the time allotted. Mr. Jackson also provides you with a sheet of questions to answer to help you determine if you are "expert" enough to teach the concept to other students. Your assignment is to find out what material (air, water, or solids) sound travels through best. Sarah, assigned to be the runner, collects the supplies (a bucket of water, a tuning fork, and a paper tube) for the activity. John, the manager, gets to experiment with the tuning forks to see when they seem the loudest. You record the results after everyone agrees upon the answers. Marcus watches the clock the whole time and announces the time at 10-minute intervals. Once each of you has become an "expert" and knows that sound travels best in solids, you are put into "learning teams" where you teach this idea to three students who had not been in your group. You engage the three new students in the same activity with the tuning fork, water, and tube of paper to demonstrate that sound travels best through the solid table, next best through the water, and least well through the air. You feel happy that you did the experiment that only John had been permitted to do when you were in your first group. Each of the three other students has become an expert in another topic and teaches you what he or she has learned. You have a great time and really understand which material sound travels through best. Although you learn some interesting ideas from what the other students taught you, you are still a bit fuzzy on others. On the test Mr. Jackson gives, you do well on questions about the type of materials sound travels through, and you do okay on the items about frequency, pitch, amplitude, and sound insulation that the other expert teams taught you.

In this example, the teacher structured the lesson by choosing the activities and providing all the directions for the students. The teacher provided students with guidance on the criteria for being an expert. In this way, each student was cognitively engaged because she needed to teach the concepts to others. This strategy seemed to work well.

Scenario 4: A Learning Community

Your class has been exploring the driving question "What causes living things to become endangered?" Ms. Nadarajah leads the class in a discussion of possible threats to plant and animal survival. After you all decide on some causes of endangerment, Ms. Nadarajah lets you choose your teammates. One group decides to investigate how pollution affects animals and plants. They conduct experiments on plants with various amounts of acids and other substances, and they research the decline of frog populations throughout the world. Another group studies the gorilla and the effect of past wars in Rwanda on the gorilla population. A third group investigates how the sea turtle in Florida has been affected by humans building along the coastline where they reproduce. Your group decides to learn more about the rainforest and causes of endangerment in the rainforests of the world. In your group, each member discusses his or her interests and ideas for investigating the rainforest. Juanita says, "I read the story called *The Great Kapok Tree* by Lynn Cherry, and I think the animals are dying because they're cutting down all the trees." Monika adds, "Yeah, but my Dad says it's because we eat a lot of hamburgers. The ranchers cut down trees to raise cows." "I don't know," Brian says. "I think it has more to do with pollution. Pollution is killing the animals." "Yeah, I saw something on TV about frogs dying because of pollution, and there's frogs in the rainforest," says Kameel. "Yeah, but I don't think frogs are the only animals dying in the rainforest," replies Monika. "And what would cutting down trees have to do with frogs dying? They live in water!" says Kameel. Ms. Nadarajah comes over to work with your group to help you figure out how you would finish the task. Ms. Nadarajah also asks you questions to get you thinking about what types of artifacts, or products to share, you might create to show what you learn.

Through collaborative planning and some guidance from Ms. Nadarajah, your group decides to divide this topic into types of endangered animals in the rainforest, causes of endangerment, and solutions to endangerment. As a team, you decide to learn about these three topics by reading books, visiting the zoo, watching a video on rainforests that you noticed in the school library, talking to scientists at the local university, and e-mailing students in South America to talk with them. You divide into smaller working pairs to answer questions about endangered animals, causes, and solutions. Over the course of a few days, you find out that deforestation is a major cause of endangerment, so you decide to conduct a small experiment to see what would happen to insect populations in a section of the schoolyard when you slowly change their habitat by removing sticks and leaves, cutting the grass, and pulling out the grass. Ms. Nadarajah helps you understand what types of insects might live in the area by teaching a lesson about insect classification and helping you locate some resources for insect identification.

At each stage, you keep data and develop artifacts to show what you are learning. You decide to take close-up photos of the schoolyard insect population. Soon after Ms. Nadarajah asks you to give a presentation, you and others in the group analyze and synthesize what you have learned and decide to present your information to classmates in the form of a poster labeled *Cause* and *Effect*. Finally, the big day comes. Your group gives its presentation to the class, and everyone applauds. The other students in the class are very interested in your project, but one team comments that you might have left out variables when comparing insect populations to the rainforest. They wonder if you could really compare the two situations because the rainforest is much more complex than a 3-foot by 3-foot area on the playground. You decide that this is a valid point and begin to wonder how to address it. The other teams present the results of their projects, and you learn that there might be other factors that you did not think about, such as acid rain, economic factors, and social factors. These ideas give you some direction for further exploring the rainforest.

In this scenario, the teacher served as an instructional planner and guide. The teacher helped students design questions, provided learning materials, and organized discussions. The driving question "What causes living things to become endangered?" was a meaningful and important question that served to organize the activities. The students in the group formed a learning community. Each negotiated with others in the group to select an investigation. Students shared the work, discussed ideas, and completed portions of the entire project. Differences in opinion were resolved amicably. Members of the community were included in the project. The Internet provided e-mail access to students in South America. The project's direction and outcome were determined by members of the group. The students were involved in "hands-on" inquiry of habitats. Students developed "artifacts" that they shared with others. Students critiqued these artifacts and the presentation in a positive manner, thereby providing the group with ideas for future exploration. In this sample scenario, students had a voice in their learning, were motivated, and were cognitively involved. Students worked together to create meaning and understanding. There weren't some "jobs" that were less attractive than others.

This fourth scenario describes a project-based science classroom in which collaborative learning is one of the major instructional features; it involves students, teachers, and members of society in trying to work through an important problem in which people are interested. Collaborative learning requires students to work with others to solve a problem or engage in inquiry.

These four scenarios illustrated different types of collaborative learning. Learning Activity 7.1 will further clarify the characteristics of different approaches.

The Nature of Collaboration

Understanding the nature of **collaboration** will help you comprehend its importance in project-based science classrooms. In this section, we will explore two questions about collaboration: *What is collaboration?* and *What are the characteristics of collaboration?*

Learning Activity 7.1

HOW DOES COLLABORATION CHANGE THE NATURE OF THE CLASSROOM?

Materials Needed:

- Two classrooms to observe

A. With the help of your instructor, a school principal, or a school curriculum director, identify two teachers and classrooms for observation. Obtain permission to observe each of these teachers. Make sure you observe at least one collaborative group setting.
B. As you observe both classrooms, analyze how collaboration changes the nature of the classroom. Try to answer the following questions:
   - How are students working together in these classrooms?
   - How does the teacher structure the group work?
   - Are students given jobs or roles to play? Are the roles or jobs meaningful intellectually, or are they only procedural tasks?
   - Are the jobs or roles necessary to complete the overall task?
   - What is the role of the teacher in the classroom?
   - What are the teacher's objectives in the lesson?
   - What instructional strategies does the teacher use?
   - How does the teacher help students work in groups (by developing interpersonal skills, encouraging students to talk with each other, resolving conflicts)?

C. What do you see as the advantages and challenges of a collaborative classroom? Write down your thoughts.
D. Discuss with peers how these challenges might be overcome.

What Is Collaboration?

Think of **collaboration** as a joint intellectual effort of students, peers, teachers, and community members to investigate a question or problem. When students work with others on a joint intellectual effort to explore a problem and to build understanding, they form a community of learners. For example, a community of learners who are collaborating to answer the question "Why does algae grow in the fish tank?" might debate the cause of algae in an aquarium. This process encourages students to challenge and support their conceptions and ideas. They work with others in the class to test their ideas and build understanding. They might set up several aquariums under different conditions, such as in the window of the classroom with sunlight, in a dark room without sunlight, in a cool place, and in a warm place. The teacher might present a lesson about algae being an aquatic organism that has some characteristics of plants (it makes its own food through the process of photosynthesis), but that lacks true roots, stems, leaves, and embryos. In the lesson, the teacher has students observe algae under a microscope so they can see these characteristics. The students might ask the owner of a local pet store how to control algae in an aquarium. In this learning community, the students, teachers, and other community members work together to find resolutions to questions or problems.

In Chapter 2, we examined Vygotsky's ideas about child development and learning. Teachers can use Vygotsky's (1986) ideas to create learning environments in which students have contact with others with different levels of expertise, that is, with multiple zones of proximal development. Vygotsky's **zone of proximal development** is described as the difference in performance between what a learner can accomplish unassisted and what he could accomplish with the assistance of a more knowledgeable or capable other. Think of the zone of proximal development as a range of cognitive tasks that you can learn with the assistance of someone who knows more. There are some tasks that you could not master, even with assistance, but there are also tasks you could complete with just a little help. For example, a 10-year-old can, over a short period of time with adult guidance, master the process of using a simple single-focus microscope (the zone of proximal development has then shifted, and the 10-year-old might next work on using a microscope with multiple magnification lenses). The same 10-year-old, however, cannot manage using a scanning electron microscope, no matter how many times or how much her teacher or a scientist assists her. So, from Vygotsky's notion of zone of proximal development, we can use the idea that providing timely and relevant support and guidance to learners can be of great assistance in their development of science knowledge and skills.

Such a learning environment can be created by lessons where more knowledgeable others provide students with basic knowledge or skills that are just beyond their level of intellectual attainment and that will enable them to expand and continue their learning. Students learning about the cause of algae in an aquarium, for example, are exposed to new concepts when the teacher gives a lesson on how to observe algae under the microscope, a skill they did not possess earlier. When the teacher supports their learning by teaching them how to use a microscope, the teacher provides that little bit of help needed to help students be able to conduct their own inquiry—something they were unable to complete before learning this skill. The students can now learn that algae make their own food through the process of photosynthesis. This concept helps the students understand that an aquarium in the sunlight will grow algae more quickly than will an aquarium placed away from a window. This same type of learning environment is created when a student explains how his family keeps algae from growing in the swimming pool or when the pet store owner explains how she keeps algae from forming on the aquariums in her store.

What Are the Characteristics of Collaboration?

Collaborative learning, unlike other types of group arrangements, encourages high levels of *equality* and *mutuality* (Damon & Phelps, 1989) as students work with others to create meaning. **Equality** refers to the similarity in the *level* of knowledge or ability that members bring to a group. In a collaborative classroom, each member should contribute equally (Cohen, 1998). This does not mean that each member of the group possesses the same knowledge or abilities—in fact, it is desirable that members have had different experiences and contribute varied prior knowledge and varied abilities. This is different from other group arrangements that encourage only cooperation—getting along or being compliant with others in a group to finish a task. It also differs from task sharing in which each person in a group completes one part of a presentation thrown together with little or no shared effort. For example, of the students studying algae growth, one has a swimming pool at home and is familiar with the techniques used to control algae; one has a fish tank in her bedroom, and she has noticed that it has less algae when she keeps it away from the window; and one thinks he has algae growing in his shower at home (where it is warm and moist), so he has theories about the cause of algae growth. All three share different, but equal, ideas with the group.

Mutuality refers to the common goals of the students. In collaborative classrooms, mutuality should be high—the students in a group should be working to answer the same question, complete the same task, reach the same goal, or find solutions to the same problem. The students studying algae growth are all investigating why algae grows on the fish aquarium in the classroom. They have a driving question that they are investigating together. They all work together to find ways to prevent algae growth.

Other group arrangements sometimes involve each student having a separate "job." The students have different goals—each task corresponds to a different goal. Look back to the first scenario in this chapter. Although the students did cooperate to present a dinosaur report to the class, the goal of one student was to make sure the report was illustrated. The goal of another student was to make sure the report was legible. Mutuality was low.

Another characteristic of collaboration is that it creates **meaning**. Think back to the four scenarios that started this chapter. In the first two scenarios, students in the groups did not work together to understand the topics. Some students did all of the work; others did not play important roles that would enable them to develop any understanding. Some students' attention wandered, and most did not cognitively engage in the learning process. In contrast, the third and fourth scenarios illustrated students collaborating. In the third scenario, students shared their expertise on various aspects of sound. In the fourth scenario, students worked together and with others (the teacher and members of a broader learning community) to create understanding about the rainforest.

The students studying the question "Why does algae grow in the fish tank?" were collaborating because they created meaning, and they experienced high levels of equality and mutuality. They all contributed knowledge and skills to the group's effort and they worked to answer the same question. Without creating understanding and establishing high levels of equality and mutuality, group work consists solely of students *being* in groups.

Types of Collaborative Learning

Collaboration can occur among all members of a community. In this section we will explore three types of collaborative learning: among students, between students and teacher, and between students and community.

Among Students

Student-to-student collaboration is a powerful method of instruction. It can occur at several different levels, ranging from one-on-one collaboration to various group configurations. In one-on-one situations, one student can tutor another. A pair of students can also work together to investigate science questions. This one-on-one collaboration can occur within a classroom or even across a geographical distance via e-mail or regular mail. Collaboration among groups of students can involve students of a single class working together on a project, students from two or more classes within a school building or district exchanging ideas, or students from one class communicating with other classes via e-mail or regular mail.

> **CONNECTING TO NATIONAL SCIENCE EDUCATION STANDARDS**
>
> **SCIENCE EDUCATION TEACHING STANDARDS**
>
> Science often is a collaborative endeavor, and all science depends on the ultimate sharing and debate of ideas. When carefully guided by teachers to ensure full participation by all, interactions among individuals and groups in the classroom can be vital in deepening the understanding of science concepts and the nature of scientific endeavors. The size of a group depends on age, resources, and the nature of inquiry (p. 31–32).

Imagine groups of fourth-grade students trying to answer the driving question "How do things with wheels help us move?" As students begin experimenting with skateboards, roller blades, bicycles, and other everyday objects with wheels to find out how they help us move, they find that they need to multiply distance by force to figure out how much work is accomplished. One student in the group does not know how to multiply. Rather than hold the entire group back while the student learns multiplication, the teacher could set up a peer tutoring situation. While the one student worked with another to learn multiplication, the entire group could move ahead with the project.

To encourage student-to-student collaboration, some teachers start by pairing students during group activities. Pairs of students can be given specific prompts to discuss ideas and make meaning from science activities. For example, teachers can ask students to compare and contrast their own ideas with those of their partners. One such open-ended prompt is "My idea is like _____'s because _____, but my idea is different from _____'s idea because _____." Over the course of a school year, teachers can begin to put students into larger groups. As students become comfortable working in groups, some teachers have them share information, debate ideas, and brainstorm solutions with students from other classes within a school building or a district. Teachers of the same grade level can encourage this level of collaboration by having their students all study the same driving question and by providing their students time to meet with members of the other classes. If meeting with other classes is not possible, e-mail or regular correspondence can facilitate collaboration. Finally, once students are skilled collaborators, a teacher can encourage collaboration with other students around the world. For example, one sixth-grade teacher's students studied the question "What's in our water?" and collaborated with students in another class via the World Wide Web.

> CONNECTING TO
> NATIONAL SCIENCE
> EDUCATION STANDARDS
>
> SCIENCE EDUCATION
> TEACHING STANDARDS—
> STANDARD B
>
> Teachers of science guide and facilitate learning. In doing this, teachers orchestrate discourse among students about scientific ideas (p. 32).

There are several reasons for using student-to-student collaboration. First, peers are not authority figures, so students tend not to feel threatened in these situations. Second, students come to school with many different prior experiences that give them different levels of knowledge about subjects. Each student has different interests, personalities, learning styles, and attention spans. As students explore solutions to driving questions, they can use these varied interests, abilities, knowledge levels, personalities, and attention spans to help their fellow students learn new ideas. Finally, deeper learning takes place when students challenge others' ideas. Alfie Kohn (1996) wrote, "To create a classroom where students feel safe enough to challenge each other—and us—is to give them an enormous gift" (p. 77). This idea is consistent with Vygotsky's zone of proximal development, because students with a little bit more knowledge or skill than their peers are providing a scaffold to help the others learn new ideas.

Scaffolding, discussed in Chapter 2, is a process in which a more knowledgeable other uses various techniques, such as modeling or coaching, to direct those aspects of the intellectual task that are initially beyond the capacity of the learner.

Between Students and Teacher

A cornerstone of collaboration is equal participation among members of a learning community. Collaboration between teachers and students is a situation in which students and teachers participate equally. Although students do not have the same level of expertise as the teacher, there are some areas in which students have more experiences or different understandings. Such experiences and understandings can become a valuable resource for the classroom.

Imagine a sixth-grade teacher whose class is studying the driving question "Are there poisons in our environment?" The teacher can generate her own questions, such as "Is insecticide a poison in our environment?" She also can suggest to students that they are experts when it comes to their own experiences and investigations. The student whose parents purchase only organically grown foods might ask, "Are food preservatives a poison?" The student who reads information on the Internet about harmful products around the home might ask, "Are cleaning products poisonous?" The student whose family just purchased a radon detector might ask, "Is radon a poison?" In such a classroom, the teacher can participate equally with the students by asking them to generate questions, rather than being the sole director of the lesson.

To enter into a teacher–student collaboration, a teacher must show students that she finds their ideas important and that she is a learner in the classroom as well. When a teacher encourages students to contribute questions and ideas to classroom projects, students become empowered. They begin to feel that they have important knowledge, and they no longer perceive the teacher as the sole source of information in the classroom. Students begin to take responsibility for their own learning and help teach the class. Once the imbalance between roles is reduced, students and teachers can begin to collaborate equally.

In establishing a collaborative relationship with students, a teacher must take on two roles—that of the collaborator who shares and debates ideas, negotiates meaning, and takes risks, as well as that of the instructor who orchestrates instructional events and keeps the class focused on

> ## CONNECTING TO NATIONAL SCIENCE EDUCATION STANDARDS
>
> ### SCIENCE EDUCATION TEACHING STANDARDS
>
> The teacher's role in these small and larger group interactions is to listen, encourage broad participation, and judge how to guide discussions—determining ideas to follow, ideas to question, information to provide, and connections to make. In the hands of a skilled teacher, such group work leads students to recognize the expertise that different members of the group bring to each endeavor and the greater value of evidence and argument over personality and style (p. 36).

learning goals. Adequately taking on both roles can be difficult. The teacher needs to establish trust among students so that students feel comfortable discussing ideas, asking questions, seeking information, and posing probable solutions. This can be accomplished by letting students know that they won't be punished or ridiculed for their ideas. The teacher can also admit to students that he doesn't know everything—that he's a learner in the classroom, too. However, the teacher also needs to maintain control of certain classroom events, maintain appropriate behavior, and keep students on task. This can be accomplished by establishing guidelines for acceptable classroom behavior, timelines for lessons and projects, limits on use of resources, and overall learning goals. (Classroom management is discussed more fully in Chapter 11.)

Collaboration between students and teachers offers several benefits. First, it helps students see classroom projects as real world problems rather than as "cookbook" lab work with a single right answer. They come to view learning as more meaningful because they begin to see that the solutions to their questions extend beyond the teacher. They come to understand that no one—not even the teacher—has all the answers, and every person must learn how to answer the questions posed. They learn that adults continue to learn and that learning is an ongoing process. Second, students gain self-esteem because they see that their ideas are valued. Third, they tend to be more motivated to learn because they are excited to find answers to their questions—especially answers that the teacher may not even know!

> ### CONNECTING TO NATIONAL SCIENCE EDUCATION STANDARDS
>
> ### SCIENCE EDUCATION TEACHING STANDARDS
>
> An important stage of inquiry and of student science learning is the oral and written discourse that focuses the attention of students on how they know what they know and how their knowledge connects to larger ideas, other domains, and the world beyond the classroom (p. 36).

Between Students and the Community

In collaboration between students and the community, students work with parents, neighbors, friends, relatives, and other community members either by meeting or by communicating over the telephone, via e-mail, or through the mail to share information, investigate ideas, or develop artifacts.

Collaboration with community members can take many forms. A student might call a person at a governmental office to request information about a given topic. A guest speaker might visit the classroom to demonstrate a concept, discuss an idea, or answer students' questions. Community members might provide resources; schools with limited budgets sometimes cannot afford materials that community members and businesses can provide. Students might correspond with community members, sharing their ideas, questions, and investigations over an extended period. Finally, community members may become mentors for students, working collaboratively with students to answer a driving question and modeling for the students specific strategies or behaviors.

To extend collaboration beyond the four walls of the classroom, a conduit for communication must first be established. There are many ways to establish this communication. A teacher might encourage students to invite to the classroom parents, neighbors, friends, and relatives who have expertise in an area being studied. A teacher might invite community resource people, such as doctors, water quality experts, chemists, and veterinarians to the classroom. As students become comfortable working with adults in the community, they can contact and interview community members on their own. Another conduit is technology. Telephones, fax machines, e-mail, and Web pages are tools that can connect students to the community. For example, in the acid rain project, students could link via the Internet to experts in the field. Using blogs (web journals) or wikis (web pages that others can edit), students can use the Internet to engage in **social computing** where others help create and react to knowledge.

Teachers also need to establish a classroom environment that encourages collaboration with community members. To accomplish this, the teacher needs first to help students understand that she and the science textbook aren't the only sources of information in a classroom—that students can learn from each other and from people in the community. Blogs can also be used to engage community members (Ray, 2006). Second, the teacher needs to allot class time for community members to visit and for students to communicate with others outside school. Third, the teacher should help students understand the different roles community members can play in teaching them science.

Individuals in the larger community represent a wealth of information for learners; they can provide resources and ideas otherwise unavailable to students. Through collaboration with the community, students see that science learned in school relates to real-world problems. They learn about people and their careers. They learn to communicate with adults in a collaborative fashion. Upper elementary and middle school students especially seem to appreciate working with adults in the community because the adults treat them as young adults capable of understanding science in the real world.

Creating a Collaborative Environment

Collaboration doesn't just happen. Teachers must work to create a collaborative environment in their classrooms. Collaboration is new to many students; it does not resemble the typical interactions that most students have experienced in schools. To understand what real collaboration is, students need experiences and guidance in collaborating. Some students will be frustrated by collaboration at first if they are not used to taking initiative and responsibility for their own education.

For these reasons, it is important to introduce students gradually to collaboration. Over the course of an entire school year, a teacher can create a foundation for collaborative learning. In this section, we will explore some considerations when designing collaborative classrooms: how to form groups, how to develop collaborative skills, and how to get all students equitably involved.

Forming Groups

Before students can collaborate in small groups, the groups must be established. Many factors need to be considered when forming groups. Should groups represent mixed academic ability or similar abilities, mixed personalities or like personalities (all shy students in the same group and domineering students in another group), and one or both genders? How many students should there be in a group? Should the teacher create the groups or allow students to select their own group members? Should groups be randomly assigned? Should friends be allowed to be in a group together? Should groups stay together throughout the entire project or should new groups be formed for each activity? As you think about forming groups in your own classroom, on what basis will you determine group composition to facilitate the best collaboration? Next we discuss some of the most important factors to consider.

One of the most important steps a teacher can take when designing a collaborative classroom is to think carefully about group composition before placing students into groups. This placement can affect students' interactions in groups (Blumenfeld, Marx, Soloway, & Krajcik, 1996; Leonard & McElroy, 2000). Collaboration works most effectively in heterogeneous groups with moderate differences in ability, personality, and prior experience and to which each student brings different strengths and weaknesses. Vygotsky's (1986) theory supports the need for situations in which discrepancies among children's views exist to promote cognitive development. Students are exposed to more zones of proximal development when they are grouped with students who may be a little more knowledgeable or have slightly different viewpoints. If differences among students are too drastic, however, problems can arise. For example, a highly competent reader in the fourth grade may become frustrated working with a peer who cannot read beyond a second-grade level. This student may even come to reject or mistreat the student who has difficulties reading. Students with extremely incompatible personalities may not work well together either. For example, an extremely extroverted student may totally dominate an extremely introverted student. Conversely, two extremely domineering personalities may only bicker with each other since they both want to be in charge of a project. Drastically different prior experiences may also inhibit students' abilities to complete a task. For example, a student who has never worked with a computer may not be a good choice for working with students who are computer fanatics. However, a student who knows a little about working with computers might make a good partner for someone who is very comfortable working with computers.

Although heterogeneous groups tend to foster collaboration, there are exceptions to this with regard to gender. Researchers studying equity issues in education have found that girls sometimes feel intimidated in groups dominated by boys (Baker, 1988; Kahle & Rennie, 1993). Boys tend to take control of conversations, activities, and decisions. Girls are often relegated to serve as scribes, taking the notes for the group or writing out the results of investigations. These educational researchers have found that girls, particularly upper elementary and middle school girls, tend to participate more in science when they work only with other girls (Baker, 1988; Kahle

> ## CONNECTING TO NATIONAL SCIENCE EDUCATION STANDARDS
>
> ### SCIENCE EDUCATION TEACHING STANDARDS
>
> Effective teachers design many of the activities for learning science to require group work, not simply as an exercise, but as essential to the inquiry. The teacher's role is to structure the groups and to teach the skills that are needed to work together (p. 50).

& Rennie, 1993). To avoid these potential problems, teachers can help make girls feel more comfortable with science by sometimes creating girl-only and boy-only groups. Then, as the girls' confidence increases, the teacher can gradually introduce mixed-gender groups.

The size of collaborative groups is also important. With younger elementary students who are in the self-centered stage of development, smaller groups—such as pairs—work best because such children have not yet learned to share attention with a large number of other people at once (Edwards & Stout, 1990). There is only one line of communication in pairs (between the two students), and no one is left out. Groups of three can be effective if there are three tasks, roles, or parts for students to take. However, there is the risk that two students will pair up, excluding the third. Groups of four seem to work better than groups of three because, even with pairing up, no one is left out. Groups of four are also physically easy to arrange—four student desks can be put together to form one larger table area for student work. When more than four students are grouped, behavior management sometimes becomes a problem, and it is more likely that students will be left out of the group dynamics.

Teachers need to decide whether or not students will be allowed to choose their own groups or whether or not the teachers will assign group membership. This decision is not always an easy one. Most teachers find that students who are not used to working in collaborative groups will select team partners who diminish effective collaboration. They tend to select their friends, and this type of group composition can sometimes cause off-task behavior and cliques that dominate activities. Some teachers randomly select members of a group because students see this as a "fair" group selection technique. Some research suggests that random selection among friends (rather than totally random assignment) leads to greater results (Mahenthiran & Rouse, 2000), and other research has found that girls have greater shared goals among friends—an important consideration for better task performance (Strough & Cheng, 2000). Other teachers make groups based on the physical arrangement of the classroom, such as by dividing the class into quadrants. Another option is to form interest groups. For example, all students who want to explore the same question can be grouped together.

Students who have learned to work in collaborative groups have skills that help them overcome the problems inherent in grouping. In fact, older elementary and middle grade students who are comfortable with collaboration will appreciate being given the responsibility of selecting their own groups.

Regardless of who selects groups, it is important to build team identity. Team identity creates team rapport and camaraderie. Young adolescents have a need for peer approval and for feeling that they belong to a group, so it is important to build team identity with upper elementary and middle school students. Some teachers have students select group names, make team banners, write team cheers, or create team slogans. The identity of a team can be related to characteristics of members of the team or to the problem it is solving. For example, a team investigating wetlands might create the team name "The Swamp Things."

The decision to keep groups together throughout a project or to form new groups presents a challenge for teachers. Some teachers find that, once the momentum has built up in a project, students want to stay together. Further, students in a group build mutual understandings and interpretations that must be built all over again each time new groupings are formed. Other teachers find that students need change to keep them interested in a project. Students may become bored working with the same group members; ideas may become stale in the group; and students eventually may learn all they can from each other. Regardless of what their final decision is, teachers find that it is only through continual monitoring of the dynamics within a group that they can make this decision.

To overcome the problem of switching groups frequently, some teachers form different levels of groups. Informal groups are used for short-term activities, such as quick brainstorming sessions. Such groups may exist for only a few minutes. Other groups may stay together for months during a project. **Home groups**, groups that are used for critiquing work and supporting each other, may be used the entire school

TABLE 7.1   Factors to Consider When Forming Groups

- Are students with different ability levels placed in the same group?
- Are the personalities of the students in the group compatible?
- Have you avoided placing students with extremely incompatible abilities in the same group?
- Have you formed heterogeneous groups?
- Has attention been paid to making sure girls are not dominated by boys in the group?
- Is the size of the group appropriate for the age of students and the task?
- For this task, do you think it better to select the students for the groups or allow students to select their groups?
- For this task, will it be important for you to select students for the groups from among friends or select them totally randomly?
- Does the physical arrangement of your room affect how you need to select groups?
- Have groups created a team identity (team name, cheer, etc.)?
- Will you need to keep the group together or break it up throughout the project?
- Have you considered creating different functional groups such as home groups, base groups, or sharing groups?

year. One option is to form base groups and sharing groups. **Base groups** are groups of students who work together on a project. They become experts in their topic area, on their driving question, or on their problem. After they have finished their project, the students move to new groups—the **sharing groups**—in which they share information about their respective projects.

Developing Collaborative Skills

To foster collaboration, teachers need to help students develop collaborative skills. Most students have not experienced collaborating, talking to each other about ideas, and working independently in small groups and may need to learn how to do so. Also, collaboration requires students to take risks. By sharing their ideas, students open themselves to criticism from their peers. Teachers need to provide a safe environment so that students are willing to take such risks by teaching students how to sensitively give feedback. Collaborative skills are essential in this type of situation, as well as in many other real-world interactions, such as completing projects with others at work or getting along as a member of a family. For effective collaboration, students need to learn the skills of decision making, task completion, trust building, communication, and conflict management.

Groups will quickly fail unless the teacher analyzes the collaborative skills needed by students and discusses, models, practices, and evaluates them. In other words, collaborative skills must be taught just as purposefully as academic skills. Before starting a lesson, many teachers discuss the skills that are necessary for collaboration. Learning Activity 7.2 explores ways to purposefully approach collaborative skills. For example, teachers frequently discuss with students behaviors necessary in groups, such as talking with appropriate voices, taking turns using science equipment, sharing science materials, acknowledging others' contributions in the group, staying on task during a project, completing work on time, helping others understand science ideas, using "I" messages (saying what one thinks rather than blaming others, such as "I wish you would redesign that artifact," rather than "You didn't design that artifact correctly"), criticizing in a positive way, listening actively, rephrasing others' science ideas, being patient, compromising, negotiating, asking for justification for science answers/responses, and probing or redirecting questions to others in the group. To discuss these behaviors, teachers ask students questions, such as "What is *negotiating?*" "Why is it important to negotiate?" "Why will your groups work better if you negotiate with each other?" and "How do you know if people are negotiating?"

A teacher might also use a T-chart to teach various collaborative skills (Johnson & Johnson, 1990). The T-chart is simply a chart in the shape of a T. Write the name of a skill above the T, descriptions of what the skill looks like to the left of the T, and descriptions of what the skill sounds like to the right of the T. Table 7.2 is a sample T-chart of the skill *negotiating*. The teacher would have students generate examples of what they would see each other doing and hear each other saying if they were negotiating. The teacher could model these behaviors or ask students in the class to model them. Next, students would practice the skill. Finally, at the end of the class period, the teacher would have students discuss and evaluate how well they

Learning Activity 7.2

HOW DO I INTRODUCE COLLABORATIVE SKILLS TO STUDENTS?

Materials Needed:

- Something to write with

A. Elementary and middle grade students need to be introduced to collaborative skills, and these skills need to be developed over time. Children don't naturally collaborate. For each of the following skills, think of questions you might ask children as a means of introduction. For example, imagine you are introducing the skill "agreeing with others." You might ask students, "What does it mean to agree with others?" (to compromise or reach the same opinion on something), "Why should members of a group agree with each other?" (so that arguments and fights are avoided), and "How will the group's ability to complete science projects improve if students are in agreement?" (it can proceed with a project without being sidetracked by disagreements).

Skills

- Talking in appropriate voices in small groups.
- Taking turns using science equipment.
- Sharing science materials.
- Acknowledging others' contributions in a group.
- Staying on task during a project.
- Completing work on time.
- Helping others understand science ideas.
- Using "I" messages (saying what one thinks rather than blaming others, such as "I wish you would redesign that artifact" rather than "You didn't design that artifact correctly").
- Criticizing in a positive way.
- Listening actively.
- Rephrasing others' science ideas.
- Being patient.
- Compromising.
- Negotiating.
- Asking for justification for science answers/responses.
- Probing or redirecting questions in a group.

B. Discussing collaborative skills is only a first step in introducing collaborative skills. Develop a collaborative lesson. Decide what you would do to reinforce these skills. How would you develop the skills?

C. Try the lesson. Evaluate how well the introduction worked with students. What else do you need to do?

D. Record your ideas.

TABLE 7.2   Sample T-Chart

Negotiating	
Looks Like	**Sounds Like**
• Heads nodding in agreement or disagreement	• "What is your idea?"
• Students making eye contact with each other	• "Can we come up with a compromise?"
• Students brainstorming more than one alternative	• "That's an interesting alternative!"
• Students not putting down any ideas	• How can we use both ideas?"

negotiated with each other. In some cases, teachers even make a video of the classroom activities and have students watch and analyze themselves to determine how well they worked together. The students would conclude by determining what areas still needed refinement. T-charts are very effective in helping students understand what behaviors look and sound like in a classroom. Without these types of discussions, children will oftentimes fail to understand what a teacher means by resolving conflicts, respecting others, disagreeing politely, maintaining self-control, showing appreciation, taking turns, encouraging others, praising others, respecting others, or accepting differences.

In Learning Activity 7.3, you will create your own T-charts to teach students skills that will be helpful in collaborative groups.

Learning Activity 7.3

MAKING T-CHARTS

Materials Needed:

• Something to write with

A. Practice developing a T-chart. Make two columns and develop your own ideas for the following ideas written at the top of the "T":
  • Resolving conflicts
  • Criticizing ideas, not people
  • Respecting others
  • Disagreeing politely
  • Maintaining self-control
  • Showing appreciation
  • Taking turns
  • Encouraging others
  • Praising
  • Respecting others
  • Accepting differences

B. Try making a T-chart of these skills with others in your group (or with children). How are their ideas different from and similar to yours? How did developing a T-chart help you communicate with the other people in your group? With children?

Decision-Making Skills

In a project-based science classroom, students take responsibility for making many decisions. Students must make decisions about division of labor (who will do what and how the work load can be equitably distributed), timelines and scheduling (when people will complete things and in what sequence), activities (what are the solutions to problems, which activities need to be completed, and what artifacts need to be made), and resources (information, supplies, and materials).

A teacher can help students learn decision-making skills by providing students with goal sheets to track division of labor, timelines and schedules to plan and sequence their time, and checklists to make decisions about resources. A goal sheet is simply a chart on which students can record their goals for the week and their plans for who will take responsibility for each goal. Table 7.3 is a sample goal sheet. After students have completed their goal sheets, they can move to developing timelines and schedules. This can be accomplished with the use of a weekly or monthly calendar, or the goal sheet can be expanded to include a plan and timeline. Table 7.4 is an example. You can use a checklist to provoke students' thinking and expand their ideas about resources needed to complete their project. A checklist option is shown in Table 7.5. Another option is to work with students to develop their own resource list.

Task-Completion Skills

Task-completion skills are necessary to teach students to stay focused on the driving question and complete projects within a reasonable time frame. Once students understand the task at hand, they need to learn how to actually complete the task. To complete a task, students must *know how* to complete it, *remember* to complete each step of the task, *stay focused,* and *monitor* their time. (Remember that it usually takes more time to cover the same objectives in a collaborative learning environment than

TABLE 7.3 Goal Sheet

What do we need to do?	Who will do it?
Find information about air pollution.	Liz
Contact the EPA.	Jeff
Figure out an experiment we can do to test for particulate matter.	Rosario

TABLE 7.4 Goal Sheet With Plan and Timeline

What do we need to do?	Who will do it?	How will it be done?	When will it be done?
Find information about air pollution.	Liz	Use World Wide Web.	By Friday, Sept. 12th
Contact the EPA.	Jeff	Call from home.	At home
Figure out an experiment we can do to test for particulate matter.	Rosario	Look in resource books.	Today, by Friday, Sept. 12th

TABLE 7.5   Resources Checklist

What?	Can we use it?	How?
Information		
Encyclopedia	Yes	Find out more about air pollution
Magazines	_____	_____
Journals	_____	_____
Books	_____	_____
CD-ROMs	Yes	Use in library
Software	No	The school doesn't have any on air pollution
Guest speaker	Yes	From the EPA
Telephone call	Yes	To the EPA
E-mail	_____	_____
Interview	Yes	Of someone at the EPA
World Wide Web	Yes	Do a search
Letter	_____	_____
Supplies and Materials		
Writing materials	_____	_____
Computers	_____	_____
Drawing materials	_____	_____
Science equipment	Yes	Check with the teacher
Calculator	_____	_____

in a traditional classroom. Students need time to discuss ideas, work together, build understanding, and complete projects.)

To help students complete tasks, some teachers use a progress report. A progress report is a modification of Tables 7.2, 7.3, and 7.4 in which students review how much they have accomplished at the end of a day or a week. You might have students turn the progress report in to you, share it with classmates, put it in a science journal, or include it in an evaluation portfolio. The main point is to engage students in the reflective practice of examining how much they have completed. This will keep students conscious of starting in a timely manner and avoiding off-task behavior.

Trust-Building Skills

Trust-building skills form a critical aspect of collaboration. Trust building makes all students feel comfortable working in small groups. Each student should be at ease giving suggestions or opinions, participating in discussions, and helping answer the driving question. Although students must feel safe when sharing their ideas, an important part of science is the critiquing of ideas; and students must be willing to have their ideas critiqued. In addition, students need to learn how to critique ideas without offending others. Students can learn to build trust within a group by including everyone in discussions. When all students feel that their ideas are wanted and considered, they will be more likely to trust their peers.

Getting to Know Others. If students in the classroom do not know each other or if the school's population is highly transient, an important first step for building trust is helping students get to know each other. Many people lack the confidence to share ideas with or take risks among strangers. Students need to introduce themselves to each other and discuss hobbies or interests. They need time to develop some common lines of communication. Teachers who neglect to facilitate students' getting to know each other will find students frequently not on task because they are taking the time themselves to build trust by talking about their lives and their interests. To help students get to know each other, engage them in ice-breaker activities. The following ice-breaker activities can help students to get to know others in the class.

Four Corners

In one ice breaker, each student writes his or her name in the center of a 5 × 8 index card. In each corner, the student writes some type of information about him- or herself. For example, children might list where they were born, personality characteristics, favorite hobbies, favorite science topics, the names of people they admire, accomplishments they are proud of, or favorite school subjects. Participants should gather in groups of four to share their cards. Participants should take on different roles—probers, recorders—to elicit more information about what is written on the cards. Each person in each group should introduce another group member to the whole class.

People Search

In another activity, teachers create a grid with one statement in each square, such as someone who likes comedy, someone who collects baseball cards, someone who loves dogs, someone who walks to school, someone who likes to read science fiction, someone who plays basketball after school, someone who dances, someone who has a chemistry kit, someone who has used a telescope, someone who is shy, someone who has visited a science museum, someone who likes to swim, someone who likes to sleep late, and someone who has lived on a farm (or in the city). The task of this ice breaker is to have students circulate among the entire group to find people who meet the criteria.

Partner Interview

Partner interviews are effective for all ages of students. Simply pair students with some probing questions, such as What is your favorite science fiction movie? What is your favorite hobby? What is your favorite sport? What science magazine have you read? What's your favorite book? Why? What do you like to do after school? What is your favorite science topic? What sports do you play? What is your favorite computer program? Who's your favorite actor or actress? What is your favorite school subject? What is your least favorite science topic? What is your favorite song? What would embarrass you? What's your favorite science experiment? Why? Where have you gone on vacation? What is your favorite restaurant? What is your middle name?

What do you think you want to be when you grow up? What science topic have you noticed in the newspaper this week? Do you have any pets? Which kind? What makes you laugh? What science topic would you like to know more about? By interviewing partners, students get to know the other students they will be working with in groups. This builds a relationship, and it helps eliminate off-task talking that naturally occurs in groups if students do not know each other well.

Interview Cube

Make paper cubes (or cover some existing cubes with paper to write on each side). On each side of each cube write the name of a topic, such as *favorite science fiction movie, favorite science topic, science-related hobby,* or *best experiment ever conducted.* Participants should form groups of four or five. Each member should "roll" the cube and respond to the topic that rolls face up. For example, if the cube lands on *science fiction movie,* the roller must tell the others the title of her favorite science fiction movie. Then members of the group must ask probing questions about the topic. For example, they might ask, "When did this become your favorite movie?" At the end of the activity, each participant should share something interesting about the person next to him or her.

Finding Common Ground

Participants should form groups of four or five and then select a topic, such as *sports, science class, TV shows,* or *school.* Each group should discuss its topic until members find an interest, dislike, or quality that *every* member of the group shares.

Finding Qualities

Participants should circulate around the room to find one positive quality about each other person. Participants should list each person's name with at least one positive quality, such as "Is always friendly toward others" or "Helps others understand science ideas."

Feeling Needed. In building trust, it is not only important for students to know each other, but also for them to feel needed by the group. One way to accomplish this is to identify situations in which not all students are participating and to encourage the nonparticipating students. You might draw quieter students into group discussions by inviting them to share their thoughts or ideas. To encourage sharing, you might make it a requirement for each student to share a given number of ideas during a lesson. You might use a "round-robin" approach in which each student must contribute one idea. Some teachers use children's literature to teach about sharing and other collaborative skills. For example, the story *The Rainbow Fish* by Marcus Pfister (1996) can teach children about making friends through sharing. *Seven Blind Mice* by Ed Young (2002) teaches children about collaborating to solve a problem and use observation skills (see Sapon-Shevin, 1999, pp. 205–207, for a comprehensive set of children's literature suggestions). To encourage equal participation, you might give each student a limited number of tokens. Each time a student contributes an idea, she "pays" a token. When she runs out of tokens, she cannot contribute more ideas until all the other students have also used up their tokens. Another approach is to have groups of students break into pairs temporarily to talk or work together before returning to the large group. While students are working in pairs, they will all participate equally. When back in the large group, they may continue to participate at a high level. Learning Activity 7.4 introduces a strategy for teaching students that they must depend on others in their group.

Offering Constructive Criticism. After students have learned about one another and have begun to feel needed by the group, they must learn how to critique others' ideas without offending. This is probably the most difficult component of trust building to teach as students often speak before thinking. You can use a number of strategies to support students in thinking about their classmates' feelings and to foster mutual respect. A T-chart, like the one in Table 7.6, can be used to help teach students how to show respect for the ideas of others.

Many children find it difficult to avoid putdowns. You might have students make lists of insulting words they should not use in a group, such as *stupid* or *dumb*. Then, students practice accepting others' ideas and avoiding the prohibited words. Another strategy involves having students create lists of acceptable words. For example, students might be tempted to say, "That's a dumb idea." The teacher can have the class brainstorm a list of respectful ways to point out the problems with an idea. Students might list, "You seem to be having trouble with this idea" or "You need to think about this again."

Students also need to be taught to avoid a tone of voice that suggests ridicule. Rules regarding making faces might need to be established. Students will probably need to practice critiquing others' work without being offensive or abusive. Finally, when students disagree with other people's ideas, they need to learn to criticize the *ideas*—not the people. To learn to criticize ideas rather than people, students can practice using "I" messages instead of "you" messages. For example, saying, "I think this might be a better way to solve this problem" is less threatening than saying, "You cannot solve this problem this way."

Communication Skills

For students to be successful at collaboration, they need communication skills. Communication skills can be broken down into three components. One, students need to learn to talk about learning. Two, they need to learn to speak clearly. Three, they need to listen carefully to what others are saying so that miscommunication doesn't arise. These three components help students communicate clearly with others, and effective communication helps prevent arguments, frustrations, and disagreements among group members.

Learning Activity 7.4

WHAT SKILLS ARE HELPFUL IN COOPERATIVE GROUPS?

Materials Needed:

- Five equal-size cardboard squares cut up as shown and sealed in envelopes[1]

A. Cut up five squares (all of the same dimensions) as shown in Figure 6.1 for each group of five people. Distribute the five squares among envelopes as shown.

B. Each member of the group should get an envelope. The goal is for each person in the group, *without communicating in any way* (without talking, gesturing, or writing notes), to form a square that is the same size as all the other members' squares. Members of the group may hand their own pieces to others in the group, but they may *not* take pieces from another person or point to pieces they want. They can only wait for a person to hand them a piece. They cannot gesture that they do not want a piece; they must accept it and give it to someone else if they do not want it. When a member gives a piece to another member of the group, she must simply hand the piece to the member and not place the piece so as to help complete a square. The activity is not timed, and it may take anywhere from a few minutes to over a half hour to complete.

C. When all the members of the group are finished, discuss what happened. What skills does it take to complete this task? Was it frustrating? Why? Was there collaboration among the members of the group? What skills would facilitate collaboration in elementary and middle grades classrooms?

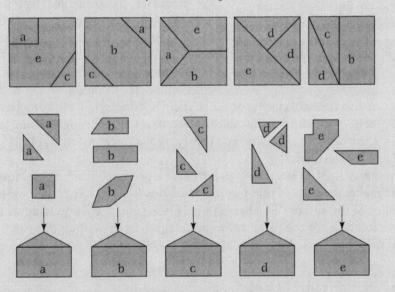

Figure 7.1   The five-square activity.

1. Adapted from Abruscato, J., & Hassard, J. (1976). *Loving and beyond: Science teaching for the humanistic classroom*. Glenview, IL: Scott, Foresman.

TABLE 7.6   Respect-Building T-Chart

Showing Appreciation	
Looks like ...	Sounds like ...
Smiling	"Thanks, I appreciate your help with this."

Talking About Science Learning. Children don't automatically know how to talk about science. They need to be taught how to ask questions and generate alternative solutions to questions. One way to teach children how to generate questions and alternative solutions is to hold a whole-group brainstorming session during which you record questions and plausible answers on the board.

CONNECTING TO NATIONAL SCIENCE EDUCATION STANDARDS

SCIENCE TEACHING STANDARDS

A fundamental aspect of a community of learners is communication. Effective communication requires a foundation of respect and trust among individuals (p. 50).

Children also need to learn how to probe for more information and request justification for ideas and evidence for conclusions. If a student says, "I think the aquarium is dirty because the fish go to the bathroom in it," other students might learn to ask probing questions like, "Why do you think that? Fish go to the bathroom in lakes and oceans too." However, when students are challenged to elaborate, they often resort to nonanswers, like "Because." To overcome this problem, teachers can have students practice responding in complete sentences and providing evidence. Evidence might be in the form of observations; data collected; or things read in books, in articles, or on the World Wide Web.

Students can also create T-charts for skills, such as summarizing information, paraphrasing others' ideas, checking for scientific accuracy, elaborating, probing for more information, advocating a position, and justifying a science answer. Table 7.7 shows a sample T-chart on elaborating.

Children also need to learn how to advocate a position. Teachers can foster this skill by engaging students in role-playing situations (such as role playing whether or not to report to the authorities a person who litters), mock trials (during which students have to argue for the prosecution or defense), letter-writing campaigns (such as to newspaper editors about a local issue), making posters (such as those advo-

TABLE 7.7   Elaborating T-Chart

Elaborating	
Looks like ...	Sounds like ...
Pointing to an investigation in progress	"Please tell me more about your project."
Holding up an artifact	"I think ... because...."

cating conservation of energy), or participation in news groups on the World Wide Web (such as those advocating bans on types of tuna fishing that endanger dolphins). As we discussed previously, developing scientific explanations and arguing for your position are an important parts of science. By teaching children skills for defending a position without being rude to others, you teach them important skills for doing scientific inquiry.

Speaking Clearly. Speaking clearly means enunciating words, using appropriate voice levels, and facing others when talking to them. Young children, of course, are still developing a repertoire of words, and they often mispronounce them. They also lose teeth and sometimes find it difficult to say certain words. In addition, children sometimes get very excited about science and, so, speak too quickly. Although these are normal characteristics of children's speech, teachers can help their students practice more effective speech by enunciating and talking more slowly so that others can understand, talking at appropriate voice levels so that people want to listen, and talking face-to-face so that others can hear them better.

Listening Carefully. Listening carefully is as important as speaking in good communication. Children like to talk, but frequently don't listen carefully to what others say. They need to be taught active listening, which includes using names and making eye contact when speaking to each other, paraphrasing what others say, asking for explanations, and summarizing conversations. Standards can be set about raising hands, taking notes when others talk, and not interrupting others in midspeech.

For example, while working on a display board for an artifact, one student might say, "I think we should do something different." A second student could interpret this to mean that they should have made a different type of display. In fact, the first student meant that they should complete their experiment before making the display board. Without understanding what the first student meant, the second student might suggest a different format for the display. This, of course, could lead to confusion and anger. To avoid misunderstanding, the second student could have said, "You said we should do something different. Do you mean that we need to have a different type of display?" The first student could then clarify his intent without causing problems.

Conflict-Management Skills

Even after students have learned decision-making, task-completion, trust-building, and communication skills, conflicts can arise among members of a group. For this reason, students also need to develop conflict-management skills. These skills include the ability to clarify disagreements, negotiate, and compromise.

Clarifying Disagreements. To clarify disagreements within a group, students need to learn to understand the points of view of others. Children frequently have difficulty understanding others' opinions. This is partly developmental (with young elementary children) and partly due to inexperience.

With young elementary students, teachers can help foster understanding others' points of view by using literature, stuffed animals, and pets. A teacher might read a story and ask children to express how the character in the story feels. The *Story of the Three Bears* is a good story to use in this way. A teacher might have students pretend that a favorite stuffed animal or pet can talk and tell stories about how the animal thinks and feels.

With older elementary and middle grade students, teachers can have students pretend (role play) that they are other people. Students can ask themselves, "What would I think if I were the other person?" Older students respond well to courtroom dramas. Teachers can read a courtroom drama or set up a courtroom simulation in which each student plays a different role. Students can discuss the thinking, feelings, beliefs, values, attitudes, desires, wants, and problems of each person in the drama. Imagine that students disagree about who should contact a guest speaker. One student wants to contact the guest speaker because he thought of the idea; another student wants to contact the guest speaker because his father works with her. By switching roles (pretending each is the other person), the students can try to understand why each believes he is the best person to call the speaker.

Negotiating. Next, students need to learn to negotiate. Negotiation does not mean that one person wins and the other loses. Negotiation results in both parties feeling as if they had won.

To teach negotiation, many teachers use T-charts or a hybrid of the T-chart, a win-win chart. On each side of the T in a win-win chart, students write the names of the disagreeing parties. The students list under each name how the person or group will win in the situation. Table 7.8 is an example of a win-win T-chart. The chart illustrates that both Matt and Luther want to feel like they contributed to the project. Matt wants the students to know that his father knows the speaker. Luther wants the speaker to know that inviting her was his idea. In order for both students to win, Luther will help Matt make arrangements for the speaker's visit. Matt will call the speaker. He will tell the speaker that inviting her was Luther's idea, and he will introduce her to Luther when she arrives.

Another way to teach negotiation is to use Venn diagrams for finding common ground. Students list differences in the outer circle and common ideas in the area of intersection of the two circles. The Venn diagram in Figure 7.2 indicates that students originally wanted to include different types of pictures in their project, but agreed upon a photograph.

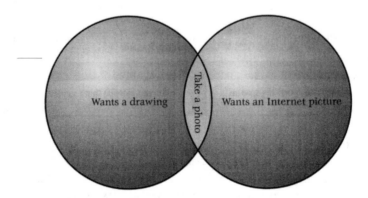

Figure 7.2   Venn diagram.

TABLE 7.8 Win-Win Chart

Who Gets to Call the Guest Speaker	
Matt	**Luther**
I can call since my dad knows the speaker.	You can mention that I suggested this.

Students can also send an e-mail about the problem to another school or student, requesting new ideas for resolving the differences. A group of students having difficulty reaching a consensus on how to build a machine to test for friction, for example, could e-mail their ideas to another school in their district and ask the students at the other school to vote on the best idea.

Compromising. If negotiation fails to produce a win-win situation, students need to learn to compromise. In a compromise, students settle differences and reach an agreement in which each "side" or person makes concessions. The resulting decision may combine the best qualities or elements of different ideas.

To teach the skill of compromise, a teacher might have students list all of the major points in their arguments. Then, the teacher instructs each side in the disagreement to give up one or two of its ideas in an exchange. Slowly, the two sides should come to an agreement that is a fair compromise. For example, students trying to decide how to best measure friction might use ideas from Table 7.9 to compromise. The compromise might turn out to be building a skateboard and measuring the friction of the skateboard across different surfaces, using a pulley.

Involving All Students Equitably

One of the essential features of collaboration involves equal participation by all members. However, frequently students feel reluctant to collaborate for one reason or another—lack of self-confidence, unfamiliarity with collaboration, and not knowing others.

Students need to understand that the products and ideas developed through collaboration must represent everyone, including them. Each student should feel that his or her contribution to the collaborative task is essential. A collaborative environment keeps all members of a group equitably involved and needed by others in the group. Some educators call this condition **positive interdependence**—the linking

TABLE 7.9 Compromise

How Should We Build a Machine to Test Friction?	
Yolanda's Ideas	**Anchee's ideas**
Make a skateboard.	Build a pulley system to measure accurately.
Test it on different surfaces.	Drag objects across different surfaces.
See if you fall off the skateboard on rough surfaces.	Chart how hard it is to move the objects.

of all group members so that one cannot succeed unless the other group members succeed.

You probably noticed in the first two scenarios at the beginning of the chapter that some students' "jobs" weren't really critical to the success of the group as a whole. Students whose efforts were not critical to the project's success quickly lost interest, got bored, or failed to participate intimately with others in the group. Positive interdependence may be built through establishing common goals or rewards, building common understanding, dividing up labor, and assigning roles.

Establishing Common Goals

Teachers can establish common goals by teaching students to move from individual ideas to group ideas. During all stages of collaborative learning (designing a driving question, investigating, developing artifacts, and sharing artifacts), students can first write down their own ideas. Open-ended prompts such as "My idea is ... ," "My plan is ...," and "My goal is ..." are helpful. After students have recorded their own ideas, they can work in small groups to share ideas. Then students can record the ideas and goals of others in the group. Finally, by comparing and contrasting goals, they can develop a common goal. Table 7.10 shows examples of charts that can be used to help students build common goals.

TABLE 7.10   Charts for Building Common Goals

Driving question			
Ideas	How the idea is like mine	How the idea is different from mine	Our group idea for a driving question
My idea			
Sara's idea			
Richard's idea			
Rachel's idea			
Inquiry			
Ideas	How the idea is like mine	How the idea is different from mine	Our group idea for an investigation
My idea			
Sara's idea			
Richard's idea			
Rachel's idea			
Artifacts			
Ideas	How the idea is like mine	How the idea is different from mine	Our group idea for a artifacts
My idea			
Sara's idea			
Richard's idea			
Rachel's idea			

Building Common Understanding

To build common understanding, teachers need to have students discuss intellectual ideas. The first step involves getting students to really listen to each other. When one student gives her idea and another student recites it back to her, two other students can add modifications to the idea; and, once all of the ideas are out, students can discuss the similarities and differences among the ideas. Understanding the similarities and differences among ideas helps students listen to each other, and it helps clarify ideas and build understanding among the members of the group.

One technique for facilitating discussion of intellectual ideas involves asking students to think about the validity of their responses. Students can be asked to justify the positions they take, and no one's idea can be included or excluded without a reason. Students can use sentences such as "We think X is better than Y because... ," "We think it is. . . because. . . ," "We don't think it is ... because. . . ," and "I think we should . . . because. . ." At first, charts like the one in Table 7.11 are useful to encourage discussions. The charts can be photocopied and distributed to other groups for discussion. Groups can discuss how their ideas are different from those of other groups. For example, the teacher can ask, "Which other group had a prediction similar to your group?" Over the course of the year, students will gain the skills to discuss ideas without the prompts in the charts.

Dividing up Labor

Division of labor occurs when students collaborate, negotiate, compromise, and interface with others to investigate a driving question. **Division of labor** does not mean *divide and conquer*. Divide and conquer strategies rarely lead to equitable distribution of knowledge, skills, or effort. Someone always seems to carry a larger workload, have a more interesting task, or contribute more information to the group. Students need to learn that true division of labor can only occur when members of the group participate equally in a task. All students need to take an active role, and no one can dominate. In a collaborative environment, it is in the division of labor that the teacher serves the role of guide—making sure students are cued to listen to one another's ideas, rather than only to work individually.

A teacher can help facilitate a true division of labor by putting students in situations in which they depend upon one another's resources. Resources include skills, abilities, knowledge, prior experiences, access to information, and friends and acquaintances. Students who have slightly different prior experiences may nonetheless contribute equally in a significant way to answering a driving question—thus dividing the labor.

TABLE 7.11   Defending Your Ideas

Idea	Why?
We think _____	Because _____
We think it is not _____	Because _____
We think_____ is better	Because _____

Imagine that students are trying to find out what factors affect plant growth. A student who has grown houseplants in an apartment complex has prior knowledge about window locations, the size of pots, and the amount of water. Another student who lives in a rural area grows food crops. This student brings to the group knowledge about farm machinery, commercial fertilizers, pesticides, herbicides, and crop yields. A third student who lives in a subdivision has a garden in her backyard. This student has experience using a compost pile as a source of fertilizer and organic gardening methods. These students can work together to divide the labor in an equitable fashion based upon their prior experiences. The first student could help the others set up an experiment to grow plants in pots of different sizes under various lighting conditions (in north, south, east, and west windows). The second student could help the others investigate how commercial fertilizers and pesticides help plants grow by setting up conditions in which some plants are treated with fertilizers and pesticides and others are not. The last student could contrast this investigation with one on organic gardening methods.

Assigning Roles

If students have difficulty working together, assigning roles can alleviate the problem. The assignment of roles ensures that all students participate and that no one dominates. It also teaches responsibility by making each student responsible for a part of the task.

It is critical that each role be important and contribute in a substantial way to solving the problem or reaching understanding. **Interpersonal roles** help ensure that members of a group work well together. Interpersonal roles include a facilitator (who makes sure everyone is participating), an encourager (who makes sure all students are participating or provides positive feedback to members of the group), a checker (who keeps notes about how well the group is doing on skills or tasks), and a noise monitor (who makes sure the group isn't too loud). **Managerial roles** include a reader (who reads directions to the group), a timekeeper (who makes sure the group finishes the task in a given amount of time), a recorder (who writes down the team's answers), and a runner (who gets materials that the group needs). Cognitive roles, or roles that require a specific type of thinking or specialized knowledge, include an executive, a skeptic or critic, an educator, and a conciliator. **Cognitive roles** can also correspond to scientific fields; such roles include a veterinarian, a botanist, a paleontologist, a geologist, an ecologist, a meteorologist, a zoologist, an oceanographer, an astronomer, and a chemist. *The Great Solar System Rescue* (1994) and *The Great Ocean Rescue* (1994) by Tom Snyder Productions are excellent examples of software programs that utilize cognitive roles to teach astronomy and oceanography.

Interpersonal and managerial roles, more than cognitive ones, tend to focus on separate jobs to get a task finished. Interpersonal and managerial roles are also less likely to be significant for the completion of a task. For example, the job of a student who is a runner is over as soon as supplies are acquired, and the student who is the encourager is likely to lose interest in the role and feel that it is phony or silly. For this reason, interpersonal and managerial roles are best used only with young children or students who lack even extremely basic skills for working together. For older elementary or middle school students (or students with more developed collaborative skills), cognitive roles are more effective. If students are studying acid rain, for example, they might take the roles of chemist, botanist, geologist, and ecologist. The chemist studies how the chemicals determine acidity. The botanist learns about how plants are affected by acid levels. The geologist researches the effect of acid rain on rocks, minerals, and soil. The ecologist looks at the overall picture, determining how acid rain interacts with the environment. With these cognitive roles, each person is needed throughout the project, and each contributes equally to the task.

Using Technology Tools To Create A Collaborative Environment

Technology tools can help teachers create a collaborative climate and enhance the quality of collaboration. Teachers can use telephones, fax machines, e-mail, and the World Wide Web to connect students with others and with the community. These technology tools create collaborative environments not otherwise possible. These applications allow students to share data, discuss and explore ideas, and talk about trends in data from various sites. E-mail, chat rooms, and listservs encourage discourse among large groups of people. Using the World Wide Web, students can create, share, and critique multimedia documents.

The quality of the collaboration is enhanced through technology because it enables students to collaborate with a large number of peers outside of their classroom, people in their community, scientists, and others around the world. By sharing ideas and data with this larger community, students and teachers take part in practices similar to those of scientists. New ideas can be quickly introduced and discussed. This extended network of collaborative partners also builds multiple zones of proximal development, because the wide range of people adds different levels of knowledge and expertise. Learning scaffolds are created when many different individuals help others explore a topic. Further, some teachers have found that many students are more comfortable asking questions or discussing topics electronically than they are in person.

A number of programs allow students to share data and ideas on the World Wide Web. Project FeederWatch (http://www.birds.cornell.edu/pfw) allows students throughout the winter months to collect and share data about birds that visit feeders in the school playground, nature area, or garden. In the project, students periodically count the highest numbers of each bird species they see at their feeders from

November through April. After each count, students report their bird counts to scientists at the Cornell University Lab of Ornithology. Students can view their results and the results of other Project FeederWatchers by viewing different maps. They can look at maps that summarize their data, or they can look at maps that present movements of bird populations. Scientists at the lab will use the data to conduct bird population research. Project FeederWatch helps scientists track movements of winter bird populations and long-term trends in bird distribution and abundance. They can also track the movement of the most common birds found at bird feeders. As members of the project, students learn more about winter birds, their movement, and how the birds fare during the winter months. By being part of Project Feeder-Watch, students take part in ongoing scientific inquiry.

In the Web-based Inquiry Science Environment project (WISE; Linn, 2005), students can share and discuss ideas with other students to explore various phenomena, such as malaria and why so many frogs are deformed (see http://wise.berkeley.edu/). Designed for students in 5th through 8th grade, students use the World Wide Web to find, share, and sythesize information with other students. Using the electronic tools available, students can share ideas and data, as well as discuss questions. Peer review is an important part of the WISE program, and it is used to help students exchange ideas and receive feedback in preparation for a final design or debate activity. Students in WISE learn how to build explanations, and these explanations are critiqued by other students. As in Project FeederWatch, students in WISE take part in scientific practices of communicating and sharing their ideas with others.

We have shared only two of the many projects that are available on the World Wide Web. Many of these programs allow students to share and discuss ideas, as well as share data. By taking part in such activities, students are participating in scientific practices and joining a much larger collaborative community to promote learning. Recent research (Tao & Gunstone, 1999) has found that collaborative learning at the computer fosters conceptual change. Computer-supported collaborative learning helped students coconstruct an understanding that enhanced learning.

Challenges That Arise When Students Collaborate in Small Groups

Students don't automatically collaborate with others, and groups do not always work well together. Students may argue, fail to communicate, or resist compromise. Students' personalities may lead to situations in which some students do all the work and others contribute virtually nothing. Sometimes higher ability students feel they are being taken advantage of by others, or they dominate. Status among members of a group can cause problems—high-status members (those with higher academic status, peer status, or social status) may take control of a group. Sometimes, members of a group may make a pact to do the least amount of work possible. Differences in diversity among students (gender, culture, physical ability, or academic ability) can also cause problems in small groups. Some students even become socialized to believe they cannot work well in a group. Students can also become socialized to believe that there is a "right answer" for everything and, thus, find it difficult to work in project-based science classrooms where students find solutions among many possibilities.

Finally, parents, colleagues, or administrators may not understand the nature of collaboration and cause problems by questioning a teacher's techniques.

A Lack of Collaborative Skills

As discussed earlier, students need to learn collaborative skills to be able to work effectively in small groups (Johnson, Johnson, & Holubec, 1986). Children do not learn these skills overnight; it can take months or years for students to learn to work well in groups. However, without these skills, students may experience unresolved conflicts, aggression, failure to challenge the viewpoints of others, and failure to make meanings explicit.

When problems arise during collaboration, try two approaches: (a) Carefully examine the types of collaborative skills students seem to be lacking and work on these specific skills, and (b) get students to try to work out their own problems. Use these two approaches before moving problem students to another group or forming new groups. If you are too quick to move students to a new team, you will only reinforce negative behaviors—students will learn that complaining results in getting to be on a team with the people they want to be with.

If students seem to argue excessively, work with them so they learn to summarize ideas in a positive fashion, request help, agree with each other, and support others' ideas. Try to get students to understand that they do not need to have arguments, give negative criticism, or overlook the positive contributions of members of the group. Have students work out their own problems, if possible. You might ask some of the following questions:

- What have you done to try to resolve the argument?
- What could you try next?
- What are some other ways to solve this?

Some teachers find that when students realize that they will work as a team for a length of time and that they must work together to be successful, they will often work out their problems and get along.

Students may still refuse to work together, however. When this happens, you may need to remove a student from the group and let him or her work alone until ready to work with others. Sometimes, isolation, while seeing team members enjoying themselves, may motivate students to want to join the team. A private meeting with an uncooperative student might also reveal reasons for negative behavior. Some teachers ask team members to think of ways to help an uncooperative student participate as a productive member of the team. Similarly, some teachers call class meetings to discuss ways to improve teamwork.

Sound levels can also become a problem in collaborative classrooms—especially if the school is not built with acoustical tiles, carpeting, and more modern ways to absorb sound. Although talking is expected and desired in a project-based science classroom (students, after all, must debate ideas, construct understanding, carry out inquiry, and make artifacts), unproductive noise can be distracting. If the teacher must

CONNECTING TO
NATIONAL SCIENCE
EDUCATION STANDARDS

SCIENCE EDUCATION
TEACHING STANDARDS

Teachers also give groups opportunities to make presentations of their work and to engage with their classmates in explaining, clarifying, and justifying what they have learned (p. 36).

say, "Shhhhh," every few minutes, something is wrong. Either the teacher's expectations for noise level are too strict, or students are too noisy. To determine which is the case, observe the classroom for a moment. Students should not be yelling at each other. Students should be able to talk and hear each other within a group. If you can see that students are struggling to hear one another, reinforce (or establish) ground rules for talking. Turn off the lights, for example, to get the room quiet, and then establish appropriate noise levels.

Finally, students may not use collaborative time effectively. During a project, some talk will be off-task. If much of the conversation focuses on off-task ideas, try to find out why. Is it that the students do not know each other and so are talking about common interests to become acquainted? Are they finished too early and trying to amuse themselves in the remaining time? Are they failing to focus on the driving question? Did they fail to make sure everyone understood the material? It is only through working *with* the groups that you will get to the heart of the problem. You may need to give them time to get to know each other (Kohn, 1996). Perhaps students need more challenging work. You may need to refocus students' attention on the driving question and create new zones of proximal development so they can sustain collaboration. You may need to show students strategies for making sure all members of the group understand the concepts. For example, students may need to create quizzes for each other or practice giving group presentations to see if each member really understands what has been learned.

Loafing

The **loafer effect** happens when one or more group members allow others to do all the work. Constant monitoring of group members' contributions to the group goal will help you determine if loafers exist. If you keep groups small (three to four students), you will notice these free riders more easily. As you move about the classroom, you can momentarily conference with each group to discuss what each student is doing. You can try to engage the loafers by directing them to tasks they can complete, ideas they can contribute, or ways they can enhance the group's work. You can also provide support mechanisms that will ensure that all students contribute to completing a project. For example, students can complete daily journals in which they record their contributions to the project. You can collect daily charts in which students document their day's accomplishments and record each group member's contribution. These types of support focus on individual and group accountability and help eliminate the loafer effect.

The Fear of Being Duped and the Dominating of Others

Sometimes a higher ability member of a group fears being "used" or **duped** by the other members and, thus, doesn't want to collaborate. Sometimes a member of a group uses others to his or her advantage, thereby gaining more than he or she would have alone. By requiring individual accountability and self-assessment, teachers can eliminate either of these effects. For example, a teacher may require each student to complete a daily journal to chronicle individual contributions and learning and to account for the accomplishments of the entire group. In this fashion, students have the opportunity to show the teacher their own work, as well as that of the group.

Status Differential

The **status differential effect** occurs when a high-status member of a group takes control. Higher status can be caused by many things—membership in a clique, socio-economic status, race, age, or academic achievement (Cohen, 1994). You can approach this type of problem from several directions. First, you can continue to stress collaborative skills so all members of the group are contributing equally and learning to accept their individual differences. Second, you can form groups carefully in an effort to avoid status differentials. Third, you can give low-status students prior training on a topic or skill so they can teach it to high-status students. Finally, you can require individual accountability within a group so all members are recognized for their efforts.

For example, a teacher may purposefully avoid placing a group of students who all play on the basketball team with a student who is not on the team. The collegiality formed among members of the basketball team may cause a clique to form that excludes the one student who is not on the team. Similarly, a teacher may teach a lower socioeconomic student who she believes will be shunned by others how to use a new computer program. This student, armed with prior knowledge that the others do not have, can teach the rest of the group how to use the computer program.

> ### CONNECTING TO NATIONAL SCIENCE EDUCATION STANDARDS
>
> ### SCIENCE EDUCATION TEACHING STANDARDS
>
> Teachers monitor the participation of all students, carefully determining, for instance, if all members of a collaborative group are working with materials or if one student is making all the decisions. This monitoring can be particularly important in classes of diverse students, where social issues of status and authority can be a factor (pp. 36–37).

Pacts

When all members of a group make a **pact** to avoid work and contribute the least amount of effort to complete a task, a poor-quality product or inadequate completion of the goal results. A teacher who knows the abilities of the students in his classroom will quickly recognize this type of behavior.

When you spot evidence of a pact, stress high standards. You might demonstrate or model what is expected at the end of the lesson by showing a sample finished product. You might establish minimum competencies so students complete an acceptable level of work. For example, you could require each student to have at least two complete artifacts to document science learning or a certain number of entries in a portfolio. Another strategy involves having students complete **learning contracts** in which they establish standards for completion of their work before they begin it. Some teachers promote peer reviews among collaborative groups. Most elementary and middle grade students are motivated to do their best when they know they are going to be required to share their work and be evaluated by other groups. Some teachers believe that friendly competition among groups helps establish high standards. While the members of a group are collaborating, the groups are competing with each other for recognition of high-quality work. However, Kohn (1996) believed that teams should never be set against each other in competition, because this practice does not build a community atmosphere in a classroom. Sapon-Shevin (1999) suggested that exclusion of certain groups of people and competition are barriers to collaboration.

Diversity

Research findings on collaborative learning show that, over time, students working in groups learn to appreciate each other's differences, including gender, racial, cultural, and physical- or mental-ability differences (Cohen, 1994; Johnson & Johnson, 1999; Manning & Lucking, 1992). With collaborative learning, students tend to accept others different from themselves, make friends with diverse members of their class, and raise the self-esteem of isolated students. However, these same differences in gender, culture, physical ability, and academic ability can cause problems in some group-learning situations. For example, research in education has shown that male students frequently dominate science activities, and female students are left behind (Baker, 1988).

You may need to directly address the differences in gender, culture, race, and physical and academic ability. For example, you can point out situations in which male students dominate group situations, or you can subtly encourage female students to participate. Sometimes, making gender-specific teams can eliminate this problem.

Problems certainly can arise when students see some group members as *different,* but they also can arise when students see some group members as *better.* The student who is interested and gifted in science and mathematics may be labeled a "nerd" in a traditional classroom environment because she is always the one raising her hand, getting called upon by the teacher, giving the right answers, and being praised by the teacher. Other students may come to resent her because her success makes them look bad. In contrast, members of a sports team do not resent a talented athlete on their team, because the talent of the star athlete is shared with the team—everyone becomes a winner when one person does well. By fostering such a team mentality with your students, diversity can come to be celebrated, and students with special abilities can come to be members of the group.

Socially Induced Incompetence

Socially induced incompetence occurs when members of a group ostracize or disparage a student to the point that he or she feels unable to contribute to the group's work. Children and adults are not always aware of how their behavior can affect others, talking in a certain tone of voice intimidates, or making faces or laughing at fellow students can create feelings of inadequacy. You should discuss these problems, model appropriate behavior, and evaluate group work so students are not socialized to believe they are incompetent. For example, a teacher can purposefully display as a representation of good work an artifact created by a student who has low self-esteem.

A Belief in the "Right Answer"

Even as early as the elementary grades, students become socialized to believe that "right answers" exist for problems or exercises. In project-based science classrooms, teachers frequently find that students rely too heavily on the notion of correct answers. You need to work with students so they come to understand that many problems or situations in real life have many possible solutions and answers. Reliance on correct answers can cause students continually to seek you out for approval, information, or reassurance. You may need to "wean" students from you, slowly teaching them that they, themselves, are sources of information, that they can help themselves find answers, and that they have the ability to find resources and information. You need to help them learn that reassurance comes from completing a project as a group, rather than from getting minute-to-minute approval from you.

Learning Activity 7.5 will help you think about what to do if collaboration fails to work well in your classroom.

Lack of Support From Parents, Colleagues, or Administrators

Parents, colleagues, or principals may not understand the benefits of collaborative learning. Colleagues or administrators with differing beliefs about learning may view a collaborative classroom as unstructured, noisy, or time consuming. These people think that teaching is not taking place when the teacher does not lecture, write on the chalkboard, or ask questions of the whole class. Principals, in particular, can become problematic if they are not supportive of the instructional techniques being used in a teacher's classroom. Pressure to conform to traditional classroom methods, such as having students sit in rows or keeping students quiet, can destroy a teacher's efforts to develop and sustain collaboration.

Learning Activity 7.5

TROUBLESHOOTING

Materials Needed:

- Something to write with
- Peers to work with
- A teacher experienced with having students work in collaborative groups

A. Think about each of the following situations. As the teacher of an elementary or middle grade classroom, what would you do in each situation? List your ideas.

- Team members are not working well together and, one student is begging you to let her out of the group.
- Two students monopolize the activities and discussions.
- The groups are too noisy.
- One shy student doesn't want to work with others.
- You have an ESL student (a student for whom English is a second language).
- Parents complain that their children are being cheated by having to work with others. One father says his child is gifted and is being held behind by working with lesser ability students.
- A student is habitually absent and, so, doesn't contribute to her group's work.
- A student in a group complains that he has done all the work.
- A student in a peer tutoring situation complains that the tutor is mean.

B. Now discuss each situation with a group of peers. How do your ideas compare?

C. Discuss these issues with a teacher experienced in using collaborative strategies. How would the teacher handle each situation?

One of the major erroneous beliefs that parents or guardians may have is that higher ability students are harmed in mixed-ability groups or inclusive classrooms. Research does not support this belief (Manning & Lucking, 1992; Sapon-Shevin, 1999). Findings suggest that higher ability students do just as well or better in mixed-ability or heterogeneous groups as they do in homogeneous groups. On the other hand, lower ability students tend to fall further behind when they are placed in same-ability groups. Parents may also believe that students are not learning when they are talking in small groups. Another erroneous belief is that collaborative learning is incompatible with the demands of standardized tests and the kinds of individual projects students must learn to complete in schools. Research has demonstrated that collaborative learning, in fact, increases student achievement, higher order thinking skills, and problem-solving skills (Bowen, 2000; Schulte, 1999; Wenglinsky, 2000). Among elementary aged children, science learning is more effective using collaborative learning environments instead of traditional ones (Marinopoulos & Stavridou, 2002).

Changing people's beliefs can be difficult. A teacher can circumvent problems by explaining to parents, colleagues, and administrators early in the school year what the class is doing and why they are doing it. For example, a teacher might write a letter to parents or to the principal explaining reasons for using collaborative learning. Another strategy is to invite parents, colleagues, or principals into the classroom to engage them in collaborative learning situations. Make them active members of the learning community, not just observers. You can put parents' careers, colleagues' hobbies, and a principal's expertise to work in your learning community. These community members can become sources of information and potential guest speakers. Very often, such a positive experience will convince them of the benefits of collaborative learning. Over time, parents and guardians will see the positive effects of collaborative learning on their son's or daughter's attitudes toward school. Learning Activity 7.6 will help you practice what you might tell a parent, principal, or peer.

Why Collaboration Almost Always Works Better Than Individual Learning

Collaboration is an essential component of project-based science. It almost always works better than individual learning for seven main reasons:

1. Collaborative learning environments create multiple, overlapping zones of proximal development so students help one another. The students also interact with the teacher and members of the community, forming additional zones of proximal development.
2. Collaborative learning has been found to be more effective than other teaching techniques for raising achievement, enhancing problem-solving abilities, and developing understanding.

Learning Activity 7.6

HOW DO I EXPLAIN COLLABORATIVE LEARNING TO COLLEAGUES AND PARENTS?

Materials Needed:

- Pen and paper or a computer

A. Imagine that you are planning to use collaborative learning in your science classroom. Think about how you will explain your teaching strategies so others will understand your goals, expectations, and management strategies.
B. Write a letter to each of the following groups to explain why you are using collaborative learning and how your classroom will function:
   - The principal of the school.
   - Parents or guardians.
   - A substitute teacher (in case you are absent).

3. Collaborative learning helps spread among members of a group the cognitive load that a task or project might demand.
4. Collaborative learning encourages students to become autonomous, self-motivated learners.
5. Collaborative learning tends to lessen anxiety about learning.
6. Girls and minorities—groups traditionally left behind in science classrooms—tend to become essential, active members in a collaborative classroom.
7. In collaborative classrooms, students develop skills necessary for real-life situations.

Lin (2006) also asserted that collaborative learning allows teachers to achieve at least three major instructional objectives listed in the National Science Education Standards (NRC, 1996).

1. It improves students' thinking and helps them construct better understandings by sharing ideas with others, refining ideas, and questioning new ideas.
2. It promotes student involvement and engagement by giving students opportunities to make their thoughts visible to others, as well as consider their own ideas and those of others.
3. It helps students learn important communication and scientific thinking skills by providing social settings where students analyze scientific ideas by interacting with others.

> ## CONNECTING TO NATIONAL SCIENCE EDUCATION STANDARDS
>
> ### SCIENCE EDUCATION TEACHING STANDARDS
>
> Working collaboratively with others not only enhances the understanding of science, it also fosters the practice of many of the skills, attitudes, and values that characterize science (p. 50).

Using Multiple Zones of Proximal Development to Provide Cognitive Support for a Variety of Learners

Trying to learn a new concept or skill without the assistance of others is often very difficult. Have you ever taken a correspondence course or tried to learn something new solely from reading a book or watching a video? Have you ever tried to learn a foreign language from an audiotape? If you have, you probably know that learning is more likely to occur in a social situation, a situation in which you can see a new skill modeled by a teacher before you try it out on your own or you can receive the assistance of someone more knowledgeable.

In collaborative classrooms, as students interact with others, they introduce new strategies for solving problems. Additional, new, or conflicting ideas crop up. Group members create learning scaffolds for one another. For example, one student might model for another how she used a motion detector to gather data about the speed that a skateboard went down a ramp. Modeling is one example of a scaffold. These collaborative interactions foster joint formation of ideas, construction of shared meaning, development of new skills, and the creation of different ways of solving problems. Collaboratively, students advance from a state of uncertainty, frustration, or confusion toward understanding.

In a traditional classroom, students compete with each other instead of helping each other learn. Low-achieving or quiet students frequently try to avoid attention and, so, don't participate or interact with others.

Yolanda, an elementary teacher, has her class answer a driving question about the effects of acid rain on the environment. Groups of students work on different subquestions. One group of students has decided to collect water samples from nearby lakes, rivers, streams, ponds, and puddles. They have measured the pH of each water sample. Now students are contemplating the reasons why the pH in the puddles, created by a fresh rainfall, had a much higher acid level than did the lake, river, stream, and pond. After all, they think, a puddle is fresh, and standing bodies of water should accumulate higher levels of acid over time. Yolanda has the group members write down what they believe are the reasons for this surprising finding. Then she has each student share his or her reason by completing the sentence "My idea is ...." One student says, "My idea is the air is very polluted and the rain picked up additional acidity when it fell." Another says, "My idea is there was something on the asphalt that was acidic—the rain was simply absorbing acid from the pavement it was sitting on." Another says, "My idea is that cars sitting on the asphalt put acid into the puddle with their exhaust." Still another student says, "My idea is that the puddle is the closest of all the water sources to a factory." Each student, with a different prior level of knowledge about acid, provides a zone of proximal development for others in the class. After students thoroughly discuss the possible causes of this phenomenon, Yolanda suggests that they test each of the ideas. She provides suggestions for testing the ideas and of people with whom they can talk. Students call the factory and find out that factory officials monitor the air and ground surface around the factor for pollution. Now they know they can obtain information about the acid level of both the air and the ground. One student contacts a local automobile dealership owner to get information about car exhaust and the effect on global warming in the environment. As the students interact with Yolanda and with other knowledgeable adults in the community, they continue to be exposed to new zones of proximal development. Each zone provides students with links to higher learning, links that would not have occurred if students had been working alone or in traditional classrooms.

Effects on Achievement, Problem Solving, and Understanding

Several research studies have shown that when students work in groups, their academic achievement improves (Bowen, 2000; Manning & Lucking, 1992; Slavin, 1992; Wenglinsky, 2000). Recent research has found that collaborative learning improves problem-solving skills and higher order thinking skills and helps students learn concepts—two important goals of the *National Science Education Standards* (National Research Council, 1996). As students interact with others, disagreements usually arise over ideas, methods to test ideas, and answers. The disagreements and their resolutions help students solve problems. In the process of resolving disagreements, students construct understanding, improve their reasoning skills, and enhance their problem-solving skills. They are also likely to form richer conceptual understandings. Attempts by students to validate their views help bring about cognitive change. Arguments that require students to defend their views help some students become dissatisfied with their own views and abandon them. Further, students are more likely to retain material learned in collaborative groups because ideas and relationships are formed and modified via communication with others. When students express ideas, they must represent ideas in new ways so stronger connections are formed. Students use cognitive elaboration—strategies that enable them to form a more detailed, thorough mental understanding—when they discuss concepts, debate ideas, and explain concepts to others in their group.

In the acid rain project, as students discuss reasons that fresh rain puddles contain more acid rain than other bodies of water, they begin to disagree and argue. Some students insist that the factory caused the acid rain, and others argue that it had something to do with the asphalt. Because the students disagreed, they decided to test out all their ideas. As a result of this exploration, students learn that the factory exceeds all EPA air quality regulations, enabling them to rule out proximity to the factory as a cause.

Spread of Cognitive Load Among Learners

Rarely in today's world is there a science topic that is simple. Most topics are complex and interdisciplinary—they overlap various science subjects, social issues, mathematics, and technology. The study of acid rain, for example, involves many disciplines. Understanding the formation of acid in the environment relies on understanding chemistry. Examining the effects of acid rain on the environment includes areas of biology, geology, meteorology, and physical science. Measuring and monitoring acid rain involves mathematics. Evaluating the sources of pollutants in the air that cause acid rain is a social concern.

Many scientific questions and problems have more than one plausible solution. For example, numerous ways exist to limit acid rain in the environment (limit sulfur-burning coal factories, add scrubbers to factory smokestacks, find ways to neutralize acid, and so on). New knowledge in science and technology is increasing at such a rapid pace that journals and textbooks, unless they are electronic, cannot keep up with it. It is nearly impossible for people to acquire, synthesize, and assimilate all the information available to answer a question that they find important.

Most children cannot on their own answer many science questions that arise. In collaborative learning situations, however, the **cognitive load** is spread among members of the group. Not every student in the class has to answer every question, grapple with every idea, obtain information from all community members, or test all possible hypotheses. The cognitive load of acquiring, synthesizing, and assimilating information is spread among members of the class, who have different learning styles and intelligences. For example, the student with strong interpersonal skills may want to interview business leaders about efforts to control acid rain, whereas the student with visual/spatial skills might build a model to demonstrate the effect of acid rain on the environment. Some students may read journals and books, while others search the World Wide Web for information. A few students can conduct a scientific experiment, while others interview people in various careers.

Promotion of Autonomous, Motivated Learning

Collaborative learning shifts the responsibility for learning from the teacher to the students. Students learn from each other as well as from the teacher and others in the community. A collaborative environment promotes students' taking active roles in their own learning, as well as in the learning of others, determining what is learned, how it is learned, and when it is learned. When students take an active role in their learning, the subject matter content becomes interesting to them, and they are motivated to learn. This type of responsibility helps students become autonomous learners who know how to ask questions and find solutions to questions.

Collaborative learning reinforces the idea that effort, not solely innate ability, begets rewards. For example, in a traditional classroom, if a student fails a test, she may conclude, "I'm just not smart enough to understand molecules!" In a collaborative environment, students are more likely to conclude that, with effort, time, and cooperation, every member of the group can learn to understand basic chemistry. This type of success is motivating.

The collaborative classroom environment is particularly powerful for preadolescents and young adolescents because it matches their developmental needs (Barnes, Shaw, & Spector, 1989). They desire peer interaction, want to be accepted by others, need to be trusted to make their own decisions, and crave meaningful learning. A collaborative environment also develops students who are intrinsically motivated to learn, thereby reducing or eliminating the need to enforce punishments or give artificial rewards.

In the acid rain project, Yolanda explores ways to make her relationship with her students more collaborative. She wants to move away from the traditional role of the teacher as holder of knowledge to one in which she is a coinvestigator with her students. She accomplishes this by becoming a learner along with her students—asking questions and exploring with students without ever being certain of the solutions that will arise. She empowers her students' status by behaving as if they have important knowledge and experiences that are useful in their exploration of the problem. The students respond by selecting resources to help them investigate the problem. They read local newspapers, consult reference books, and talk with individuals who work in related fields. Some students choose to investigate the topic on the World Wide Web. Others set up experiments to measure acid rain levels and the effect on plant life. The students become responsible for their own learning. The classroom is a democratic one—both the students and the teacher share in the decision making. In Learning Activity 7.7, you will explore how the role of the teacher changes in a collaborative classroom.

Learning Activity 7.7

HOW DO TEACHERS' ROLES CHANGE IN A COLLABORATIVE CLASSROOM?

Materials Needed:

- Two teachers

A. With the help of your instructor, a school principal, or a school curriculum director, identify two teachers—one who does not use much group work and one who uses many collaborative techniques in his or her classroom. Obtain permission to observe and interview each of these teachers.

B. Observe each teacher. Analyze the likenesses and differences between the teachers in terms of
   - The academic goals for students.
   - The structure of daily lessons.
   - The style of classroom management.
   - The reactions of students during lessons.
   - The level of interaction among students in the class and between the teacher and students.
   - The methods of evaluation in each class.

C. What do you see as the advantages and disadvantages of collaborative learning? How do teachers' roles change in collaborative classrooms? Record your ideas.

Reduction of Anxiety About Learning

In collaborative classrooms, **anxiety,** or uneasiness and apprehension, about learning science is reduced for several reasons (Sheridan, Byrne, & Quina, 1989). First, students share ideas without fear of being put down or ridiculed. Peers typically will not feel threatened by one another, so they can engage in quality discussions, share thoughts, and debate ideas without anxiety. Second, collaborative work tends to be supportive rather than competitive. Collaborative learning classrooms form caring environments in which students show kindness, fairness, and responsibility. Third, in collaborative classrooms, the teacher and the students become allies in the learning process. The teacher isn't an authority, but, rather, a guide in the learning process. In addition, the teacher is not the sole disseminator of information; students play an important role in asking questions, finding information, and sharing what they have learned. Fourth, the collaborative classroom stresses that there aren't only short, correct answers but possibly many answers, interpretations, or solutions to questions. In this atmosphere, students know that there are often many ways to answer a question or solve a problem and that their ideas are as valid as those of others.

The students learning about acid rain feel very comfortable asking questions. Neither the other students in the class nor their teacher, Yolanda, will make them feel intimidated for asking questions. In fact, students are encouraged to ask as many questions as possible, since it is part of the learning process to answer questions. As students take control of their own learning and become more comfortable in the project-based classroom, their self-esteem increases. Their attitudes toward learning about the effects of acid rain are very positive. Yolanda finds that, over the course of a school year, she is able to get students more involved. All students participate equally, and they don't seem anxious about learning science. Students seem to like school, enjoy science class, and are concerned for others in their group. This caring environment reduces or eliminates the anxiety frequently observed in a traditional, competitive classroom.

Inclusion of Groups Traditionally Left Behind in Science

Research indicates that girls and minorities lag behind others in their pursuit of learning about science, developing science-related interests, selecting science electives in high school, and embarking on careers in science (Kahle, 1985; Pollina, 1995; Rosser, 1990; Seymour, 1995). Collaborative learning has been demonstrated to be a positive teaching strategy for including girls and minorities in science (Baker, 1988; Harwell, 2000; Manning & Lucking, 1992; Mayberry, 1998; Strough & Cheng, 2000). There are several explanations for why collaborative learning helps girls and minorities. First, collaborative learning removes the competitive atmosphere that prevails in some science classrooms in which some girls and minorities feel intimidated. Second, because collaborative learning demands that all students participate equally in investigating a question, students who may not perceive science as a strength do not "fall through the cracks." They must participate in discussions, inquiry, and decisions. Third, collaborative learning develops the social and interpersonal skills of all students, making them better able to work with students from backgrounds different from their own and making them more accepting of others and their opinions or ideas.

American Indians, for example, are underrepresented in science, typically drop out of high school, and often do not attend college. Although differences exist among Native American tribes, Native Americans hold certain core values. For example, Native Americans respect all life and encourage respect for individual people (Soldier, 1989). They teach their children to solve their own problems and look to their elders for wisdom. Many Native American children are part of extended families in which they learn to cooperate, share, and live in harmony with others. Traditional classroom environments that stress competition for grades; individual work, rather than group work; and the authority of the teacher conflict with basic Native American values. Collaborative classrooms, on the contrary, promote cooperation, group work, sharing, and respect for all individuals in a group. Further, collaboration in a project-based science classroom encourages student-directed learning; the teacher is not the authority but, rather, the guide who might be sought out for his or her wisdom—a concept of teacher that is consistent with American Indian culture.

In the acid rain project, a quiet girl named Maria is confident enough to share her ideas with three other classmates, but she tends to fail to participate in whole-class situations. Maria's family, originally from West Virginia where coal mining is a major industry, believes very strongly in the benefits of burning coal to produce energy. As a result of sharing her experiences with classmates about the economy in West Virginia, others in the group learn to accept her and have greater tolerance for her ideas. They begin to explore ways that factories might still use sulfur-burning coal and not damage the environment.

Development of Real-Life Skills

Collaborative learning helps develop the abilities and skills students need in real life. The Labor Secretary's Commission on Achieving Necessary Skills (Brock, 1991) identified five minimum competencies (called the *new basics*) necessary for all adult citizens. These are the abilities to work with others; acquire and use information; identify, organize, and allocate resources; understand complex interrelationships; and work with a variety of technologies. To work with others, for example, a person must be able to participate as a member of a team, contribute to a team's efforts, negotiate with others, resolve differences, and work with people of diverse backgrounds. These types of skills are developed in collaborative learning situations.

As students in the acid rain project investigate the driving question, they develop skills they will use throughout their lives. They need to work effectively with other students in their group. This project requires them to communicate with others, compromise, develop new ideas together, and resolve differences. Although they come from diverse backgrounds and have different opinions, they must learn to work together. To acquire and use information, the groups of students need to read the local newspapers, identify appropriate journals and books to read, and synthesize information from these readings. They figure out how to obtain related information from the World Wide Web. To conduct their plant experiments, they need to identify, organize, and allocate such resources as pH meters, plants, sources of acid rain, and locations for taking samples. To understand the effect of acid rain on the environment, they have to make sense of complicated interrelationships, such as the effects of biological factors on plants, the physical effects of acid rain on buildings and other surfaces, the chemical relationship between sulfur-burning coal and the formation of acid rain, and the geological processes that take place in the atmosphere and hydrosphere. They use scientific equipment and technological tools that include pH meters, microscopes, computers, fax machines, and telephones.

In Learning Activity 7.8 you will learn about available curriculum resources that use collaborative strategies.

Chapter Summary

In this chapter we discussed the nature of collaboration among students, teachers, and members of the community. Because collaborative learning doesn't just happen in a classroom, the chapter focused on ways to create a collaborative *environment*, including forming groups, developing collaborative skills (such as decision-making, task-completion, trust-building, communication, and conflict-management skills), and getting all students equitably involved (through establishing common goals,

Learning Activity 7.8

WHAT CURRICULUM RESOURCES USE COLLABORATIVE STRATEGIES?

Materials Needed:

- A variety of curriculum resources listed in the following activities

A. In recent years, a variety of curriculum materials that use various collaborative strategies have emerged on the market. Obtain several of the following curriculum materials and investigate them. If possible, use the materials with a class of students:
    - *Science for Life and Living: Integrating Science, Technology, and Health.* 1992. BSCS, 830 N. Tejon, Suite 405, Colorado Springs, CO 80903, or Kendall/Hunt Publishing Company, 4050 Westmark Drive, P.O. Box 1840, Dubuque, IA 52004-1840. An elementary science textbook series.
    - *Decisions, Decisions Series.* 1990, 1991, 1993. Tom Snyder Productions, 80 Coolidge Hill Rd., Watertown, MA 02172-2817, 1-800-342-0236. Different simulation programs about colonization, immigration, revolutionary wars, environmental issues, and more.
    - *The Great Solar System Rescue.* 1994. Tom Snyder Productions. Simulation in which students work to find space probes lost in the solar system.
    - *The Great Ocean Rescue.* 1994. Tom Snyder Productions. Simulation in which students work to solve environmental problems about the oceans.
    - *Fizz and Martina.* 1991. Tom Snyder Productions. Cooperative learning activities for mathematics.
    - *The Geometric Supposer.* 1987. Education Development Center, Sunburst Communications, Inc., 39 Washington Ave., Pleasantville, NY 10570, 1-800-431-1934. Cooperative learning activities to teach about geometry.

B. How are these materials different from other curriculum materials you've seen for teaching science to children?
C. How well do the materials utilize the following philosophical framework for project-based science?
    - *Focus on a driving question.* Students investigate meaningful and important questions that orchestrate activities and organize concepts and principles.
    - *Investigations.* Students investigate the driving question by asking and refining questions, making plans, designing experiments, debating ideas, collecting and analyzing information and data, drawing conclusions, and communicating ideas and findings to others.
    - *Generation of products or artifacts.* As a result of performing inquiries, students develop a series of artifacts or products that represent their knowledge in a variety of ways.
    - *Learning communities and collaboration.* Students, teachers, and individuals outside the classroom collaborate to investigate the driving question.
    - *Use of cognitive tools.* Learners represent and share ideas and support their research efforts with cognitive tools, such as computers.

building common understanding, and allocating labor and roles equitably). We discussed ways teachers can overcome challenges that arise during the implementing of collaborative groups. Specifically, we discussed what to do when students lack collaborative skills and how to deal with particular problem situations, such as students' failing to work in groups or not being accepted by peers. We discussed how to work with parents, colleagues, and administrators who may not be familiar with collaborative group work. Finally, we discussed reasons collaboration is almost always better than individual learning, such as the development of zones of proximal development; improvements in achievement, problem solving, and understanding; the spread of cognitive load; the promotion of motivated learning; the reduction of anxiety about learning; the inclusion of groups traditionally left behind in science; and the development of real-life skills.

Chapter Highlights

- Collaborative learning is a key feature of project-based science.
- There are three types of collaborative arrangements:
  - Among students.
  - Between students and teacher.
  - Between students and the community.
- Teachers are responsible for creating a collaborative environment.
- Collaboration must be planned. Planning includes
  - Forming groups.
  - Developing collaborative skills among students in the class.
  - Involving all students equitably.
- Students' collaborative skills need to be developed. These include:
  - Decision making.
  - Task completion.
  - Trust building.
  - Communication.
  - Conflict management.
- Teachers face many challenges when implementing collaborative groups. These include:
  - A lack of collaborative skills.
  - Loafing.
  - The fear of being duped and the dominating of others.
  - Status differential.
  - Pacts.
  - Diversity.
  - Socially induced incompetence.
  - A belief in the "right answer."
  - Lack of support from parents, colleagues, or administrators.
- Collaboration almost always works better than individual learning. There are numerous reasons:
  - Multiple zones of proximal development are created among students that help scaffold learning.
  - It positively affects achievement, problem solving, and understanding.

- The cognitive load is spread among students.
- It promotes autonomous, motivated learning.
- Anxiety about learning is reduced.
- Groups traditionally left behind in science are more likely to be included.
- Real-life skills are developed.

Key Terms

Anxiety	Home groups	Positive interdependence
Base groups	Interpersonal roles	Scaffolding
Cognitive load	Learning contracts	Sharing groups
Cognitive roles	Loafer effect	Social computing
Collaboration	Managerial roles	Socially induced incompetence
Division of labor	Meaning	Status differential effect
Duped	Mutuality	Trust-building skills
Equality	Pact	Zone of proximal development

References

Abruscato, J., & Hassard, J. (1976). *Loving and beyond: Science teaching for the humanistic classroom.* Glenview, IL: Scott, Foresman.

Baker, D. (April 1988). Teaching for gender differences. *Research Matters to the Science Teacher, 30,* 3. National Association for Research in Science Teaching (NARST).

Barnes, M. B., Shaw, T. J., & Spector, B. S. (1989). *How science is learned by adolescents and young adults.* Dubuque, IA: Kendall/Hunt.

Blumenfeld, P. C., Marx, R. W., Soloway, E., & Krajcik, J. S. (1996). Learning with peers: From small group cooperation to collaborative communities. *Educational Researcher, 25*(8), 37–40.

Bowen, C. W. (2000). A quantitative literature review of cooperative learning effects on high school and college chemistry achievement. *Journal of Chemical Education, 77*(1), 116–119.

Brock, W. E. (1991). Continuous training for the high-skilled work force. *Community, Technical, and Junior College Journal, 61*(4), 21–25.

Cohen, E. G. (1994). *Designing groupwork: Strategies for the heterogeneous classroom.* New York: Teachers College Press.

Cohen, E. G. (1998). Making cooperative learning equitable. *Educational Leadership, 56*(1), 18–21.

Damon, W., & Phelps, E. (1989). Critical distinctions among three approaches to peer education. *International Journal of Educational Research, 13,* 9–19.

Edwards, C., & Stout, J. (1990). Cooperative learning: The first year. *Educational Leadership, 47*(4), 38–41.

Harwell, S. H. (2000). In their own voices: Middle level girls' perceptions of teaching and learning science. *Journal of Science Teacher Education, 11*(3), 221–42.

Johnson, D. W., & Johnson, R. T. (1990). Social skills for successful group work. *Educational Leadership, 47*(4), 29–33.

Johnson, D. W., & Johnson, R. T. (1999). *Learning together and alone: Cooperative, competitive, and individualistic learning.* Boston: Allyn and Bacon.

Johnson, D. W., Johnson, R. T., & Holubec, E. J. (1986). *Circles of learning: Cooperation in the classroom.* Edina, MN: Interaction Book Company.

Kahle, J. B. (1985). *Research matters to the science teacher: Encouraging girls in science courses and careers.* Cincinnati, OH: National Association for Research in Science Teaching.

Kahle, J. B., & Rennie, L. J. (1993). Ameliorating gender differences in attitudes about science: A cross-national study. *Journal of Science Education and Technology, 2*(1), 321–334.

Kohn, A. (1996). *Beyond discipline: From compliance to community.* Alexandria, VA: Association for Supervision and Curriculum Development.

Leonard, J., & McElroy, K. (2000). What one middle school teacher learned about cooperative learning. *Journal of Research in Childhood Education, 14*(2), 239–245.

Lin, E. Sumer (2006). Cooperative learning in the science classroom. *The Science Teacher,* 35–39.

Linn, M. C. (2005). WISE design for lifelong learning — Pivotal cases. In P. Gardenfors & P. Johansson (Eds.), *Cognition, education, and communication technology.* Mahwah, NJ: Lawrence Erlbaum.

Mahenthiran, S., & Rouse, P. J. (2000). The impact of group selection on student performance and satisfaction. *International Journal of Educational Management, 14*(6), 255–264.

Marinopoulos, D., & Stavridou, H. (2002). The influence of a collaborative learning environment on primary students' conceptions about acid rain. *Journal of Biological Education, 37*(1), 18–24.

Manning, M. L., & Lucking, R. (1992). The what, why, and how of cooperative learning. In M. K. Pearsall (Ed.), *Relevant research.* Washington, DC: National Science Teachers Association.

Mayberry, M. (1998). Reproductive and resistant pedagogies: The comparative roles of collaborative learning and feminist pedagogy in science education. *Journal of Research in Science Teaching, 35*(4), 443–459.

National Research Council. (1996). *National science education standards.* Washington, DC: National Academy of Science.

Pfister, M. (1996). *Rainbow fish.* New York: North South Books.

Pollina, A. (1995). Gender balance: Lessons from girls in science and mathematics. *Educational Leadership, 53*(1), 30–33.

Ray, J. (2006, Summer). Blogosphere: The educational use of blogs (aka edublogs). *Kappa Delta Pi Record,* 175–177.

Rosser, S. V. (1990). *Female-friendly science.* New York: Pergamon Press.

Sapon-Shevin, M. (1999). *Because we can change the world: A practical guide to building cooperative, inclusive classroom communities.* Boston: Allyn and Bacon.

Schulte, P. (1999). Lessons in cooperative learning. *Science and Children, 36*(7), 44–47.

Seymour, E. (1995). The loss of women from science, mathematics, and engineering undergraduate majors: An explanatory account. *Science Education, 79*(4), 437–473.

Sheridan, J., Bryne, A. C., & Quina, K. (1989). Collaborative learning: Notes from the field. *College Teaching, 37*(2), 49–53.

Slavin, R. E. (1992). Achievement effects of ability grouping in secondary schools: A best evidence synthesis. In M. K. Pearsall (Ed.), *Relevant research.* Washington, DC: National Science Teachers Association.

Soldier, L. L. (1989). Cooperative learning and the Native American student. *Phi Delta Kappan, 71*(2), 161–163.

Strough, J., & Cheng, S. (2000). Dyad gender and friendship differences in shared goals for mutual participation on a collaborative task. *Child Study Journal, 30*(2), 103–126.

Tom Synder Productions. (1994). *The great ocean rescue.* Watertown, MA: Author.

Tom Synder Productions. (1994). *The great solar system rescue.* Watertown, MA: Author.

Tao, P. K., & Gunstone, R. F. (1999). Conceptual change in science through collaborative learning and the computer. *International Journal of Science Education, 21*(1), 39–57.

Vygotsky, L. (1986). *Thought and language.* Cambridge, MA: MIT Press.

Wenglinsky, H. (2000). *How teaching matters: Bringing the classroom back into discussions of teacher quality.* Princeton, NJ: Milken Family Foundation and Educational Testing Service.

Young, E. (2002). *Seven blind mice.* New York: Scholastic.

Chapter 8

Instructional Strategies That Support Inquiry

Introduction

This chapter focuses on various instructional strategies that teachers can use to help students learn science. The topic of instructional strategies raises a number of questions: *What instructional strategies do teachers use in elementary science classrooms? Are some strategies more effective than others in helping students develop meaningful understandings? How can I improve on my instructional techniques?* This chapter will address these and other questions. Since teaching skills, like good questioning, are essential in promoting student thinking, we end the chapter by discussing various techniques that improve questioning and discussions.

Chapter Learning Performances

- *Compare and contrast the reasons for using various types of strategies to teach science.*
- *Explain the value of using transformational instructional strategies to teach science.*
- *Describe several techniques that are used to make presentation of material more meaningful to students.*
- *Justify why graphic organizers, such as concept maps, are useful, and create a concept map for several related science concepts to be taught.*
- *Distinguish among various types of questions.*
- *Justify the use of community resources and field trips in elementary/middle grades science.*
- *Clarify why certain instructional skills, such as wait-time and probing, are important.*

Before beginning our discussion of instructional strategies, consider the following three scenarios on the presentation of scientific ideas. These scenarios will illustrate the importance of various instructional strategies in your teaching.

Scenario 1: Reading About Insects

Mrs. Patterson, your fifth-grade teacher, starts class like this: "Okay, let's open our science textbook and turn to page 28. Today, we are going to read about the variety of insects that live in our world. Karen, would you please read the first paragraph?" At the end of each paragraph, Mrs. Patterson asks a different student to read the next section and writes new vocabulary words on the board. After about 20 minutes of reading and summarizing, Mrs. Patterson says, "Now turn to page 41 and answer the first four questions about insects. Make sure to use complete sentences and check your spelling." You turn to page 41, take out a piece of paper, and start to answer the questions.

You follow most of the reading, but occasionally you drift off and think about playing soccer after school. You are happy that you answered the questions in the amount of time Mrs. Patterson gave you. You aren't sure why you were reading about the insects. Some of the pictures of the insects are pretty cool, and you wonder if you have insects like that in the school yard or in your backyard.

Sound familiar? Chapter 1 referred to this method of science instruction as *reading about science*. Although important new terms were introduced to students with this instructional method, students experienced science as a passive reading activity. The concepts of science were not linked to everyday life, and the teacher did not use strategies to engage students in the reading. Reading is an important component of science learning; but, when reading about science is not made active or tied to a larger picture that has meaning for students, it can become a routine and uninteresting way of learning.

Scenario 2: Using a Microscope

Mr. Morales, your fifth-grade science teacher, starts class by saying, "Today we're going to find out about how to use a microscope." Mr. Morales gives a number of cautions about handling a microscope and then gives a demonstration. For the remainder of the class period and the next day, you and your partner practice using a microscope and look at some "awesome" stuff under the microscope. Mr. Morales wants you to draw what the objects look like when viewed through a microscope. You examine and draw pictures of onion skin, human hair, and a fly wing. You hear "oohs" and "aahs" during class as students look at the various objects under the microscope. You and your partner do not fully understand why you learned to use the microscope, but you remember Mr. Morales mentioning that scientists use it to make discoveries.

In Chapter 1, you learned that this kind of science teaching is called *process science teaching*. Although it was interesting things under the microscope and the activity generated excitement in the classroom, learning to use the microscope wasn't tied to a larger purpose. Mr. Morales used the activity to meet one of the district's curriculum objectives: use scientific tools to make more precise observations. He used a demonstration, followed with practice using the microscope. He had students make drawings of their observations. Although students learned how to make observations using a microscope, the observations were not connected to new ideas or to answering questions. Although at the elementary and middle school level it is important to develop students' skills, skills learned in isolation are often not transferred to new situations. As we discussed in Chapter 4, you don't learn to play tennis by learning a backhand return; the skills must be practiced in context.

Scenario 3: Observing Insects

Your class is exploring the question "When do various insects appear on our playground?" As part of your project activities, you are observing when various insects appear on the playground and in the small wooded area behind the school. As part of your work, you decide to draw the various insects and to write down the date when you first see them. A number of the children in the class ask your teacher, Ms. Cheng, how they could better observe the insects. Ms. Cheng, noticing your activities, decides that it would be beneficial for the class to learn how to make more careful observations with either a microscope or a magnifying glass. She also realizes that a lesson on microscopes and magnifying glasses would help meet several of the district's objectives: using scientific tools to make more precise observations and teaching students about the taxonomy, structure, and function of parts of insects. She decides to plan a lesson on how to use a magnifying glass and a microscope to observe and identify insects and learn about their taxonomy, structure, and function. Ms. Cheng starts the next science class by explaining what she observed about the students and what questions students asked. She then says, "How can we better observe the insects we see on the playground?" After several students' responses, Ms. Cheng summarizes and elaborates on what the students said: "Scientists often use various tools to help them in their work. Tools that can help people better see the insects are magnifying glasses and microscopes." She demonstrates how to use a microscope, shows the class how to make slides to observe various parts of the insects, and gives a number of cautions about how to handle a microscope. She also demonstrates how to use a magnifying glass. For the remainder of the class period and the next day, you and your partner practice using a microscope and look at some "awesome" stuff under the microscope. You look at onion skin, human hair, and a fly wing. You hear "oohs" and "aahs" throughout the class as students look at the various objects under the microscope. Once you know how to use the microscope and understand the cautions about using it, Ms. Cheng allows you and your partner to use your new tool to continue your observations of insects and to learn more about insects' body parts and the structure and function of the parts. In your science journal, you make drawings and keep notes about your observations as they related to the driving question about insects. Near the end of the week, Ms. Cheng takes your class on a field trip to the local university to see a scanning electron microscope that scientists use to conduct research. At the end of the week, your class has a discussion about what you learned to summarize the week's activities. Science is fun; you do some new activities; and you learn how to use a tool to help you answer questions you have about insects.

This third scenario illustrates how a teacher might use a variety of instructional strategies to help students further their work in the project, as well as meet some curriculum objectives. The students learned how to make observations with a microscope, but their learning was connected to new information that would help them answer their questions. The students learned about the taxonomy, structure, and function of insects' body parts through active observation of them. The teacher used a variety of instructional strategies, such as demonstration, inquiry, journals, field trips, and discussions. The lessons exposed the students to new ideas, allowed them to interact with materials, held their interest, and were tied to a driving question to derive meaning.

An Overview of Instructional Strategies

The teachers in the scenarios above used a variety of instructional strategies, or teaching techniques, to help students learn. In Chapter 2, you learned how children construct knowledge, and you discovered that teachers use *receptional* (strategies that transmit information to students) or *transformational* (active strategies that support students' making sense of material) approaches to teaching. Although there are reasons for using both approaches, a constructivist classroom tends to use more transformational approaches that help children develop integrated and meaningful learning. Teachers need to be very purposeful in selecting instructional strategies. Effective teachers carefully select instructional strategies to help students make connections between ideas. This way, students build deep understanding about concepts. One way of thinking about the broad array of instructional strategies that teachers can employ in a classroom is to classify them into direct, indirect, experiential, and independent strategies. Many direct receptional strategies can be supplemented with other instructional strategies to make them more engaging for students.

Direct instructional strategies are teacher-directed and are used primarily to convey information or teach step-by-step science skills. Common direct instructional strategies include such techniques as **lecture** (oral delivery of content or presenting information), **drill and practice** (memorization or repetitive skill attainment), **passive reading** (reading without use of strategies to engage students during the process), and **demonstrations** (showing something). Teacher-directed activities can help students learn important concepts or skills associated with a project. For instance, during a project focused on exploring whether the community has acid rain, a class could explore the concept of acidity by first investigating the acidity of various household materials. Next, students might explore how acids affect different materials. Such activities would need to be teacher-directed since they would involve potentially dangerous chemicals and since students would not necessarily be able to envision how to carry them out on their own.

Direct instructional strategies can also show children how to use various scientific instruments, such as microscopes, graduated cylinders, and rulers. The teacher might lead a classroom activity in which students learn how to use a pH meter by measuring the pH of various household chemicals, such as bleach, ammonia, shampoo, and detergent. The teacher in the third scenario above used direct instruction to demonstrate how to use a microscope, and practice strategies were used to perfect students' skills.

Teachers need to think carefully about the use of direct instructional strategies, because they are commonly passive and receptional. Children's attention spans may be short with passive activities. For a teacher-led instructional strategy to have meaning for the learner, it must be connected to the project's driving question. Students must understand that learning how to use a pH meter, for example, is related to their driving question about acid rain. However, there are ways to integrate direct instructional strategies with other, more transformational approaches to engage children's thinking. For example, a teacher can effectively use questioning strategies throughout a lecture or demonstration to provoke children's thinking and help develop meaningful thinking. A well-crafted demonstration, although teacher directed, can be very educational for students.

Indirect instructional strategies are more student-centered than direct instructional strategies and include such techniques as case studies, debates, reflective discussions, and active reading and writing. Used appropriately, these strategies can be transformational. **Case studies** are short scenarios or stories given to students that typically pose a realistic problem that must be explored. Oftentimes, this strategy is used when a teacher wants to provide a real-life problem that cannot logistically happen in the context of the school. For example, the teacher may set up a scenario where students are provided with a hypothetical environmental disaster situation that may, in real life, be too dangerous for students to engage in. In exploring the case study, students engage in critical thinking, problem solving, and research. **Reflective discussions** where students talk about something they have read, seen, or experienced can develop integrated learning. For example, teachers oftentimes use this strategy along with a more direct technique, such as demonstrations. **Active reading strategies** help students interpret and understand the material they are reading. Oftentimes, elementary teachers use children's literature and elementary-oriented magazines to motivate and interest students. It is important for students to **write** and interpret written material in science classrooms. Additionally, use of **graphic organizers**, such as **concept maps**, can help students develop understandings in science (see http://scienceideas.org).

Experiential strategies are student centered and transformational. These activities form the basis of many techniques used to teach science. They include such strategies as **inquiry investigations**, use of **community resources, field trips, simulations,**

games, **role playing**, and **debates**. Because of the central role that **investigations** play in science, we devote Chapter 4 to this topic. Many new technologies are experiential and provide educational simulations, games, virtual field trips, and investigations. Because of the importance to these technology-based strategies, we devote Chapter 6 to this topic. **Community resources** (such as parks, hospitals, police stations, courts, radio stations, universities, and businesses) and **field trips** help make science meaningful to real life and, therefore, are important strategies to use in science teaching. **Role playing** (playing the part of a particular role) can help students learn important content and understand various points of view. Similarly, **debates** (organized arguments around a topic) enable students to develop organized communication skills and argumentation skills to defend a point of view. This technique is particularly effective for examining controversial topics, such as global warming.

Independent study strategies are designed to help foster self-reliance or individual work. There are many techniques that can be considered independent study strategies, including writing **essays**, keeping a **journal**, keeping a **blog**, writing **reports**, working at **investigation centers**, completing **independent research projects**, as well as many forms of **homework**. Oftentimes these strategies are also used as forms of assessment. Thus, we will address some of these strategies in more detail in Chapter 10 on assessing student understanding. In this chapter, we highlight investigation centers (independent inquiry activities).

Throughout the rest of this chapter, we focus on a few examples of instructional strategies in each of these categories that are particularly effective for teaching science. We also discuss ways to make passive instructional strategies more engaging. By highlighting a few strategies, we draw your attention to techniques you will find useful as an elementary or middle grades teacher. Throughout your career, you will find that there are many more strategies that can be employed to teach science.

Direct Instructional Strategies

Lectures

At times, you will find it necessary to lecture, or present information to students to further the work of a project. We all learn by listening to what others say. However, there is strong evidence that merely presenting facts will not help children develop understandings that they can apply and build upon. Some techniques exist that can be used to present information in powerful ways. Analogies and metaphors, graphs, movies or videos, educational television programs, and diagrams or pictures represent other ways to powerfully present information. Guest speakers often can present information in interesting ways that tie to a project.

Metaphors, Similes, and Analogies

Metaphors, similes, and analogies actively involve students in constructing meaningful knowledge by linking new ideas or concepts to prior knowledge or previous experiences. Educational researchers (Vosniadou & Brewer, 1987) suggested that metaphors, similes, and analogies can help students develop understandings by making connections between students' prior knowledge and the target understandings.

A metaphor is an implicit comparison between two things that helps you understand something about one of them. For example, a teacher might say, "The heart is a pump" or "Don't eat garbage food." The first metaphor helps students picture the heart (an unfamiliar object) as a pump (a familiar object). In the second metaphor, the use of the word garbage compares something with negative connotations (garbage) to unhealthy snack foods.

A simile, a type of metaphor, makes a comparison between two things, using the words like or as. For example, a teacher might say, "An exoskeleton of an insect is like a shield of armor," "The iris in the eye is like a shutter in a camera," "Dots and dashes in Morse code are like DNA in genes," or "The arteries of the body are like the plumbing of a house." Each of these similes compares a familiar object to an unfamiliar object. The similes also help students understand relationships between ideas. For example, understanding that a camera shutter opens and closes helps students understand that the iris in the eye opens and closes.

An analogy, a comparison between seemingly unlike things, points out a similarity between them and thus implies that they might be alike in other ways as well. For example, "The circulatory system is to the body what a transportation system is to a country" helps people envision that the circulatory system transports things for the body. Once a person has begun to think of the body as a country, it is a small leap to think about the nervous system as analogous to a country's information system. The skeletal system might then be compared to the economy of a country, and the immune system in the body can be compared to a country's military. Such analogies help students make connections among all the things they are learning, improve their retention of new information, and encourage them to anticipate new information, which increases motivation.

Diagrams, Graphs, and Pictures

A picture says more than a thousand words. This old adage is perhaps more pertinent in today's media-driven society. Showing students diagrams, graphs, and pictures that represent key ideas helps students understand the ideas and form links between them, helping learners build deeper understandings.

Imagine a class exploring why the land around the school has the form that it does. Water that cut through the school's neighborhood made a meandering stream. A teacher could describe to students what a meandering stream looks like; however, showing them aerial views of meandering streams or perhaps even an aerial view of the particular stream would do much more for understanding than a description. After showing the students the meandering stream, a teacher could have them describe what they saw.

New technologies, such as DVDs, CD-ROMs, and the World Wide Web, are excellent sources of a variety of diagrams, graphs, and pictures. The images available on the Web can bring real meaning into the classroom. A CD-ROM might be the better source for high-quality illustrations of a meandering stream; however, the World Wide Web might offer pictures and diagrams not yet available on CD-ROM. Imagine that a class is completing an activity on "Can life exist on Mars?" Some excellent, current pictures of Mars that are available on the World Wide Web at the NASA site are not available elsewhere.

Movies, Videos, and Educational Television

If a picture says more than a thousand words, then a video says more than a book. Although videos certainly can be overused, when used judiciously, they provide numerous beneficial educational experiences.

Teachers can use videos to set the context of a project, and they can use them to enhance the context. A teacher completing a project on air pollution could use a video to show children the influence of acid rain on the environment. Time-lapsed photography would give students the chance to watch the changes that occur on a rock as a result of acid rain. Showing and discussing a video would help contextualize the project for students. The *Adventures of Jasper Woodbury* series developed by Vanderbilt University (Cognition and Technology Group at Vanderbilt, 1992), and *The Great Ocean Rescue* (1994) and the *Rainforest Researchers* (1996) developed by Tom Snyder productions are excellent examples of video-based material available on compact disc that contextualizes an instructional unit. For more information, see http://peabody.vanderbilt.edu/projects/funded/jasper/Jasperhome.html and http://www.tomsnyder.com/).

Videos also enable children to experience phenomena that would be too dangerous or costly to experience firsthand. Imagine that students are doing a project on weather and exploring the driving question "What will it be like outside tomorrow?" Videos about tornadoes can represent for children the power of this natural phenomenon without requiring their actual involvement. Similarly, videos of earthquakes and volcanoes show children the power of natural phenomena in a safe, but dramatic manner. *National Geographic*'s DVD and video store is a good source for high-quality videos about natural phenomena (http://shopngvideos.com/?utm_source=ngcomglobal&utm_medium=link).

Videos can also be used to raise issues or help pull together a number of issues raised throughout a project. Imagine students are exploring the driving question "Why do we need to protect the environment?" A number of excellent videos could be used in this project. One of the best is the Dr. Seuss video *The Lorax* (Geisel & Geisel, 1971), which raises a number of issues regarding conservation and recycling in a style that both young and old children (and adults) find entertaining. The movie, *An Inconvenient Truth*, addresses the issue of global warming in a PG-rated film that may be appropriate for middle grades students.

Guest Lecturers

Teachers want their students to see them as people who like to learn, who are knowledgeable, and who care about their learning; but it would be unreasonable to expect teachers to be experts on all subjects. Guest lecturers can easily fill in gaps in expertise. Guest lecturers offer informed perspectives, share in-depth information, create excitement, and present ideas in new ways. Guest lecturers also share careers, serve as role models, bring in resources not normally found in schools, and build community ties. Don't be shy when it comes to asking community members to come into the classroom; many of them will respond to the request enthusiastically.

Imagine that students are completing a project exploring the question "How do we care for our classroom pets?" This is a perfect opportunity to invite in a local veterinarian to explain to the class how to care for pets. The veterinarian will further explain some of the concepts discussed already in class and introduce many new ideas that will help the children develop better understandings of nutrition, health needs, and anatomy.

Using Demonstrations

Think of a demonstration as showing something to someone. In a project-based science classroom, the purpose for showing something to students is to prepare them to find solutions to the driving question or to conduct their own investigations. Imagine that a fifth-grade class is exploring a stream that runs next to the school. One of the questions that the students want to explore is how the temperature and clarity of the stream change throughout the school year. Before sending students to the stream, the teacher needs to demonstrate how to use a thermometer to measure the temperature and how to use a Secchi disk to measure water turbidity.

> ### CONNECTING TO NATIONAL SCIENCE EDUCATION STANDARDS
>
> #### SCIENCE CONTENT STANDARDS
>
> In partial inquires, [students] develop abilities and understanding of selected aspects of the inquiry process. Students might, for instance, describe how they would design an investigation, develop explanations based on scientific information and evidence provided through a classroom activity, or recognize and analyze several alternative explanations for a natural phenomenon presented in a teacher-led demonstration (p. 143).

The Purpose for Demonstrations

Demonstrations can be a particularly effective strategy for teaching science. They can be used to teach a difficult concept, model a new skill, show something that would be too difficult or too dangerous for students to do on their own, or encourage children to question and wonder.

A teacher can use a demonstration to illustrate a difficult concept associated with a project. For instance, during a weather project exploring "When are skies blue?" a teacher might demonstrate a number of properties associated with weather, including air pressure. In addition, a teacher might demonstrate a laboratory technique or skill or show how to use an apparatus. For instance, if students wanted to measure pH, a teacher would need to demonstrate how to use a pH probe or pH paper.

Although students need to experience science for themselves, depending on their age and maturity, some activities may be too dangerous for children to do on their own. Sometimes, the teacher needs to perform a demonstration for safety purposes. For instance, in a project on "What chemicals are found in my home?" the teacher might demonstrate how some household substances, like vinegar (acetic acid) and "Drano" (sodium hydroxide), can generate very high temperatures if mixed together. Such a demonstration would illustrate to students how dangerous some chemical reactions can be, demonstrate safety techniques (such as the use of a safety shield) for some investigations, and teach students about the properties of certain chemicals.

A final purpose for demonstrations is to have students begin to question and wonder. In a project about "How safe is our water to drink?" a teacher might demonstrate what happens when various substances, such as cooking oil, car oil, and lemon juice, are added to water and how some substances mix and others don't. Such a demonstration could provoke some students to ask, "How do we know what is dissolved in water?" or "How do we remove materials that sink to the bottom but don't dissolve?"

Presenting Demonstrations to Learners

What techniques should a teacher use to present a demonstration, and how can teachers engage students during demonstrations? Teachers can conduct demonstrations in an interactive manner that stresses inquiry and engages students. In any demonstration, teachers can ask students to make predictions about what will happen and list the predictions on the board. For instance, the teacher can start by asking students what they think will happen when cooking oil is added to water. Next, he can ask students to describe what they see and to compare their observations with their predictions. Again, the teacher can list their observations on the board. Do all students agree? Finally, the teacher can engage students in a discussion to explain their observations. Predicting, observing, and explaining should be central features of all demonstrations because they help students practice important inquiry skills and keep students cognitively engaged.

Besides engaging students in demonstrations, teachers need to make sure that all students in the class can see what occurs during the demonstration. Nothing is worse than sitting in the back of the room and not seeing what is going on. Make sure demonstrations are visible to all by standing in an appropriate place in the room, moving students' desks to help them see, or removing obstructions. You may want to refer to Chapter 11 for more information about classroom arrangements.

Make certain that demonstrations do not become dry, boring information sessions. Students typically find demonstrations more interesting and motivating when the teacher is exciting. This can be your chance to ham it up. Some teachers tell jokes, act silly, or dress up for demonstrations.

Cautions in Giving Demonstrations

Demonstrations can be exciting, as well as informative, but they can also be dangerous. Fire comes out of a test tube, we hear a loud bang, a top blows off a can, a can is crushed when placed in water. Because of potential hazards, a demonstration is sometimes preferred to student activity. However, there are safety precautions that you, too, must follow when leading a demonstration:

- Practice good laboratory technique by always wearing safety goggles. At times, you might need a safety shield. Protect your eyes and the eyes of students.
- Students frequently want to crowd around the teacher during a demonstration. Although this is never good practice, it is especially poor practice during a demonstration that involves safety hazards. Teachers should make sure students are at all times at a safe distance from the demonstration area. Make sure students are not standing in the way of objects that might project toward them.
- Whenever a flame of any type is used in a classroom, whether from a candle or a propane torch, make sure that a fire extinguisher is at hand. Also, before using a flame in a classroom, think if an electrical form of heat would work as well. Remember when handling hot objects to use heat-resistant gloves or hot pads.
- As with any instructional technique, teachers should practice demonstrations before doing them in front of the class. Practice will help you understand what precautions to take and make you aware of what can go wrong. Practice will also help ensure that the demonstration will work when it is performed in front of the class.

Discrepant Events

Discrepant events comprise a special class of demonstrations. A discrepant event consists of a grabber—an event that goes against what students expect and thus sparks an open-ended question to stimulate student thought. The unexpected event captures the attention of the class, and the open-ended question engages students intellectually. Demonstrations featuring discrepant events can be used to contextualize a project, stimulate curiosity and interest, and provide focus on central concepts.

Discrepant events can be used to grab student attention at the start of an investigation or a new project. For instance, in a project with the driving question "How can I stay on a skateboard?" a teacher can start off the project by demonstrating the center of mass of an odd-shaped object (that appears to defy gravity by balancing in an awkward position). A discrepant event can also be used as the central focus of a lesson to help students learn important concepts linked to the project. For instance, in a project on weather forecasting with the driving question "Will we have blue skies?" the teacher might demonstrate the force that air pressure can exert by crushing a pop can with air pressure.

Two excellent sources of discrepant event ideas are *Invitations to Science Inquiry* by Tik Leim (1981) and *Teaching Science to Children: An Inquiry Approach* by Alfred E. Friedl and Trish Yourst Koontz (2004). Learning Activity 8.1 will familiarize you with some discrepant events.

In Learning Activity 8.1, you probably did not expect the result you observed; you probably thought that water would come rushing out of the cup. This discrepant event probably caused some dissonance between what you observed and what you thought should occur. As a result, you are probably curious about the reasons behind what you observed. Your interest is raised and you are more likely to want to learn more. Just like you, after students have seen a discrepant event, they invariably ask,

Learning Activity 8.1

THE INVERTED JAR

Materials Needed:

- A glass jar
- Water
- A note card
- A group of elementary or middle school students to work with

A. Slowly pour water into a jar. Fill the jar to the edge. Place a note card over the mouth of the jar. Now hold the card in place and turn the jar and card over. Ask students to predict what will happen if you let the card go. Let the card go. What happens?
  - *Variation 1:* Stretch two layers of cheesecloth over the mouth of a jar and hold them in place with a rubber band. Pour water into the jar slowly. What do students think will happen if you quickly turn the jar over? Try it. What happens?
  - *Variation 2:* To add some excitement, have a student volunteer to sit in a chair and turn the jar and card over the student's head (do this only after you have practiced a few times).

B. What concepts are covered in this discrepant event that serve as a central focus in a lesson? For what driving question might this discrepant event provide a lesson?
C. Interview the students. What are their questions? What do they want to learn more about after having seen the discrepant event?

"Why?" and they are motivated to find an answer. Students are very likely to remember discrepant events for a long time because they engage them intellectually.

Student Demonstrations

Demonstrations don't always have to be performed by teachers. Students can also perform demonstrations. Several positive results occur when children lead demonstrations. First, when children make a presentation, most other children find it extremely motivating. Presenting a demonstration is a good time for students to "show off" in a productive manner. Second, when students have to present and explain the material to others, they, themselves, develop a better understanding of the concepts involved. Many experienced teachers recognize the value of explaining a concept for the first time and have experienced the "ah ha!" phenomenon when teaching a concept for the first time. In the process of teaching the concept, they come to finally understand it. Third, by making a presentation, students practice presentation skills. Fourth, presenting to a live audience can be a very motivating experience—presenters want to do well so they work hard to prepare. Fifth, by presenting a demonstration, students build ownership of their project. A word of caution is in order: A teacher should never let a student give a demonstration without first reviewing the content and format. This lets the teacher clear the demonstration for safety issues.

An example of a student demonstration that serves as a lesson to help answer the driving question "When do various insects appear on our playground?" is a demonstration of how to use a magnifying glass. Early elementary students can demonstrate how to hold the magnifying glass, focus it on an insect, and clean it without scratching the lens.

> **CONNECTING TO NATIONAL SCIENCE EDUCATION STANDARDS**
>
> **SCIENCE CONTENT STANDARDS**
>
> Students should present their results to students, teachers, and others in a variety of ways, such as orally, in writing, and in other forms—including models, diagrams, and demonstrations (p. 192).

Indirect Instructional Strategies

Case Studies

Case studies are stories or scenarios that teachers present to students to engage them in problem solving, analysis, and discussion. Cases are commonplace in law and medicine, where students examine a scenario based on a real-life event (oftentimes a dilemma or problem) and analyze the story to arrive at a solution. Usually, cases give students enough detail to be able to picture the situation and analyze it from different points of view.

In science classes, cases can be created for many different issues or problems. For example, students are presented with a dilemma, such as disposal of toxic waste. Enough information is given in the case that students are provided with the problem, possible disposal solutions, and the pros and cons of each solution; then they are asked to study the case to make a decision on the best course of action for the situation.

Reflective Discussions

Reflective discussions are a key component of inquiry-based science classrooms. These discussions help students pull together ideas and arrive at shared understanding of a project. Imagine that a third-grade class is completing a project on the behavior of pillbugs. Students have just finished making observations of pillbugs, estimating their size, and looking for behavior patterns. The teacher now wants to synthesize these various observations and determine how many of the children agree to them. A classroom discussion is an appropriate strategy for accomplishing this synthesis.

Value of Discussions

Throughout a project, a teacher may want to hold numerous large-group discussions. These discussions do not have to be long; they can last for as few as 10 minutes and still provide valuable information to both the teacher and the students. First, discussions can provide teachers with feedback regarding how the project is going and let them know if they need to provide any additional lessons. Second, a discussion can help students synthesize ideas. For instance, during a project with the driving question "Is our water safe to drink?" a discussion about the various water quality indices (such as the effect of turbidity on the amount of dissolving oxygen and animal life) will serve not only as a review of the indices, but also as a way to synthesize the ideas.

Discussions offer additional advantages. Students will find discussions engaging because they enable them to share their ideas and findings. Discussion is also an excellent instruction technique for building *student-to-student interaction*. Students can also hear *different viewpoints* during discussions, which can help scaffold the learning of new ideas.

Leading a Discussion

Good discussions don't just happen. Engaging students and helping them to express themselves is a difficult task. Leading a discussion that fosters thinking and student interaction will take much thought on your part. Discussions must be carefully planned out beforehand. Beginning teachers often mistakenly blaze into class and lead a discussion without preparation. Learning how to lead an excellent discussion will come about only with much practice, preparation, and thought.

How do you start the discussion? How do you help students discuss ideas among themselves and not just with you? How can you help students to express their ideas more clearly? One way to plan a discussion is to prepare beforehand a set of guiding questions. Asking open-ended questions is critical. Such questions promote in-depth responses and foster student–student interaction. "What" and "how" questions tend to be effective: "What observations did you make?" "What other observations were made?" "Who doesn't agree with this and why?"

> ## CONNECTING TO NATIONAL SCIENCE EDUCATION STANDARDS
>
> ### SCIENCE TEACHING STANDARDS
>
> Teachers encourage informal discussion and structured science activities so that students are required to explain and justify their understanding, argue from data and defend their conclusions, and critically assess and challenge the scientific explanations of one another (p. 50).

Reading

Reading is a critical aspect of science teaching and learning. Through reading, students can learn valuable background information, find information related to their projects, and learn new ideas associated with the project work. Reading, abstracting information, and evaluating what is read play central roles in project work.

Many teachers new to project-based science think that students don't need to read in a project environment. How wrong they are. It is true that in project-based science students don't read from a single textbook, but they do read from a variety of sources to obtain information related to the project. As they explore their question, they read from textbooks, magazines, trade books, newspapers, and World Wide Web postings. For example, during a project related to air quality, sixth-grade students might search the Web for related articles, read the local paper, and read trade books about pollution. In a project about insects in the neighborhood, first graders might look through insect picture books to identify insects they found.

Children's Literature

Children's literature and trade books can serve as powerful resources in a project-based science environment. With its illustrations and high accessibility, children's literature tends to capture the attention of students in a way that textbooks do not. Children's literature prompts children to *imagine* science concepts, and the writing often relates these concepts to their lives. Biographies written for children are interesting ways to learn about people, such as Thomas Edison. In addition, this literature enables the teacher to match children with appropriate reading levels.

Imagine that a class explores the driving question "Will humans always be on earth?" As part of the project, the teacher needs to teach a lesson on dinosaurs. Dinosaurs are one of those topics that are difficult to teach since dinosaurs are not around for students to observe. However, children's literature provides an excellent resource for teaching about these magnificent creatures. *Dinosaur Dreams* (1990) by Dennis Nolan (New York: Macmillan Publishing) helps children imagine what life would be like if dinosaurs still lived today. *My Visit to the Dinosaurs* (1969) by Aliki Brandenburg (New York: Harper & Row) tells children about characteristics of dinosaurs. *Dinosaur Babies* (1991) by Lucille Recht Penner (New York: Random House) explains how dinosaurs hatched from eggs and how they survived. *Digging up Dinosaurs* (1981) by Aliki Brandenburg (New York: Harper & Row) teaches children how archaeologists find dinosaur fossils and learn about dinosaurs from the fossil remains. *The Day of the Dinosaur* (1987) by Stan and Jan Berenstain (New York: Random House) tells children all about dinosaurs. *How Big Is a Brachiosaurus?* (1986) by Frederic Marvin (New York: Platt & Munk Publishers, a division of Grosset & Dunlap, Inc.) helps children conceptualize the size of a brachiosaurus by relating it to the size of more familiar objects. Give children's literature a try; many teachers find that it is a great resource for the science classroom.

If you are not familiar with children's literature, your first stop should be the children's section of a local library or bookstore. Children's librarians are usually very eager to assist you in locating books on a given topic. The National Science Teachers Association's journals, *Science and Children* and *Science Scope*, publish annual lists of recommended children's literature that can be used to teach science. These journals also often feature reviews of new books. The number of children's stories and trade books available today is enormous, so these reviews are excellent ways to keep up on what is being published. Indiana University resource librarians compiled a listing of children's literature for teaching science (http://www.indiana.edu/~reading/ieo/bibs/childsci.html). Other good sources of ideas for using children's literature in science include the following:

Brainard, A., & Wrubel, D. H. (1993). *Literature-based science activities: An integrated approach.* New York: Scholastic.

Butzow, C. M., & Butzow, J. W. (1989). *Science through children's literature: An integrated approach.* Englewood, CO: Teacher Ideas Press.

Cerbus, D. P. (1991). *Connecting science and literature.* Huntington Beach, CA: Teacher Created Materials.

Gertz, S. E., Portman, D. J., & Sarquis, M. (1996). *Teaching physical science through children's literature.* New York: Learning Triangle Press.

Mayberry, S. C. (1994). *Linking science with literature.* Greensboro, NC: Carson-Dellosa Publishing Company.

Staton, H. N., & McCarthy, T. (1994). *Science and stories: Integrating science and literature—Grades K–3.* Glenview, IL: Goodyear Books.

Staton, H. N., & McCarthy, T. (1994). *Science and stories: Integrating science and literature—Grades 4–6.* Glenview, IL: Goodyear Books.

Magazines and Periodicals for Children

Many magazines and periodicals written especially for children make great resources in a project-based science environment. Teachers can use these resources to broaden the context of a project, present new information, or review ideas explored previously in project work. Just like literature books, children's magazines and periodicals are written to capture children's attention, and they are filled with wonderful illustrations and photographs. Because magazines and periodicals are printed monthly or quarterly, they contain more up-to-date information than do books. Learning Activity 8.2 will familiarize you with some of these resources and help you identify ways to use them in the classroom.

Written Information

Children refer to many different kinds of written information in science classrooms. New technology-based resources, such as those found on the World Wide Web, are a rich source of information for students as they explore their driving question. Resources found on the World Wide Web have several advantages over text-based materials. First, the content is more current. On the Web, students can obtain information that may be just minutes old. For example, weather data are continuously updated on the Web. Second, the Web offers a great deal of primary-source material, or firsthand information. For example, weather data is provided by the U.S. National

Learning Activity 8.2

HOW CAN I USE A CHILDREN'S SCIENCE MAGAZINE?

Materials Needed:

- Pencil and paper or a computer
- One or more of the following children's magazines (visit your local library, if necessary): *Ranger Rick, My Big Back Yard, Science World, The Curious Naturalist, World, Zoo Books*

A. Read through as many magazines as you can. Identify ways you could use these resources in a project-based science environment. What would be the driving question of the project? How could you use the magazine in a lesson?

B. If possible, have an elementary or middle school student look through each of the magazines. Interview the student about his or her opinion of the magazines. What do students like and dislike about them?

Oceanic and Atmospheric Administration (NOAA); the primary source of these data are NOAA's Web site. Scientists, too, use NOAA's databases for this reason. Third, Web content is comprehensive. Typical libraries used by children offer only subsets of popular and scholarly material on a given subject.

The Web expands the range of content enormously, thereby giving students access to an unprecedented range of information sources. For example, although students will find plenty of information about weather in library books, the Web offers real-time weather maps, satellite images of hurricanes, and historical data on weather patterns. Fourth, Web resources are represented in various formats, including digital form, which students can then easily manipulate. Information can also be conveyed through video and sound—imagine dynamic views of the ozone holes and the sounds of a tornado. These new ways of conveying information are particularly helpful for students who are visual–spatial learners. Fifth, students can publish their own work online, sharing it with a wide audience. Many students find this sharing valuable and motivating, because immediate contact with others makes their work more meaningful. Sixth, Web content is readily accessible. Information on the Web is gathered in a single source, with a single point of access (the computer).

Clearly, the World Wide Web offers a wide range of learning opportunities. However, some of the Web's advantages have the potential to become disadvantages. Although advanced learners might find it valuable to read from the same sources that scientists do, such material may confuse young learners. The materials available on the Web are sometimes too comprehensive and dense for a young student to sift through. However, teachers can point students to web sites designed with young learners in mind, such as National Geographic Kids (http://www.nationalgeographic.com/kids/index.html) or Yahooigans, where there is a science link (http://yahooigans.yahoo.com).

Making Sense of Written Material

Many readers, including many adults, find making sense of written text very difficult. Students can use a number of active reading strategies to help them construct meaning from text. One strategy is to ask a variety of questions before, during, and after reading to make the reading process active (Jones & Leahy, 2006; Yopp & Yopp, 2006). Before a student starts to read, he might ask himself, "What do I want to learn from reading this text?" During reading, he might ask, "What are the author's goals and purposes?" After reading, he might ask, "How can I summarize what I just read?" Writing down answers to these questions lets students examine their own learning. Table 8.1 lists a variety of questions that students can ask to make meaning of text.

Another strategy to help reading is to monitor what is read. A learner might ask, "Do I understand what I just learned?" Evaluation is another critical habit that good readers employ consistently. A learner might ask, "Do I agree with arguments made in the text?" Table 8.2 lists questions that learners can ask themselves to help monitor and evaluate the materials they are reading.

Using Graphic Organizers to Make Sense of Learning

Many teachers use graphic organizers to help students make sense of their learning. Graphic organizers (pictorial representations of learning) can help students under-

---

TABLE 8.1   Constructing Meaning From Text

Before Reading
- What are my goals for reading this text? What do I want to learn?
- How can I use the text structure to help me learn? (headers, side-bar questions, key words)
- Which parts of this text will most likely contain relevant information (for my goal)?
- What are my initial ideas about what the text is about?
- How might this text or reading relate to what I already know?

During Reading
- What is the important information?
- Where are the key words?
- Where are the topic sentences?
- What are the writer's goals and purposes?
- What was the last sentence or section about?
- What do I think will happen next?
- What were my initial ideas about the materials? Were they correct? If not, what are some new ideas?

After Reading
- Am I rereading sections that were unclear?
- What important information is in the text?
- How did my ideas work out? Why?
- How can I summarize the material?
- Do I understand what I read?

TABLE 8.2 Monitoring and Evaluating Text Materials

Monitoring

- Am I finding appropriate information for my goal?
- Do I reread sections when I need to?
- Do I agree with what the author said?
- Am I finding definitions to unfamiliar terms?

Evaluating

- Do I agree with arguments made in the text?
- Are the examples clear?
- Is the writing clear?
- Is the content trustworthy? Can I believe the material?
- What are my reactions to the text?

stand their own thinking, processes, concepts, or tasks (Hyerle, 1996; Parks & Black, 1992). Graphic organizers are visual tools that make abstract ideas more concrete by organizing them or illustrating relationships among them. Although there are many kinds of visual tools, we illustrate five examples: a concept map, a compare and contrast diagram, steps designed to understand a task, Venn diagrams, and storyboards.

Concept Maps. Joseph Novak, a professor at Cornell, created the **concept map** as a tool to assess the changes in conceptual learning that were occurring in the science students he was studying over a 12-year span of schooling (Novak & Gowin, 1984). Since then, concepts maps have been be used in a variety of ways to help children learn science. Using concept maps before a project starts elicits students' initial understandings, providing a baseline. During a project, students can use concept maps to track the concepts they are learning and to integrate them with prior understandings. As the project continues, students make new concept maps, helping them form links among concepts. By comparing earlier versions of their concept maps with later versions, students see how their conceptual understandings are developing. Additionally, students can compare their concept maps with those of other students in order to see the connections formed by other students, sparking new connections for themselves. Concept maps developed at the end of a project help students tie together all the concepts explored and serve as a form of assessment (see Chapters 9 and 10 for more assessment information). For additional ideas on how to use concept mapping with young children see Novak and Gowin (1984).

A concept map is an educational tool used to tap into a learner's cognitive understanding and to externalize that understanding. On a typical concept map, each word representing a **concept** is enclosed by an oval, circle, or rectangle. The concepts are connected by **lines** and **linking words** to represent relationships among concepts. Together, the linked relationships form a **network.**

A very simple concept map would consist of just two concepts connected by a linking verb to form a **proposition** or single relationship. For instance, the concept *chemical reactions* could be linked with the concept *products* by the verb *form* to

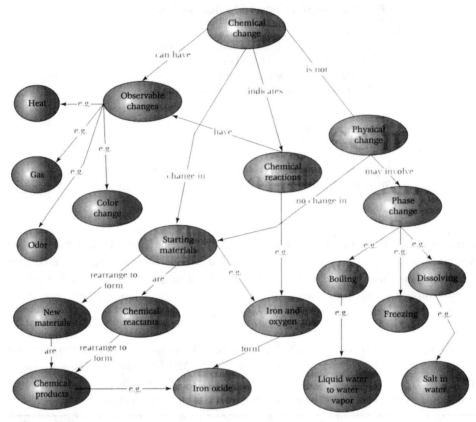

Figure 8.1   Concept map of chemical change.

make the proposition *chemical reactions form products*. However, most concept maps are much more complex. Figure 8.1 shows an example of a concept map drawn with a computer program. Although there are some advantages to using software programs to draw concept maps, teachers and students can quite easily make concept maps with paper and pencil, index cards and tape, or even poster board and yarn.

In a concept map, the relationships among concepts are **hierarchical**. The more general (**superordinate**) concepts—*chemical change, chemical reaction,* and *physical change*—are located toward the top of the hierarchy, and the more specific (**subordinate**) concepts—iron oxide and salt in water—are located below, to reflect the degree of their generality. Linking words connect the various concepts. Cross-links show the **interrelationships** among the concepts included on the map.

The mental images you see when you think about *dog* are your concepts of *dog*. You might picture a golden retriever, a French poodle, or a mutt. You probably picture other associated concepts as well, such as a dog snuggling up to you on a chair or a dog barking and chasing you down the road. All affiliated images form your conceptual map (framework) of *dog*. When you hear the terms *oxidation* and *stomata,* the images that come to your mind probably are not as rich. In science teaching, we want to create understandings that are as rich and detailed as those that occur when you think of *dog*. One goal of science teaching is to develop concepts that have numerous associated concepts and that can grow and change as students learn more information.

A good way to start developing a concept map is to brainstorm all the concepts related to a topic of the project. You can leaf through books to help you identify concepts. Once you have developed your list of concepts, identify which concepts are superordinate (the most general) and which are subordinate (the most specific). Identify the most superordinate concept and link it to subordinate concepts. Remember that, on a typical concept map, each concept word is enclosed by an oval, circle, or rectangle and that the concepts are connected by lines and linking words. If you are like most learners, you will find it difficult to determine the exact hierarchy on your first attempt. When first developing a concept map, you might write your ideas on sticky notes, small index cards, or scraps of cardboard so that you can make changes easily. Place one concept on each index card, sticky note, or piece of cardboard. The sticky notes, index cards, or pieces of cardboard can easily be moved around. You also can easily use computer programs such as Inspiration (Helfgott & Westhaver, 2000), Kidspiration (Inspiration Software, 2000), or Pico Map, which is a Palm application program downloadable for free from the hi-ce Web site (http://hi-ce.org) to make concept maps and modify them.

Linking words are used with concept words to construct sentences with meaning. Table 8.3 shows a number of different types of linking words. The first row in Table 8.3 shows general linking words. The second row shows linking words that signal an illustration. The third row contains linking words that show a relationship among concepts. The fourth row shows process linking words, which are used to show how one concept affects another concept.

In conclusion, concept maps provide visual external representations of the relationships among concepts. Concept mapping empowers learners by making them aware of their own thinking. By connecting or linking new ideas and experiences with existing ones, concept maps help structure information into long-term memory, developing meaningful understanding (Eggen & Kauchak, 1992). Each link in the network increases the meaningfulness of the concept, because it represents a connection to another related topic. Through this networking, the learner develops a working schema. Once information is stored, mapping aids in the retrieval of the information from long-term memory and facilitates through a greater number of associations the transfer of a new idea to another setting. Through mapping, learners can take charge of their learning—in essence, they can learn how to learn.

TABLE 8.3 Linking Worlds

Types of linking words	Examples
Common linking words	Are, where, the, is then, with, such as, as in the, by the, has
Illustration linking words	Is for example, is needed by, is made of, can be, is in a, comes from, is in, determines, depends on, is the same as
Relational linking words	Is bigger than, is faster than, contains, live in, is part of, leads to, helps, divides into, is based on, is done to, occurs when, is essential for, involves, depends on, describes, is a kind of
Process linking words	Causes an increase, produces, consumes, changes, uses, results in, aids in, employs, is formed from, comes from, goes into, washes away

Issues to Consider When Using Concept Maps

You need to consider a number of issues when introducing students to concept maps:

- Most learners, including many adults, find making concept maps a very difficult cognitive activity. Don't be surprised to find that students dislike making concept maps at first. One way to help students learn about concept maps and see their value is to have them start by creating concept maps of familiar things—the grocery store, movies, songs, or sports. The maps they create about what is familiar will flow easily and show students how rich their ideas are. This activity also isolates learning about concept maps from learning about a particular concept, focusing students' attention and energy on one learning task at a time. If students don't know how to create a concept map and they are trying to map new ideas related to decomposition, for example, they will struggle with both making the map and understanding the links among the new concepts.
- Students don't always recognize all of the concepts that can go in a concept map. You need to help students identify additional concepts. One way to do this is to have students brainstorm concepts and ideas. For example, for a concept map of *home, you* might have students first brainstorm rooms in a home, such as bedroom, bathroom, garage, kitchen, den, and basement. Then, they might brainstorm as many things as they can that might go in a garage. By brainstorming as a class, students usually exhaust most concepts that should be included on a map. Some teachers keep a running list of concepts and ideas that they have taught. Students can be encouraged to add to the list when they find new concepts. This makes them aware on a daily basis of the new concepts they are learning, and it provides them with a prompt of ideas for mapping. Finally, some teachers help students identify concepts that should be included in a map by having them refer to field trips, guest speakers, textbooks, books, activities, the Web, and other project materials and experiences. The act of remembering all that has been experienced during the course of the project often will help students remember additional concepts.
- Students will show reluctance to creating concept maps with a hierarchy. You need to encourage students to search for inclusive concepts and order less-inclusive concepts under more-inclusive ones. A useful strategy to help students learn how to develop hierarchy is to pick a familiar concept, like a home, and elaborate on all the subconcepts associated with it. First, have students identify rooms in a home. Second, have students list what is found in these various rooms. Students will be able to understand that the items in a room (such as tissue paper, soap, shampoo, and toothpaste) are subordinate to the rooms in a home (such as a bathroom). Record each concept on sticky notes, index cards, or scraps of cardboard so that students can easily move the concepts around. Finally, have students share their ideas with partners to generate a hierarchy. By working with others, students usually see new and more detailed ways to link ideas.
- Students sometimes fail to use linking words to connect concepts in a concept map. Linking words are critical in clearly communicating hierarchical relationships. Stress to students the importance of selecting linking words. At first, you will need to help students choose good linking words. You might provide students with potential linking words to use. You can also pair students and have them discuss the relationships among the concepts. As one student tells the other about the relation-

ships among the concepts, the other writes down the linking words on the lines connecting the ideas. A conversation between students in a pair might sound like this:

"I connected the rooms, like bathroom, garage, Mom and Dad's bedroom, my sister's bedroom, kitchen, and family room, all to the word house because they are all in the house."
"So these rooms are all in a house?"
"Yeah."
"So, we could say the link is "contains a"—the house contains a bedroom and a bathroom."
"Yeah, that's right. Write down "contains a" on each of those lines."

- Students frequently don't use arrows on their links. You need to encourage them to do so, however, because arrows show the direction of relationships; and it is crucial for students to indicate the direction in the relationship. Pairing students, as described, can help students use arrows. Another way to stress the value of arrows is by asking questions of students as they work on their maps. For example, you could ask, "So, are you saying that a bathroom is made of toothpaste?" Children will laugh and say, "No. Toothpaste is found in a bathroom!" You might then say, "Oh, so what direction should the arrow go so that I don't get confused about that?"
- Students are unlikely to produce good maps on the first attempt. To encourage better maps, have students revisit and redraw their maps whenever possible throughout a project. Students very often will be surprised to compare their initial maps to their final project maps. Filing maps in portfolios is an excellent way to track how students' understandings change throughout a project. Show that you value the process of revision by giving students time to work on their maps. Show that you value the revisions themselves by having students compare, in class discussions, how their maps have changed or by acknowledging improvement on maps (either with verbal praise, points, or grades).

Compare and Contrast Diagrams. Compare and contrast diagrams (Parks & Black, 1992) are visual tools that illustrate the relationships among the characteristics of two different objects. Learning the classifications of various animals, for example, requires students to understand similarities or differences among animals. Figure 8.2 is a sample compare and contrast diagram that depicts the similarities and differences between mollusks and arthropods.

Steps. Steps provide a simple pictorial representation of a task, helping students to visualize it. Figure 8.3 is a simple pictorial representation of the steps to focus a microscope.

Venn Diagrams. Venn diagrams, oftentimes used in mathematics, can be used to illustrate relationships between different groups of things. Overlapping circles are used to illustrate characteristics in common, whereas unrelated characteristics fall within separate circles and outside of the overlap. For example, the Venn diagram pictured in Figure 8.4 could be used to show children how plants and animals are alike and dissimilar.

Storyboards. Storyboards are a sequence of pictures, usually with illustrations drawn within boxes, which tell a story. Comic strips are probably the most common type of storyboards. This type of graphic organizer is very useful in science, because it can be used to summarize an investigation, explain a sequence of events (such as

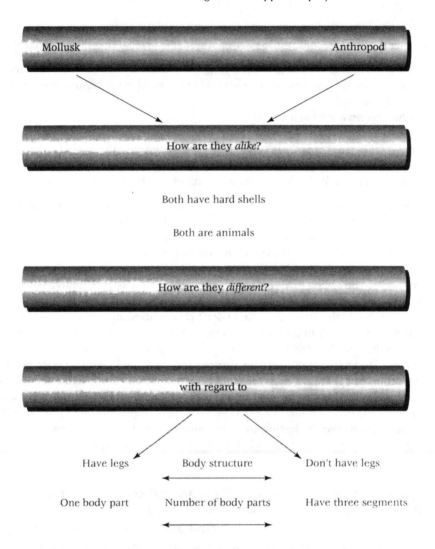

Figure 8.2   Compare and contrast diagram to represent a concept.

stages of metamorphosis, plant growth, or water cycle). The technique is useful with young children and older students, because it allows for use of pictures, words, or both. The storyboard shown in Figure 8.5 illustrates the drawings students might make when studying frog metamorphosis.

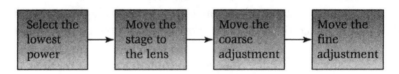

Figure 8.3   Visual tool to represent steps in a task.

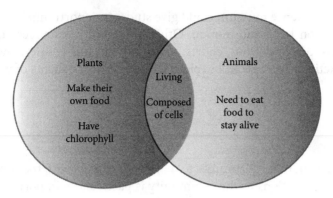

Figure 8.4 Venn diagram.

Experiential Instructional Strategies

Community Resources

Teachers also have opportunities during a project to make use of community resources. Research has shown that connecting learning to students' lives is critically important for children living in urban environments and children of poverty (Haberman, 1995). Community resources provide valuable sites for field trips and sources of educational materials and experiences. During a project exploring the driving question "Will we have blue skies?" a class could take a trip to a local weather station. The local meteorologist could explain how to predict the weather using a variety of different scientific instruments. Although the meteorologist might explain some concepts the class has already explored, he or she would also introduce new ideas, such as the influence of weather on crop growth and local economies, that will help the students understand their topic in greater detail. The review will also help the students develop further understandings of how weather affects daily life. Students will also be able to see how new technologies and new visualization tools are used to predict the weather.

Science museums provide students with hands-on displays, interactive experiences, and special programs. Many science museums have wonderful interactive exhibits of the principles of motion, such as large, swinging pendulums. During a project related to motion that addresses the question "Why do I need to wear seat

Figure 8.5 Storyboards.

belts?" a trip to a science museum could give students opportunities to interact with various exhibits on force and motion. Other community resources are local zoos, parks, hospitals, police stations, courts, radio stations, universities, and businesses. In Learning Activity 8.3, you will explore the educational benefits of various educational resources.

Field Trips

During a project, many opportunities to make connections with the community will arise. Making connections to the community is particularly important for children

---

Learning Activity 8.3

WHAT EDUCATIONAL VALUE DO COMMUNITY RESOURCES HAVE?

Materials Needed:

- Pencil and paper or a computer

A. On a sheet of paper or on a computer, make two columns—one marked *Community Resource* and the other marked *Scientific Educational Value*. For each of the following community resources, identify a scientific educational value. For example, a radio station can teach students about sound production and broadcast and expose them to a science-related career.

Community Scientific	Resource Educational Value
Radio station	• Teach students about careers • Teach students about sound production and broadcast

Zoo
Park
Hospital
Police station
Fire station
Library
Water treatment plant
Sewage treatment plant
Court
Universities
Retail businesses:
- Photocopy store
- Grocery store
- Pet store
- Bookstore
- Welding business
- Furnace and air-conditioning business
- Restaurant
- Gas station
- Auto dealership
- Mechanics shop

B. Select one community resource and an elementary or middle school in your geographic area. Pretend you are teaching at the school. How would you answer the following questions?
 1. What is educational about a field trip to this location? Is there a clear purpose for going there?
 2. What travel arrangements could you make?
 3. Who would handle the arrangements?
 4. Is there any legal liability? Do you need parents' permission? Does the community resource have any particular rules concerning student age, safety, or behavior?
 5. What is the cost?
 6. How long would the trip take?
 7. What is the area of expertise of the people at the community resource?
 8. Do these people know how to handle and talk to children?
 9. What arrangements will you make to visit the location yourself in advance so that you know what the students will encounter?
 10. What is the phone number of the resource person to contact?
 11. How much lead time is required to arrange a trip to this resource?
 12. How will you keep track of the students when you are there? Will you need parent assistance?

C. If possible, contact this community resource and find out the answers to the questions you were unable to answer on your own. What have you learned about community resources?

living in urban and poor communities, because it helps teachers build on students' interests (Haberman, 1995). One project-related opportunity for investigating the question "What birds live near school?" is a visit to a zoo to learn about birds that live in different environments. Students might take a field trip to a local nature preserve or wildlife refuge to see what other types of birds live there. Field trips, like other strategies for presenting information, should be related to project activities. Such related activities help students build a deeper understanding of the concepts and principles being explored in the project.

Field trips in the community do not need to be elaborate. Often the best trips are those around the school to streams, ponds, fields, woods, or playgrounds. Have students mark off a 1-meter by 1-meter area of the playground and record everything they see and hear in the area during a 5- or 10-minute period. They will learn a great deal from this simple field trip about the animals and plants that live in the area, and they will learn about people from the litter they find.

Make sure children know the purpose of the field trip. To prepare students for a trip to a community resource, use the KWL strategy (Ogle, 1986). Ask students what they know (K) about the location you are visiting. List their ideas on the board or have students record them. Next, have students tell what they want to learn (W). Again, record these questions or ideas. After the trip, summarize what was learned (L). Record these ideas, too. Another technique is to give students a list of questions they need to answer during the trip or on the bus home from the trip. Providing students the questions ahead of time gives them an advanced organizer for focusing their attention during the trip.

Games

Games can be very educational and motivating for students (Marek & Howell, 2006). Games include board games, card games, and variations of popular outdoor games, such as tag. Some good examples of games can be found in *Project WILD* (CEE, 1999). For example, in one game entitled Quick Frozen Critters, students play freeze tag to lean about predator-prey relationships. In another game called Oh, Deer! students take on roles of deer, habitat, food, and water to explore factors in the environment (such as shortage of food for a large deer population) that limit populations. Simple card games can be made to teach science concepts. For example, a game of matching allows students to find playing cards that match some criterion, such as parasitic relationships (the parasite and host are matching cards). Games modified from popular television shows, such as Jeopardy, can be used to review science concepts.

Technology has made the use of games very popular in schools since there are a huge number of educational online games and software programs. Commercial Suppliers, such as Delta Education (http://www.delta-education.com), sell games that teach about the human body, geology, and insects.

Role Playing

Role playing can raise excitement, as well as help students develop deep understanding of the issues, concepts, and principles of a project. Through role playing, students may explore subject matter and their attitudes and values, which are brought out in these simulated situations. Role playing can also help students develop greater comfort in expressing their feelings. Role playing allows learners to put themselves in the roles of experts—scientists, doctors, local politicians, or teachers. Role playing can also help children imagine themselves in situations that they can't actually experience. For example, students cannot see digestion as it occurs, but they can role play the human digestive system digesting a hamburger.

Placing children in these role-playing situations, allows them to "act out" or represent their understanding. Howard Gardner has written a number of books about **multiple intelligences** (1983, 1993, 1999) in which he asserts that people have many types of intelligences (verbal/linguistic, logical/mathematical, visual/spatial, bodily/kinesthetic, musical/rhythmic, interpersonal, intrapersonal, and naturalist). Role-playing is a powerful method of reaching students with strong visual/spatial and bodily/kinesthetic skills, enabling them to learn in a way that is consistent with their forms of intelligence. Because children act out their understandings during role playing, it can be considered another form of embedded assessment. Chapters 8 and 9 address this issue in greater detail, and Chapter 12 provides an in-depth discussion of Gardner's theory of multiple intelligences.

Imagine a class is exploring the driving question "Why do I need to wear a seat belt?" Students could engage in a role-playing activity about revoking the law that mandates the wearing of seat belts. During a mock court trial, one or several students could take on the role of physicians' explaining why seat belts can *prevent* injuries. Others might role play scientists' explaining how seat belts can *cause* injuries. Students can also play the parts of lawyers, judges, community members, and insurance agents.

Debates

Debates are organized arguments around a topic. They help students develop communication and argumentation skills. The word *argumentation* probably conjures in your mind a negative picture, such as two people fighting. However, in science, argumentation is actually a useful skill for defending a point of view (of course, as long as it is conducted in a civil manner!). Debates are particularly effective for examining controversial topics, such as global warming. Debates can be set up between individuals (as one would see on television between two people running for a public office) or between groups of students.

In science classes, debates can be structured around almost any controversial issue. As a teacher, you will need to decide the appropriateness of controversial topics in your school or district with the age group that you teach. Such subjects as evolution, stem cell research, mandatory no-smoking policies, or global warming make for good debates where students must analyze and defend various points of view. A recent example in Ohio is a state voting ballot with two proposed nonsmoking laws. One proposal would ban smoking in all public places, and the other would ban smoking in most public places (exempting some restaurants, bars, and entertainment places, like bowling alleys). In a debate, students could study the pros and cons of each item on the voting ballot, research cancer and smoking, and defend how they would vote on the two issues.

Independent Instructional Strategies

Writing

Writing is an important process of learning science. In science classes, students engage in a variety of writing activities, including journaling, writing essays, writing lab reports, and keeping a blog. Writing is an important part of communicating and demonstrating scientific understandings. A recent publication by the National Science Teachers Association (Tierney & Dorroh, 2004) focused on writing to learn science. Many online sites provide tools that teachers and students can use to become better writers. For example, Online Writing Assistant (http://www.powa.org) helps with a number of writing strategies. LabWrite (http://www.ncsu.edu/labwrite/index_labwrite.htm) is an interactive site to help students write laboratory reports.

Investigation Centers

Investigation centers, self-instructional science learning activities, help small groups of students learn particular science concepts, principles, or laboratory skills associated with a project. Investigation centers typically provide a set of detailed instructions for students to follow on their own, without need for further direction and science materials needed to complete the activity. An investigation center need not be elaborate—it can be as simple as a shoe box containing the materials, instructions, and maybe a list of understanding performances, a glossary of scientific terms, and a posttest for assessing learning. Usually, the center requires the learner to submit a product (such as observations, a posttest, or a self-evaluation checklist) to the teacher.

Imagine a class investigating how many birds visit the classroom bird feeder. One possible instructional objective is for students to notice subtle differences among birds so they can identify them. The teacher might set up an investigation center where students examine subtle differences among several everyday objects, such as sea shells, buttons, nuts, bolts, and screws. By investigating differences (such as size, shape, color, and luster), young students improve their observation skills (Bradekamp, 1987). After this activity, the teacher might set up an investigation center near the window with the bird feeder where students use a bird guide to identify the types of birds coming to the feeder. After completing the independent activities, students might turn in observation sheets, describing the different types of birds they saw and a checklist of birds they saw at the feeder.

Instructional Skills

In this section, we focus on skills needed to ask good questions and carry out effective discussions. Because these skills transcend most instructional strategies, we distinguish them from the classifications of instructional strategies described earlier in the chapter. For example, teachers need to ask good questions while giving a demonstration, during investigations, on field trips, or to engage students in active reading. Thus, regardless of the instructional strategy, the skill of asking good questions is important to an effective lesson.

Using Questions

Questioning is a central aspect of science teaching. Why is this so? Children need to manipulate real materials and experience phenomena, but manipulating materials and experiencing phenomena is not an end in itself. Hands-on science is a means to an end. Project-based science seeks to cognitively engage students in hands-on activities. By using probing and open-ended questioning, teachers can cognitively engage children.

Science teachers seek out and use students' questions and ideas to guide lessons and projects. They encourage students to discuss, elaborate upon their responses, and challenge others' ideas. Questioning techniques help the teacher identify prior conceptions, understandings, and possible misconceptions. These techniques also help students create, refine, and elaborate upon their understandings.

Types of Questions

Over the last few decades, many researchers have examined questioning strategies used in classrooms. Researchers have found that teachers ask many questions. Unfortunately, most questions are too low level and factually oriented (Gall, 1984; Wilen, 1987). This means that teachers ask questions that only require students to *tell about, define, recite, list,* or *identify.* Few questions ask students to *analyze, differentiate, contrast, imagine, create, prove,* or *evaluate.* In other words, most teachers are simply stressing rote memorization of factual material. In a project-based science classroom, higher level questions need to be stressed.

The types of questions teachers ask will greatly affect the quality of the answers they receive from students and the character of classroom discourse. Educators classify questions as **divergent** and **convergent** or **higher order** and **lower order**. Here we will put questions into three general classifications: higher level, medium level, and lower level. Higher level questions require complicated thinking, such as evaluation, synthesis, or application. Medium level questions require understanding of concepts. Lower level questions require recall and can often be answered without understanding of ideas.

Consider this question: "What is the best way to clean up an oil spill?" The question has more than one answer, requiring students to think about multiple possibilities and evaluate the best one. The question takes more intellectual thought than just, for example, recalling an explanation given in a textbook. Now consider this question: "What is a vertebrate?" This question is a lower level question that narrows the thinking of the learner to one scientifically acceptable answer. A student could also simply memorize a definition for the word *vertebrate* without having a true understanding of the concept.

Teachers want to plan the major questions they will ask in a lesson. Although a lesson should always be flexible—lessons would probably resemble drills if teachers did not allow for deviation from planned questions—few effective teachers let all questions surface serendipitously during a lesson. They usually plan in advance a list of higher level questions to cover during the lesson.

Table 8.4 presents a classification scheme for lower level, medium level, and higher level questions. You will notice that this table uses the same classification scheme that we presented earlier when we discussed learning performances. These clue words in Table 8.4 will help you plan lessons, form questions, and facilitate higher level

---

CONNECTING TO NATIONAL SCIENCE EDUCATION STANDARDS

SCIENCE TEACHING STANDARDS

In the science classroom envisioned by the *Standards,* effective teachers continually create opportunities that challenge students and promote inquiry by asking questions (p. 33).

TABLE 8.4   Question Classification Scheme

Level of question	Clue words
Lower level questions	
Remember (recognizing, identifying, recalling, and retrieving)	Give me the definition of... Locate the word for... List examples of... Name as many...
Medium level questions	
Understand (interpreting, clarifying, translating, exemplifying, classifying, summarizing, generalizing, inferring, concluding, comparing, contrasting, explaining)	Describe what happens when... Retell what happened when... Explain what... What are some...
Higher level questions	
Apply (executing, carrying out, implementing, using)	Make use of the information... Implement an experiment to show... Carry out a procedure to show how you could... Apply what you learned to...
Analyze (differentiating, distinguishing, organizing, attributing, deconstructing)	Differentiate between... Break down details in this situation to... Compare and contrast the... Give an alternative way to...
Evaluate (checking, monitoring, critiquing, judging)	Decide if you would... Critique the design of... Judge the circumstances in... Evaluate the situation and...
Create (generating, planning, producing, designing, constructing)	Invent an alternative way to... Design an experiment to... Produce a way to... Construct an argument to show...

Source: Modified from L. W. Anderson, and D. R. Krathwohl (2001).

thinking. Although at times you will want to ask lower level questions, the most effective questioning you do will be questioning that sparks higher order thinking.

In Learning Activity 8.4, you will determine the levels of questions and practice writing your own higher-level questions.

Teachers should be very concerned with developing their questioning skills. Many of us went through school experiencing the bombardment of low-level questions, focusing on giving "correct responses" to questions in a designated textbook or a teacher's lecture. This recitation method of questioning was short and fast paced. It went something like this:

Teacher: "What is a molecule?"
Student: "A building block."
Teacher: "A building block of what?"
Student: "Of everything."

Learning Activity 8.4

IDENTIFYING LEVELS OF QUESTIONS

Materials Needed:

- Pencil and paper
- An interesting newspaper article

A. Using the information you have learned about questioning, as well as Table 8.4, identify each of the following questions as either lower level, medium level, or higher level.
   1. What are the causes of animal extinction?
   2. Explain what happened in your investigation.
   3. What is a molecule?
   4. Why would or wouldn't you want a low-level toxic waste site in your town?
   5. What causes oil to float on top of the water?
   6. What might life be like on Mars?
   7. How are frogs different from toads?
   8. How effective is your state at handling solid-waste problems?
   9. Do you think that the spotted owl should be protected? Why or why not?
   10. What is evaporation?
   11. What causes evaporation?
   12. How might you stop water from evaporating from a swimming pool?
   13. How might you increase water evaporation from a wet towel?
   14. What are the characteristics of a reptile?
   15. Who invented the light bulb?
   16. How is sedimentary rock formed?
   17. What is a fossil?

B. After you have identified each of these questions as higher level, medium level, or lower level, meet with a small number of classmates to discuss your answers. Come to a consensus on the level of each of the questions.

C. Form teams. In your team, locate an interesting article in the local newspaper that covers an important current event that might be discussed in an elementary or middle school classroom. Try to find a different article from the other teams. Write three lower level questions, three medium level questions, and three higher level questions for your article. After you have accomplished this task, pair up teams and trade articles. Then ask your paired team your three higher level or medium level questions and take notes about their responses. Do not ask them the lower level questions.

D. Analyze the answers to the higher level questions. What did you notice? What types of thinking occurred? Did team members demonstrate knowledge of the lower level questions when they answered the higher level questions or, even, the medium level questions? If so, what does this tell you about the lower level questions?

Teacher: "Good. What are all molecules made of?"
Student: "Atoms."
Teacher: "Right. What are atoms made of?"
Student: "Protons, neutrons, and electrons."
Teacher: "Right."

This type of recitation leads students to believe that science questions always have one "correct answer." Such questioning does not demand much thinking, and it does not lead to discussion. Because of teachers' prior experiences and beliefs about teaching, multiple answers, uncertain responses, tentative answers, opinions and judgments, and divergent answers may seem frightening. Because many teachers have limited science preparation, they fear open-ended and divergent questions, questions to which there are no answers written in the textbook. Consider this type of questioning:

Teacher: "What did you find out about landfills from your research?"
Student 1: "I found out that landfills are filling up, and the state has laws prohibiting the opening of new ones unless you go through a lot of regulations and red tape."
Student 2: "We also found out that our city doesn't allow grass clippings or tree branches in the garbage."
Student 3: "Yeah, and my subdivision doesn't allow composting, so what are we supposed to do?"
Student 1: "What do you think we should do?"
Student 3: "I don't know. Maybe write to the city."
Teacher: "What do others in the class think?"
Student 4: "I think we should make posters and hang them up in town to tell about this problem."
Teacher: "Do you think this would be beneficial?"
Student 4: "Yeah, it would make people aware of the problem, and maybe they'd do something about it."
Teacher: "Is there anything else we could do, and is there any 'red tape' that we'd encounter?"

As you can see in this example, the teacher asks very open-ended questions. Students can respond in a variety of ways, and the teacher may not know the answers to the questions asked. Remember: Science is a process and a way of thinking. Use community resources, guest speakers, the Web, and other resources to find the answers you don't know. You don't need to be afraid. In a project-based classroom, everyone participates in finding solutions. Learning Activity 8.5 will help you analyze your own questioning style.

Techniques for Improving Classroom Discussions

Several techniques can help teachers improve discussion techniques. Techniques, such as wait time, probing, and redirecting, encourage better dialogue among stu-

Learning Activity 8.5

ANALYZING YOUR QUESTIONING STYLE

Materials Needed:

- A video camera
- A video
- A prepared lesson

A. Prepare a short lesson, no more than 10- or 15-minutes long, on a topic of your choice. Teach it to a small group of your peers. Make a video of yourself teaching.

B. Watch yourself on video and analyze your lesson:
   1. What was the average number of questions you asked per minute?
   2. What percentage of your questions were lower level? Medium level? Higher level? Procedural (just part of organizing the lesson)?

C. After analyzing your lesson, list in your portfolio ways that you could improve your questioning strategies. What are you doing well?

dents, engage more students in discussions, facilitate the flow of discussions, and increase student interaction.

Wait Time. Wait time refers to the amount of time teachers wait to call on specific students after asking a question and to the amount of time that elapses between a student's answer and the teacher response. Mary Budd Rowe, a former professor at Stanford University, and Pat Blosser, a former professor at Ohio State University, researched questioning in classrooms and found that, with increased wait-time, students' responses improved in duration, quantity, and quality; students' confidence increased; the number of questions among students increased; responses from students who typically didn't respond increased; and more students became involved in answering questions. These results were evident when teachers waited for between 3 and 5 seconds (Blosser, 1990; Rowe, 1996; Swift, 1983; Wilen, 1987). This amount of time allows the students to think about, reflect on, improve, and elaborate on answers. Therefore, one way to improve your class discussions is simply to wait after asking a question.

When you first try waiting 3 to 5 seconds, it may seem as though an eternity is passing and as if you are wasting time. At first, you may feel a bit nervous about the "empty time." Science education researchers have found that teachers fear wait time for three reasons: (a) they feel as if they have too much material to cover to slow down, (b) they think fast-paced questions keep students motivated, and (c) they fear that discipline problems will occur if they wait too long between questions (Swift, Gooding, & Swift, 1996). Research does not, however, support these teacher fears. When too much material is covered, students aren't really learning it; fewer students participate in fast-paced drills; and because fewer students are engaged, it is more likely that classroom discipline problems will erupt with fast-paced questioning. Don't worry that wait-time will cause problems or waste time. Try it. To make sure you really wait a few seconds after asking a question, silently count to 10 before calling on someone. Most teachers find that with longer wait-times, more students get involved in class discussions and give more thoughtful responses.

Probing. Teachers also find probing a useful technique for improving classroom discussions. **Probing** involves asking students to elaborate on their answers. This technique is especially useful for identifying students' prior conceptions and possible misunderstandings. It is also useful to help students clarify understandings and form new conceptions. Imagine a teacher just dropped a toothpick into a gallon of water and asked, "Why did the toothpick float on top of the water?" If an elementary child answered, "Because it is lighter," a teacher might assume that the child understood that the toothpick was less dense than the water and move on to the next question. However, the child could mean that the toothpick is lighter than air, not water. The teacher probes the response by asking the child, "What do you mean, it is lighter?" The child answers, "Well, it's lighter. It floats because it is small compared with the gallon of water." This response tells the teacher that the child is comparing the tiny toothpick to a gallon of water and concluding that, because of their relative sizes, the toothpick must be lighter. The child is not comparing the density of the material—the amount of mass in a given area. The teacher probes the child's answer further and, through questioning and further explorations, directs the child's thinking to the idea that the toothpick would float on a much smaller amount of water. The teacher could also use counterexamples, asking the students to predict if a thin nail would float. Teachers also can probe by asking for more information: "Tell me a little more about what you mean by that?" or "How did you come to that answer?"

Redirecting. Many teachers do all the talking in a classroom. They ask all the questions and provide feedback to all student answers. Sometimes, in the rush for an answer, teachers even answer their own questions. In fact, many students have been conditioned to wait to answer the teacher's questions and *not* to respond to their neighbors' questions or comments. In a project-based science classroom, teachers strive to have students evaluate themselves and their peers. They make every effort to have students discuss with other students, not just with the teacher.

How do teachers accomplish this? If students are not talking to each other, asking each other questions, or critiquing each other's responses and ideas in classes, teachers might start by **redirecting** students' answers to other students in the class. For example, if a child cannot elaborate on her answer or express why she thinks the toothpick is lighter, the teacher might redirect the question to another student in the class: "Anissa, could you provide more information to Jessica's answer?" "Is there anyone else who could help provide information?" This redirecting is not meant to ridicule Jessica for her inability to elaborate on her answer. Care should be taken to make sure that there is a sense of trust in the classroom and that students feel free to give answers without worrying about being wrong or being perceived as failures (see Chapter 11 for more information about class climate). The teacher could ask Jessica to respond to Anissa's answer. This type of redirecting encourages students to talk with each other. It stimulates student-to-student discussions.

TABLE 8.5 Tracking Teacher's Verbal Comments During Discussions

Initiatory (Talking)
- Lectures (presents information) or gives directions (for example, tells the students to get out a pencil, paper, and magnifying glass).
- Makes statements or asks rhetorical questions (for example, says, "The magnifying glass is a good way to observe the insect in your jar").

Initiatory (Questioning)
- Asks short-answer questions (for example, says, "What kind of insect do you think you have in your jar?").
- Asks extended-answer questions (for example, says, "How are the insects in your jar alike or different from those found by other students in the class?").

Responding (Teacher Centered)
- Rejects students' comments, answers, or questions (for example, says, "No, that isn't quite correct").
- Accepts students' comments or answers (for example, says, "Yes. That's correct!").
- Confirms students' comments or answers (for example, says, "John's ideas regarding how insects are alike and different are right on target!").
- Repeats students' answers or comments (for example, says, "Did you hear Lesley's ideas about insect behavior? She said she thinks insects behave this way to protect themselves").
- Clarifies or interprets what students said (for example, says, "Do you mean to say that insects are protecting themselves against predators that might eat them?").
- Answers students' questions (for example, student asks, "Do all insects have six legs?" and teacher responds, "Yes, that is true").

Responding (Student Centered)
- Asks students to clarify or elaborate (for example, says, "Could you please explain what you mean by 'protect themselves'?").
- Models questions he or she wants students to ask each other (for example, says, "You will want to ask your partner a question that makes him or her use observations of the insects, such as 'What color is your insect?'").
- Uses students' questions or ideas (for example, says, "That idea of protection is a good one. How could we test this idea further?").
- Asks for other ideas (for example, says, "Does anyone else in the class have an idea as to why this insect is holding up its tail in the air?").

Source: Modified from Vince Lunetta (1977).

Keeping Track of Your Comments

During a discussion, keep track of your comments so you can evaluate how well you are using discussions. The scheme in Table 8.5 was developed by Vincent Lunetta, a professor from the Pennsylvania State University, to track teachers' verbal commentary. This table will help you complete Learning Activity 8.6.

Chapter Summary

In this chapter, we introduced instructional strategies, which can be classified as direct, indirect, experiential, and independent. We considered strategies to make teaching more active, such as role playing, investigation centers, and field trips. We discussed

Learning Activity 8.6

LEADING A DISCUSSION

Materials Needed:

- Paper and pencil or a computer
- A video camera

A. Prepare a short discussion, no more than 10- or 15-minutes long, on a topic of your choice. Situate the discussion in a project. What is the driving question of the project?

B. Lead the discussion with a small group of your peers. Make sure to create a video of the session.

C. Watch yourself and analyze the discussion. Use the categories in Table 8.5 to track your verbal behavior:
   1. What was the average number of questions you asked per minute?
   2. What kind of questions did you ask?
   3. What is your pattern of asking questions?
   4. Did you use wait-time, probing, and redirecting?

D. How would you like to change your questioning? What other issues might come up in a science class?

ways that teachers can use literature, including children's books, magazines, and the World Wide Web. We also discussed ways that teachers can help students make sense of textual material. Finally, we discussed teaching skills that improve questioning and discussions.

Chapter Highlights

- Instructional strategies can be classified as:
  - Direct
  - Indirect
  - Experiential
  - Independent
- Direct instructional strategies are teacher directed and are used primarily to convey information or teach step-by-step science skills.
- Indirect instructional strategies are more student centered.
- Experiential strategies are student centered and transformational.
- Independent study strategies are designed to help foster self-reliance or individual work.
- Metaphors, similes, and analogies can be used to make content more meaningful to students.
- Key ideas can be clarified by using diagrams, graphs, and pictures.

- Content can be presented, using movies, videos, and educational television.
- Guest lecturers help fill gaps in expertise and contextualize learning.
- Demonstrations are an effective classroom strategy when used well.
- Discrepant events capture students' attention by stimulating curiosity.
- Discussions help students pull together ideas and develop shared understanding.
- Reading is an important aspect of project-based science.
- Teachers can use a variety of materials to use reading effectively.
  - Children's literature
  - Magazines and periodicals for children
- Writing is an important aspect in science classrooms.
- Teachers can help students make sense of written material by using graphic organizers.
- Graphic organizers are visual depictions of knowledge.
- Some graphic organizers include:
  - Concept maps
  - Compare and contrast diagrams
  - Steps
  - Venn diagrams
  - Storyboards
- Community resources help students connect learning to life.
- Field trips are important for connecting the community to science teaching.
- Role playing helps students explore science topics, attitudes, and values.
- Debates help students develop important skills, such as communicating and argumentation.
- Investigation centers are useful independent instructional strategies that promote inquiry.
- Questioning is an important instructional strategy in inquiry.
- Different types of questions affect the quality of responses from students and classroom discourse:
  - Divergent
  - Convergent
  - Higher order
  - Lower order
- Teachers can use several techniques to improve discussions:
  - Wait time
  - Probing
  - Redirecting

Key Terms

Active reading strategies
Analogy
Blog
Case studies
Children's literature
Compare and contrast
   diagrams

Feedback
Field trips
Games
Graphic organizers
Hierarchical
Higher order
Homework

Metaphor
Multiple intelligences
Network
Passive reading
Primary-source
   material
Probing

Community resources
Concept
Concept maps
Convergent
Debates
Demonstration
Direct instructional
   strategies
Discrepant event
Divergent
Drill and practice
Essays
Experiential instructional
   strategies

Independent research
   projects
Indirect instructional
   strategies
Inquiry investigations
Instructional strategies
Interrelationships
Investigation centers
Journal
Lecture
Linking words
Lines
Lower order

Proposition
Questioning
Redirecting
Reflective discussions
Reports
Role playing
Simile
Simulations
Subordinate
Superordinate
Steps
Venn diagrams
Wait-time

References

Anderson, L. W., & Krathwohl, D. R. (Eds.). (2001). *A taxonomy for learning, teaching, and assessing: A revision of Bloom's taxonomy of educational objectives.* New York: Longman.

Berenstain, S., & Berenstain, J. (1987). *The day of the dinosaur.* New York: Random House.

Blosser, P. (1990, March 1). *Using questions in science classrooms. Research matters to the science teacher.* National Association of Research in Science Teaching.

Bradekamp, S. (1987). *Developmentally appropriate practice in early childhood programs serving children from birth through age 8.* Washington, DC: National Association for the Education of Young Children.

Brandenburg, A. (1969). *My visit to the dinosaurs.* New York: Harper & Row.

Brandenburg, A. (1981). *Digging up dinosaurs.* New York: Harper & Row.

Cognition and Technology Group at Vanderbilt. (1992). The Jasper series as an example of anchored instruction: Theory, program description, and assessment data. *Educational Psychologist, 27,* 291–315.

Council for Environmental Education. (1999). *Project WILD.* Baltimore: United Book Press.

Eggen, P., & Kauchak, D. (1992). *Educational psychology: Classroom connections.* New York: Macmillan.

Friedl, A. E., & Koontz, T. Y. (2004). *Teaching science to children: An inquiry approach.* Boston: McGraw-Hill.

Gall, M. (1984). Synthesis of research on teachers' questioning. *Educational Leadership, 42,* 40–47.

Gardner, H. (1983). *Frames of mind: The theory of multiple intelligence.* New York: Basic Books.

Gardner, H. (1993). *Multiple intelligences: The theory in practice.* New York: Basic Books.

Gardner, H. (1999). *The disciplined mind.* New York: Simon & Schuster.

Haberman, M. (1995). *Star teachers of children in poverty.* West Lafayette, IN: Kappa Delta Pi.

Helfgott, D., & Westhaver, M. (2000). *Inspiration 6.0.* Portland, OR: Inspiration Software (Version 6.0) [Computer software].

Hyerle, D. (1996). *Visual tools for constructing knowledge.* Alexandria, VA: Association for Supervision and Curriculum Development.

Inspiration Software. (2000). *Kidspiration.* Portland, OR: Author.

Jones, J., & Leahy, S. (2006). Developing strategic readers. *Science and Children, 44*(3), 30–34.

Leim, T. (1981). *Invitations to science inquiry.* Lexington, MA: Ginn Custom Publishing.

Marek, E., & Howell, B. (2006). Game time! *Science and Children, 44*(3), 48–50.

Marvin, F. (1986). *How big is a brachiosaurus?* New York: Platt & Munk Publishers.

Nolan, D. (1990). *Dinosaur dreams.* New York: Macmillan.

Novak, J. D., & Gowin, D. B. (1984). *Learning how to learn.* Cambridge, MA: Cambridge University Press.

Ogle, D. (1986). A teaching model that develops active reading of expository text. *The Reading Teacher, 39*(2), 564–570.

Parks, S., & Black, H. (1992). *Organizing thinking: Graphic organizers.* Pacific Grove, CA: Critical Thinking Press and Software.

Penner, L. R. (1991). *Dinosaur babies.* New York: Random House.

Rowe, M. B. (1996, September). Science, silence, and sanctions. *Science and Children, 34*, 55–37. (Reprinted from March 1969 issue)

Seuss, Dr. (Geisel, T. S. & Geisel, A. S.). (1971). *The Lorax.* New York: Random House.

Swift, J. N. (1983, November). Interaction of wait time and questioning instruction on middle school science teaching. *Journal of Research in Science Teaching, 20*, 721–730.

Swift, J. N., Gooding, C. T., &. Swift, P. R. (1996, October 23). Using research to improve the quality of classroom discussions. *Research Matters to the Science Teacher.* National Association of Research in Science Teaching.

Tierney, B., & Dorroh, J. (2004). *How to write to learn science,* (2nd ed.). Arlington, VA: National Science Teachers Association.

Tom Snyder Productions. (1996). *Rainforest researchers.* Watertown, MA: Author.

Tom Snyder Productions. (1994). *The great ocean rescue.* Watertown, MA: Author.

Vosniadou, S., & Brewer, W. F. (1987). Theories and knowledge restructuring in development. *Review of Educational Research, 57*(1), 51–67.

Wilen, W. W. (1987). *Questioning skills for teachers.* Washington, DC: National Education Association.

Yopp, H. K., & Yopp, R. H. (2006). Primary students & informational text. *Science and Children, 44*(3), 22–25.

Chapter 9

Assessing Students in Science

Introduction

This chapter focuses on the purpose of assessment in a project-based science class and in other school settings. You may have many questions about assessment: *Why is assessment such an important topic today? How do I assess science understanding? When should students be assessed? What kinds of learning should be assessed? Are some ways to assess better than others? How do I know that all students are learning?* This chapter answers these and other questions about assessment. We start by discussing the purpose of assessment in science and the ways today's assessment differs from the past. Five different types of assessment are reviewed, and we explain why

---

Chapter Learning Performances

- *Explain why assessment is a critical component of project-based science.*
- *Compare immediate, close (embedded), proximal, distal, and remote assessments.*
- *Distinguish between classroom assessment and large-scale, high-stakes assessment.*
- *Explain the value of using various assessment techniques to promote learning for populations of students traditionally at risk in science learning.*
- *Compare traditional assessment techniques to current assessment techniques.*
- *Create learning performances and corresponding assessments for content, procedural, and metacognitive knowledge across different cognitive dimensions.*
- *Justify why a variety of different assessment approaches should be used during different time frames in science classrooms.*
- *Identify ways that teachers can network with colleagues to improve their assessment practices.*

assessment is such an important topic for educators today. We cover many topics related to making assessment fair and consistent with today's educational goals. We consider what to assess in science classrooms and when to assess student learning. Finally, we introduce the idea of using technology tools to network with colleagues to improve assessment practices. Before we begin deliberating the purpose of assessment, let's examine three classroom scenarios that focus on assessment.

Scenario 1: Traditional Paper-and-Pencil Tests

Mr. Isarov says, "Now that we've finished grading the pop quiz in science, I have a few reminders for you. Remember, boys and girls, next Friday is the last day of the quarter, so you will be taking tests each day next week that will be included in this grading period. On Monday, you will have a test on Chapter 3 of the social studies book, which covers the Revolutionary War. Tuesday's test in math will cover division of decimals. Wednesday, we will have a test on Chapter 4 of the science book on simple machines. On Thursday, you will take a test on possessive nouns, and Friday's test will be the regular week's spelling list. Oh, I almost forgot. Make sure you take home the letter I gave you this morning about the Iowa Test of Basic Skills that you'll be taking in 2 weeks. Study hard and have a nice weekend; we'll be starting to learn all new topics next quarter!"

Do these words bring back memories of school testing? In the past, teachers commonly administered paper-and-pencil tests as the sole form of assessment. Students were expected to pass these tests, which for the most part determined their letter grades in each school subject. These tests were usually multiple choice, true/false, matching, or essay; and they were administered to the whole class after the class covered a chapter or some other unit of study. After completing the designated chapter or unit, the teacher rarely brought up the topic again for the rest of the year. *Assessment* **was synonymous with** *test,* **and** *test* **meant** *grades.* **The whole idea of testing caused nervousness, anxiety, and perhaps sweaty palms.**

Scenario 2: Tests Used in a Project-Based Environment

Mr. Corry, a science teacher who teaches a combined fourth- and fifth-grade science class, hopes his students will develop understandings of two National Science Education Standards:

Science Content Strand K–4; Physical Science; Motion and Force.
The position and motion of objects can be changed by pushing or pulling. The size of the change is related to the strength of the push or pull.

Science Content Standard 5–8; Physical Science, Motion and Forces.
If more than one force acts on an object along a straight line, then the forces will reinforce or cancel one another, depending on their direction and magnitude. Unbalanced forces will cause changes in the speed or direction of an object's motion.

To help his students develop these ideas, he plans a project on simple machines that is based on the FOSS Levers and Pulleys curriculum that his school district adopted. Mr. Corry decides that the driving question "How can I move big things?" will allow his students to engage in a number of activities to help them develop understandings identified in the standards. In his planning, he realizes that the project will also allow him to meet a number of National Science Education Inquiry Standards, such as *develop descriptions, use explanations, make predictions*, and *use models for evidence*.

Mr. Corry is pleased with how the project proceeds. His students seem engaged during the activities, and the work students turn in is of high quality. As students design experiments to explore ideas, such as "Why do you use ramps to move heavy objects onto a truck?" he walks around asking the students questions. Although he is pleased with most of the responses, he wonders if all students would respond appropriately to basic questions that focus on the main ideas of the project. He decides to give his students a short test that focuses on the main ideas explored in the project. He tells the class that they will have a test on Friday that will ask them to use ideas they explored in the project. He spends part of the day reviewing with the students what they learned through their investigations.

Mr. Corry writes a test consisting of short-answer responses that match the content objectives of the project. Some of the questions ask students to recall basic definitions, but others have students consider ideas in everyday-life situations. Some of the questions follow:

- How would you define *force*?
- When a book is sitting on a desk, what forces are acting on the book?
- What do you need to do to slide a box of bricks along the floor? Explain what is happening using scientific ideas.

Over the weekend, Mr. Corry scores the test using a rubric he created. He is surprised that some of the students who did well in class and wrote strong reports did not do so well on the test. Some of the students simply did not explain their ideas clearly enough. Mr. Corry writes notes on the tests asking the students to explain their ideas further. He uses the results of the test to reinforce some of the major ideas he intended students to learn in a class discussion on Monday when he goes over the test and helps students build even further understanding. He contrasts various responses to help students in the class develop a better idea of what constitutes a complete response on a short-answer test.

When used and designed appropriately, tests can serve as important vehicles for giving both the teacher and the students feedback. Both Mr. Corry and his students obtained important information about the understanding that was developed. Perhaps even more important, Mr. Corry used the results of the test to further help his students develop understanding, both of the content he wanted students to learn and also on how to write complete thoughts to explain one's ideas. Using paper-and-pencil tests in this manner is consistent with a project-based environment. In addition, Mr. Corry knows that these concepts are related to the National Science Education Standards, which were adopted as a curriculum framework for his state. Therefore, these concepts will be tested on his state's achievement tests given to students each year.

Scenario 3: Embedded Assessment

Students in a third-grade class ask, "Why do pumpkins decay after Halloween?" To answer this question, they plan several investigations. Some students think that pumpkins rot when they freeze outside because they noticed that fresh fruit is mushy after being frozen. That group investigates how fast a pumpkin decays when it is frozen and when it is left inside the building. Another group of students investigates whether "germs" have anything to do with the decay. They wash one pumpkin with an antibacterial soap and leave another one alone to investigate which decays first. Another group buries pumpkins in some leaves and soil to see what effect this has. Some students compare carved and uncarved pumpkins, while another group leaves its pumpkins in the sun and in the shade. Between team investigations, the teacher, Mrs. Molina, teaches several lessons on bacteria, decay, and mold growth. The class also visits a compost site where the city dumps leaves to generate fertilizer for sale. Some students write to grocery store managers to see how they keep fruit and vegetables from decaying.

Next, Mrs. Molina has students design a poster (an artifact) depicting what they learned as they completed their investigation. The poster is divided into two sections labeled "Things that cause a pumpkin to rot" and "Things that stop a pumpkin from rotting." Students draw pictures and take photographs of the things listed on each side of the poster to illustrate it. Students include with the poster a short essay on the benefits of decomposition in their daily lives.

This third scenario is also characteristic of project-based science. What did the students learn in this investigation? They learned to ask questions and devise investigations. In the process, they learned about the effect of freezing on plant cells, about bacteria and using soap to kill bacteria, and about composting and the relationship of sunlight and oxygen to decomposition. They used communication skills to write letters, and they learned about community efforts to compost leaves. Their learning did not occur in a single, discrete step, but unfolded gradually through a variety of activities in a learning community. Not all students were required to do the same activities at the same time.

As you can imagine, assessment for this project environment would need to differ from the traditional assessment described in the first scenario to accurately measure what students had learned. Students could take a single multiple choice test at the end of the unit with questions like "What is decomposition?" and answers like "a. dead plants and animals, b. rotting, c. separation of matter into its basic components, d. breaking up of elements, e. none of the above." However, one test at the end of the entire unit with questions like these would fail to assess the multitude of ideas students learned. Such a test wouldn't measure students' ability to work as a team. Tests serve an important role in assessing student learning, but they often fail to identify the most important ideas students learned. They would not show how students could apply their knowledge and skills to everyday life. They would not show that they could plan investigations or interpret data. The use of paper-and-pencil tests as the sole measure of understanding is not consistent with the philosophy of project-based science. It does not measure the kinds of understandings students gain while in the process of pursuing answers to driving questions, and it is not used as a tool to give students information that will help them gain further understandings.

Assessment techniques in a project-based science environment differ considerably from the technique depicted in the first scenario. First, various forms of assessment (not just tests) are used so that a wide variety of understandings, skills, and attitudes are measured. Second, assessment takes place during instruction, as well as after. Third, higher level cognitive skills, such as asking questions, designing investigations, interpreting data, and drawing conclusions, are assessed, as are affective outcomes, such as curiosity, skepticism, and open-mindedness. Fourth, students are involved in assessment decisions along with their teachers. Fifth, assessment is a continuous process, embedded in learning, not an end in itself.

Although some teachers stubbornly refuse to move beyond end-of-term paper-and-pencil assessment practices, most teachers have learned to use a variety of techniques to assess their students throughout a unit of study. After reading this chapter, you should understand why the opening scenario does not exemplify assessment practices recommended today. Before we move to the next section, Learning Activity 9.1 will help you think about the purpose of assessment.

The Purpose of Assessment

Assessment can be thought of as any method used to judge or evaluate an outcome or help make a decision. We use assessment data for a variety of purposes in educational settings today: to gather evidence of student learning; assist learning; guide teaching; measure individual achievement; give grades; allocate resources; evaluate programs; and inform local, state, and national policy. Used for these purposes, assessments tell us how students are learning; improve education; and give feedback to students, educators, parents, policymakers, and the general public about how students are doing in science (Pellegrino, Chudowsky, & Glaser, 2001). In fact, the National Science Education Standards state, "Research shows that regular and high-quality assessment in the classroom can have a positive effect on student achievement."

Assessment may be classified into five categories (Ruiz-Primo, Shavelson, Hamilton, & Klein 2002): Immediate assessments, close or embedded assessments, proximal assessments, distal assessments, and remote assessments. Although all of these types of assessment are aimed at measuring student achievement, they differ in their relationship to classroom teaching. **Immediate assessments** are those assessments that are a direct measure of curriculum implementation. For example, a teacher may

Learning Activity 9.1

WHAT IS THE PURPOSE OF ASSESSMENT?

Materials Needed:

- Pencil and paper or a computer

A. Think about why we assess students. What do you think is the purpose of assessment? On a sheet of paper or on a computer, make three columns. Label the first column *Know,* label the second column *Want to know,* and label the third column *Learned* (Ogle, 1986). Take a few moments to list in the first column as many things as you know about why we assess students.

B. Interview a classmate to see why he or she thinks we assess students. How does his or her viewpoint compare with yours? Find another classmate (or a teacher) with a different viewpoint. Compare and contrast these two colleagues' ideas about assessment.

C. Interview an elementary or middle grade student (preferably a student in grade four, five, six, or higher in which testing becomes more frequent), and ask the student why he or she is assessed or tested in school. How does the student's view of assessment compare with the teacher's? With yours?

D. After conducting these interviews, list as many things in the second column that you can think of that you want to know about assessment. At the end of the next chapter, you will assess your own learning by filling in the third column with what you have learned about assessment (see Learning Activity 10.7).

have students keep a journal related to the investigations conducted in the classroom. As such, the journal is immediately and directly related to the curriculum being used in the classroom. **Close assessments** (sometimes called embedded assessments) are those that teachers use in classrooms that are related to the curriculum being implemented, but are at a higher level. For example, after studying a project-based unit on "How can I move big things?" that included concepts of force and motion from a FOSS kit, Levers and Pulleys, a teacher may have students design a machine that could be used to solve a real-life problem, such as moving furniture from the school for a remodeling project. In designing the machine, students use their understandings of force and motion, but the assessment goes beyond the actual FOSS curriculum materials. **Proximal assessments** are those that measure knowledge or skills in an entirely new way than that introduced in the curriculum. For example, while learning about concepts of force and motion, students also learn to manipulate variables and test hypotheses. The skills of manipulating variables and testing hypotheses can be applied to an entirely new topic, such as causes of pollution in the local environment. **Distal assessments** are large-scale state and national assessments focused around a curriculum framework. For example, in Scenario 2, the teacher was concerned with several understandings identified for fourth and fifth graders in the National Science Education Standards. This teacher was using the school's adopted FOSS curriculum on levers and pulleys, but he was also concerned with teaching the concepts that would be tested on the state's achievement tests. As such, the state's tests are distal, or large-scale assessments related to the curriculum framework in the National Science Education Standards. **Remote assessments** are national level standardized tests, such as the Advanced Placement Test, Scholastic Achievement Test (SAT), and American College Testing Program (ACT). For purposes of this book, we narrow down these five categories to two: classroom assessments and high-stakes assessments. Immediate, close (embedded), and proximal assessments are those we place in the category of **classroom assessments**. Distal and remote assessments are **large scale** and oftentimes also **high stakes**.

Classroom Assessments

Classroom assessments are used to help teachers evaluate students' progress during a unit of study and make day-to-day decisions regarding curriculum and instruction. These types of day-to-day assessments, sometimes called formative assessments, are found in many formats, such as short quizzes and tests, observations of students, and artifacts. Teachers use these assessments to provide immediate or short-term feedback to students with the goal of helping them learn. There is also research to indicate that, by improving formative assessment, student achievement can be raised (Black & Wiliam, 1998a). Black and Wiliam (1998b) stated "The research reported here shows conclusively that formative assessment does improve learning. The gains in achievement appear to be quite considerable...amongst the largest ever reported for educational interventions" (p. 61).

Assessment used to improve curriculum and instruction determines what knowledge and skills children bring to science lessons and identifies what students know and can do following instruction. Scenarios 2 and 3 both used assessment to help guide and improve instruction. Assessment for this purpose is matched to instructional goals and conveys expectations to students and their parents in a way that motivates and helps students to learn. Assessment used in this manner also assists teachers in making decisions about what instruction has not been effective and how it should be modified. This type of assessment is an ongoing process that occurs during, as well as after, project-based science instruction and helps students, teachers, and parents monitor individual student's learning.

> CONNECTING TO NATIONAL SCIENCE EDUCATION STANDARDS
>
> CLASSROOM ASSESSMENT AND INQUIRY
>
> Assessment data can be used to plan a lesson, guide a student's learning, calculate grades, determine access to special programs, inform policy, allocate resources, or evaluate the quality of curriculum or instruction. In the breadth of its application, assessment merges seamlessly into considerations of the curriculum and teaching (NRC, 2000, p. 76).

Large-scale assessments are used to evaluate school curriculum, school district progress, and programs. An example of a large-scale assessment often used to evaluate programs is the National Assessment of Educational Progress (NAEP) in which student achievement can be compared across states and with previous generations of students (NAEP testing began in 1969). Sometimes, large-scale assessments are used to compare a student's individual achievement with that of peers. Assessment of individual achievement in the form of end-of-unit tests, letter grades, and large-scale assessments is called summative assessment; some happen in the context of the classroom, and some are delivered as large-scale assessments. More and more, high-stakes assessments are used to make important decisions regarding student advancement to the next grade level and graduation from high school. They are also used to determine which students or schools receive special services (such as gifted programs, special education programs, and tutoring services), evaluate teachers and administrators (and sometimes determine how they are paid), determine how resources are allocated to schools, and determine whether schools get rewards or sanctions from the state (Pellegrino et al., 2001).

Large-Scale and High-Stakes Assessment

Assessment is a fervent topic in education today, and it is scrutinized at the public level and within our educational system. At the public level, it seems that we face almost daily issues pertaining to assessment. This increase in attention paid to educational assessment is due primarily to the focus in the last 2 decades on high standards and measuring students' progress toward standards. Journalists have written articles about how our nation's students compare on standardized tests with previous generations (National Assessment of Educational Progress) or with students in other countries (TIMSS). Policymakers, community leaders, parents, and school administrators are demanding that students be held to certain levels of performance and understanding. As a result, most states have debated and implemented new educational standards that include large-scale or high-stakes testing, such as proficiency testing or high school graduation qualification testing. Many people believe that these externally mandated tests inform the public about how our schools are doing (Klassen, 2006). They are intended to measure the nation's educational achievement.

Since the passage of the No Child Left Behind (NCLB) Act in 2001 (http://www.nochildleftbehind.gov), much greater emphasis has been given to stronger accountability toward student achievement. Under the Act's provisions, each state must describe how they will close achievement gaps and make sure all students learn. Annual state and district reports are required, and schools are subject to losing funding if their school does not make yearly progress toward these goals. Starting 2006–2007, the Act requires states to assess science.

The increase in focus on large-scale and high-stakes assessment has sparked debates about what should be tested and how it should be tested. Within our educational system, many people are debating the relevance of using only traditional paper-and-pencil tests, commonplace in schools and used for most large-scale assessments, and are looking for alternative means of assessment. For example, Mislevy (1993, p. 19) wrote, "It is only a slight exaggeration to describe test theory that dominates educational measurement today as the application of twentieth century statistics to nineteenth century psychology." Advances in cognitive science (how people learn) and psychometrics (how we measure learning) have taught us a great deal in the last several decades about how people learn, and this has led to recommendations for changes to assessment to include measuring different types of knowledge: content, procedures, and metacognition. New types of measurement instruments and statistical procedures make it easier to examine data in different ways. However, some testing techniques, such as looking at students' products after they conduct an investigation, limit widespread use of these instruments and procedures because of cost, feasibility, and concerns about subjective interpretation of students' answers.

Societal, economic, and technical changes have also prompted demands for changes in assessment by illustrating the need to focus on a wider range of knowledge and competencies, such as applying knowledge in a meaningful way (McAfee & Leong, 2002; Pellegrino et al., 2001). For example, a student who can repeat a memorized definition of *density* may possess little to no understanding of density, whereas a student who can use scientific equipment to figure out the density of an object has in-depth knowledge of the concept. It is this in-depth knowledge that societal, economic, and technical changes demand. In Scenario 2, the teacher not only asked students for the definition of a concept on a paper-and-pencil test, but also asked them to apply the concept to something in everyday life. However, use of an item, such as this, on a large-scale assessment would require large numbers of people to read and score students' answers—a process too time consuming and costly for widespread use.

Many educators argue that large-scale assessments are only one-shot methods (they don't show progression or growth) that may not measure what is taught in the local school's curriculum. Since our state and local district curriculum can vary dramatically, a "one size fits all" testing policy may not be fair. Thus, policymakers, who see these large-scale assessments as a way to measure achievement and change what is going on in schools, may be using the assessment data inappropriately. It is more fair to view large-scale assessments as targets for educators to pursue than as perfect measurements of students' learning and growth. All assessments are flawed and only estimates of student learning.

Some educational leaders view large-scale assessments as an insufficient means of meeting the assessment standards found in the *National Science Education Standards* (NRC, 1996). Most of the large-scale assessments do a reasonable job of measuring knowledge of facts and procedures, but they fail to capture the breadth and richness of knowledge. For example, Pellegrino et al. (2001) pointed out that current tests don't measure students' organization of knowledge, problem representation, use of strategies, self-monitoring skills, and individual contributions to group problem solving. As a result of pressure on schools to do well on high-stakes tests, many teachers are "teaching to the test" or teaching to the low-level questions on the test, rather than focusing on the intent of educational reforms. To meet the assessment standards found in the *National Science Education Standards* (NRC, 1996, p. 100), there needs to be a greater emphasis on "assessing what is most highly valued" (as opposed to what is easily measured); "assessing rich, well-structured knowledge" (as opposed to discrete knowledge); "assessing scientific understanding and reasoning" (as opposed to only knowledge); "assessing to learn what students do understand" (as opposed to what they do not know); "assessing achievement and opportunity to learn" (instead of only achievement); and "engaging students in ongoing assessment of their work and that of others" (as opposed to end-of-term assessments given by teachers).

Finally, many criticize the fact that, although teachers and those who work in schools are judged by the results of high-stakes assessments, the assessments do little to inform teachers' day-to-day decisions, modify instruction, or change the curriculum. Frequent testing doesn't necessarily improve instruction. An analogy we once heard compared this phenomenon to the frequent weighing of a cow to ensure its growth. To grow, a cow needs proper nutrition, medical care, and habitat. Weighing it alone won't make it grow. Similarly, frequent testing without changes in curriculum or instruction won't improve students' learning. However, many states' standards and, as a result, the high-stakes tests used to measure attainment of the standards vary in their focus and are often too vague to serve as a useful blueprint for curriculum, instruction, and assessment. The results of tests reported to the schools and teachers are often limited—providing only general information about how students did relative to their peers. Furthermore, these scale assessments are usually given only once a year. There is a large lag until the results of the tests come back and, therefore, they are seldom used to help students or teachers in day-to-day curricular or instructional decisions that affect learning. Pellegrino et al. (2001) pointed out the need to create a better match between classroom practices and large-scale assessments: "A vision for the future is that assessments at all levels—from classroom to state—will work together in a system that is comprehensive, coherent, and continuous" (p. 9).

At this time, complete Learning Activity 9.2 on high-stakes assessment to explore this topic more thoroughly.

The Nature of Classroom Assessment

Although there are many reasons we assess students in schools today, in this book, we focus primarily on the nature of *classroom assessment*. Just as instruction in project-based science has certain characteristics that distinguish it from traditional science teaching, assessment best suited for a project environment has special characteristics that distinguish it from traditional assessment, as described in the first scenario. Assessment that is appropriate for project-based science has many different labels: *active assessment, direct assessment, performance assessment, alternative assessment,* and many others. Let's examine one label, **authentic assessment**. The word *authentic* means "genuine or justifiable." This suggests that assessments that most closely measure what happens during instruction and inform instruction are more genuine and justifiable. This also suggests that we align learning goals and assessments to measure what we value most in our schools.

Learning Activity 9.2

HIGH-STAKES TESTING

Materials Needed:

- Reference materials

A. High-stakes testing is testing that has a significant consequence for students—testing that is required for high school graduation, placement in "remedial" or "advanced" classes, or for national and international comparisons. Find out more about this type of testing by researching your state's testing policies; national tests, such as NAEP, SAT, and ACT; and international tests, such as the International Science Study. Report your findings.

B. Why do some people feel these tests are important? Why do others dislike them? What are the advantages and disadvantages of this type of assessment? Why do you think it has become more popular in recent years? Will high-stakes testing improve our educational system? Include in your report your opinion of high-stakes testing.

C. Find out what other countries do in terms of assessment. Are there high-stakes testing programs? How do teachers assess within classrooms? Countries you might want to investigate include Japan, Germany, England, Singapore, and Australia. Good sources of information on this topic are the TIMSS Web pages at http://nces.ed.gov/ and http://wwwcsteep.bc.edu/timss.

D. Record your findings.

Assessment appropriate for today's science classrooms has the following features:

1. It is responsive to context.
2. It is a continuous process embedded in instruction.
3. It is multidimensional.
4. It engages students in the assessment process.
5. It is valid and reliable.
6. It matches today's educational goals.
7. It accommodates cultural diversity.
8. It is consistent with learning theory.
9. It measures meaningful understanding.

CONNECTING TO NATIONAL SCIENCE EDUCATION STANDARDS

EXECUTIVE SUMMARY

It is necessary to align assessment in the classroom with externally developed examinations, if the goals of science education are to be consistent and not confuse both teachers and students (NRC, 2001, p. 2).

Current Assessment Is Responsive to Context

As depicted in the opening scenario, assessment traditionally has been thought of as mainly a measure of students' understanding of content knowledge at the end of a unit of study. It does not help students reflect on their own learning. It is not used to determine what should be taught. It does not influence teachers' day-to-day lessons or provide a means for revising lessons and teaching techniques. On the other hand, assessment in a project-based science environment is context-responsive. It is used to help plan instruction, guide day-to-day interactions with students, and, if necessary, revise instruction on a moment-to-moment basis. In short, assessment is aligned with learning goals.

Teachers in a project-based environment use assessment techniques to diagnose students' prior knowledge of science and any problems they may have with science learning. Teachers also examine students' interests. These diagnoses assist teachers in planning and carrying out lessons. For example, if students are having difficulty making meaning of the effects of seasonal change on the behavior of animals and the growth of plants, a teacher would revisit the concepts in a different way and plan new experiences to help students better understand the concepts. If students are extremely interested in the Venus flytrap, the teacher might use this interest as an avenue for teaching about types of plants, including carnivorous ones.

Assessment techniques also provide teachers with information that may alter their moment-to-moment and day-to-day decisions. Teachers using assessment to monitor understanding, skills, and attitudes are less likely to blindly continue lessons without regard to how students are progressing. In other words, teachers using this type of assessment are **reflective practitioners**. They think about, analyze, and mentally debate what should be done to most effectively teach particular students at any given moment. Such teachers continually ask themselves questions, such as "Are the students interested in the lesson?" "Was the student misbehavior a result of my management and questioning?" "Did I choose the best instructional strategy to get across the concept?" "Was my pacing okay?" "Did I call on the same students or many different students?" "Did I wait long enough after I asked questions?" "Did I give good examples?" "Were my transitions good?" "Did I end the lesson in a way that summarized what was learned today?" Such ongoing reflection frequently results in teachers' changing the direction of their questioning, instructional strategies, choice of curriculum, and classroom environment.

Current Assessment Is a Continuous Process Embedded in Instruction

In the opening scenario, traditional assessment was portrayed as something that always happens at the *end* of instruction. Such assessment is a discrete event separate from instruction. The assessment was not necessarily aligned with the learning goals. Assessment in a project-based science classroom is consistent with the instruction and aligned to learning goals. It is a *continuous process* that is embedded in instruction. Treagust, Jacobowitz, Gallagher, and Parker (2001) and Wenglinsky (2000) concluded that assessment embedded in teaching does lead to improved science learning. It is contextualized and realistic, rather than contrived or staged. It does not occur under artificial testing conditions; rather, it matches classroom instruction and real life. Teaching, learning, and assessment are thought of as reciprocal. Assessment continuously monitors student learning; it is not solely an end product of learning.

Assessment with these characteristics looks more like the assessment discussed in Scenarios 2 and 3.

Imagine that you are on a month-long vacation to Australia; you have only brought black-and-white film; you only have enough film to take one photograph per week; and your camera has only one setting. How well will your photographs portray your vacation experiences? How could you have improved the documentation of your vacation? First, you might have brought more film so you would not be limited to one picture per week. You might also have purchased color film, and you might have used a 35 mm camera so you could change the focus. If you had a wide-angle lens and a zoom lens, you could take panoramic pictures and close-up photographs. You might even have brought a video camera so you could document continuous sights as well as sounds; and you might have carried a diary so you could write down your impressions of Australia. Finally, you might have gathered souvenirs to take back home with you.

> CONNECTING TO NATIONAL SCIENCE EDUCATION STANDARDS
>
> AN OVERVIEW
>
> Ideas about assessments have undergone important changes in recent years. In the new view, assessment and learning are two sides of the same coin. When students engage in assessments, they should learn from those assessments (pp. 5–6).

This analogy is very similar to a teacher's efforts to document students' progress in school using assessment techniques. Traditional forms of assessment, when used appropriately, supply the teacher with only quick monochrome snapshots of what students know, can do, and are like. Today's forms of assessment are more analogous to the use of a digital video camera, a 35 mm camera, and a journal and the gathering of souvenirs—strategies that gather a wide range and depth of information. In a project-based classroom, teachers and students will continually collect information during the instructional process. The "souvenirs" of science, which are the artifacts of a project, are especially important because they are often the most remembered and, therefore, the most meaningful for students and teachers. Students will create many different types of "science souvenirs" or artifacts that can be used to assess their knowledge, skills, and attitudes.

Current Assessment Is Multidimensional

Assessment in a project-based science classroom aims to make fair and accurate judgments about student progress, and the most legitimate judgments come from diverse sources of information to document student learning. Traditional paper-and-pencil assessment techniques can provide teachers with important information, but, if used alone, can ignore differences in students' learning styles, interests, attention spans, or types of intelligences. For example, on traditional tests, a student might not correctly identify the definition of *metamorphosis* from five choices provided. However, the student might be able to demonstrate understanding of *metamorphosis* by drawing a diagram illustrating stages of growth or choreographing a performance depicting changes occurring in an animal's life cycle. A student who demonstrates understanding of *metamorphosis* in one of these ways has exhibited understanding of the concept as much, if not more, than the student who identified the term from a list of choices.

For teachers who work in schools that require letter grades, the alternative assessment techniques used in project-based science can be of great assistance. Using diverse forms of assessment helps teachers make more valid decisions about students' grades. Students also tend to learn concepts in a more in-depth manner, so they are more likely to score well in other types of assessments. We address different assessment techniques and grading in more detail in Chapter 10.

Current Assessment Engages Students in the Assessment Process

In a project environment, teachers and students engage collaboratively in assessment. At each stage of the assessment process, teachers and students work together to collect data, make decisions about individual progress, document progress, and set goals. This collaborative process helps students become self-reflective, self-monitoring learners who regulate and take responsibility for their own learning. This is supported by the National Science Education Standards (NRC, 2001).

Imagine that students are engaging in an investigation. During the investigation, they make numerous observations and record this information in their science notebooks. While students engage in this process, the teacher makes notations in his journal about the students who are having trouble making observations and recording information. Later, while students work in collaborative groups to make decisions about the artifacts they will present to the class, the teacher assembles information from the students' notebooks to file in their portfolios. The teacher judges each student's depth of understanding from the items contained in his or her portfolio. After the artifacts are presented to the class, the entire class engages in a critical dialogue. Later, the teacher discusses with each student the items contained in his or her portfolio. The teacher reviews individual progress with each student, and together they devise strategies to improve the student's academic growth. Students are involved in every step of the assessment process, which has been designed to help them become motivated, self-regulated learners.

> **CONNECTING TO NATIONAL SCIENCE EDUCATION STANDARDS**
>
> **EXECUTIVE SUMMARY**
>
> Student participation is a key component of successful assessment strategies at every step. If students are to participate effectively in the process, they need to be clear about the target and the criteria for good work, to assess their own efforts in light of the criteria, and to share responsibility in taking action in light of the feedback (NRC, 2001, p. 1).

Current Assessment Is Valid and Reliable

Good assessment is both valid and reliable. **Valid assessment** is appropriate and allows the teacher to make accurate generalizations from it about a student's understanding. Invalid assessment does not permit a teacher to make accurate decisions about what students know and what they are ready to learn. Each one of us can provide examples of tests or assessments from our own educational experience that we felt weren't fair. We agonized over the results because we knew that they didn't represent what we really knew or what we were capable of doing. Therefore, any generalizations made about us were incorrect or, at best, only partially correct. When assessment focuses on more than simply the attainment of content knowledge, uses diverse sources of information, and involves the student in the assessment process, it is less likely that it will result in biased conclusions or errors.

Reliable assessment provides consistent results across different trials. In other words, a reliable assessment tool will yield similar results on different occasions or will result in different assessors coming to similar conclusions about the student being assessed. Think about your car. If it starts one day, but not another under similar weather conditions, you would say that the car isn't reliable. The results are not consistent, so the car isn't trustworthy. Of course, if this happened only once, you could not blame your car for being unreliable. If, however, your car repeatedly starts one day and not the other, you could assuredly conclude that it isn't reliable. With traditional end-of-term paper-and-pencil assessment techniques, a single "snapshot" of students' abilities is taken at the "conclusion of learning." Many factors, such as testing conditions, the assessment items themselves, and a student's frame of mind, can influence the reliability of these snapshots; and, so, the results are not always trustworthy. Assessment techniques are more reliable because they give students many opportunities and ways to demonstrate their abilities, thus eliminating variables that may alter results.

Current Assessment Matches Today's Educational Goals

It is widely recognized today that learning should be active and integrated; the knowledge explosion has made it futile to ask students to memorize large quantities of facts, as they will quickly become outdated. In our information age, labor force experts are calling for students who can solve problems, make decisions, learn, collaborate with others, and manage themselves (*SCANS Report*, 1991, in Packer, 1992). Educators think it is important for students to be able to understand how they have arrived at an answer, transfer knowledge to real-life situations, think critically, document and communicate information, and analyze and synthesize information.

New state and national educational goals have required changes in the way we teach. Therefore, assessment practices need to match educational goals. If a teacher were trying to measure content knowledge, or knowledge that is concerned only with what central concepts, principles, and theories students have acquired, selected response assessments, such as true/false and multiple choice tests, would measure a small part of students' understandings. However, as we discussed in Chapter 2, teachers today are trying to measure much more: **content knowledge**, **procedural knowledge**, and **metacognitive knowledge** as related to out-of-school experiences. Students are assessed on their abilities to solve problems, make decisions, and collaborate. Not only do assessment techniques *match* today's goals, but they also help to *meet* them. Sometimes teachers criticize assessments for forcing them to teach to the test. However, if tests measure what we value in intended learning goals and the test items are of good quality, there should be nothing wrong with teaching to the test. To help align assessment with learning goals, Project 2061 is developing strategies and tools for evaluating the alignment of K–12 assessments in science with national and state standards and benchmarks (http://www.project2061.org/research/assessment.htm). Project 2061 developed a set of criteria for determining if assessment items are aligned with standards (AAAS, 2004). These analysis criteria include **necessity** (the knowledge or skill measured in the assessment item is specified in the learning goal), **sufficiency** (the knowledge or skill is outlined in the learning goal *enough* to answer the assessment item), **comprehensibility** (students can understand the assessment item), **clarity of expectations** (do students understand what is expected of them in the assessment item), **appropriateness of context** (the assessment task is familiar and engaging to the student), and **resistance to test-wiseness** (students are unlikely to get the answer correct by guessing or using general test-taking strategies).

Learning Activity 9.3 asks you to compare assessment practices with several reform documents that have established today's educational goals.

Learning Activity 9.3

HOW DOES SCIENCE REFORM AFFECT ASSESSMENT PRACTICES?

Materials Needed:

- Reference materials, including *Benchmarks for Science Literacy* (AAAS, 1993); *Project 2061: Science for All Americans* (Rutherford & Ahlgren, 1990); *National Science Education Standards* (National Research Council, 1996); and *Classroom Assessment and the National Science Education Standards* (NRC, 2001).

A. On a sheet of paper, list as many answers as you can to the question "What is a scientifically literate person?"

B. How could you assess whether a person had met this criteria? List ways to assess each criterion.

C. How many of the items did you choose to assess with traditional multiple-choice, true/false, and essay questions? How many required different approaches to assessment? Why is this so?

D. Compare your list of characteristics of scientific literacy to those in major national reform reports such as *Benchmarks for Science Literacy, Project 2061: Science for All Americans, National Science Education Standards,* and *Classroom Assessment and the National Science Education Standards.* How do your criteria for scientific literacy compare with the criteria in these policy reports? What effect do you think these reports will have on curriculum, instruction, and assessment?

Current Assessment Accommodates Cultural Diversity

Policy reports in science education have called for changes that will attract all students to science learning (National Research Council, 1996; Rutherford & Ahlgren, 1990). When considering the diverse needs, interests, and abilities of students in our country—particularly girls, minorities, students with disabilities, and those with limited English proficiency—it becomes clear that traditional or narrow assessment techniques sometimes discriminate against some populations of students. They overlook the prior experiences, learning styles, multiple intelligences, and interests of our diverse population. Use of a variety of assessment techniques, on the contrary, allows for students' unique differences.

For example, a student with limited English proficiency might have difficulty answering essay questions about the process of metamorphosis. This test, however, would more likely be assessing her ability to write English sentences than her understanding of metamorphosis. An assessment that asked her to draw pictures of the stages of life that an insect goes through from egg stage to adult stage would more likely measure her understanding of the concept.

> **CONNECTING TO NATIONAL SCIENCE EDUCATION STANDARDS**
>
> **ASSESSMENT IN THE CLASSROOM**
>
> Assessments should be equitable and fair, supporting all students in their quest for high standards (NRC, 2001, p. 58).

A study (Lawrenz, Huffman, & Welch, 2001) found significant differences among ethnic groups (Caucasians, Asian Americans, Hispanic/Latino Americans, and African Americans) for different test formats. For example, Caucasians performed better on multiple-choice and open-ended tests, while Asian Americans fared best at hands-on stations. Atwater (2000) pointed out that many test items fail to measure the content taught in African-American classrooms, don't match the language used by African-American children, and are incongruent with many minority children's life experiences. Solano-Flores and Nelson-Barber (2001) noted that sociocultural differences can affect the validity of assessment items. Sociocultural influences include students' values, beliefs, experiences, communication patterns, teaching and learning styles, and socioeconomic status. Similarly, Fusco and Calabrese-Barton (2001) pointed out the need to assess science understanding of diverse students. McAfee and Leong (2002) outlined several guidelines for teachers to be more sensitive to cultural and social differences among students:

- Assume that there are cultural differences in every classroom that will affect assessment.
- Learn the difference between social, cultural, ethnic, language differences, and disabilities that require special services.
- Use multiple assessment measures.
- Involve parents and the community in the assessment process.
- Design assessment measures so that they match students' interests, experiences, and prior knowledge.

Current Assessment Is Consistent With Learning Theory

In Chapter 2, we examined how students construct understanding, the social nature of learning, and the need for science to be anchored in students' daily lives. Social constructivist theory has several implications for assessment. One major tenet is that learners actively construct knowledge and connect new learning to prior knowledge that enables them to integrate knowledge in new ways and to solve problems. Cognitive theories focus on how people develop structures of knowledge, including concepts associated with subject matter expertise and procedures for reasoning and problem solving (Pellegrino et al., 2001). One implication for assessment, for example, is that assessment techniques that measure only content knowledge in short-term or working memory (and sometimes only inert knowledge) are inadequate. What follows are a number of learning principles and the assessment practices that derive from them (adapted from Herman, Ashbacher, & Winters, 1992, and Pellegrino et al., 2001).

People generate knowledge through an active process of creating personal meaning by integrating new information with prior knowledge. This principle implies that students must actively generate meaning for themselves. For example, in project-based science, some students might create posters to demonstrate what they have learned, while others might create a video documentary. This principle also implies that assessment should measure how students relate new learning to personal experiences and prior knowledge. Finally, assessment should measure how well students apply what they have learned. For example, students investigating why the water in the classroom faucet doesn't flow well might present their data to the city water department, creating a perfect opportunity for assessment.

Students of all ages and abilities can think and solve problems. This principle implies that assessment techniques should engage students in problem-solving activities and investigation. Because learning does not happen in discrete, small steps, assessment should measure holistic knowledge and abilities. When students present artifacts, they are demonstrating a wide variety of knowledge, skills, and attitudes learned—not isolated facts. More than one right answer should be encouraged.

Children differ in learning styles, attention spans, memory, and developmental rates. This principle implies that assessment techniques should be varied and multidimensional. Students should be given more than one chance to demonstrate their competence, and they should have time to complete tasks. Students should be allowed to revise work and show improvement over time. Finally, students should be able to demonstrate their various abilities, whether they be academic, artistic, verbal, or social. Student production of artifacts presents an opportunity for appropriate assessment because students can construct the products over time and present information in whatever format best matches their learning styles. After presenting their artifacts to a public audience, students receive feedback, which can then be used for revision or further development of the artifacts.

People need to know when to use knowledge and how to adapt it. They also need to learn how to manage their own learning. This principle suggests that students should be able to participate in their assessment, consulting with the teacher, reflecting upon their own work and progress, and helping set their own learning goals.

Motivation and effort affect learning and performance. This principle implies that students should play a part in identifying what they will do, how they will do it, and how it will be evaluated, because motivation and effort are enhanced when learners are able to set their own goals and when the criteria for assessment are demystified. Learners also need clear guidelines about expectations, and they need to see the connection between their efforts and their results. When students determine what artifacts to include in their portfolios for assessment, their motivation increases. They play active roles in selecting, developing, and giving presentations that will be assessed. They participate in assessing the presentations and artifacts of their classmates.

Learning occurs within a social context. This principle implies that assessment of group work should be included in evaluation. As students work together, they construct understanding and find solutions to questions and problems. As a result, it is critical to assess the products that emerge from students working together.

Current Assessment Measures Meaningful Understanding

Do you remember memorizing answers for a test and not understanding what you were memorizing? What is Avogadro's Number? What is the equation that represents the process of photosynthesis? What type of lens focuses light to a single focal point and then inverts it to enlarge the image? What are the names of the three bones in the ear? What is a vacuole? What are the six simple machines? If you cannot answer these questions, it's probably because you learned them by rote memorization in order to regurgitate them on a test at the end of a chapter or quarter. It is likely that you have never used or applied these concepts since and that you lack a meaningful understanding of them.

Paper-and-pencil tests are typically true/false, multiple choice, matching, short essay, fill-in-the-blank, or circle the correct picture. Although these types of tests can be structured to measure meaningful understanding of concepts, they frequently are

Directions: Circle the best picture to answer the question.
1. Which plant needs the least amount of water?

a.   b.   c.

Figure 9.1   Sample test item

not (NRC, 2000). Students can easily memorize terms and guess at correct answers based on shallow and disconnected understanding of material. Sometimes ambiguous words, unclear questions, or vague sentences make it difficult for students even to respond during these types of tests. Imagine a question that instructs children to look at pictures of plants to identify which plant needs the least amount of water. Two of the plants pictured are a cactus and a head of cabbage. The "correct" answer is supposed to be the cactus. However, students could reason that since the head of cabbage is dead (since it has been cut from its roots), it no longer needs *any* water to live. Choosing the head of cabbage actually demonstrates a deeper, more meaningful understanding of the concept in question. Such questions fail to measure what students know; instead, they measure what students do not know.

Teachers need to write paper-and-pencil tests so they measure higher level thinking, rather than low-level, factual answers. True/false, sequencing, and multiple-choice tests can limit student answers, but can be expanded to let students answer open-ended follow-up questions as well. Students often take paper-and-pencil tests on statewide assessments. For this reason, some teachers like to use paper-and-pencil tests in their classrooms. For more information regarding the construction of good paper-and-pencil tests, see Chapter 10.

Authentic assessment measures meaningful understanding by stressing open-ended answers rather than a single "correct" one. Students engage in higher level reasoning through discourse with others (teachers, peers, members of the community). Creating artifacts is more complex and engaging than is selecting answers from a multiple-choice list. Artifacts can demonstrate higher level thinking skills, such as planning, inventing, and making conclusions. Students construct knowledge, rather than simply repeat memorized information. Finally, students engage in self-reflection, which develops self-regulation and personal responsibility.

Imagine that your students have spent considerable time investigating the driving question "Why does flooding cause so much damage?" You could try to measure students' understanding of water pressure on levees by asking a multiple-choice question, such as "What is water pressure? a. the amount of force a given amount of water gives on an object, b. the amount of water that presses down upon the earth, c. the amount of pressure on water from the air above it." Or you could assess understanding of water pressure on levees by using a performance-based assessment technique. You could provide students with 2-liter soda bottles, each with three holes about 2 inches apart—one near the top, one in the middle, and one near the bottom. Instruct students to tape the holes and then pour water into their bottles. Ask students to explore what happens as they remove the tape from each hole. Ask them to explain the results of the activity and then to compare the results of this activity to the breaking of levees during Mississippi River floods of many towns in the Midwest or during wet El Niño weather in the winters. Question them about their opinions on building levees along a river.

Which of these examples of assessment measures meaningful understanding? Although paper-and-pencil tests can measure meaningful understanding, in the aforementioned example, a student could memorize a definition of *water pressure* from a textbook and really not understand the concept. Although not all performance assessments are necessarily well constructed to measure meaningful understanding, in the aforementioned example, a student must have a meaningful understanding of

Learning Activity 9.4

WHAT ARE THE PITFALLS OF TRADITIONAL TESTING?

Materials Needed:

- Traditional textbooks and accompanying commercial tests

A. Reread Scenario 1 at the beginning of this chapter. Make a list with two columns labeled *Traditional assessments* and *Contemporary assessments*. Under each of these headings, make two additional columns labeled *Advantages* and *Disadvantages*. Identify as many advantages and disadvantages as you can for each type of assessment.

B. With a few classmates, compare and discuss the advantages and disadvantages. Make a list of what you all agree are the most important advantages and disadvantages. How do the advantages of each category help students learn, help teachers administer assessment, help parents understand student progress, and help students understand their own progress?

C. Discuss whether the advantages of today's assessment techniques outweigh the disadvantages. How can the disadvantages be overcome?

D. Examine the commercial tests that accompany a traditional elementary or middle grade textbook. How many of the following can you find in the tests or assessments?

1. A variety of testing formats, including whole class, small group, and individual
2. A variety of formats including artifacts, teacher observations, essays, student-produced products, student self-evaluation, and concept maps
3. Assessment throughout the learning process
4. Measurement of higher level cognitive outcomes
5. Measurement of skills
6. Monitoring of affective outcomes
7. Student monitoring of his or her own progress and time to continue to learn content until it is mastered
8. Measurement of relationships, the nature of science, critical thinking, problem solving, and interrelationships between school science and life outside of school
9. Measurement of student understanding of the subconcepts within the test
10. Feedback beyond a letter grade or percentage score

E. As a classroom teacher, what steps would you need to take to make a traditional test accompanying a textbook series more valid in its assessment method?

water pressure to explain the results of an investigation. The student must construct meaning to explain why the water in the bottom hole flows out faster and at a different angle. The student must also have a deep, meaningful conceptual framework to discuss how the results of the investigation resemble the river flooding and whether levees are ultimately helpful or harmful.

In Learning Activity 9.4, you will compare traditional assessment techniques with contemporary assessment techniques and examine the pitfalls of traditional testing in order to better understand the merits of today's assessments.

What to Assess

When you were in school, your teacher probably focused almost solely on measuring your attainment of scientific facts. You may remember asking the teacher, "Is this going to be on the test?" You and your classmates were probably keenly aware that tests measured certain content, and you wanted to make sure you memorized the particular answers that you would be asked to give. It is unlikely that your teacher paid much attention to whether you had misunderstandings about scientific ideas. The teacher probably did not focus on procedural knowledge—whether you were able to ask and refine questions, plan and design investigations, collect data, make sense of data, or report your findings. It is even less likely that your teacher was concerned with examining your metacognitive knowledge, your attitudes and dispositions toward learning science, or your motivation to learn. Teachers using project-based methods might still use paper-and-pencil tests from time to time, but the focus of their tests is different. These tests measure what students learned by doing investigations. The focus is not mastery of inert knowledge, such as vocabulary terms for the sake of learning the terms. Rather, understandings that can be measured on a test are the outcome of investigations. As a result, assessment is aligned with important learning goals. The test, as shown in the second scenario, may also be used to help students monitor their own progress or to help the teacher adjust his or her teaching.

Traditionally, students were expected to know vocabulary words, such as vertebrate, invertebrate, crustacean, mollusk, and reptile. For almost all students, this type of knowledge didn't have much relevance outside of school. It was thought that knowing such vocabulary words would build a foundation for understanding science. Today, science educators see the teaching of such concepts as a much more complex process than simply requiring students to memorize definitions and recognize examples. Today, teachers are concerned with examining students' prior beliefs about the scientific world and their understanding of its major concepts. Science educators also see knowledge as something that helps students analyze a question, solve a problem, or conduct an investigation in a way that is meaningful to them. This type of knowledge is not superficial. For example, rather than focusing on the definition of invertebrate, crustacean, and reptile, a teacher might have students analyze how a crustacean (with a hard outer covering) and a reptile (with a hard shell) are alike and different. This knowledge of similarities and differences would help students set up a functioning aquarium to investigate a driving question about how some crustaceans and reptiles can coexist in a saltwater tank with poisonous invertebrates. If a group of students actually assembled a functioning aquarium, they would be applying their knowledge and skills. As a result, the assessment of such an activity would extend beyond measuring inert knowledge. This assessment is more consistent with today's assessment goals (NRC, 1996, 2000, 2001; NSTA, 2002).

Learning Activity 9.5

HOW CAN I CHOOSE A GOOD ASSESSMENT TECHNIQUE TO MEASURE UNDERSTANDING?

Materials Needed:

- Paper and pencil or a computer

A. What follows are several learning performances that might be contained in a typical elementary or middle school curriculum. Form a team of three students. Read each learning performance and decide what would be the most beneficial ways to measure whether students have gained understanding of the topic. Keep in mind that good assessment should:
   1. Be responsive to context.
   2. Be a continuous process embedded in instruction.
   3. Be multidimensional.
   4. Engage students in the assessment process.
   5. Be valid and reliable.
   6. Match today's educational goals.
   7. Accommodate cultural diversity.
   8. Be consistent with cognitive learning theory.
   9. Measure meaningful understanding.

LEARNING PERFORMANCES

- After investigating simple machines, each student will show how machines help us do work.
- After investigating what animals need to eat, each student will be able to distinguish between a consumer and a producer.
- After investigating what types of trees are in our environment, each student will be able to identify two types of trees.
- After investigating where garbage goes, each student will be able to tell why it is important to recycle.

B. As a team, develop a lesson to teach one of the learning performances. Develop two different ways of assessing the objective after you have taught it. Teach the lesson to another team in your class and try the two different methods of assessment.

C. Ask the team you taught which method of assessment best measured what they learned. How do that team's beliefs about good assessment techniques compare to yours?

In Learning Activity 9.5, you will develop your own assessment procedure for measuring understanding.

The National Science Teachers Association (NSTA, 2002) released a position statement about assessment in science. The NSTA stressed assessment of understanding of science content and process; ability to think critically and solve problems; ability

CONNECTING TO NATIONAL SCIENCE EDUCATION STANDARDS

CLASSROOM ASSESSMENT AND INQUIRY

In the context of inquiry, assessments need to gauge the progress of students in achieving the three major learning outcomes of inquiry-based science teaching: conceptual understandings of science, abilities to perform scientific inquiry, and understandings about inquiry (NRC, 2000, p. 75).

to design scientific experiments, analyze data, and draw conclusions; ability to recognize relationships between science topics and real-world issues; and skill in using mathematics as a tool to learn science. Similarly, the National Research Council (2000) stressed that we need to assess science content, science process, and understandings about inquiry.

Current assessment also monitors the affective side of learning. Students are encouraged to ask questions, become excited about an investigation, remain open-minded about new ideas, be thorough in their investigation, stay fair in their conclusions, and be honest in their presentation of evidence. Imagine that a group of students is investigating the driving question "Where does all the garbage go?" The students could become aware of landfill problems, discuss their attitudes about the situation, and create artifacts that depict the problem without exhibiting positive dispositions or taking action, two other objectives of the project. It would be better if students responded to the landfill problem by recycling aluminum cans and encouraging others to recycle. It would be even better if students organized a schoolwide recycling program for all possible school products—glass, aluminum, steel cans, paper, cardboard, and plastic. By focusing assessment on the different types of knowledge (content, procedural, and metacognitive), skills, and attitudes of students, it broadens the purpose of assessment and treats it as a process that matches instructional goals and life outside of school.

Table 9.1 illustrates how we can assess different knowledge domains.

When to Assess

Research shows that assessment can help students learn science (Black & Wiliam, 1998b). For assessments to help students learn, however, they need to inform practices in the classroom. If used effectively, both formative and summative assessments can inform classroom practices and help students learn. **Formative assessments** are those that help teachers make day-to-day decisions about instruction or help students learn. These assessments are used in the context of instruction to assess prior knowledge and are also embedded into instruction to guide teaching and learning. Using the classification introduced earlier, immediate, close (or embedded), and proximal assessments tend to be formative in nature. **Summative assessments** occur at the end of a unit of instruction or time period (such as upon completing a grade level) to determine achievement, issue grades, promote students, or demonstrate accountability (NRC, 2001). As discussed earlier, summative assessments are often external tests

TABLE 9.1 Assessing Learning Performances

The cognitive process dimension

	Remember	Understand	Apply	Analyze	Evaluate	Create
The knowledge domain	*Definition:* to recall something from the memory	*Definition:* to comprehend and be able to explain the meaning of a concept or process	*Definition:* to develop solutions to familiar or new problems	*Definition:* to examine a concept or process in detail, to learn more about it	*Definition:* to make judgment about value, quality, importance, or condition	*Definition:* to produce something
Factual knowledge *Definition:* Knowledge of details and facts	*Learning performance:* State that matter is made up of atoms	*Learning performance:* Use the idea of conservation of atoms to explain how the mass after a chemical reaction compares to the mass before the reaction	*Learning performance:* Carry out an experiment to illustrate the conservation of mass	*Learning performance:* Given a table of data, be able to discriminate the relevant from irrelevant data needed to answer a question related to conservation of mass in a chemical reaction	*Learning performance:* Judge appropriate if conclusions were drawn regarding a set of date collected from a series of chemical reactions	*Learning performance:* Design an experiment to illustrate the conservation of mass
	Assessment: Have students select the correct definition of matter on a multiple choice test item	*Assessment:* Have students complete a short-answer test item asking how the mass after a chemical reaction compares to the mass before the reaction	*Assessment:* Examine a student's science journal to see how he or she carried out an experiment	*Assessment:* Interview a student to probe what he or she means by relevant and irrelevant data; ask what the relevant data are	*Assessment:* Examine student's daily journals for evidence of appropriate conclusions	*Assessment:* Examine student's notes from a science journal to determine how the experiment was designed

(Continued)

TABLE 9.1 Continued

	Remember	Understand	Apply	Analyze	Evaluate	Create
Conceptual knowledge *Definition:* Knowledge of interrelationships, principles, theories, and models	*Learning performance:* Define flow of energy in an ecosystem *Assessment:* Have students write a poem that illustrates flow of energy in an ecosystem	*Learning performance:* Explain how sunlight is transformed through photosynthesis to provide food for plants, herbivores, and carnivores *Assessment:* Have students write a song that explains how sunlight is transformed through photosynthesis to provide food for plants, herbivores, and carnivores	*Learning performance:* Implement an experiment to show the influence of sunlight on the growth of plants *Assessment:* Have students make a video of their experiment and evaluate the ability to implement the experiment from watching the video	*Learning performance:* Analyze an ecosystem to show how extinction of a small carnivore might influence the population of a herbivore *Assessment:* Have student draw diagram that illustrates how extinction of a small omnivore might influence the population of an herbivore	*Learning performance:* Evaluate how a new government policy might influence various species in ecosystem *Assessment:* Have students write a position statement regarding how a new government policy might influence various species in an ecosystem	*Learning performance:* Create an alternative EPA policy to improve the ecosystem of the Great Lakes *Assessment:* Have students write to their local newspaper to suggest an alternative EPA policy to improve the ecosystem of the Great Lakes

	Remember	Understand	Apply	Analyze	Evaluate	Create
Procedural knowledge *Definition:* Knowledge about how to do something, conduct inquiry, or use a skill	*Learning performance:* Recognize the parts of a microscope *Assessment:* Have students label the parts of a microscope from a drawing	*Learning performance:* Clarify to a fellow students how to focus a microscope *Assessment:* Have a student teach a peer how to focus a microscope	*Learning performance:* Use a microscope to observe single-celled pond organisms *Assessment:* Observe students while they are focusing a microscope to determine if they can focus it to see a pond organism	*Learning performance:* From observations of single-celled organisms, organize a data chart to distinguish the organism *Assessment:* Have students create a chart (with appropriate labels) of single-celled organisms	*Learning performance:* Determine whether conclusions drawn from observations of single-celled organisms are appropriate *Assessment:* Have students record their conclusions in a science journal and defend how the conclusion logically follows the observations	*Learning performance:* Devise a procedure to determine if the number of single-celled organisms in a pond sample changes over time *Assessment:* Have students present their procedures to the class for determining if the number of single-celled organisms in a pond sample changes over time; rate the quality of the procedure
Metacognitive knowledge *Definition:* Self-knowledge, knowledge of one's own cognition	*Learning performance:* Recall that you find writing in a science journal challenging *Assessment:* Interview a student about his or her strengths or weaknesses recalled from the past science experiences	*Learning performance:* Summarize which tasks you find most difficult in science classes and why they are difficult for you *Assessment:* Have students summarize in a portfolio which tasks they find most difficult in science classes and why	*Learning performance:* Implement a procedure to help you study for a science test *Assessment:* Have a student reflect in a journal entry on his or her learning and design a way to study	*Learning performance:* Break down the components of a science test that you will need to focus on to be successful *Assessment:* Have students send you an e-mail message, starting how they will change their study habits to better understand the material	*Learning performance:* Critique whether you have improved you own test-taking skills *Assessment:* Have students summarize their improvement on test-taking skills by making an entry in a portfolio	*Learning performance:* Construct a portfolio and organize it in a manner that would be most compatible with your own learning style to show what you have learned *Assessment:* Discuss with a student why his or her portfolio is arranged as it is

Source: Modified from L. W. Anderson & D. R. Krathwohl (2001).

used to determine placement, promotion, achievement, or program success (accountability of the schools). Thus, distal and remote assessments tend to be summative in nature. Summative assessments can also inform instruction and learning if used as illustrated in the second scenario. Here, a summative assessment (the state's achievement test that was aligned to the National Science Education Standards) was used to assess attainment of important understandings of force and motion in the state's curriculum framework. When used in this fashion, the distinction between formative and summative can become somewhat blurred—the assessment can help students revise and redo their work or help teachers revise their subsequent teaching.

The National Science Education Standards assert that effective formative assessment has a framework of three guiding questions:

1. Where are you trying to go? (identify and communicate the learning and performance goals).
2. Where are you now? (assess or help the student self-assess current levels of understanding).
3. How can you get there? (help the student with strategies and skills to reach the goals) (NRC, 2001, p. 14).

Using this framework, a teacher clearly communicates expected learning or performance goals to students. For example, the teacher might explain to the

---

CONNECTING TO NATIONAL SCIENCE EDUCATION STANDARDS

THE RELATIONSHIP BETWEEN FORMATIVE AND SUMMATIVE ASSESSMENT

Tests that are given before the end of a unit can provide both teacher and student with useful information on which to act while there is still opportunity to revisit areas where students were not able to perform well. Opportunities for revisions on tests or any other type of assessment give students another chance to work through, think about, and come to understand an area they did not fully understand or clearly articulate the previous time (NRC, 2001, p. 62).

CONNECTING TO NATIONAL SCIENCE EDUCATION STANDARDS

ASSESSMENT IN THE CLASSROOM

Formative assessment refers to assessments that provide information to students and teachers that are used to improve teaching and learning. These are often informal and ongoing, though they need not be. Data from summative assessments can be used in a formative way. *Summative assessment* refers to the cumulative assessments, usually occurring at the end of a unit or topic coverage, that intend to capture what a student has learned, or the quality of the learning, and judge performance against some standards. Although we often think of summative assessments as traditional objective tests, this need not be the case. For example, summative assessments could follow from an accumulation of evidence collected over time, as in a collection of student work (NRC, 2001, p. 25).

students that they will understand the concept of *friction* after they finish investigating their subquestion "How can we make roller blades go faster?" The teacher uses a variety of techniques to help students assess their current level of understanding. For example, the teacher might have students keep a science journal where they record their investigations involving roller blades rolling across different surfaces (rough, smooth, wet, dry, etc.). The teacher would provide students with timely **feedback**. For example, a teacher might sit down with a student to discuss his or her journal and provide suggestions for testing additional ideas related to friction against the roller blades. If the students in the class are having a difficult time understanding the concept of friction, the teacher might modify his or her teaching to introduce a new lesson to teach about friction.

Regardless of technique used (such as written remark or conferences with students), it is the teacher's responsibility to implement the science standards and provide students with regular feedback regarding how they are progressing toward desired learning performances. It is also the teacher's responsibility to keep good records so that he or she can give each student feedback and report progress to parents. Teachers should give students opportunities to revise their work and improve it based on feedback. It is the student's responsibility to use feedback to take action to learn the desired knowledge or skill. Many techniques can be used to give students feedback, such as journals, and to help students take responsibility to learn, such as self-evaluations. We devote Chapter 10 to discussing numerous assessment techniques.

Using Technology Tools to Examine Assessment

One way that teachers can improve assessment is to network with other teachers. By networking, teachers make time to discuss their teaching and assessment practices with others. Often this discussion focuses on students' work to determine whether the students understand the expected learning performances. Technology tools can help you become part of a community of learners who are interested in this type of discourse. For example, teachers who have little time to meet together might exchange ideas with each other over e-mail, discuss particular assessment concerns in an online chat room, or share assessment ideas in a threaded discussion.

> **CONNECTING TO NATIONAL SCIENCE EDUCATION STANDARDS**
>
> **AN INTRODUCTION TO ASSESSMENT IN THE SCIENCE CLASSROOM**
>
> Teachers need time and assistance in developing accurate and dependable assessments. Much of this assistance can be provided by creating settings in which teachers have opportunities to talk with one another about the quality of student work (NRC, 2001, p. 9).

The National Science Teachers Association (NSTA) has a section at its Web site entitled *NSTA Community*. This section of the NSTA Web site includes a discussion board where you can talk with others. The NSTA *Building a Presence for Science* section is a networking initiative designed to improve the teaching and learning of science. The specific goal of this initiative is to break down the isolation some teachers face and develop a professional network of colleagues. In this project, NSTA connects key leaders around the country to school districts to provide electronic communication and support for teachers. Similarly, many state science professional organizations have Web sites where teachers can interact with others to discuss curriculum and assessment. These Web sites often enable teachers to talk with others about their state's standards and high-stakes assessment. Finally, the Eisenhower National Clearinghouse Web site (http://goENC.com) contains a number of links to places where teachers can communicate and collaborate with others.

Chapter Summary

In this chapter, we discussed the purpose and nature of assessment. We explored the advantages and disadvantages of various types of assessments. We considered features of high-quality assessments that are responsive to context, embedded in instruction, and multidimensional. Current assessments also engage students in the assessment process, are valid and reliable, match today's educational goals, accommodate cultural diversity, are consistent with learning theory, and measure meaningful understanding. The chapter examined what to assess by focusing on measurement of learning performances across a variety of knowledge domains. We discussed both formative and summative assessments. Finally, we discussed how technology tools can help teachers network with others to examine their assessment practices.

Chapter Highlights

- Assessment is any method used to judge or evaluate an outcome or help make a decision.
- Assessment data are used for a variety of purposes in educational settings today:
  - Assisting learning
  - Guiding teaching
  - Measuring individual achievement.
  - Giving grades
  - Allocating resources
  - Evaluating programs
  - Informing local, state, and national policy
- Research shows that regular, high-quality assessment can have a positive effect on student achievement.
- Assessments can be classified as immediate, close (embedded), proximal, distal, or remote.
- Large-scale assessments are used to evaluate students' achievement, school curriculum, school district progress, and programs.
- Many educators argue that large-scale assessments are only one-shot methods and may not measure what is taught in a local school's curriculum.
- Some educational leaders view large-scale assessments as insufficient for meeting the assessment standards found in the *National Science Education Standards* and advocate for multidimensional assessments.

- Assessment appropriate for today's science classrooms has several features:
  - It is responsive to context.
  - It is a continuous process embedded in instruction.
  - It is multidimensional.
  - It engages students in the assessment process.
  - It is valid and reliable.
  - It matches today's educational goals.
  - It accommodates cultural diversity.
  - It is consistent with cognitive learning theory.
  - It measures meaningful understanding.
- Assessments measure science content, science process, and understandings about inquiry.
- Assessment also monitors the affective side of learning.
- The cognitive process domain can be used as a framework to assess student understanding.
- Formative assessment helps teachers evaluate students' progress during a unit of study and make day-to-day decisions regarding curriculum and instruction.
- Effective formative assessment has a framework of three guiding questions:
  - Where are you trying to go?
  - Where are you now?
  - How can you get there?
- Summative assessment occurs at the end of a unit of instruction or time period to determine achievement, issue grades, promote students, or demonstrate account-ability. Both formative and summative assessments can inform classroom practices and help students learn.
- Technology tools can help teachers become part of a community of learners who are interested in improving assessment practices.

Key Terms

Appropriateness of context	Immediate assessments
Assessment	Large-scale assessments
Authentic assessment	Metacognitive knowledge
Clarity of expectations	Necessity
Classroom assessments	Procedural knowledge
Close assessments	Proximal assessments
Comprehensibility	Reflective practitioners
Content knowledge	Reliable assessment
Distal assessments	Resistance to test-wiseness
Embedded assessment	Remote assessments
Feedback	Sufficiency
Formative assessments	Summative assessment
High-stakes assessment	Valid assessment

References

American Association for the Advancement of Science. (1993). *Benchmarks for science literacy.* New York: Oxford University Press.

American Association for the Advancement of Science. (2004, Summer/Fall). Assessment with precision. *2061 Today, 14*(1), 1–8.

Anderson, L. W., & Krathwohl, D. R. (Eds.). (2001). *A taxonomy for learning, teaching, and assessing: A revision of Bloom's taxonomy of educational objectives.* New York: Longman.

Atwater, M. M. (2000). Equity for black Americans in precollege science. *Science Education, 84*(2), 155–179.

Black P., & Wiliam, D. (1998a). Assessment and classroom learning. *Assessment in Education, 5*(1), 7–74.

Black, P., & Wiliam, D. (1998b, October). Inside the black box: Raising standards through classroom assessment. *Phi Delta Kappan, 139*–148.

Fusco, D., & Calabrese-Barton, A. (2001). Representing student achievement in science. *Journal of Research in Science Teaching, 38*(3), 337–354.

Herman, J. L., Ashbacher, P. R., &. Winters, L. (1992). *A practical guide to alternative assessment.* Alexandria, VA: Association for Supervision and Curriculum Development.

Klassen, S. (2006). Contextual assessment in science education: Background, issues, and policy. *Science Education, 90*(5), 820–851.

Lawrenz, F., Huffman, D., & Welch, W. (2001). The science achievement of various subgroups on alternative assessment formats. *Science Education, 85*(3), 270–290.

McAfee, O., & Leong, D. J. (2002). *Assessing and guiding young children's development and learning.* Boston: Allyn and Bacon.

Mislevy, R. J. (1993). Foundations of a new test theory. In N. Frederiksen, R. J. Mislevy, & I. I. Bejar (Eds.), *Test theory for a new generation of tests..* Hillsdale, NJ: Erlbaum.

National Research Council. (1996). *National science education standards.* Washington, DC: National Academy Press.

National Research Council. (2000). *Inquiry and the national science education standards.* Washington, DC: National Academy Press.

National Research Council. (2001). *Classroom assessment and the national science education standards.* Washington, DC: National Academy Press.

National Science Teachers Association. (2002). NSTA position statement on assessment. *NSTA Reports, 13*(3), 14–15.

Ogle, D. (1986). A teaching model that develops active reading of expository text. *The Reading Teacher, 39*(2), 564–570.

Packer, A. H. (1992). Taking action on the SCANS report. *Educational Leadership* 49, 27–31.

Pellegrino, J. W., Chudowsky, N., & Glaser, R. (2001). *Knowing what students know: The science and design of educational assessment.* Washington, DC: National Academy Press.

Ruiz-Primo, M. A., Shavelson, R. J., Hamilton, L., & Klein, S. (2002). On the evaluation of systemic science education reform: Searching for instructional sensitivity. *Journal of Research in Science Teaching, 39*(5), 369–393.

Rutherford, F. J., &. Ahlgren, A. (1990). *Science for all Americans.* New York: Oxford University Press.

Solano-Flores, G., & Nelson-Barber, S. (2001). On the cultural validity of science assessments. *Journal of Research in Science Teaching, 38*(5), 553–573.

Treagust, D. F., Jacobowitz, R., Gallagher, J. L., &. Parker, J. (2001). Using assessment as a guide in teaching for understanding: A case study of a middle school science class learning about sound. *Science Education, 85*(2), 137–157.

Wenglinsky, H. (2000). *How teaching matters: Bringing the classroom back into discussions of teacher quality.* Princeton, NJ: Educational Testing Service.

Chapter 10

Assessing Student Understanding

Introduction
Assessment of Student Understanding
Another Look at the Advantages of Educational Assessment

Introduction

This chapter focuses on ways that we can assess student understanding in a project-based science class. You may have many questions about assessment: *How do I assess science understanding? Are some ways to assess better than others? What are portfolios, and how do I use them in science classes? How do I know that all students are learning?* This chapter answers these and other questions about assessment. In this chapter, we consider numerous strategies for gathering assessment information, such as conducting observations, using concept maps, writing test items, and implementing performance-based assessments. We discuss artifacts as ways to form lasting memories in students' minds, strategies for assembling information into artifacts, and ways that students can present information. We consider information about using scoring rubrics to evaluate assessment information. We discuss how to use assessments for determining grades. Before we begin, let's examine three classroom scenarios that focus on stages of assessment: gathering, assembling, and making judgments.

Chapter Learning Performances

- *Explain the steps a teacher can use to assess student growth and achievement.*
- *Create assessments, using different strategies to measure content, procedural, and metacognitive knowledge.*
- *Compare and contrast the reasons for using various types of assessment.*
- *Describe how a teacher can determine if a test or quiz is well constructed.*
- *Justify why artifacts and portfolios should be consistent with goals in a project-based environment.*
- *Design a scoring rubric to evaluate a performance.*
- *Explain the value of assessment for teachers, students, and parents.*

Scenario 1: Gathering Assessment Information

Mrs. Miller tells her fourth-grade class, "Please pass your papers to the front of the row. I will be grading these tonight and putting them down in the grade book for this quarter. Also, remember to study tonight for the test tomorrow on animal classification. This will be the last test that will go toward determining your science grade for this quarter." You tense up a bit with the thought of tomorrow's test on animal classification, because you really don't have time to study tonight—it is the big soccer game night. Oh well, you know that Mrs. Miller will put down the test and homework score and average it with the rest from the semester. You think you did well on the other tests this quarter, but there are a few that Mrs. Miller hasn't passed back to the class yet. If you are correct, your average (and, thus, your grade for the quarter) shouldn't be affected much, even if you don't do as well on the test tomorrow.

In this scenario, Mrs. Miller viewed assessment as solely for the purpose if summative evaluation (determining achievement from end-of-unit tests and homework). She used the average of scores on homework and tests to issue a letter grade at the end of the quarter. She did not always hand back homework or tests in a timely manner; and she did not use tests or homework to help students improve their learning, set goals, or monitor their own progress. In short, it would appear that Mrs. Miller has not aligned assessment with her instruction. It is also difficult to determine whether she has assessment tightly matched with the learning goals.

Scenario 2: Assembling Assessment Information

Ms. Mominee, a second-grade teacher, sits down with one of her students, Rachael, to discuss her drawings of her plant in the plant portfolio she turned in at the end of the previous week. Ms. Mominee says, "Rachael, I see that you have five drawings here in your portfolio that show how your plant has looked in the last few weeks. Tell me something that you learned about plants." Rachael responds, "I learned that they need water or they will wilt like this one did." Ms. Mominee probes, "How do you know that this one wilted because of lack of water?" Rachael says, "Because I know the bell rang to go to lunch that day when I was watering my plants, and I stopped to go to lunch. I forgot to water that one." "Oh," responds Ms. Mominee, "are there any other things you learned about what plants need to grow?" Rachael says, "Dirt!" Ms. Mominee probes further to determine Rachael's understanding of the need for sunlight, and she helps Rachael come to understand that some plants grow without dirt. She finishes the discussion by asking Rachael how well she thinks she did on the plant project. After she finishes talking with Rachael and other students about their plant portfolios, she makes some written comments on the portfolios, records a grade in her grade book for this project, and sends the portfolios home for the students' parents to see.

In this scenario, Ms. Mominee used an immediate assessment technique associated directly with the curriculum to help determine progress and grades. She used more than one method of classroom assessment (self-evaluation, portfolios, and a formal clinical interview). However, she provided the students with feedback, but only at the end of the project. The teacher did not meet with students to provide them with feedback while they were growing their plants, an activity in the project "How do plants grow in our environment?" Thus, the students could not use the feedback to modify their project, and the teacher could not use the feedback to reflect on her teaching and possibly change her instructional plans.

Scenario 3: Making Judgments About Assessment Information

Students in Mr. Iwamoto's sixth-grade class are investigating the question "How do plants grow in our environment?" Several students set up an investigation to examine the effect of organic insect-control methods (as opposed to using pesticides) on plant growth. As students plan their investigations, Mr. Iwamoto asks them to keep a journal about their plans. The students set up a few procedures to test out some homemade greenhouses. Mr. Iwamoto sits down with the group to discuss their plan and give the students feedback. He says, "Tell me why you decided to construct a small greenhouse for your investigation." Lei responds," Well, we think that we need to make sure that the temperature is controlled so that it doesn't affect our results. The insects need to be in the same temperature so that it doesn't affect if they live or die." Mr. Iwamoto says, "Will one greenhouse let you compare the organic and pesticide-sprayed plants?" Lei and her classmates say, "Yes." Mr. Iwamoto realizes that he hasn't clearly taught about controlling variables and students may need a lesson on pesticides. His students don't seem to realize that they have two separate variables involved (organic and chemical uses to control for insects), and they aren't aware that pesticides placed in close proximity to the organically controlled plants could run off to the organic plants and contaminate the investigation. Thus, Mr. Iwamoto adapts his instruction and teaches about controlling variables and pesticides in several subsequent lessons.

The next week, Mr. Iwamoto gives a small quiz to his class about topics they are studying. The quiz covers concepts of photosynthesis, plant form and function, and environmental factors that affect plant growth. He returns the quiz to students and discusses the results. He asks the students whether they need to modify anything in their investigations based on what they have learned so far. One group of students decides that they need to be able to measure the temperature of the soil where their plants are growing in order to have accurate data regarding use of organic compost soil.

When students are finished with their study of plants, Mr. Iwamoto has the groups of students present their artifacts to their classmates. Some students have photographs of their investigations. One group gives a PowerPoint presentation to the class. Students in the class are encouraged to critique each other's presentations and offer suggestions for further study. Finally, Mr. Iwamoto uses a scoring rubric to provide students with summative feedback and give them a grade (score).

In this scenario, Mr. Iwamoto used both formative and summative assessment procedures. He used formative assessments to provide students with feedback that helped them learn and to modify his teaching. Mr. Iwamoto also used a variety of immediate and close classroom assessment techniques and did not rely only on the quiz he administered. He used summative assessment to judge students' achievement and issue a grade.

Assessment of Student Understanding

In the previous chapter, we defined **assessment**, and we discussed the nature and benefits of assessment. In this chapter, we will discuss how teachers assess student understanding in a project-based science environment. As we discuss these, you will see how teachers can use various techniques to align learning goals, instruction, and assessments.

The process of assessing student understanding might be thought of as a three-step procedure. First, you and your students gather information that will help in forming generalizations about students' learning to determine if your learning goals are being achieved. Second, after securing information during the instructional process, you and your students assemble and present the information in some fashion so that it is recorded. Third, you and your students evaluate the assessment information. The purpose of evaluation is to make judgments about student growth, to set goals, and to report information to students and their parents.

Imagine a third-grade teacher who has the learning goal: *Students will learn how different animals grow and change. Insects go through different stages of growth called metamorphosis.* The teacher's instructional plans have students investigating how insects grow and change. Students could record in their notebooks information about the length, mass, and appearance of the insects at different stages of growth. During inquiry, the teacher could observe the students and keep information about their progress on a checklist. To assemble and present information, students might draw pictures or take photographs of the insects at each stage and share them with classmates. The teacher could place copies of these items into each student's portfolio. Finally, to evaluate progress, the students might keep a journal in which to enter their personal reflections. The teacher could use a scoring rubric (described later in this chapter) to determine levels of understanding or skill attainment on a per-

CONNECTING TO
NATIONAL SCIENCE
EDUCATION STANDARDS

ASSESSMENT IN
SCIENCE EDUCATION

Assessment is a systematic, multistep process involving the collection and interpretation of educational data (p. 76).

formance-based assessment. Assessment that can provide feedback to students and to the teacher requires multiple sources of information and methods of evaluating. The following sections will discuss specific methods of collecting, assembling, and evaluating assessment information.

Gathering Assessment Information

Classroom teachers have many assessment options. No matter which options they choose, teachers need to be able to collect sufficient information so they can make generalizations about students' learning and make good planning decisions. The information needs to be relevant to teaching and the goals desired in the classroom. Therefore, the information-gathering techniques presented in this section are only options, not recommendations. It is up to you, the expert on your classroom, to continually reflect upon your learning goals, adopted curriculum, instructional methods, and students' performance in order to select and develop methods for gathering information that meet the particular needs of your classroom.

Observation-Based Assessment

Observation is a great technique for gathering data. Observations can help teachers make valid decisions about curriculum and instruction and determine how much progress students are making. Observations play a key role in a teacher's minute-to-minute decision making during a project.

Teachers often make informal observations of students. For example, you might notice that a particular student raises her hand often during social studies but very rarely during science, indicating that she understands social studies topics but is having difficulty with science concepts. A puzzled look on a student's face might provide you with the evidence that a student is having difficulty with a concept or task. Off-task behavior by students might indicate problems with methods of instruction. These types of observations can help you make "in-flight" curricular or instructional decisions that refocus a lesson.

However, observations can also be planned and formal. For example, you may intentionally ask groups of students questions about the driving question and their investigative plans. You can also purposefully listen to and watch a particular student while he solves a problem, reads aloud, or creates a project to identify problem areas, as well as areas of mastery.

Observations are inevitably subjective to some degree. When you observe a particular behavior, you interpret what you see. For example, a college student wore an iPod with ear buds during his science methods class. During the first week of class, the professor interpreted his behavior as rude and inattentive, and she decided that he was a mediocre student. However, after talking with him about the iPod, she discovered that he was taking sixteen quarter hours and working two jobs to pay for his tuition. The science methods class was his last class of the day, and he was listening to music to stay awake during class. Incidents like this demonstrate why it is preferable to use a variety of observation techniques that provide varied information than to rely on a single technique. There are several observation techniques that you can use to ensure that you gather varied information. These include discussion-based observations, anecdotal records, checklists, and clinical interviews.

Discussions. During the instructional process, discussions can provide a rich array of information about students' understanding of important science concepts. Discussions can also be used to identify students' process skills and thinking patterns during investigations. Finally, discussions can be used to check students' attitudes. Using discussions to extract information about students' knowledge, skills, and attitudes is also less threatening to most students than are traditional forms of testing and, consequently, invites broader student participation.

Unfortunately, teachers' lack of discussion skills often limits the information they glean from students during a discussion. They sometimes ask too many superficial questions, give students too little time to formulate answers to their questions, or call on the same students all the time. To obtain better answers from more students during a discussion, teachers should learn the techniques of **wait-time**, **probing**, and **redirecting**. These techniques were discussed more thoroughly in Chapter 8. With good questioning technique, teachers can find out from students how well they really understand science concepts.

Teachers can also engage students in discussions about their thought processes and their investigative procedures. For students to discuss their thought processes, teachers sometimes have them "think aloud"—that is, talk about what they are thinking during an investigation. Ensuing discussions can then focus on their reasoning, beliefs, and misconceptions.

Discussions in project-based classrooms can also be used to detect students' attitudes. For example, students investigating why frogs are disappearing in their state might discover that sulfur-burning factories in the area have increased the acid in the air to a level that is toxic to frogs. Students can then discuss whether they would vote for stricter air pollution laws. This type of value-oriented discussion is good for making students confront their own beliefs and encouraging them to apply those beliefs in the area of social policy.

Anecdotal Records. In elementary and middle grade classrooms with 20 to 30 students, teacher's observations can easily be forgotten or associated with the wrong students. For these reasons, many teachers choose to keep anecdotal records of their observations. **Anecdotal records** include written notes that describe student behaviors made at or near the time that the behaviors occurred. These records can be kept in a teacher's notebook, in folders, on index cards, on sticky notes or labels, or on the computer. After a class period, you might make notations about students' understandings, types of questions they asked, and possible misunderstandings demonstrated. Some teachers have all students jot down their responses on paper before calling on anyone to give an answer. This technique gives each child the time to think about and compose his or her response before the question is discussed, and it gives the teacher a written record of students' thinking. Hand-held computers or personal digital assistants (PDAs) permit teachers to easily make observations *during* instruction, instead of after.

We find keeping a file with an index card on each student to be a valuable technique. At the end of each day, take a moment to write down any significant observations about students on their cards. You will find these cards very beneficial when you write up a student's progress report or have discussions with parents at parent-teacher conferences.

Checklists. To keep track of students' observations, teachers often keep **checklists**. These checklists can be used to evaluate knowledge, skills, or attitudes. When used for assessment, checklists can count or tally the frequency of something. For example, on a class roster, you can easily tally the frequency of students' participation in classroom discussions. Or you might maintain a checklist of important concepts and check off students who demonstrate understanding of the concepts. For example, such a list might include "understands why machines make work easier," "sees why we lubricate machines to reduce friction," and "knows how energy is transferred in a system."

You can assess skills by observing students and recording on a checklist which skills they use. During a lesson, walk around the room with your checklist while students are working. Choose a few students to observe and mark on the checklist the skills in which they seem to be proficient. Imagine your students working on a project that explores the kinds of trees in the neighborhood. As part of the project, they observe leaves, classify them into groups, and graph the numbers of trees in each classification. You might use the checklist in Table 10.1 for this type of classification project.

Checklists save teachers time. Teachers can rapidly mark off on a checklist evidence of a skill or the meeting of an objective. You might keep an observation sheet in a three-ring binder for each student, making a functional grade book for tracking student progress. This method of observation has the added benefit of keeping you focused on your lesson objectives. It also helps you make sure that no one falls between the cracks and goes unnoticed. Finally, it provides documented evidence of progress for grade cards or progress reports or for discussions with students and their parents.

Checklists can be either project specific, as was the leaf-classification checklist, or generic. Table 10.2 shows a generic checklist designed to gather information about

---

TABLE 10.1   Project-Specific Checklist

1. Observation

_____ 1.1 Can identify major characteristics of the leaves

_____ 1.2 Notices similarities among the leaves

_____ 1.3 Notices differences among the leaves

2. Classification

_____ 2.1 Can determine a method of grouping leave into two groups

_____ 2.2 Can determine a way to group leaves within groups

_____ 2.3 Can explain why leaves are classified

_____ 2.4 Can classify a new leaf into the existing classification scheme

TABLE 10.2   Generic Checklist

Questioning	Gathering data to explore questions
The student can:	The student can:
___ Explore ideas	___ Observe properties
___ Form a question	___ Identify ways to measure data
___ Remain curious	___ Seek additional information
___ Design an investigation	___ Use science equipment
___ Formulate a hypothesis	___ Measure accurately
___ Note discrepant events	___ Estimate answers
___ Notice problems	___ Examine more than one variable
___ Challenge ideas	___ Select variables
___ Remain skeptical	___ Name an object
Examining data to answer questions	**Answering questions**
The student can:	The student can:
___ Classify objects into groups	___ Communicate results
___ Make predictions	___ Summarize information
___ Analyze results	___ Make revisions
___ Describe observations	___ Make decisions
___ Make comparisons	___ Demonstrate knowledge
___ Graph information	___ Identify limitations
___ Order or sequence data	___ Make inferences
___ Search for patterns	___ Describe discrepancies
___ Tabulate information	___ Establish criteria
___ Identify errors	___ Offer evidence
___ Interpret data	___ Verify results
___ Control variables	___ Solve problems
___ Calculate	___ Generate new questions
___ Make drawings or diagrams	___ Establish relationships

general skill development. In Learning Activity 10.1, you will try your hand at developing a skills checklist.

Checklists can also provide you with a list of affective attributes to look for in a project-based science classroom. Table 10.3 is an inventory of dispositions to be used

Learning Activity 10.1

DEVELOPING A SKILLS CHECKLIST

Materials Needed:

- Materials to teach a lesson developed in this investigation

A. Design a lesson that requires a great number of skills, such as observing, measuring, classifying, inferring, and concluding. Teach this lesson to a team of classmates after you design a checklist of all the skills you might observe while your classmates are engaging in the investigation. Try to keep track of your classmates' performance while they participate in your lesson.

B. Share the checklist with your classmates. Discuss whether it was an accurate measure of the skills they used. How might it be improved?

TABLE 10.3 Affective Attributes of Students in Project-Based Science

Ability to compromise with others	Flexibility with ideas	Responsibility to project
Ambition to investigate	Honesty in artifacts	Satisfaction with artifacts
Cooperation with others	Independence	Self-confidence
Curiosity about the world	Objectivity	Self-discipline
Dependability	Open-mindedness	Self-reliance
Disciplined thinking	Patience with others	Sensitivity to others
Enthusiasm to continue	Persistence with a task	Skepticism about results
Excitement about science	Precision	Thoroughness
Fascination with findings	Questioning attitude	Tolerance for change
	Respect for evidence	Willingness to change

as a guide when observing students. In Learning Activity 10.2, you will develop an assessment for monitoring affective attributes.

Clinical Interviews. Interviews with people seeking jobs can tell an employer a great deal that résumés, grade transcripts, or applications cannot. For example, an interview can provide insight into a candidate's motivation and rapport with people. Also, during an interview, a candidate has time to elaborate on specific items in his or her résumé, transcript, or application. Interviews are also common for candidates seeking Master's or Doctoral degrees. During these 1- or 2-hour interviews, university committee members can ask candidates unanticipated questions, and candidates can demonstrate in greater depth what they know.

Classroom interviews, or **clinical interviews**, with your students can serve similar purposes. In an interview, students can explain in greater detail what they understand, how they're progressing, what problems they're having, and what steps they or you might take to improve learning. In an interview, you can get to know students, work on individual student goals, and clarify misunderstandings and concerns. Interviews provide a depth of information that other forms of assessment cannot. They also provide opportunity for clarification of classroom observations (Southerland, Smith,

Learning Activity 10.2

ASSESSING AFFECTIVE ATTRIBUTES

Materials Needed:

- Pencil and paper

A. Evaluate a science lesson (one you participated in during this course or one you taught to children). What affective attributes (attitudes or dispositions) do you think would be important to the lesson? List them.
B. Create a method to monitor the dispositions. Would it be a formal attitude inventory, an interview, a journal?
C. Some educators do not believe that monitoring attitudes or dispositions is an important teacher activity. Write an essay presenting your opinion on the matter, or debate the idea with others in your class.

& Cummins, 1999). For example, you might observe that a student never raises her hand during class and conclude that she's having difficulty with the topic. However, an interview with the student may reveal that she is simply bored, tired, or shy.

Interviews with students can be either informal or formal. Students find **informal interviews**, which resemble teacher-student discussions (written notes can be made after the discussion), less stressful than other forms of assessment. Imagine that in an informal interview about a water quality project you ask, "Why did you use the balance to measure the amount of water?" and the student answers, "I wanted to see which weighed more." To probe for more in-depth understanding and to ascertain why the student is only interested in a general level of measurement ("more water" versus "less water") rather than an exact measurement in milliliters, the teacher might say, "Tell me what you mean by 'more.'"

Formal interviews may seem more stressful, but they don't have to be. What differentiates formal and informal interviews is the greater structure of formal interviews. You might set up formal interviews, for example, to elicit the same information from each student. A list of guiding questions will keep you focused on your objectives. Some guiding questions you might ask are "What is your driving question?" "What makes that an appropriate driving question?" "What is the design of your investigation?" "How does the design help you answer your questions? "What materials or equipment are you using?" "What have you accomplished so far?" "Are you having any problems?" "What kinds of results have you obtained?" "What are your conclusions so far?" and "What do you intend to do next?" However, guiding questions only provide the foundation to a formal interview. You will need to ask additional questions to clarify student answers and elicit in-depth answers.

Open-ended questions are a great way to start either formal or informal interviews. These are discussed in Chapter 8 as well, where we call them divergent and convergent questions or open and closed questions. An open-ended question gives a child a cue to explain what she thinks. Compare the following starting questions: "Lisa, how many stages did the butterfly go through?" "Lisa, what can you tell me about butterfly stages?" and "Lisa, what have you learned about butterflies this week?" The first question squelches conversation and limits the child's response. She will have answered the question if she only says, "Five." The teacher could just as easily have obtained this amount of information with a written test. The second question, while still focusing on butterfly stages, permits a

---

CONNECTING TO NATIONAL SCIENCE EDUCATION STANDARDS

A CASE FOR STRENGTHENING ASSESSMENT IN THE SCIENCE CLASSROOM

Portfolios (collections of student work, regular self-reflection and peer assessment, assessment conversations, journals, projects, class discussions, performances, well-planned quizzes, and tests)—any combination of these assessment activities—can support improved science learning (NRC, 2001, pp. 18–19).

richer explanation. However, the third question gives Lisa the greatest opportunity to elaborate on what she knows and doesn't know about butterflies, and it may lead to some unexpected responses. The teacher can then ask additional probing questions to see if Lisa has any misunderstandings and to find out how much she knows about butterflies and their stages of growth.

Another effective technique for either informal or formal interviews is the "What if?" question. For example, a teacher might ask, "What if the balance were used to measure the helium balloon? What would you see?" This type of question probes students' understanding even further by forcing them to think about new variables and different situations.

Keeping records of both informal and formal interviews provides a rich source of information for portfolios, parent-teacher conferences, and future discussions with students. *After* an interview, jot down your thoughts about students' understandings and misunderstandings. If you take notes *during* interviews, some children will become intimidated and stop talking. Also, jotting down a note about each question makes the interview process too time-consuming.

Assessment Based on Concept Maps

In Chapter 8, we learned that concept mapping can be used to link new ideas or experiences with earlier ones. **Concept mapping** can also be a powerful method of assessing individual student's conceptual understandings or the understandings being formulated among groups of students (Edmondson, 1999).

Examine the concept maps shown in Figures 10.1 and 10.2. If a fourth-grade student had completed the first map to show prior understanding about magnets, what would you be able to say about the student's understanding? What major concepts are missing? First, magnets are not attracted to all types of metal, so the map demonstrates an incomplete understanding. Second, opposite poles attract and like poles repel, but the student depicts both the north and south poles attracting to the north pole. Third, the map fails to include some major concepts, such as temporary and permanent magnets, natural and human-made magnets, and magnetic fields. In short, you could use this concept map to diagnose the student's knowledge of magnets and adapt instruction accordingly. If the same fourth-grade student later completed the second concept map following a project that investigated magnets, you could conclude that the student's conceptual understanding of magnets had improved dramatically.

Some teachers use concept maps with groups of students during discussions, interviews, and group projects. A concept map drawn on the board or projected on an over-head transparency or with a computer projector can form the basis for a class discussion about the relationship among ideas that surface while investigating a driving question. Concept maps can be used to probe student interpretation of investigation findings. Finally, students can collaborate on their own concept maps to illustrate the relationships among their findings.

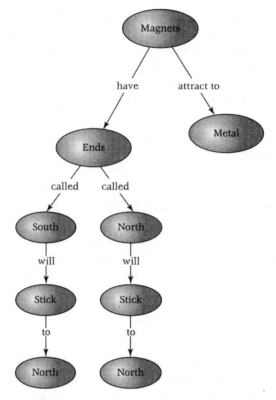

Figure 10.1   Limited concept map.

Concept maps can also be evaluated. Students' concept maps are commonly scored according to the complexity of the conceptual relationships they illustrate (Novak & Gowin, 1984). For example, students might be awarded one point for every hierarchical level in a concept map, quantifying the complexity of their thinking. A point could also be given each time a concept was branched into new categories. For example, in the second concept map, the student has branched the idea of magnetic fields into two subcategories, north and south. Likewise, students may be awarded points for cross-links such as those between human-made magnets and electromagnets and between natural magnets and lodestone. You might award more points for branches and cross-links that occur at higher hierarchical levels.

Assessment Based on Classroom Tests

Because assessment should be integrated throughout the teaching and learning process, at times you might want to write some classroom tests or quizzes to assess the progress of your students. Quizzes and tests can give both you and your students information regarding their basic understanding of concepts and use of process ideas. Tests and quizzes provide another opportunity for students to demonstrate

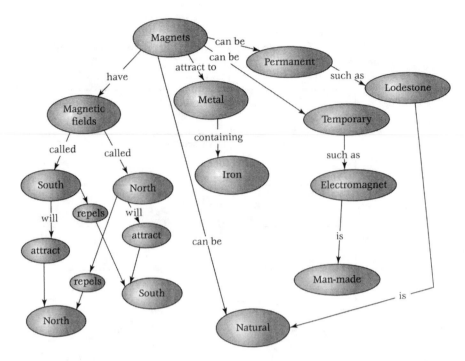

Figure 10.2   Complex concept map.

what they understand. Tests provide students and teachers with feedback. You can also use the results of the tests and quizzes to modify your classroom instruction. Their effectiveness depends on how you use them. Tests that provide students and teachers with feedback and that are used to enhance instruction are a positive example of test use.

Creating Tests and Quiz Items. You can construct test items to measure both content and process understanding. Consider the following problem:

> *Test/Quiz Item 1:* While riding your bike, you drop a quarter out of your hand. Someone watching you would see the quarter hit the ground _____
>
> A.   in front of you.
> B.   beside you.
> C.   behind you.
> D.   there's no way to tell.

This item would measure a student's understanding of content knowledge related to Newton's First Law of Motion. You might recall that Newton's First Law states that an object will stay in motion unless acted upon by an outside force.

Now, consider the following item:

Test/Quiz Item 2: Which of the following distance-time graphs shows an object with the greatest speed?

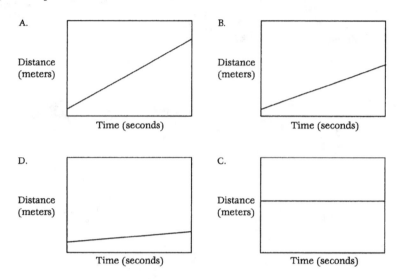

This item would measure a student's understanding of interpreting graphs, a process skill.

Test and quiz items can also be constructed across several cognitive levels—low, medium, and high. Low cognitive items would measure a child's understanding to recall, recognize, identify, or retrieve information. We discussed levels of cognitive items in Chapter 8. The following item would be an example of a low-level cognitive domain item:

Test/Quiz Item 3: The movement of soil from the side of a river bank is an example of which process?

   A.   Runoff
   B.   Deposition
   C.   Erosion
   D.   Absorption

To respond to this item, a student needs only to recall the definition of erosion, so we classify it as a low-level cognitive item. In contrast, medium-level cognitive items require students to interpret, clarify, translate, exemplify, classify, summarize, generalize, infer, conclude, compare, contrast, or explain. Test/quiz items 1 and 2 are examples of medium-level cognitive demand items. To answer test item 1, a learner needs to clarify a situation by applying understanding of Newton's First Law of Motion.

High-level cognitive demand items require a student to apply, analyze, create or evaluate. Item 4 illustrates a high-level cognitive demand item:

Test/Quiz Item 4: Design an experiment to show the relationship between plant growth and the amount of sunlight a plant receives.

To answer item 4, a student would need to design an experimental procedure to show how sunlight influences plant growth. The student would need to define and control variables, discuss the importance of trials, and describe how to measure plant growth. Such an item is very open-ended with several possible solutions. Therefore, we would classify it a high-level cognitive item.

Types of Paper-and-Pencil Items. In the previous examples, we used primarily two types of techniques: selection and supply (Airasian, 2001). **Selection** techniques require students to select a response from choices provided on the test. **Supply** techniques require a student to construct a response to a question or problem.

Items 1, 2, and 3 are examples of multiple choice selection items that require a student to select the best answer. **Multiple choice** items include a stem and various options. The *stem* presents the problem or question to the student, and the *options* present the various choices from which a student can select a response.

True and false, matching, and short-answer items are other types of selection items. **True and false** items require a student to classify an item as true or false, correct or incorrect, or yes or no. As such, true and false items are typically classified as low-level or medium-level cognitive demand items. Test/quiz item 5 *is* an example of a true and false item.

> Test/Quiz Item 5: As the amount of nitrates increases in a lake, the lake becomes less likely to support a variety of different kinds of living organisms:
> a. True
> b. False

To respond to test/quiz item 5, a student needs to recall basic information (what nitrates are) and then make a decision if the statement (regarding the effect on living organisms) is true or false. As such, we classify the item as a medium-level cognitive demand question.

Matching requires students to connect premises made in one column with responses in another column. Typically, matching items are low-level cognitive demand items. An example would be a list of words in one column to be matched to their definitions in the second column.

Supply techniques require students to construct a response to a problem. **Short answer** items typically require students to respond with only one or two words. Most often, short response items are low level and require a student merely to recall information. Test/quiz item 6 illustrates a short answer item:

> *Test/Quiz Item 6:* As water flows in a river, soil may be transported downstream where it is "laid down." The process of the soil being "laid down" can be described as _____.

Test/Quiz Item 4 is another example of a supply item. **Essay** items, such as this, require students to construct a response that typically measures high-level understanding. Essay tests can be more cognitively challenging for students than other types of supply items because, often, they require students to explain and interpret. For example, students might be asked on an essay test to explain weather data (date, temperatures, barometer readings, wind speed, and location of high and low fronts) found on a map and predict (with explanations and justifications for their predictions) the weather one might expect the next day.

Each type of test/quiz item offers advantages and disadvantages over the others. For instance, true-false, matching, short answer, and multiple choice formats are easy to score and are considered more impartial (not open to interpretation by the person grading the item). However, many teachers find it very difficult to write good, high-level cognitive items for such tests. Students can also guess correctly on these types of items. Essay items, on the other hand, are very time-consuming to score and are open to interpretation. Moreover, teachers are more easily able to write challenging essay items. However, very verbal students might be able to bluff their way through an answer. Essay and performance items are also more time-consuming for students to complete and, as a result, teachers may not be able to use them to test all of the intended learning performances.

Issues to Consider When Writing Test/Quiz Items. Constructing good test items can be challenging. Although you might be able to write low-level cognitive items fairly easily, you may find it very difficult to write medium-level and high-level items. We will discuss several issues that need to be considered when writing good test and quiz items.

When writing tests, make sure items match and measure the learning performances you expect of the students. The items you write should test the content you hope students will learn and allow students to perform at the level you expect. For instance, if you ask students to recall the definition of *force,* the item might match the content of the unit; but, because it only requires low-level cognitive demands, it does not match the level at which you expect students to perform. Project 2061 calls this **necessity** (the knowledge or skill measured in the assessment item is specified in the learning goal).

Although the items you write might correlate to the learning performances you designed for the unit, you need to make sure you actually stressed these learning performances in your teaching. Often, what we hoped we would get accomplished in our teaching did not really occur. Be careful not to write items that measure learning performances not included in your instruction. Even if you covered an idea in class, you might not have had students engage in the level of difficulty intended. For example, you may have had students describe an idea rather than use it in an everyday-life application. Hence, make sure the level of difficulty of the item matches what happened during instruction. The knowledge or skill that is outlined in the learning goals should be *enough* to answer the assessment item. Project 2061 calls this **sufficiency.**

Teachers also take into consideration students' reading levels and language. A science test is not fair if it measures the student's ability to read, rather than understand science concepts. Make sure that vocabulary words on the test are ones students recognize and are able to read. Be careful that language, especially for students whose first language is not English, does not limit their ability to express what they know in science. Project 2061 calls this **comprehensibility**.

Teachers need to make sure directions are clear on a test. Students should be able to understand what is expected of them in the test item. Project 2061 calls this **clarity of expectations**.

Next, teachers need to consider students' cultural backgrounds. They need to think about students' beliefs, values, prior experiences, and other factors that may influence how students respond on assessments. For example, urban students may not respond well to a test item asking about the acidity of soil for growing soy beans. Even if they might be able to reason through such a question, they might not even try because they believe they don't know anything about growing soy beans. The same item could be more culturally sensitive if rephrased to ask about the acidity needed to grow house plants. Project 2061 calls this **appropriateness of context** (the assessment task is familiar to the student).

> **CONNECTING TO NATIONAL SCIENCE EDUCATION STANDARDS**
>
> **ASSESSMENT IN SCIENCE EDUCATION**
>
> Assessment tasks must be appropriately modified to accommodate the needs of students with physical disabilities, learning disabilities, or limited English proficiency (p. 85).

Finally, teachers should construct items so that students cannot easily guess the answer or arrive at a correct answer by using general test-taking strategies, such as eliminating items without having to know the science content. Project 2061 calls this **resistance to test-wiseness**. Sometimes teachers use **two-tiered** assessments to help reduce test-wiseness (Sampson, 2006). They will have students answer the test item, and they also require students to write an open-ended response that explains their thinking behind the answer.

In summary, ask yourself these questions when writing a test:

- Is the knowledge or skill measured in the assessment item specified in the learning goal?
- Is the knowledge or skill outlined in the learning goal *enough* to answer the assessment item?
- Can students understand the assessment item?
- Do students understand what is expected of them in the assessment item?
- Is the assessment task familiar and engaging to the student?
- Are students unlikely to get the answer correct by guessing or using general test-taking strategies?

If you can answer "yes" to each of these questions, you will have written a fair test. A fair test is **valid**. A valid test measures what the students have been prepared to understand and makes accurate generalizations about students' learning.

A second concern you need to consider is reliability. **Reliability** refers to how consistently a test measures a student's performance. To illustrate the idea of reliability, Airasian (2001) used the example of weighing yourself. If you step on the scale and it reads 121 lbs. and then you step on it again a minute later and it reads 126 lbs., you probably won't trust either reading. Now let's say you step on it a third time after another minute, and it reads 119 lbs. You would conclude that something is wrong with your bathroom scale because it isn't measuring your weight consistently. This is an example of an instrument that is valid (it is measuring what it is suppose to be measuring—your weight), but it isn't doing so consistently (it isn't reliable). A reliable

CONNECTING TO
NATIONAL SCIENCE
EDUCATION STANDARDS

ASSESSMENT IN
SCIENCE EDUCATION

Assessment tasks and the methods of presenting them provide data that are sufficiently stable to lead to the same decisions if used at different times (p. 84).

bathroom scale would read the same weight consistently if you stepped on and off it during a particular time period. As classroom teachers, we strive to build classroom tests on which a student could achieve a similar result if she took the test again within a short time period.

Writing clear items that do not include vague or ambiguous terms helps to improve the reliability of test items. Clearly written items help ensure that each time the student reads them, he interprets them in a similar manner. You can also improve the reliability of classroom tests by including on them two or more items that measure the idea you are trying to assess. If a student misses one of the items, you can look at how he or she responded to the other item that measures the same concept or skill to determine the student's understanding.

Although you might try very hard to write items that reflect what your instruction stressed, a student's score on the test doesn't necessarily reflect what he or she really knows. Some students experience **test anxiety**. Most of us have experienced some test anxiety, particularly before a very important examination, such as the Scholastic Aptitude Test (SAT) for admission to college. Some students, however, consistently experience high-test anxiety, which affects how they perform on a test. In severe cases, students freeze and simply cannot respond. In other cases, students cannot recall even easy answers. Test anxiety represents one threat to the validity of a test.

Sometimes even well-intentioned teachers promote test anxiety by stressing the importance of a test. We saw this in the opening scenario of Chapter 9. Although we want students to take assessment seriously, we don't want to overemphasize the importance of a classroom test. As a classroom teacher, you can help students with test anxiety by creating multiple ways to assess students in a classroom. If a student realizes that her evaluations on other classroom assessments are acceptable, she might not be as nervous when it comes time to take a test. Multiple assessment techniques help students build positive self-images. You can also reduce test anxiety by teaching effectively, bringing closure to lessons so that students understand what they were to have learned, giving students ample opportunities to prepare for a test, and practicing sample test items. Including items at different cognitive levels on tests also helps reduce anxiety. Some teachers like to start the test out with relatively low-level items so that students can gain some confidence before facing more stressful items.

CONNECTING TO
NATIONAL SCIENCE
EDUCATION STANDARDS

ASSESSMENT IN
SCIENCE EDUCATION

An individual student's performance is similar on two or more tasks that claim to measure the same aspect of student achievement (p. 84).

The following guidelines can help a teacher write test items that are valid and reliable (modified from Airasian, 2001):

- Write items clearly at the level of the students' reading level.
- Try to avoid statements that are vague.
- Use language that students in your class will understand.
- Use appropriate vocabulary. Be mindful of your students and their reading level.
- Make sure there is only one correct answer for each item (unless it is intended to be an open-ended response item with more than one possible answer).
- Make sure you develop rubrics for open-ended items (rubrics are covered more thoroughly later in this chapter).
- Make sure the items match the instruction students experienced.
- Make sure that the items match the level of difficulty for which students were prepared.
- Make sure that several items attempt to measure the same understanding.

> ## CONNECTING TO NATIONAL SCIENCE EDUCATION STANDARDS
>
> ### ASSESSMENT IN SCIENCE EDUCATION
>
> Assessment practices must be fair. Assessment tasks must be reviewed for the use of stereotypes, for assumptions that reflect the perspectives or experiences of a particular group, for language that might be offensive to a particular group, and for other features that might distract students from the intended task (p. 85).

Try writing one of your own tests now by completing Learning Activity 10.3.

Performance-Based Assessment

Performance-based assessment refers to methods of directly examining students' knowledge, skills, and dispositions. Through these methods, teachers collect assessment information while students perform a targeted activity. The activity can be a natural part of a lesson (observing students as they look under a microscope), or it

> ### Learning Activity 10.3
> #### WRITING A CLASSROOM TEST
>
> *Materials Needed:*
>
> - The list of guidelines for writing test items
>
> A. Write a classroom test. Make sure the test includes items that measure both content and process that are written at different cognitive levels. Also, be sure to include a variety of formats.
> B. Classify each of the items as content or process and also state the cognitive level of the item.

can be a prompt given to provoke an observable action (providing equipment and a set of directions asking students to assemble a slide and focus a microscope).

In project-based science environments, students must integrate knowledge and skills to carry out their investigations, create artifacts, and present their projects to others. Performance-based assessments involve similar integration and, consequently, are a good fit with project-based science. For example, students who have been investigating how to set up and maintain an aquarium may have learned about acids and bases (pH), nitrogen cycles, life cycles of fish, salt- and freshwater, and classifications of animals. They may also have used a variety of science skills, such as *observing* the behavior of fish, *measuring* the pH or temperature of the water, *classifying* the types of fish that can get along together, *making inferences* about fish that die, and *forming conclusions* about appropriate conditions for the fish. A performance-based assessment might consist of giving students a pH test kit and five water samples. A prompt tells the students that certain tropical fish can only survive in pH levels ranging between 7 and 8. Students are then asked to determine which of the water samples would be safe for the fish. This question could be extended to measure problem-solving skills by asking students what they could do to make the inappropriate water samples livable for the fish. This performance-based assessment item integrates students' abilities to measure water accurately, interpret a chart, understand pH, and draw a conclusion.

To summarize, good performance-based assessment seeks to put students in realistic situations in which they must integrate relevant knowledge and skills in order to perform a target activity. A good performance-based test item usually includes the following characteristics:

- **It reflects important curriculum targets.** Good performance-based assessment reflects those curriculum goals that have been the objects of classroom instruction. For example, if students have been investigating the weather, they needed to learn how to use a thermometer. It is, thus, more *realistic* in an assessment to have students use a real thermometer to measure and record the temperature of air than it is to have them identify the temperature on a picture of a thermometer.

- **It asks students to create a product (artifact).** For example, students who make their own barometer to measure air pressure show their understanding of the purpose of a barometer, of how it works, and of the effect of barometric changes on the weather.

- **It supports students in collaboration and use of resources.** The performance-based test asks students to work with others and with equipment, materials, or resources to solve a problem and create a product. For example, students might work in a team to solve a problem about what types of ground cover collect the most heat in the summer. The team might decide to use a thermometer to measure and a computer spreadsheet to record the temperature of four different materials—stone, grass, asphalt, and sand.

- **It encourages active investigation by students.** Students engage in exploration of ideas and questions rather than in passive activities, such as listening to the teacher, watching a movie, reading about a concept, or watching a demonstration.

- **It permits students to apply multiple approaches to solving a problem.** For example, students might be asked to construct an "ear" that would be a good sound collecting device.[1] There is no one answer to this task. Students could design a cone-shaped ear that looks like a megaphone. They could make a long, narrow ear like a rabbit's. They could draw pictures of their ears and explain the design of the human ear in accompanying text.

- **It requires students to integrate ideas.** If students create an "ear" that is cone shaped and has "hairs" in it, they have combined the idea that funnel shapes gather sound waves with the idea that hairs (which are solids) carry sound waves better than does air. They also have used their artistic skills to design the ear and their literacy skills to explain the design.
- **It requires students to apply higher order thinking and to understand concepts, rather than memorize individual facts.** For example, students are asked to demonstrate how air pressure affects the weather or how well sound travels through various objects, as opposed to regurgitating definitions memorized of *air pressure* or *frequency* and selecting the "correct answer" from among a list of possible answers.
- **It promotes students' interest.** Good performance-based assessment is inherently more interesting to students than traditional paper-and-pencil tests. Making a barometer or an "ear" is more challenging and motivating than selecting from a list of definitions. When students enjoy the learning process involved in assessment, they feel less anxiety, yielding more reliable assessment results.
- **It is developmentally appropriate.** Performance-based assessment emphasizes the characteristics promoted as developmentally appropriate for both early childhood and young adolescent students (NAEYC, 1986; NMSA, 1995). These characteristics include the use of concrete materials. Performance-based assessment stresses problem-solving processes over memorization of abstract ideas. It does not punish students who are nonreaders or poor readers. It engages students in direct, purposeful experiences, rather than in contrived assessment tasks. It integrates subject areas. By encompassing multiple approaches, it builds self-esteem and allows for differences in learning types, ability, and development. Finally, it is interesting and motivating.

How do you go about designing a performance-based assessment? You follow these five steps:

1. Decide what science content to include in the performance. For example, the learning performance might be "Students will demonstrate how energy is transferred in an electrical system."
2. Determine what skills or complex processes to observe while students engage in the activity. For example, students assembling an electrical system will classify objects according to whether they conduct electricity or not, make inferences about the type of energy used in a dry cell, and make conclusions about the transfer of energy from the dry cell to the wires and the light bulb to give off light.
3. Generate the activity for the performance assessment. For example, you might decide to set up an activity in which students are given light bulbs, dry cells, wire, and a small motor and asked to make the motor work, using only the items provided.
4. Test the assessment item by letting students try out the activity.
5. Revise the performance-based task based upon trial runs.

In recent years, implementing performance-based tasks has become such a popular technique for collecting assessment information that large assessment companies, such as the Educational Testing Service (ETS), are now developing and adopting these types of tests. This should not be surprising. In all other aspects of our lives, we make evaluations on the basis of actual performance. We want our automobiles to run, our doctors to rid us of our ailments, our dentists to fix our cavities, and our plumbers to fix our sinks. We expect artists to generate emotions, entertain, and challenge and athletes to amaze us with their physical abilities. Why, then, do we expect our students only to regurgitate answers and not also to do what it is that they should be able to do? Learning Activity 10.4 asks you to develop a performance-based assessment.

Learning Activity 10.4

DEVELOPING A PERFORMANCE-BASED ASSESSMENT

Materials Needed:

- Materials to teach a lesson developed in this investigation

A. Following are two ideas an elementary or middle grade teacher might want a student to develop in a project-based environment. These require a complex combination of knowledge and skills. Form a team of three. Read each idea and decide what would be the most genuine ways to measure whether students have gained the pertinent knowledge and skills.

  - Each student will be able to observe the differences between two types of soils and determine which one is the best in which to plant a cactus.
  - Each student will be able to read a weather map and determine what the weather will be like tomorrow.

During your decision making, consider the following criteria of good performance-based assessment:

1. It reflects important curriculum targets.
2. It is consistent with instruction.
3. It asks students to collaborate and use resources.
4. It encourages active investigation.
5. It permits multiple approaches.
6. It integrates ideas.
7. It requires higher order thinking.
8. It is interesting.
9. It is developmental appropriate.

B. As a team, develop a lesson to teach one of the two learning performances. Develop two different ways of assessing the performance after you have taught it. Teach the lesson to another team in your class and try the two different methods of assessment.
C. Ask the team you taught which method of assessment best measured what they learned. How did their beliefs match yours about good techniques for assessing complex combinations of knowledge and skills?
D. How do these findings fit with your state science proficiency model?

Assembling and Presenting Assessment Information

After teachers and students have collected assessment information, they must assemble it into a presentable form for others to see. In project-based science, there are two main ways to assemble and present information: artifacts and portfolios.

Artifacts are the products that result from an investigation: writing samples, journals, physical or computer models, drawings, videos, or multimedia documents. Artifacts, which represent students' work and emerging understandings, are a concrete product that can be assessed. They have long been a natural part of assessment in the arts, but they are now being recognized for their potential contribution to science education.

What do you think of when you look at the shells you collected at the ocean, the turquoise jewelry purchased in the Southwest, the miniature lighthouse you bought in Maine, the alligator-shaped mug you collected in Florida? What do your parents think of when they look at your baby pictures, that Mother's Day card you drew in first grade, or that graduation tassel? Like souvenirs and keepsakes, artifacts that have been assembled for the purpose of assessment form lasting memories in our minds. Students are more likely to remember what they learned in science if it culminates in some type of public presentation: a presentation about landfills given to younger students, a poster about recycling displayed at the mall, or a video about weather that becomes a tool for future instruction.

Why would students more likely remember artifacts than abstract knowledge? There are six good reasons:

1. Through the creation of artifacts, understanding is constructed in students' minds. As students reflect upon what they have learned and develop artifacts, they actively manipulate science ideas and, thereby, generate understanding.
2. The fact that learning does not occur in linear, discrete steps argues against assessment items that are constructed around small, discrete bits of information. Artifacts, on the other hand, enable students to display their learning in a fashion consistent with real-life learning, which unfolds as a continuous process.
3. All students learn differently; students can display their learning in ways that are compatible with their learning styles. Students who are interested in drama might role play their understanding. Those interested in music might write songs to describe what they learned. Artistically inclined students can create models, posters, and murals. Talented writers can compose reports and papers.
4. Students remember what is meaningful to them. Do you remember an obscure lesson from fifth grade that involved calculating the area of a 4-inch by 3-inch square? Probably not. However, if you ever had to measure a floor or wall to purchase carpeting or wallpaper, you might remember making those calculations because there was a real purpose to them. Likewise, when students develop artifacts to share with an audience, they are more likely to remember the experience since it has social meaning.
5. Artifacts developed as part of a collaborative process provide social experiences students will remember.

> ## CONNECTING TO NATIONAL SCIENCE EDUCATION STANDARDS
>
> ### THE RELATIONSHIP BETWEEN FORMATIVE AND SUMMATIVE ASSESSMENT
>
> ...many aspects of the portfolio and the portfolio process provide assessment opportunities that contributed to improved work through feedback, conversations about content and quality, and other assessment-relevant discussions. The collection also can serve to demonstrate progress and inform and support summative evaluations (NRC, 2001; p. 61).

When you think back to your own education, you probably have vivid memories of friends you played with or collaborated with on a school play. Do you have similar memories of routine classroom lessons? Social experiences form lasting memories that we tend to cherish. Because project-based science stresses development of student artifacts in collaborative settings, students are far more likely to have lasting memories of these group experiences than they are of typical school lessons that involve little social interaction.

Portfolios, collections of student artifacts, can be thought of as both objects and methods of assessment. As objects, they are a place for holding materials, such as papers, photographs, or drawings, that are representative of students' work and progress. As methods of assessment, portfolios provide ways for teachers to continuously collect and assess student work. Haberman (1995) pointed out that exemplary teachers of children in poverty (star teachers) spend little time on tests and grading. They focus more on students' effort and accomplishments, and portfolios are much more in line with the proclivities and predispositions of students who live in poverty.

Assembling Artifacts

Students can assemble information to place into artifacts in a variety of ways. We will focus on writing samples, daily journals, physical products, drawings, music, videos, and multimedia documents.

Writing Samples. Students can demonstrate understanding in a number of ways, using writing samples, and there are several good reasons for using this technique to assemble information (Champagne & Kouba, 1999). First, **writing samples** provide a creative and accurate measure of student understanding. Second, writing is a curriculum-based task; it matches goals of instruction and integrates the language arts into science. Third, writing engages students in an active, rather than a passive, activity. Fourth, writing allows students to provide rich interpretations of their investigations. Fifth, writing targets students with verbal/linguistic abilities, enabling them to show what they know in a way that will enable them to succeed. Finally, students consider writing a worthwhile task.

As a teacher, you will need to show students how to write a report because, when children write reports, several benefits accrue. Having children write reports assesses their ability to obtain and synthesize information and to communicate through reading and writing. Also, students can learn a great deal from writing independent reports on given topics.

Students can create many other types of writing samples that can be used for evaluating understanding. Students can write stories, create poetry, make television commercials, produce plays or puppet shows, create newspaper articles, write documentaries, draft letters, compose essays, keep diaries, write computer presentations, or create comic strips. Students also can write reactions to guest speakers, films, videos, or software programs; they can develop biographies or autobiographies; and they can create annotated bibliographies. Each of these samples can be a powerful way to assess science understanding.

Imagine that fifth-grade students write stories about parasitic and symbiotic relationships. One boy writes a clever story about a flea named Billy. Billy wanted a new world away from the familiar scene he had on the back of an old, outdoor dog. The story tells of the flea's adventures in the big city—almost getting run over by a car, being fumigated by a pest control sprayer, and being stepped upon by a pedestrian. The flea, hungry, scared, and tired, decided that there was no place like home and returned to the back of the dog where he was safe and ate well. Of course, the dog wasn't too happy about Billy the Flea's return![2] Using a story format, this student explained most of what he knows about this particular parasitic relationship. This same student might have failed a test item that asked him to identify the correct definition of *parasite*. Would the teacher accurately have measured the boy's understanding with a test item? Would he have been able to elaborate on the parasitic relationship with the depth and richness that he displayed in this story?

Here are some other writing ideas that can be used in science classes. Have students take the viewpoints of other people to illustrate their understandings of particular concepts. For example, a student might present a lesson on evolution from the point of view of Charles Darwin—how would Darwin argue about evolution? Or imagine that students have launched an investigation into the changes over the last century in Lake Erie, Lake Okeechobee, or the Great Salt Lake. After learning about the history of the lake, they might create a travel brochure telling about the lake's geological and environmental features and the impact of humans. Students could draw maps, diagrams, or illustrations of the area for the travel brochure.

Daily Journals. Daily **journals** record the personal events, experiences, and reflections of children. They offer several advantages over other forms of assessment. First, they enable students to assemble some of their own assessment information. Second, they permit the teacher to conduct assessment at the most convenient time. Third, journal writing encourages students to connect science to their daily lives. Fourth, journals encourage candid teacher-to-student or student-to-student communication.

The typical elementary or middle grade teacher's day is very busy and hectic. Teachers are usually responsible for teaching all subjects to 25 to 30 students of varying ability levels and cultural backgrounds. When students make their own observations in a journal about what they have learned from a particular activity and about what continues to present difficulty, the teacher is relieved of some of the assessment burden. Because teachers can read journals after school hours when things aren't so hectic, journal assessment helps ease the assessment burden even further.

Journal writing also encourages students to associate science with their everyday lives. For example, a teacher might ask students to keep track in their journals of the science-related literature stories they read, of science-related news items, and of science-related shows they see on television. They could also comment in their journals on what they learn in science class and how it is related to their lives.

Journal writing enhances communication between students and teachers because students often open up in journals and write things that they would not say publicly. Students are often more willing to write than talk about what they learn; and they commonly describe their feelings, attitudes, and fears in a journal. When teachers respond to what students have written, what results is an ongoing dialogue that helps guide instruction. For example, if a student were to share in her journal that she is nervous about presenting in front of her classmates, the teacher could take steps to alleviate this fear— perhaps by having the student present with a partner.

Some teachers encourage students to communicate with each other by trading journals. By writing back and forth with a friend, students can discuss what they did in science class, construct understandings in a social context, and compare what they are learning to what others are learning. Other teachers ask students to keep electronic journals on the computer and send e-mails to others about their learning.

Physical Products

If you wanted to buy a new automobile, would you rather read about and see the car on a Web page or see the real product? What's the difference? The real product provides you with a *realistic* assessment of the car's riding comfort, sound quality, road handling, acceleration rate, turning ratio, workmanship, and design. Reading about it might only give you the opinions of others. Similarly, to see what students really know and can do, teachers need to make evaluative judgments of actual student products. The finished work of students provides excellent evidence of the knowledge and skills applied to create the product.

A **product** can be a book, a physical model of some kind, or a working apparatus that a student makes to demonstrate what she has learned. For example, imagine that students were interested in the human lung. They could develop a project in which they researched how human lungs worked; write a report about lungs; develop a physical 3-D model of the lung using ordinary household objects, such as straws, balloons, sponges, and cups; and give a presentation on how smoking is harmful to lungs. When finished, students would have demonstrated their ability to collect and synthesize information, to communicate and to solve problems, and to create a product.

Teachers should always inform students about their expectations for a project and how they will be evaluated on it. Expectations and evaluation criteria can be communicated verbally or in the form of checklists. Checklists are particularly effective because they can be shared with parents and because teachers easily can convert them into scoring rubrics to evaluate products. Figure 10.3 shows an example of a parent letter and checklist for the human lung project.

Dear Parents,

Your son/daughter has been learning about the human body in science. So far, we have learned about the skeletal system and the circulatory system. The students have become interested in how the lungs work and are now investigating this topic. The project they are working on will show what they learned from their investigations and will be shared with classmates on November 15th.

The project is designed so that your son/daughter will work as part of a collaborative team investigating this topic. However, you can help by taking him/her to the library or helping obtain materials that could be used to develop a project. The project may require some common household materials, such as straws, balloons, sponges, paper, cups, and glue, but you should not spend any significant amount of money on these materials.

When your son/daughter shares the project on November 15th, it will be evaluated on the following criteria:

1. The report must answer the question "How do our lungs work?"
2. The report should be thorough enough to answer the question completely.
3. Teams are expected to use at least two references in the report, with only one of the references being an encyclopedia.
4. Pictures or diagrams should be used to help communicate this information.
5. Teams should construct a physical model that demonstrates how the lungs work.
6. Each team member should be able to explain how the model represents a real lung.
7. Each team member should be able to explain why smoking harms the lungs.

If you have any questions about this project, please contact me at school.

Sincerely,

Mrs. Czerniak

Figure 10.3   Parent letter.

Drawings. Drawings give students a chance to represent scientific understanding in diverse forms, such as murals, bulletin boards, posters, and bumper stickers. For example, students investigating the types of animals and plants in their neighborhood might create a mural of the neighborhood. Students might create posters of what they found in their school trash that would end up in a landfill. Finally, they might create bumper stickers, using permanent markers on strips of contact paper, to communicate their opinion of local recycling policies.

Drawing, as a form of assessment, offers the following advantages to students. First, it lets visual/spatial learners demonstrate their knowledge in a way that is comfortable for them. Second, drawings are sometimes a more accurate reflection of understanding because children can depict concepts, such as the water cycle, more completely through illustration than they can through exposition. Third, drawings can be used to express attitudes or feelings about science. Fourth, drawings are particularly effective for assessing science understanding in early elementary grades when student aren't able to read and write well. The following examples illustrate some of the benefits of assessments based on drawings.

Imagine that you are teaching seventh-grade students how to use a microscope to observe animal cells (a common activity in which students observe prepared animal cells on a slide) and onion cells (an activity in which students place a single layer of onion skin on a slide) because they are investigating how plants and animals differ. Your learning performances for the students might include learning to use a microscope correctly (a skill) and learning the differences between animal and plant cells (knowledge). While an elementary student could probably describe to you the differences between cells, if you want students to notice that animals cells are rounded with a cell membrane and plant cells are stiff and rectangular with a cell wall, then having the students draw what they observe will give you a good way to measure achievement of your objectives.

Imagine a third-grade classroom in which students are learning about how animals see. The teacher taught students how the eye works, and students participated in a teacher-led activity in which they used flashlights to observe how classmates' pupils expanded or contracted in the presence or absence of light. To assess students' understanding, the teacher asks them to draw pictures that explain how their eyes work. One child draws a picture of rays shooting out of the eyes toward an object and then returning to the eye. This drawing tells the teacher that the child has an inaccurate conception of sight and thinks that we see because of rays that come from our eyes. Drawings help teachers identify when children are having difficulty understanding science concepts.

Imagine a classroom in which students are drawing pictures of scientists. This technique, called *draw a scientist,* examines students' beliefs and attitudes about science (Barman, 1996; Bodzin & Gehringer, 2001; Boylan, Hill, Wallace, & Wheeler, 1992; Finson, Beaver, & Cramond, 1995; Huber & Burton, 1995; Rampal, 1992; Sumrall, 1995). Researchers have found that students commonly depict scientists very stereotypically, as white, balding men wearing glasses and white lab coats. Of course, the scientist is usually surrounded by glassware out of which vapors are rising. Frequently, he has a disheveled or evil appearance. This activity can be used to monitor students' beliefs about who can become scientists and about what scientists can do in their careers.

Music. Students who possess strong musical–rhythmic intelligence can demonstrate their understanding of science by creating songs, raps, jingles, or cheers. They can engage in choral readings or set plays to **music**. A number of commercial science songs provide students with stimuli for their own writing. Songwriter Raffi, the Banana Slug String Band, and other bands have created science-related songs, such as "Baby Beluga" and the "Water Cycle Boogie," that are popular among elementary students. These albums are commonly available through online services, such as Amazon.com or in local music stores.

Videos. Students can record the images and sounds of their investigations on **video**. Almost all schools now have access to video cameras and VCRs. Most are now purchasing digital video cameras. Videos are useful forms of assessment for several reasons:

1. Creating a video is easy. The technology has advanced to the point at which even young elementary students are able to use it.
2. The videos can be edited to reflect changing understandings.
3. They permit subsequent playback or broadcasting to public audiences. Since videos can be played back for broadcast, they are records that can be used for assessment and for instruction.
4. They conveniently document long-term projects, because video can be filmed in segments, edited, and stored until additional filming is done. An investigation about why leaves drop from trees in the fall might span an entire school year. Video footage could be collected over a long period of time to document this year-long process. Moreover, it is very easy for students to create iMovies using digital video to illustrate their long-term projects. iMovies can also be converted to podcasts for students to share with others who can play them on an iPod.
5. Videos form vivid memories in students' minds. Video recording a vacation, for example, provides continuous sight and sound documentation. In a fifth-grade classroom, one of the authors of this book made an 8 mm movie of William Tell. To this day, many years later, the author remembers more about Swiss history than about the history of any other European country.
6. The act of creating a video can force students to focus on the main ideas learned. If, for example, students are required to present their science project to classmates in a 15-minute video, they must focus on only the most important concepts and best examples. As a teacher, you will need to help students focus on ideas and not on just the way the product looks. This editing process can help students sift out the most important information from a mass of facts or unrelated data.
7. Videos can be used to generate unique and complex presentations, vividly capturing what students know. For example, older elementary students can connect video cameras to microscopes and computers to add new dimensions and edit video and audio footage. Digital video cameras can document images and scenes that can be added to electronic documents, such as PowerPoint presentations, Web pages, and electronic portfolios.

Presenting Artifacts

Any of the artifacts (writing, journals, products, drawings, music, videos, and multimedia documents) already discussed can be used in presentations to groups of younger students, peers, parents, or community members. Presentations do not necessarily have to involve artifacts, however. For example, students who possess a strong bodily–kinesthetic ability may enjoy engaging in debates, simulations, plays, and role plays. Whatever the format, presentations can motivate students and enhance their self-esteem.

Students are more motivated to complete investigations and put forth their best effort to develop artifacts when they see a purpose to the activity. Too often in elementary and middle grade classrooms, students complete work that is only seen by the teacher. Presenting an artifact to a real audience provides an investigation with a real, meaningful purpose and gives students the experience of dialoguing with others about their work.

Developing Portfolios

Portfolio assessment is a process of collecting representative samples of students' work over time for purposes of documenting and assessing their learning. Typically, four types of documentation are contained in portfolios: artifacts, reproductions, attestations, and productions (Collins, 1992). As already discussed, artifacts include writing samples, journals, physical products, drawings, music, videos, and multimedia documents produced in the normal course of a science investigation.

Reproductions are documents attesting to the learning process—not the actual items produced. They include photographs of projects, videos of projects, or audiotapes of presentations. The difference between artifacts and reproductions is that artifacts are created during the investigative process, whereas reproductions are created afterward to document a project. A student's photographs of trees taken during the year to track seasonal changes are artifacts. A teacher's photographs of students' leaf collections is a reproduction.

Attestations are testimonials about a student's work prepared by someone other than the student, such as the teacher, a parent, or a peer. Some examples are a critique by a peer, a letter from a parent, and a report card from the teacher.

Productions are documents that help explain the contents of a portfolio. They include such items as goal statements, personal reflections, captions, and descriptions of what the items in the portfolio represent.

While studying neighborhood trees, students might take photographs of trees throughout the seasons, write reports about trees, collect samples of leaves or needles, map the locations of the trees, and draw pictures of the seeds of trees. Some or all of these artifacts could be placed in a portfolio. Since the actual leaf collection would not hold up well in a portfolio, a photograph of the set (a reproduction) could be added to document the learning. The portfolio might also include a testimonial (an attestation) from a parent about a child's trip to a local park to identify different types of trees. Students could reflect on their experiences and add to the portfolio a letter to the teacher (a production) telling the most important idea they learned about trees in the neighborhood.

Portfolios as Containers of Artifacts

A portfolio can be contained in a file folder, a box, or any other suitable container for student work. Some middle grades teachers are even having students create *electronic portfolios*. However, a portfolio is not just a random collection of student work carelessly thrown together. It is a carefully planned collection that includes representative samples of student artifacts, reproductions, attestations, and productions collected over the entire school year and organized so as to clearly represent learning for the viewer. Like documentation of a vacation that includes photographs, diaries, videos, and personal mementos, a good portfolio contains a carefully assembled sample of student work that encourages reflection.

There are many questions and issues to think about when implementing portfolios in a project-based science classroom. First, the purpose of the assessment should be established before starting the portfolio. Working together, the teacher and student should decide what kinds of documents best demonstrate the student's knowledge and abilities. Will the portfolio contain only the "best" samples of work or samples of varied quality? Must the evidence be individual student products, or should the work be completed and documented through a collaborative effort of several students? How much evidence should be in the portfolio? What criteria will be used to judge each piece of work? Teachers frequently construct their criteria, scoring rubrics, checklists, or other methods of judgment prior to selecting the student work to be included in the portfolio. Teachers must also decide if all students will be evaluated by the same criteria; determine how often and by whom the portfolio will be evaluated; decide where the portfolio will be kept and what access students will have to it; help students select samples of their work that accurately represent their abilities; and periodically critique the portfolio for its effectiveness in helping make decisions and in determining what students know, can do, and are like.

Making Judgments From Assessment Information

The purpose of evaluation is to make judgments about student growth, set goals, and report information to students and their parents. In the past, teachers used assessment mostly for reporting grades, and only the teacher had a voice in determining how well students were doing. Today, assessment techniques give students a voice in the assessment process. In this section, we will discuss how scoring rubrics can be used to assess artifacts, performances, and portfolios. We address the processes of self-assessment and peer assessment. Finally, we discuss grading.

Scoring Rubrics

One helpful technique for evaluating an artifact, performance, or portfolio is to use a scoring rubric. A **scoring rubric** is a brief, written description of different levels of quality for student performance that can be used to rank or rate a level of performance. It often uses the same set of observable criteria that go into a checklist. Typically, rubrics are classified as either holistic or analytic.

Holistic rubrics measure the overall quality of an artifact, performance, or portfolio (Arter & McTighe, 2001). They evaluate criteria, such as creativity, relevance to real life, impact, clarity, completeness, content accuracy, and organization. For example, the following holistic rubric contains four general criteria. Although the rubric provides a general framework for scoring students, it does not provide *specific* criteria, such as the focus of the scientific content or the characteristics that would make the project creative, complete, and clearly presented:

- *3 points:* The student makes logical inferences supported by data collected in the investigations, gives rationales for the inferences, and supports the inferences with diagrams and drawings.
- *2 points:* The student makes inferences supported by data collected in the investigations but does not support them with rationales.
- *1 point:* The student makes a logical inference, but it is not supported by the data collected in the investigations.
- *0 points:* The student fails to make a logical inference.

Analytic rubrics measure artifacts, performances, or portfolios in a quantitative manner by assigning or taking away points for specific traits, dimensions, or criteria that are present or missing (Arter & McTighe, 2001). For example, a teacher who is evaluating the investigative process and students' abilities to make inferences might use the analytic scoring rubric shown in Table 10.4. It contains five observable criteria: logical inferences, scientific accuracy, supporting data, supporting rationale, and supporting diagrams and drawings.

Imagine that students have been investigating how sound travels and how animals use their ears to collect sounds. They have learned that sound travels best through solids, next best through liquids, and least well through air. They have discovered that sound can be collected or projected in a cone-shaped object. Students create a model of an animal ear to demonstrate what they have learned. How will the teacher assess this model ear? The following rubric might be used to accomplish this task:

- *5 points:* The students designed an ear that clearly demonstrates all four sound concepts learned from the investigations—sound travels best through solids; sound travels better through liquids than air; sound travels least well through air; and a funnel–tunnel shape can be used to collect and amplify sound waves so they sound louder. The students can explain why the ear would help an animal obtain food and avoid predators, and they provide evidence from their investigations to support their design.
- *4 points:* The students designed an ear that demonstrates at least three of the four sound concepts learned from the investigations—sound travels best through solids; sound travels better through liquids than air; sound travels least well through air; and a funnel–tunnel shape can be used to collect and amplify sound waves so they sound louder. The students can explain why the ear would help the animal obtain food and avoid predators, but they fail to support their design with data from their investigations.

TABLE 10.4   Analytic Scoring Rubric

	Excellent	Good	Satisfactory	Poor
Uses logical inferences	4	3	2	1
Is scientifically accurate	4	3	2	1
Has supporting data	4	3	2	1
Has supporting rationale	4	3	2	1
Has supporting diagrams and drawings	4	3	2	1

- *3 points:* The students designed an ear that demonstrates at least two of the concepts learned from the investigations—sound travels best through solids; sound travels better through liquids than air; sound travels least well through air; and a funnel–tunnel shape can be used to collect and amplify sound waves so they sound louder. The students can explain why the ear would help an animal obtain food and avoid predators, but the explanation is not linked directly to their investigations.
- *2 points:* The students designed an ear that demonstrates at least one of the concepts learned from the investigations—sound travels best through solids; sound travels better through liquids than air; sound travels least well through air; and a funnel–tunnel shape can be used to collect and amplify sound waves so they sound louder. The students can explain why the ear would help an animal obtain food and avoid predators.
- *1 point:* The students designed an ear and can give a logical explanation about how it would help an animal obtain food and avoid predators. However, the students do not demonstrate any of the four sound concepts covered in the investigations.
- *0 points:* The students designed an ear, but are unable to explain how it would help an animal obtain food and avoid predators; and the design is not logical or consistent with concepts learned about sound.

By including evaluative criteria in the scoring rubric, the teacher establishes an objective basis for judging the artifact, performance, or portfolio. Scoring rubrics also provide students with feedback or data supporting grades and evaluations. When students know the criteria on which they were judged, they can take steps to expand their knowledge and improve their skills. Additional reasons for using scoring rubrics are outlined by Andrade (2000): Rubrics make it easy to explain students' progress to others, make teachers' expectations clear, support learning and the development of skills, and support the development of understanding and good thinking.

The criteria used in scoring rubrics depend upon the teacher's goals and learning performances for the lesson. Some criteria will be mandated by state and local school districts' curricula guidelines, while others are likely to follow national trends or curriculum frameworks, such as the *National Science Education Standards* (National Research Council, 1996, 2000, 2001) or *Benchmarks for Science Literacy* (AAAS, 1993). In project-based science classrooms, the following criteria may be used to design scoring rubrics:

- *Understanding of concepts:* Determine whether students developed shallow, inert knowledge or deep, meaningful understanding by examining the level of detail in the artifact, performance, or portfolio. Note missing links and levels of differentiation or completeness in the explanations.
- *Use of higher order thinking:* Determine whether students used higher order thinking by examining such factors as their ability to formulate and answer questions; interpret or explain decisions; and discuss assumptions that underlie the artifact, performance, or portfolio. Also assess their ability to explain relationships and formulate new problems or questions and apply information to new situations or problems.
- *Ability to answer driving questions:* Judge how well students answered the driving question. Was there a weak or strong relationship to the driving question? Did students specify the relationship in clear, specific terms?

- *Relatedness to the world:* Determine if students made strong connections between the artifact, performance, or portfolio and the real world. Did students apply information to a real-life example in a strong, convincing manner?
- *Level of collaboration:* Does the artifact, performance, or portfolio show interaction with others? Is the sharing of ideas and the use of other resources, especially community resources, evident?
- *Level of creativity:* Does the artifact, performance, or portfolio reveal a high level of creativity? Does it construct new ideas or connect existing ideas with new ones? Does it use ideas in novel ways or translate ideas from other subject areas?

> **CONNECTING TO NATIONAL SCIENCE EDUCATION STANDARDS**
>
> **ASSESSMENT IN SCIENCE EDUCATION**
>
> The process of scoring student-generated explanations requires the development of a scoring rubric. The rubric is a standard of performance for a defined population (p. 93).

- *Presentation:* Did students thoroughly explain the key ideas that answered the driving question? Were their classmates engaged? Did they answer classmates' questions? Did they use enough detail to support their conclusions?
- *Use of cognitive tools:* Did students use cognitive tools, such as print media, computers, peripherals, software applications, and telecommunication? Determine the level of use of these types of tools.

After you have determined the criteria for your scoring rubric, you need to create the rubric using a series of general steps (Andrade, 2000; Arter & McTighe, 2001):

1. Gather samples of student work and performances and sort them into groups by quality. Examine samples of related student work to get a feeling for the range of possible answers and responses.
2. Determine how an "expert" would perform in the situation. List the knowledge, skills, or dispositions an "expert" might display.
3. Identify the observable differences between "excellent" and "poor" performance. For example, a student who has mastered an understanding of sound waves is able to explain the related science concepts, how the ear is designed to collect and amplify sound, and how the ear helps an animal obtain food and avoid predators. In contrast, the student who has a poor understanding of sound may be able to create an "animal ear" and explain how it helps an animal avoid predators, but is unable to explain the science concepts related to the design of the ear.
4. Turn the "good" and "poor" performance into a range of possible performances that exemplify the traits for a variety of levels. You should use no more than three to six levels or ranges that will become *points* on a rubric (Arter & McTighe, 2001). For example, students are able to identify all four concepts, three of the concepts, two of the concepts, one concept, or none of the concepts.
5. Try to assess students with the range of performances you identified.
6. Revise the criteria as needed.

There are also useful Web sites for creating rubrics. Two good examples include: http://rubistar.4teachers.org/ (to create rubrics for project-based activities) and http://4teachers.

Learning Activity 10.5

DEVELOPING A SCORING RUBRIC

Materials Needed:

- Paper and pencil
- The results of Learning Activity 9.5

A. After completing Learning Activity 9.5 (or another performance-based activity), develop a scoring rubric to evaluate the performance. Use the following steps:
   1. Collect samples of student work.
   2. Determine how an "expert" would perform in the situation. List the knowledge, skills, or dispositions an "expert" might have.
   3. If possible, examine work that students have completed to see the range of possible answers and responses.
   4. Identify the observable differences between "excellent" and "poor" performance.
   5. Turn the "good" and "poor" performance into a range of possible performances.
   6. Try to assess students with the range of performances you identified.
   7. Revise the criteria as needed.

B. Work in teams to try to score the performance using the rubrics created. Did the rubrics seem accurate and fair? How might they be revised? If you need more information, see Arter, J., and McTighe, J. (2001). *Scoring rubrics in the classroom: Using performance criteria for assessing and improving student performance.* Thousand Oaks, CA: Corwin Press.

org/projectbased/checklist.shtml (checklists for project-based work). In Learning Activity 10.5, you will develop your own scoring rubric.

Assessment of Portfolios

After students have collected a wide range of items for their portfolios, they need to reflect upon each item and sort out those that best represent what they have learned. Some teachers have students think about "added value." What evidence will an item add to the portfolio? They ask open-ended question stems, such as, "Looking at everything in my portfolio, I like this piece most because..." or "This artifact shows that I...." If nothing is gained by including an item, it should be left out. For example, while investigating why pumpkins rot, students might produce a short report on decomposition, photographs of their investigation, notes, audiotapes of interviews with a grocery store manager, transcripts of the recorded interview, drawings of the various stages of pumpkin rot, graphs of decomposition over time, and journal entries. After reflecting upon their accumulated evidence, they would probably decide that the journal entries and the recording of interviews did not depict what they had learned and omit these items from the portfolio.

After they have selected items to include in the portfolio, students need to reflect upon their progress and overall growth. This reflection can occur in different ways. Teachers frequently have students write captions for each artifact to explain what learning it depicts. Some teachers have students write letters describing all the items in their portfolios and explaining why each artifact is evidence of learning. Others have students create tables of contents to organize their portfolios in a meaningful way. A table of contents might be arranged in a linear fashion to show progression, or it might be arranged around themes to show the relationships among concepts learned.

Students find "portfolio reviews" with the teacher, parents, or peers meaningful. The portfolio review gives students and teacher an opportunity to discuss progress, achievement, and goals. Students can share their portfolios with parents or peers, either in small groups or before the whole class. This final step encourages positive critiques from significant others, and it helps students celebrate with others their progress and success.

Self-Assessment

Self-assessment can encourage students to think about such things as learning styles, what was learned, the quality of that learning, and personal goal setting. When students think about their own learning styles and preferences, they reflect on how they best learn (metacognitive knowledge)—whether through reading, hands-on manipulation of physical objects, or working collaboratively with others, for example. When students reflect on what they learned, they also think about how they feel about their learning. When they think about the quality of their learning, they examine how well they are doing and how much they have improved and their strengths and weaknesses. Using such self-assessment information, students can then decide what steps to take to improve. Many teachers use self-assessment as one step in a process that encourages students to rethink and revise their work. The teacher has the students reflect on their progress, provides feedback to the students, and gives the students an opportunity to redo their work. As students progress further down the path of self-regulated learning, their intrinsic motivation to learn also grows; and this, of course, is what

> ## CONNECTING TO NATIONAL SCIENCE EDUCATION STANDARDS
>
> ### INTRODUCTION
>
> Student participation is a key component of successful assessment strategies at every step. If students are to participate effectively in the process, they need to be clear about the target and the criteria for good work, to assess their own efforts in light of the criteria, and to share responsibility in taking action in light of the feedback (NRC, 2001, p. 9).

teaching and learning are all about—the production of highly motivated, self-regulated learners.

Several useful techniques can be used to encourage self-evaluation. You might have students keep journals, detailing what they learn in their investigations. However, some students fail to open up in journals, and they make useless entries, such as "I did my math homework after school. Then I worked on my science investigation." This type of information doesn't reveal much about a student's thinking, learning, or problems. To encourage true self-assessment, you might provide students with guiding questions. Students can answer these questions directly; or they can answer them as part of a journal, portfolio, or interview. Guiding questions might include "Look back over your science report—what did you find easy about writing the report?" "What was difficult for you?" "What was the most important idea you learned?" "Why?" "How does this relate to the driving question?" "What do you like about your artifact?" "What do you feel are its strengths and weaknesses?" "What would you like to improve or change?" "What skills would you like to work on?" "Are you satisfied with your progress in answering the driving question?" "What else do you need to investigate?" and "In what areas of science do you think you have improved the most?"

You might choose to base self-assessment on open-ended statements. For example, you could have students complete the following statements:

- The things I am still wondering about are ...
- I discovered ...
- I'm beginning to wonder why ...
- I learned ...
- I could improve by ...
- I never realized that ...
- I was surprised that ...

You might assess curiosity by determining how many things a student is wondering about. You might use "attitude inventories" that ask students open-ended questions about their interests and beliefs:

- The thing I like best about science is ...
- My favorite topic to study is ...
- My favorite subject is ...

These open-ended self-assessment techniques also encourage students to continue wondering or questioning—affective attitudes desired in a project-based science classroom.

Finally, students can assess their own attitudes about science with a formal assessment instrument. For example, if you are interested in measuring how excited the students are about studying natural disasters, you could ask them the following questions:

a. How excited are you about learning about hurricanes, floods, tornadoes, and earthquakes? (1) very, (2) somewhat, (3) not at all.

b. How important was your investigation in terms of answering the driving question? (1) very, (2) somewhat, (3) not at all.

Whatever self-assessment method you choose, it will foster **metacognitive knowledge** that makes students more aware of what they are learning, how they are investigating, and how they are feeling about science.

Peer Assessment

In **peer assessment**, students give feedback to other students in the class. Most students respect the opinions of their peers and value their input. Peer assessment, therefore, can be an effective motivator for many students. In addition,

> ## CONNECTING TO NATIONAL SCIENCE EDUCATION STANDARDS
> ### ASSESSMENT IN SCIENCE EDUCATION
>
> Students need the opportunity to evaluate and reflect on their own scientific understanding and ability. Before students can do this, they need to understand the goals for learning science. The ability to self-assess understanding is an essential tool for self-directed learning. Through self-reflection, students clarify ideas of what they are supposed to learn (p. 88).

when students know that they will be presenting artifacts to a real audience, they are more likely to see a purpose for their work.

In collaborative learning groups, students can assess the contribution each member made to successfully completing a task (Strom, Strom, & Moore, 1999). Teachers frequently have students evaluate such outcomes as how well their group members were able to work together, carry out assigned tasks, stay on task, share materials, monitor their own time, listen to each other, ask questions, generate alternative answers, contribute ideas, take different perspectives, summarize information, encourage each other, show respect for one another's ideas, and challenge ideas.

Be careful, however, when you first start to use peer assessment in a project-based science environment. Students need to trust each other before they can function well in such an environment. Before trying this method of assessment, build a trusting classroom environment. As discussed in Chapter 7, you can create this trusting environment by developing interpersonal skills and encouraging collaborative group work. Also be careful to protect the privacy of students' grades. A recent court case challenged the practice of having students grade their peers' papers and seeing peers' grades as a violation of the Family Educational Right to Privacy Act (Buckley Amendment), but the court did not uphold this challenge.

Figure 10.4 summarizes what you have learned about the assessment process.

Grading

We discussed assessment in this chapter as a process of collecting information about students' progress, assembling the information, and making judgments about prog-

Stage	Techniques	Advantages
Gathering information	• Observations of students during instruction—e.g., watching students while they conduct an investigation	• Helps in minute-to-minute and day-to-day decisions • Can be formal or informal
	• Focused questioning of students —e.g., asking students why they completed an investigation in the manner they did	• Provides teachers with in-depth information about students' understanding • Can be formal or informal
	• Anecdotal records—e.g., notes about student performance jotted on index cards	• Is helpful for determining progress at the end of a term • Is useful for parent conferences • Helps teachers remember information
	• Checklists—e.g., lists of concepts (understand causes of weather), skill (can record data accurately), and dispositions (appreciate usefulness of weather predictions)	• Can be used while students are working • Cuts down time charting assessments information • Keeps teacher focused on obtaining data from all students
	• Clinical interviews with students —e.g., sitting down with each student and asking, "What is the design of your investigation?" and "What have you accomplished so far?"	• Provides teachers more detailed information • Enables teachers and students to know each other • Can clarify any misinterpretations made of students while observing them
	• Concept maps—e.g., 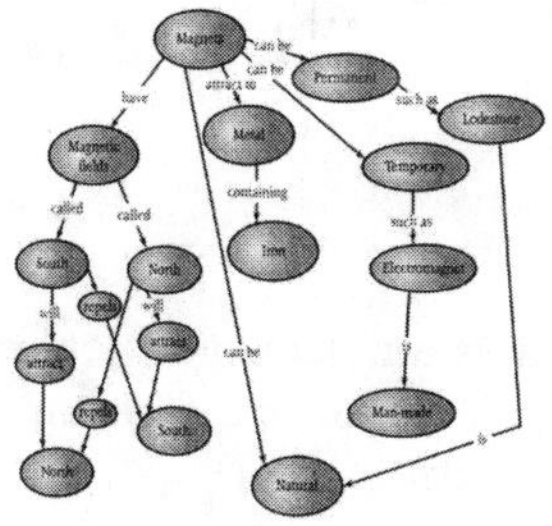	• Lets students demonstrate complex understanding of concepts • Can diagnose misunderstandings
	• Test items—e.g., Explain why it isn't a good idea to build a house on the bank of a river that flows quickly.	• Easy to grade • Can measure a variety of levels of understanding
	• Performance-based assessment —e.g., providing students a prompt to complete a skill-based activity	• Is based on actual classroom activity • Measures complex skills and depth of knowledge • Is linked to instruction • Is concrete and active

Figure 10.4   The assessment process.

ress or achievement. **Grading** is an interpretation or judgment made from data collected and recorded and, thus, is an extension of evaluation. To issue grades, a teacher must think about all of the various techniques used to gather and assemble assessment data. For example, do you want to include only written products? Do you want to include observations? Self-assessments? Once you make these decisions, you need to develop a logic or set of operating rules for converting this information to a grade.

Assigning grades is no easy task if the school uses letter grades. All assessments are flawed in some ways and only an estimation of student learning. A single letter grade can be problematic, because it reduces students' achievement in a variety of areas, such as content knowledge, process skills, decision-making skills, motivation, and team and collaborative skills to a single evaluative statement. You need to determine what you think is most valued, whether some things will be weighted more than others, and whether you will give grades based on progress and effort or only end products or final achievements. After you make a decision on the logic or rule you will use, you need to look for patterns and changes and compare them with your intended learning outcomes. From this, a teacher makes his or her best professional judgment about a grade.

Not all schools use a single letter grade for grade reports, and some allow teachers to add additional materials to the grade card, such as narrative reports in which statements about progress or achievement can be included. Other school district grade cards look more like scoring rubrics. In these districts, procedures resembling those described earlier for creating rubrics can be used for determining grades. If you find yourself in a situation in which single-letter grades are the only method of reporting to parents at the end of a term, we encourage you to supplement the grades with writ-

---

Learning Activity 10.6

CREATING A REPORT CARD

Materials Needed:

- Pencil and paper or a computer

A. Think about a report card you received as an elementary or middle grade student. What kinds of things did it tell about your performance in science? What do you wish it had included?

B. Imagine you are on a committee to design a new report card. It is your task to design the section for reporting science learning in a project-based classroom. Design a report card that you think best reflects what students and parents should know. Share the report card with others in your class.

C. Would your report card have letter grades? Debate with others whether or not letter grades should be obsolete.

D. What are the major implications of constructivist theory that would influence your report card?

E. How would you explain to parents that your assessment procedures and report card are effective?

ten narratives or other more detailed explanations of students' progress and achievement. Learning Activity 10.6 helps you answer this question.

Using Technology Tools To Enhance Assessment

Earlier, we discussed how artifacts can be meaningful for students and result in thorough and useful assessment. Technology tools can be used to create **multimedia documents** that combine writing, illustrations, photographs, videos, recordings, computer applications, and other media to enrich conceptual understanding and assessment even further. Multimedia documents give students an opportunity to present their understandings more fully than words alone ever could. In addition, they give students the opportunity to depict learning in ways that are most consistent with their learning styles.

For example, student photographers in your classroom might enjoy documenting their learning with photographs. Computers enable students to combine images and edit them into their written documents. Students can take photographs with digital cameras or cell phones, and these digital pictures easily can be added to documents. Students can readily collect data and create graphs using computer programs. Using a digital video camera and Quick-Time or Macintosh's iMovie, students can add short clips of footage to their writing or Web pages or create a podcast (sometimes called vodcast when in video format). PowerPoint and Web design programs, such as Front Page or Page Mill, enable students to combine writing, audio, and visual images into a single document.

Technology tools have also begun to improve the assessment process. Assessments no longer need to consist solely of pencil-and-paper tests, and they don't need to be done only by the teacher. Many school districts offer computerized exams, quizzes, and performance items on which students simply input answers into the computer. These assessments are then scored by computer. Technology can also provide rapid feedback to teachers and students, thereby improving the teaching and learning process. Some teachers are using blogs (Ray, 2006) or wikis to document student learning. A **blog,** short for weblog, is a journal-like electronic entry of ideas on the Internet. A blog can contain text, video, and links to other electronic documents. Students can keep blogs related to their learning a particular topic or their work during a project. A **wiki** is collection of Web pages that others can edit. The Wikipedia is probably the best known wiki. Teachers can have collaborative groups of students (or even the entire class and community experts) create a wiki related to the learning goals in their project.

Technology can enrich the nature of science instruction by providing multimedia, probes, simulations, and interactive programs for examining and measuring students' problem-solving skills, actions, and decisions. Such technology allows teachers to measure in-depth, meaningful learning, rather than discrete facts and inert knowledge. For more information about assessment, see NSTA's SciLinks Web site (http://www.scilinks.org/).

Finally, Project 2061 is developing high-quality online assessment tools, including multiple-choice and open-ended questions, for middle grades teachers (see http://

www.project2061.org/research/assessment.htm). Project 2061 researchers are identifying state and national science content standards and key ideas students should learn. Then, they are developing assessment tools to measure these ideas.

Another Look at the Advantages of Educational Assessment

Assessment in a project-based science classroom, regardless of the method used, should be designed to help the teaching and learning process—not humiliate, demean, or trick students. Good assessment is aligned with learning goals and matches instruction. It helps diagnose, monitor, and evaluate students' acquisition of concepts, skills, and attitudes. Assessment is both formative and summative. It is an ongoing process that guides curriculum selection and instruction. Teachers in project-based science think of assessment as a responsibility shared between teacher and student for the purpose of informing students, parents, and teachers.

> **CONNECTING TO NATIONAL SCIENCE EDUCATION STANDARDS**
>
> **SCIENCE TEACHING STANDARDS**
>
> Teachers of science engage in ongoing assessment of their teaching and of student learning. In doing this, teachers use multiple methods and systematically gather data about student understanding and ability (p. 37).

A college student who took an ornithology class learned about different types of birds and how to identify them by color, plumage, shape, size, and song. He learned where they live, what they eat, and whether they are endangered. The final exam consisted of 50 pictures of birds' legs. It asked students to identify the birds only by their feet. The student, frustrated and exasperated, told the professor that the test was unfair as it did not measure what had been taught in the course and that he wasn't going to take it. The professor replied, "That is fine. I will just record a zero in my grade book for this test. Now, what is your name, young man?" The student pulled up his pant leg and said, "I'm not going to tell you. See if you can identify me by my legs!"

This story, reworded from a story tole at a conference (author unknown) makes a point very well. In project-based science classrooms, teachers purposefully, carefully, and thoughtfully design and implement assessment procedures to provide information to help students grow and also to learn about students' progress. When assessment is carefully crafted and executed, there are numerous advantages for teachers, students, and parents.

Advantages for Teachers

The assessment techniques used in project-based science classrooms encourage teachers to become more **reflective** in their practice. They force them to think about their learning goals, the design of their curriculum, and their instructional practices. It helps teachers think about ways to revise and adapt their lessons so that all students learn the intended goals. They also encourage teachers to take students' comments into account when planning their instruction. Teachers and students thus become allies in the learning process. Assessment is not an end in itself, but a means to a mutually sought end—student learning.

Teachers come to know their students better when using assessment consistent with project-based science. All students become important, and it is less likely for students to fall between the cracks. Because each student works with the teacher, the teacher can discern each child's capacity, style of learning, and rate of learning. Teachers use more active learning techniques in their practices because they arrive at a deeper understanding of what students know and how they have come to know it. If a school requires letter grades, a variety of assessment strategies provide teachers with a more honest and valid appraisal of students' learning on which they can base letter grades.

Advantages for Students

Most educators feel that assessment in a project-based environment encourages students to become involved in their own learning and helps them become reflective, self-regulated learners. Teachers and students work together to determine which pieces of work represent students' abilities and how the work will be evaluated. Students can analyze the strengths and weaknesses of their own work and establish their own goals for improvement. This **self-monitoring** behavior and metacognitive knowledge helps students develop intrinsic motivation to learn and strengthens their relationship with their teachers.

Assessment in project-based science also allows for individual differences in student abilities, since the focus is on student improvement, rather than on comparison with others. Students are encouraged to chart their own improvement over time. Unlike **criterion-referenced tests**, which grade all students according to a single set of preestablished standards, assessment that focuses on individual student improvement tends to promote self-esteem.

Effective assessment techniques also promote collaboration with peers. Students work with classmates to develop and share artifacts and portfolios, and they learn to seek suggestions for improvement. During this process, students develop important social skills. They learn to support and coach others and to work collaboratively and cooperatively. Finally, assessment techniques provide a less-threatening environment for evaluation since evaluation is an ongoing part of instruction rather than an occasional, anxiety-producing situation.

Advantages for Parents

Parents and guardians play a crucial role in the education of their children. Many people believe that the assessment techniques used in project-based classrooms accomplish for parents what traditional forms of assessment alone cannot.

Standardized (normed) test scores (usually distal or remote methods of assessment), while important for standardized comparisons of all the other children who have taken the test, do not provide the depth of understanding stressed in the *National Science Education Standards* (NRC, 1996, 2000, 2001). A 50th percentile score on mathematics, for example, means only that half of the students who took the test did better and half of the students who took the test did worse. This tells the parents how their child compared to other children taking the test. However, this kind of information tells parents little about children's capability, progress, strengths, and weaknesses. It tells nothing about children's motivation to improve.

Traditionally, letter grades did not offer much more information than standardized test scores because most were based on memory-oriented, criterion-referenced tests. These tests were often administered at the end of a chapter or other unit of study, and a child's single score would determine his or her letter grade. For example, a C grade would tell the parent only that the child answered approximately 75% of the questions correctly. Was the test valid and reliable? Did the child learn more things about science than the test measured? Is the child interested in science? Has the child developed scientific-thinking and problem-solving skills? Formal criterion-referenced tests provide only a partial answer to such questions in parent-teacher conferences.

Contemporary assessment techniques offer to parents a strong, multidimensional approach to understanding their children's learning. By viewing samples of work over time, they gain a better understanding of the entire learning process and of a child's improvement. Since many schools require criterion-referenced letter grades, these alternative assessments are often used in conjunction with other methods.

It is the job of teachers in project-based classrooms to communicate with parents about their children's progress and explain why project-based science requires different forms of assessment. Parents may need to be won over to these new ideas through involvement in the assessment process.

To involve parents in the assessment process, teachers can, for example, have parents help select some of their children's work to be included in a portfolio. Parents might also play the role of teacher by holding their own conferences with their children about their work, progress, strengths, weaknesses, and future goals. Parents can also supply valuable observations of their children's learning at home. For example, a parent might supply information about a child's trip to a science museum and resulting interest in building paper airplanes and rockets or Lego structures. Parents might even help fill out checklists of science skills, attitudes, or abilities that they have observed at home. Most teachers who have used alternative assessment techniques and kept in close communication with parents find that the parents appreciate the depth of understanding the teacher has about their children. In short, assessment in project-based classrooms brings together all the stakeholders (students, teachers, and parents) in the learning process.

In the last learning activity, we will have you assess your own learning about assessment from Chapters 9 and 10 by completing Learning Activity 10.7.

Chapter Summary

In this chapter, we discussed the purpose of assessment, and we explored a three-phase method of assessing student understanding: gathering information, assembling

Learning Activity 10.7

WHAT HAVE YOU LEARNED ABOUT ASSESSMENT?

Materials Needed:

- The KWL list from Learning Activity 9.1

A. In Learning Activity 9.1, you made three columns on a piece of paper or with a computer word processing program. You listed in the first column as many things as you knew about assessment, and in the second column you listed what you wanted to know. Reread your second column. Think about what you learned about assessment in Chapters 9 and 10, both of which were devoted to assessment. In the last column, list everything you learned in these chapters that answered the questions you had before you read the chapters.
B. How did your beliefs about teaching elementary and middle grades change after reading these chapters?
C. Record your ideas.

and presenting assessment information, and evaluating assessment information. We explored numerous techniques for gathering information, including administering tests/quizzes, making observations, keeping anecdotal records, using checklists, using interviews, using concept maps, and conducting performance-based assessment. We considered techniques for improving observations and discussions and examined types of tests and testing methods. We discussed how to construct valid and reliable test and quiz items. We also considered a number of methods for assembling and presenting information, including student writing samples, daily journals, physical products, drawings, music, videos, and multimedia documents. We explored the use of portfolios and artifacts to present information, including different methods to document work in portfolios—artifacts, reproductions, attestations, and productions. The chapter introduced four techniques that teachers can use to evaluate assessment information: scoring rubrics, portfolios, self-assessment, and peer assessment. We considered the difference between holistic and analytic scoring rubrics, and we discussed procedures for creating rubrics. We examined how assessment information is used to give grades and make decisions, and we examined how technology tools can enhance the assessment process. Finally, we looked at the advantages of assessment for teachers, students, and parents.

Chapter Highlights

- The purpose of evaluation is to make judgments about student growth, to set goals, and to report information to students and their parents.
- The process of assessing student understanding might be thought of as a three-step procedure:
  - Gathering information that will help in forming generalizations about students' learning

- Assembling and presenting the information so that it is recorded
- Evaluating the assessment information
- Teachers can make formal or informal observations of student progress.
- Discussions can be improved by practicing wait-time, probing, and redirecting.
- Concept maps can be used to assess students' conceptual understandings or the understandings being formulated among groups of students.
- Tests and quizzes can be used to measure students' understanding of content knowledge and process skills.
- Tests and quizzes can be written at low-, medium-, and high-cognitive levels.
- Paper-and-pencil items can use two techniques: selection and supply.
- Selection formats include:
  - Multiple choice
  - True and false
  - Matching
- Supply formats include:
  - Essay
  - Short answer
- Some formats (such as multiple choice and matching) are easy to score and more objective. Other formats (such as essay) take time to answer and grade, and they are open to interpretation.
- Teachers need to take care to write good test/quiz items.
- Test items can be analyzed for necessity, sufficiency, comprehensibility, clarity of expectations, appropriateness of context, and resistance to test-wiseness.
- Good teachers help alleviate text anxiety by teaching well, bringing closure to lessons, reviewing lessons, and practicing sample test items.
- *Performance-based assessment* refers to methods of directly examining students' knowledge, skills, and dispositions.
- After teachers and students have collected assessment information, they should assemble it into a presentable form for others to see.
- In project-based science, there are two main ways to assemble and present information: artifacts and portfolios.
- Portfolios, collections of student artifacts, can be thought of as both objects and methods of assessment.
- Artifacts include writing samples, daily journals, physical products, drawings, music, videos, and multimedia documents.
- Artifacts can be can be used in presentations to groups of younger students, peers, parents, or community members.
- Portfolio assessment is a process of collecting representative samples of students' work over time for purposes of documenting and assessing their learning.
- Typically, four types of documentation are contained in portfolios:
  - Artifacts
  - Reproductions
  - Attestations
  - Productions
- The purpose of evaluation is to make judgments about student growth, set goals, and report information to students and their parents.
- A scoring rubric is a brief, written description of different levels of quality for student performance that allow a teacher to rank or rate a level of performance.
- Scoring rubrics can be holistic or analytic.

- There are six steps for creating a rubric:
  1. Gather and sort student work by quality;
  2. List the knowledge, skills, or dispositions of an "expert";
  3 Identify differences between "excellent" and "poor" performance;
  4. Identify performance levels;
  5. Assess students according to performance levels; and
  6. Revise criteria as needed.
- After students have selected items to include in the portfolio, it is essential that they reflect upon their progress and overall growth.
- Self-assessment can encourage thinking about learning styles, what was learned, the quality of that learning, and personal goal setting.
- In peer assessment, students give feedback to other students in the class. Most students respect the opinions of their peers and value their input.
- Grading is an interpretation or judgment made from data collected and recorded.
- Technology tools have begun to improve the assessment process.
- When assessment is carefully crafted and executed, there are numerous advantages for teachers, students, and parents.

Key Terms

Analytic rubrics	Holistic rubrics	Redirecting
Anecdotal records	Informal interviews	Reflective
Appropriateness of context	Journals	Reliability
Artifacts	Metacognitive knowledge	Reproductions
Assessment	Matching	Resistance to test-wiseness
Attestations	Multimedia documents	Selection
Blog	Multiple choice	Scoring rubric
Checklists	Music	Self-assessment
Clarity of expectations	Necessity	Self-monitoring
Clinical interviews	Observation	Short answer
Comprehensibility	Open-ended questions	Sufficiency
Concept mapping	Peer assessment	Supply
Content knowledge	Performance-based	Test anxiety
Criterion-referenced tests	assessment	True and false
Discussions	Portfolio assessment	Validity
Drawing	Portfolios	Videotape
Essay	Probing	Wait time
Formal interviews	Product	Wiki
Grading	Productions	Writing samples

Notes

1 Credit for this idea goes to our colleagues in science education, Dr. Andrew Lumpe at Seattle Pacific University and Dr. Jodi Haney at Bowling Green State University.
2 Credit for this story goes to one of Charlene's former fifth-grade students, Robert Coggin.

References

Airasian, P. W. (2001). *Classroom assessment: Concepts and applications.* New York: McGraw-Hill.

American Association for the Advancement of Science. (1993). *Benchmarks for science literacy.* New York: Oxford University Press.

Andrade, H. G. (2000, February). Using rubrics to promote thinking and learning. *Educational Leadership, 57*(5), 13–18.

Arter, J., & McTighe, J. (2001). *Scoring rubrics in the classroom: Using performance criteria for assessing and improving student performance.* Thousand Oaks, CA: Corwin Press.

Barman, C. (1996). How do students really view science and scientists? *Science and Children, 34*(1), 30–33.

Bodzin, A., & Gehringer, M. (2001). Breaking science stereotypes. *Science and Children, 1,* 36–41.

Boylan, C. R., Hill, D. M., Wallace, A. R., & Wheeler, A. E. (1992). Beyond stereotypes. *Science Education, 76*(5), 465–476.

Champagne, A. B., & Kouba, V. L. (1999). Writing to inquire: Written products as performance measures. In J. J. Mintzes, J. H. Wandersee, & J. D. Novak (Eds.), *Assessing science understanding.* San Diego, CA: Academic Press.

Collins, A. (1992). Portfolios for science education: Issues in purpose, structure, and authenticity. *Science Education, 76*(4), 451–463.

Edmondson, K. M. (1999). Assessing science understanding through concept maps. In J. J. Mintzes, J. H. Wandersee, & J. D. Novak (Eds.), *Assessing science understanding.* San Diego, CA: Academic Press.

Finson, K., Beaver, J. B., & Cramond, B. L. (1995). Development of and field-test of a checklist for the draw-a-scientist test. *School Science and Mathematics, 95*(4), 195–205.

Haberman, M. (1995). *Star teachers of children in poverty.* West Lafayette, IN: Kappa Delta Pi.

Huber, R. A., & Burton, G. M. (1995). What do students think scientists look like? *School Science and Mathematics, 95*(7), 371–376.

National Association for the Education of Young Children. (1986). *Developmentally appropriate practice in early childhood programs serving children from birth through age 8.* Washington, DC: Author.

National Middle School Association. (1995). *This we believe: Developmentally responsive middle level schools.* Columbus, OH: Author.

National Research Council. (1996). *National science education standards.* Washington, DC: National Academy Press.

National Research Council. (2000). *Inquiry and the national science education standards.* Washington, DC: National Academy Press.

National Research Council. (2001). *Classroom assessment and the national science education standards.* Washington, DC: National Academy Press.

Novak, J. D., & Gowin, D. B. (1984). *Learning how to learn.* Cambridge, UK: Cambridge University Press.

Rampal, A. (1992). Images of science and scientists: A study of school teachers' views. I. Characteristics of scientists. *Science Education, 76*(4), 415–436.

Ray, J. (2006, Summer). Blogosphere: The educational use of blogs (aka edublogs). *Kappa Delta Pi Record,* 175–177.

Sampson, V. (2006, February). Two-tiered assessment. *Science Scope,* 46–49.

Southerland, S. A., Smith, M. U., & Cummins, C. L. (1999). "What do you mean by that?" Using structured interviews to assess science understanding. In J. J. Mintzes, J. H. Wandersee, & J. D. Novak (Eds.), *Assessing science understanding*. San Diego, CA: Academic Press.

Strom, P. S., Strom, R. D., & Moore, E. G. (1999). Peer and self-evaluation of teamwork skills. *Journal of Adolescence, 22*, 539–553.

Sumrall, W. J. (1995). Reasons for the perceived images of scientists by race and gender of students in grades 1–7. *School Science and Mathematics, 95*(2), 83–90.

Chapter 11

Managing the Science Classroom

Introduction

Project-based science creates many unique management challenges in the elementary or middle grade classroom. To be successful, teachers need to create a learning environment of trust and self-responsibility, a task that may raise numerous questions: *What does a teacher need to do to establish a positive classroom climate? How do I organize a classroom? How do I structure the school day? What can be done to make sure students are safe? How do I manage student behavior?* In this chapter, we will explore how teachers can create learning environments that support science teaching. We will focus on classroom climate, classroom organization, and management strategies.

In the classroom climate section, we will consider ways to establish a positive learning environment and examine a framework for thinking about classroom climate in

Chapter Learning Performances

- *Explain how the constructivist framework is useful for thinking about classroom management in project-based science.*
- *Describe how a teacher can create an effective classroom climate.*
- *Evaluate classrooms to determine their atmosphere.*
- *Compare and contrast the reasons for using various types of classroom management strategies.*
- *Explain the value of planning classroom management before, during, and after instruction.*
- *Justify why equitable classroom practices are necessary for managing project-based science.*
- *Distinguish among various types of safety precautions.*

the context of constructivism. We will discuss the roles of teacher and student, the relationship between teacher and student, and classroom interactions.

In the classroom organization section, we will review how to arrange a science classroom, structure the school day, and maintain a safe classroom. We will explore how to prepare for a lesson and what to do when off schedule.

In the section on management strategies, we will examine ways to foster positive student behavior, anticipate problems, distribute materials, make transitions between classes, deal with disturbances, and reinforce good behavior. We also will consider ways that teachers can handle multiple groups of students' working on the same activities, as well as groups of students' working on different activities at the same time. The chapter outlines techniques for dealing with students with diverse abilities and students who finish activities at different rates.

Although this chapter offers realistic management strategies, these strategies are not panaceas for problems in the classroom. Building a classroom learning environment that supports project-based science takes a great deal of hard work. The trade-offs for the hard work are the positive results seen in student learning and motivation.

What follows are several scenarios of lessons about pesticides in a food chain. As you read each scenario, focus on various features of the instructional setting. What are the students doing? What does the classroom climate seem to be like? What is the role of the teacher? What is the relationship between the teacher and the students? What instructional supports does the teacher provide? How has the teacher managed the instruction? How is the room arranged?

Scenario 1: Reading About Science

Sixth-grade students are sitting at individual desks. They are about to read a section of their science textbook on pollution in a food chain. Ms. Jung decides to use the KWL method to focus the students' reading (see Chapter 8 for more information on KWL). Before students begin reading, Ms. Jung asks the students what they already know about food chains. She lists this information under the *K* column, representing what students know. Next, she asks students what they want to know more about. She lists these responses under the *W* column, representing what students want to learn. Ms. Jung then selects one student at a time to read a paragraph in the book. When the section is completed, she asks the students to answer in writing the two questions at the end of the chapter. The two questions will be used to focus a discussion about what students learned—the *L* on the KWL chart. A few students seem eager to begin reading the assignment because they want to complete an investigation tomorrow that focuses on pesticides in a food chain. Most of the students begin to answer the two questions: "What is pollution?" and "How does pollution affect the animals at the top of a food chain?" Ms. Jung walks around as students are answering the questions to check that each student is on task and quiet. April, who is sitting at the front of the class, is playing with something in her desk and laughing with another student. Ms. Jung reminds April that the classroom rules require students to be quiet when they are reading or writing. After 5 minutes, Ms. Jung notices that Robbie has not started his assignment. She asks him if he has any questions, and she helps him focus on the topic of the paragraph—pesticides. Dr. Sylvia Brown, the principal, walks in to ask Ms. Jung a question and smiles as she sees the students working diligently. Ten minutes later, Ms. Jung collects the students' answers so she can give them feedback. She also holds a discussion with the class about what they learned from the reading. She lists these items under the *L* column on the chart at the front of the room. She informs the students that they will begin to design an investigation the next day about pollution found in a food chain. The students seem excited about this and begin to talk about some ideas that they have. After a few ideas have been shared, Ms. Jung instructs the students to take out their mathematics books.

Scenario 2: Direct Science Instruction

In a sixth-grade classroom down the hall, Mrs. Lochbihler is teaching the same topic. She has 10 students wear signs made of envelopes labeled *mice*, 10 students wear signs made of envelopes labeled *snakes,* and 5 students wear signs made of envelopes labeled *hawks*. The 25 students are told that mice eat grains, snakes eat mice, and hawks eat snakes or mice. Next, Mrs. Lochbihler says, "We will be going outside to play a game. We will be running around on the playground during this game. How can we make sure that everyone learns and that no one gets hurt during this activity?" The students decide that they need to stay away from playground equipment, and they should have a discussion after the game is over to see what people learned from it. Mrs. Lochbihler and the students go outside. The game begins by the mice eating grain represented by an assortment of white and colored bits of paper. The mice put the bits of paper into their envelopes as they "eat" it. Then the snakes are allowed to tag or "eat" the mice, and the hawks are allowed to tag mice or snakes. As the students tag each other, they collect the envelopes of their "prey." They put these envelopes inside their own. Students are laughing and enjoying the game of tag. Mrs. Lochbihler cautions a few students that they are near the playground equipment.

After a number of mice and snakes are "eaten," Mrs. Lochbihler asks the students to return to the classroom. The students are excited by this outdoor activity, and they arrive at their classroom in a noisy fashion. Mrs. Lochbihler models for the students how they should get quiet as they reenter the school, and the students seem to do this quickly. Dr. Sylvia Brown, the principal, is waiting for Mrs. Lochbihler as she returns to the room with her students. The principal is happy to see that the students are enjoying science, and she joins the class so she can see what all of the excitement is about. The students begin to count how much "food" they "ate" and how many white and colored bits of paper are contained in the envelopes they collected. Mrs. Lochbihler informs the students that the colored bits of paper represent food that is contaminated with a pollutant called DDT. If an animal ate any food that included any colored bits of paper, the animal will become sick, and its offspring might be deformed. If an animal's food supply consisted of at least as many colored bits of paper as white, the animal will die from the pollution. Students complete the task of counting their bits of paper, and Mrs. Lochbihler engages the students in a discussion about the effects of DDT pollution. She asks them to think about the effect that DDT may have on other animals, particularly humans if they eat animals that have been exposed to it. Finally, she instructs the students to compare how the mice, snakes, and hawks were affected by DDT. The students notice that the animals at the top of the food chain, the hawks, collected a lot more DDT than did the other animals. Mrs. Lochbihler finishes the lesson by having students draw a picture of how the pollution was passed through the food chain. A few students finish early, so Mrs. Lochbihler asks them to go to the computers to see if they can learn more about DDT. When students finish their pictures, they file them in their portfolios. The students who finished early share the information they found on the World Wide Web about DDT. Students in the class are so interested in this topic that they keep talking about the fact that DDT is still found in our environment, even though it was banned in the United States years ago. Mrs. Lochbihler is surprised that this lesson has run a half hour longer than she planned. At the end of the lesson, Mrs. Lochbihler helps the students determine how they will structure their investigation tomorrow.

Scenarios 3: Process Science

A few students in Mr. Morales's class bring to class an article about the level of PCBs found in fish from a local lake. In the article, students find that PCBs are reported to be in the fish caught in the lake. The article refers to *ppm*, and students wonder what *ppm* means. Mr. Morales approaches the topic of PCBs in food chains by using a laboratory activity as a lesson. Following directions, the students fill small plastic cups with red food coloring. The red food coloring is a 1 part per 10 solution. Next, the students take one drop of this red food coloring and transfer it to a clean plastic cup, and they add 9 drops of clear water. The solution becomes a 1 part per 100 solution of red food coloring. After they have done this, they take one drop from this cup and transfer it to a clean plastic cup and add 9 drops of clear water. This becomes a 1 part per 1000 solution of red food coloring. The students continue this procedure until they have mixed up a 1 part per million (1 ppm) solution. Mr. Morales notices that Martin is not following directions, and he reminds him about the contract they made for him to complete assignments in class. Next, Mr. Morales tells the students that pollution is frequently measured in parts per million (ppm) or parts per billion (ppb), and he tells the students that these tiny amounts of pollution can enter the food chain through many different sources (through plants that we eat, animals that we eat, or plants that animals we eat have eaten). The students in the class seem excited to learn this information; they now understand what the term *ppm* means. Tomorrow they will conduct an investigation in which they will measure pollutants from a local stream and report them in ppm.

Scenario 4: Multiple Investigations

Mrs. Kimble is teaching her sixth-grade class about food chains. Students in her class are investigating the question "How do chemicals affect an animal's food supply?" Several groups of students have decided to investigate different topics, so Mrs. Kimble sits down at her desk to grade some papers.

One group of students is studying the effect of pollution on the bald eagles in the Great Lakes region of the United States. Today these students are working on the computer, and they are communicating with a group of students in California who are studying a similar question: the effects of pollution on the condor. The students from both schools are comparing the pollutants found in their regions.

A second group of students is reading an article from the local paper about consumption of fish caught in the Great Lakes. They learn that there are recommended limits for consumption of fish from some of the lakes because of pollutants called *PCBs* and *mercury,* which are found in the fish. These students are discussing whether they could set up an experiment to test the effects of pollution on fish in an aquarium in their classroom. The discussion is getting very loud and is disturbing other groups. Mrs. Kimble gets up from grading papers and walks over to this group to question the students about their plan and helps the students understand procedures and regulations governing research conducted on vertebrate animals. She also points out that they need to keep their voices down.

A third group of students is studying the topic of biotechnology, investigating whether there are any known effects on humans who drink milk from cows given chemicals to increase their milk production. These students have decided to invite a guest speaker to come to their classroom to talk about research findings on dairy production, and they are excited at the prospect of calling the speaker.

Another group of students is studying the effects of pesticides and herbicides on fruits and vegetables. These students are trying to find out whether these chemicals are harmful to humans who eat the foods. They want to locate two farmers with different opinions about the use of pesticides and herbicides: one who uses organic farming practices, and one who uses the chemicals. They are asking other students in the class if they know any local farmers who use these methods.

Like the scenarios you read in Chapter 1, the first three scenarios you just read represent what is called *reading about science, direct instruction,* and *process science teaching.* Each type of science instruction can be used in a project-based environment as long as it supports the intended learning goals and student investigations. In these scenarios, the *read about science, direct instruction,* and *process science teaching* lessons are used to give students basic information about food chains and pesticides that will enable them to conduct investigations later. The last scenario illustrates students working on separate activities. Each scenario demonstrates different classroom management characteristics. The *climate* (prevailing feeling or state of mind of members of the class), *organization* (how the teacher structures classroom activities and space), and *management style* (how the teacher sets, models, and reinforces classroom behavioral expectations) of these four classrooms vary, but the management objective is the same in all of them. In this chapter, you will be learning about managing a project-based science classroom. You'll start by analyzing in Learning Activity 11.1 the opening scenarios to glean from them some basic aspects of managing a project-based science classroom.

Learning Learning Activity 11.1

MANAGING A SCIENCE CLASSROOM

Materials Needed:

- Pencil and paper or a computer

A. Compare and contrast the four classrooms described in the opening scenarios. How are they alike and different with respect to climate, organization, and management? What are the advantages and disadvantages of the management of each classroom? Use the following list of questions in your comparison.

Climate
- What are the classroom goals and expectations? Do students have a role in making the rules? What happens when someone doesn't follow the rules?
- Does the teacher smile a lot or frown? Do students seem to be enjoying themselves—or do they seem unhappy?
- Are students encouraged to take initiative and be autonomous, or are they expected to follow the directions of the teacher?
- Do students talk with each other and with the teacher, or is the classroom discourse only between students and teacher?
- Do students' ideas and responses help drive instruction and behavior in the classroom, or does the teacher determine the topics, activities, and behavioral expectations?

- Do students share responsibility for classroom decisions with the teacher, or does the teacher make all decisions?
- Does the teacher focus on "correct answers" to science questions, or are students encouraged to come up with many different answers?
- Are students encouraged to critique the teacher and give suggestions for improving instruction that will help them learn?
- How does the teacher assure that each student's self-esteem is improved?
- Does the teacher believe that all students can learn science? How does the teacher show this? What does the teacher do to ensure that all students are successful?
- How are girls and minorities treated in science classes? What does the teacher do to encourage all students in science?

Organization

- How are students' desks arranged? How is the furniture in the classroom organized?
- Are students free to move about the classroom? Do they sit at their desks?
- Where are materials stored? How are they set up for student use?
- How does the teacher structure the day?
- How many minutes are given to each subject?
- How does the teacher structure a lesson? How are materials introduced and used? What comes first and last?
- What does the teacher do to make a transition between subject areas?
- Is the classroom set up to enhance students' safety? How?

Management Strategies

- What is acceptable student behavior? Who establishes this—the teacher or the students and teacher working together?
- Does the teacher use contracts to establish and reinforce good behavior?
- How does the teacher create a climate for good behavior?
- Does the teacher anticipate problems that might occur and plan for them?
- How are materials distributed?
- How does the teacher deal with disturbances?

B. What do you think an elementary or middle grade classroom should look like? What should be going on in the room? What should the teacher and the students be doing? Imagine looking through a window or a door into a project-based science classroom. Describe your view of a classroom with respect to organization and climate.

C. Compare your views with those of others. How are each of your views similar and different? How might you work with colleagues with differing viewpoints on classroom management? How would you work with a principal or administrator who had different ideas of management?

D. If possible, visit an elementary or middle grade science classroom. Make a video of the classroom and take detailed notes about the climate, organization, and management of the room. How would you characterize the teacher's management style? What is your opinion of the classroom environment?

As you completed Learning Activity 11.1, you probably discovered that the objective of each lesson in the scenarios was the same—students will be able to define pollution, and they will understand the effect that pollution has on a food chain. During your comparison, you probably grappled with many questions: *What should a classroom look like? How noisy or quiet should it be? Should students be moving around or sitting down? Should they be laughing and talking, or should they be quiet and attentive? Should they read from the textbook? Can they learn if they are playing a game? What kind of classroom environment best facilitates learning? Will they learn more if they are having fun?* It is not always easy to answer these questions. Numerous factors influence how teachers organize their classrooms and affect classroom climate.

In the next sections, we will discuss three basic aspects of managing a project-based science learning environment: classroom climate, classroom organization, and management strategies. Although some of the topics and techniques introduced here are applicable to any subject area, each of these sections is designed to help teachers deal with some of the unique concerns and challenges associated with an inquiry-based science environment.

Classroom Climate

The opening scenarios in this chapter depicted very different classrooms. One way they differed was in terms of **climate**, or prevailing conditions and environment in the classroom. What factors determine classroom climate? Why is it important to facilitate a good classroom climate? In this section, we will examine a constructivist rationale and framework for thinking about classroom climate. We consider ideas about establishing a climate in which students solve classroom problems. We also discuss some basic concerns and challenges associated with classroom climate and the responsibilities of the teacher to create an effective classroom climate. We will consider how the teacher must be a role model, select good curriculum, promote a positive attitude toward science, enhance positive affective factors, balance the relationship between student and teacher, and ensure equality. Before we begin to discuss these topics, think about your own prior K–12 classroom experiences. What memories do you have of negative classroom experiences? What specifically made the classroom climate so unpleasant? Conversely, what are your best classroom experiences? What made the classroom climate so positive?

Your memories of positive classroom experiences probably included classrooms in which teachers boosted your self-esteem and made you feel important and capable. Your ideas were valued, and you were pushed to try harder, take risks, and think for yourself. The teachers probably had senses of humor and cared about students. They did not favor some students over others and did not discriminate against students based on gender, socioeconomic status, or race. The classroom was a safe haven, and students enjoyed being there. There was a sense of belonging. Lessons were interesting and motivating. The teachers seemed to enjoy teaching the subjects. There were few classroom management and student behavior problems. You probably still think of those teachers as role models.

A Constructivist Rationale and Framework for Thinking About Classroom Climate

The work of Alfie Kohn (1996), *Beyond Discipline: From Compliance to Community*, provides a rationale and framework for thinking about climate in a constructivist classroom. Kohn asserts that most unwelcome classroom behaviors can often be traced to the larger classroom context and the curriculum. Kohn argues that many discipline plans try to do things *to* children rather than *with* them and are, therefore, not constructivist in nature. He argues that making students act "appropriately" is not consistent with constructivist classrooms:

> My argument is that the quest to get students to act "appropriately" is curiously reminiscent of the quest to get them to produce the right answers in academic lessons. Thus, the constructivist critique, which says that a right-answer focus doesn't help children become good thinkers, also suggests that a right-behavior focus doesn't help children become good people. (p. xv)

Kohn compared a right-behavior focus to a traditional model of teaching:

> This approach is strikingly similar to the traditional model of academic instruction, where information or skills are transmitted to students so they will be able to produce correct answers on demand. For anyone who understands the limits of the "right answer" approach to learning, it can be illuminating to see that classroom management is basically about eliciting the "right behavior." This analogy also may help us to think about what we could be accomplishing instead. (p. 66)

Kohn argued that a constructivist classroom with a positive climate is not one characterized by forcing or coercing children to comply with the teacher's demands. Rather, it is one in which students are asked to reflect on what they should do and to solve problems together.

If you asked a group of teachers to think about the long-term goals they have for their students (what they would want students to know, be like, or act like long after they had been their teachers), what would they say? Probably, they would hope that the students would be responsible, caring problem solvers. Most classroom management programs, however, are totally inconsistent with this long-term goal. Kohn (1996) wrote, "It is unsettling because it exposes a yawning chasm between what we want and what we are doing, between how we would like students to turn out and how our classrooms and schools actually work" (p. 61). He added, "No one says, I want my kids to obey authority without question, to be compliant and docile" (p. 61). He argued that there is conflict among our ultimate goals, short-term goals (classroom management), and methods (coercion, threats, punishment, bribes, and so on). In order for there to be cohesion between the goals of a project-based curriculum, which stresses inquiry and collaboration, and classroom management, the management system too must stress inquiry and collaboration.

Creating an Effective Classroom Climate

It takes hard work to establish a positive classroom climate. Creating an effective classroom climate requires a comprehensive effort on the part of the teacher. The teacher must give real choices, be a role model, select curriculum carefully, promote a positive attitude toward science, promote positive affective factors, maintain a balanced teacher-student relationship, and ensure equality.

Giving Real Choices

Real choices help teachers create positive learning communities in which students respect and care about each other (Horsch, Chen, & Nelson, 1999). Kohn (1996) gave several other reasons that students should have real choices:

- People of any age ought to have a say in what happens to them.
- If they have a say, it is more likely that they will do essentially what we want.
- Misbehavior will diminish when children feel less controlled.
- Children are more respectful when their need to make decisions is respected.
- Children become more self-disciplined when given choices.
- Choices help children grow into ethical and compassionate people.

What are **real choices**? Real choices are ones that are not contrived by the teacher. A good way to have students make real decisions in a classroom is to ask them, "What do you think we could do to solve this problem?" The teacher depicted in Scenario 2 asked the students to work out how to play a game on the playground. With this question, students were given a chance to reason through a real problem related to their behavior, analyze possibilities, and negotiate solutions. Asking, "How do you want to line up to come in from recess?" however, does not present a real choice. A real choice would be "Should we line up?" or "How can we best come into the room from the playground?" The question "Do you want to raise your hands during the discussion or do you want to take turns?" does not offer a real choice. Asking, "What do you think would be the best way for us to share our results with minimal problems?" does offer a real choice.

CONNECTING TO
NATIONAL SCIENCE
EDUCATION STANDARDS

SCIENCE EDUCATION
TEACHING STANDARDS

Teachers of science design and manage learning environments that provide students with the time, space, and resources needed for learning science. In doing this, teachers engage students in designing the learning environment (p. 43).

Children may not be used to making real choices, however, and a teacher may have to work with them throughout the year to get them ready. For example, at the beginning of the year, students might be asked to select one method from a list of ways to control noise during investigations. Later, the teacher should move students toward making their own list of possibilities. Or a teacher might have students vote on the best way to take turns at the computer. Later, the teacher should move students toward coming to a consensus. Or a teacher might hold a class meeting to discuss ways for students to critique classmates' artifacts without insulting them. Class meetings have been found to be an effective technique for fostering a positive school climate (McEwan-Landau & Gathercoal, 2000). Later, the teacher should move students toward making their own decisions about behavior continually throughout the day.

Being a Role Model

Teachers who are **role models** in a project-based science class exemplify the behavior appropriate for inquiry science. Such teachers seek answers to questions and model how to find these answers. For example, a teacher might say, "Class, I read in the newspaper last night about some people over on Waterford Street who were getting sick from their drinking water. I was wondering if I could find out what was causing the problem. I think I could probably investigate this." The teacher models curiosity and questioning and shows the students how to find answers through reading, conducting investigations, or contacting members of the community. The teacher might say, "You know, I think I could find out what is causing the water problem by reading about water pollution on the Web." The teacher models using technology by logging on to the Internet. Teachers show students that they, themselves, are members of a learning community. The teacher might say, "I found a lot on the Web, but I don't know where to start to solve this problem. Maybe I need to call someone from the Environmental Protection Agency." Finally, teachers model the sharing of artifacts. The teacher might share with the students a synthesis of articles she found on the Web, notes from her inquiry, and the conclusions she formulated.

Selecting Good Curriculum to Promote Positive Classroom Climate

An essential contributing factor to classroom climate (and the way students behave) is instructional planning. One critical component of instructional planning is the curriculum. Kohn (1996) wrote, "When students are 'off task,' our first response should be to ask, 'What's the task?'" (p. 19; italics Kohn's). He added, "If discipline programs studiously refrain from exploring whether an adult's request was reasonable and, more generally, how the environment created by the adult might have contributed to a student's response, their most salient omission must surely be the curriculum. A huge proportion of unwelcome behaviors can be traced to a problem with what students are being asked to learn" (p. 18). The curriculum is often too simple, boring, or too difficult. Students need rich curriculum that extends thinking, elicits curiosity, and helps students answer questions that are important to them (Kohn, 1996, p. x). Kohn suggested

> **CONNECTING TO NATIONAL SCIENCE EDUCATION STANDARDS**
>
> **SCIENCE EDUCATION TEACHING STANDARDS**
>
> Teachers of science plan an inquiry-based science program for their students. In doing this, teachers develop a framework of yearlong and short-term goals for students (p. 30).

that we should ask if the curriculum is worth doing, meaningful, and relevant (p. 19). Similarly, Haberman (1995) stressed the importance of using project work with urban children in poverty because project-based work is meaningful and important to their daily lives.

Promoting a Positive Attitude Toward Science

Attitudes are one of the strongest predictors of behavior (Ajzen & Fishbein, 1980; Bandura, 1986; Pajares, 1992), and a teacher's attitude can influence the learning environment (Rutherford & Ahlgren, 1990). It is critical that elementary and middle grade teachers exude a positive attitude toward science. They should enjoy teaching science, give it a substantial amount of time in the curriculum, and encourage students to explore science topics.

A teacher needs to display genuine interest and enjoyment in science by talking about science topics and being excited about teaching the subject. A teacher's enthusiasm about science will rub off on the students. The *National Science Education Standards* (NRC, 1996) stressed that enthusiastic teachers who demonstrate the wonders of science help students develop positive attitudes. Imagine that a third grader is excited about his new pet guinea pig and wants to bring it to school the next day. A teacher who wants to promote a positive attitude toward science will build upon this interest and show excitement about the topic. This could result in a visit from the guinea pig to the classroom or some other type of investigation about pets.

A teacher who wants to promote a positive attitude toward science will also spend time teaching science. Research findings (Nelson, Weiss, & Capper, 1990; Nelson, Weiss, & Conaway 1992; Weiss, 1978, 1987) indicate that science is often the subject that gets the least attention in the elementary curriculum. Elementary teachers must give science equal priority with other subjects. Science cannot be considered a subject that "we'll get to if we have time after everything else has been taught." In the next chapter, we discuss ways that science can be integrated throughout the curriculum and be given sufficient time in the school day.

CONNECTING TO NATIONAL SCIENCE EDUCATION STANDARDS

SCIENCE EDUCATION TEACHING STANDARDS

Teachers who are enthusiastic, interested, and who speak of the power and beauty of scientific understanding instill in their students some of those same attitudes (p. 37).

Finally, a teacher who promotes a positive attitude toward science will encourage students to explore science topics. The third grader who comes to school excited about his new guinea pig might, for example, be encouraged to investigate the types of foods his guinea pig needs and likes. Teachers in a project-based science classroom listen to students, ask questions, and seek information about students' interests so they can encourage science inquiry. Students' ideas and responses help drive instruction in the classroom; the teacher is not the only person who determines the topics and activities.

Promoting Positive Affective Factors

Affective factors include curiosity, excitement, persistence, enthusiasm, flexibility, skepticism, and open-mindedness. Teachers promote positive affective factors when they, themselves, display them and when they encourage them in students. Rather than discouraging multiple answers in science and emphasizing a single right answer, a teacher who is fostering open-mindedness will seek more than one solution to a question and encourage students to do the same. Table 11.1 provides a few illustrations of ways to promote positive affective factors.

Maintaining a Balanced Teacher–Student Relationship

Teachers in an inquiry environment need to maintain a delicate balance between being the ones in charge and being members of the collaborative group. The teacher cannot be too authoritative or the collaboration will be squelched, because students will tend to look to the teacher for answers rather than collaborating with peers. However, if the teacher is too relaxed, she may fail to achieve basic curricular outcomes required by the school district or state. Teachers need to continually reflect on desired learning goals and outcomes, daily lessons, students' academic progress, students' skills, and classroom climate. Through this reflection, teachers can analyze where collaboration is succeeding or failing and take steps to sustain it.

A second factor to consider in maintaining a balanced teacher–student relationship is whether or not students are encouraged to critique the teacher and give suggestions for improving instruction to enhance their learning. Many educators using constructivist theories (Brooks & Brooks, 1993; Taylor, Fraser, & White, 1994) believe that students should have a "voice" in establishing classroom practices. This means sharing some decisions that are made in the class. Sharing decision making with students (such as about class rules, norms, topics to be studied, group arrangements, and methods of assessment) requires a delicate balance between teacher as adult with authority and student autonomy. When this balance is achieved, students are encouraged to take initiative and be autonomous, and the classroom learning environment fosters inquiry.

Teachers also balance teacher–student relationships by using various questioning strategies that encourage students to talk to one another. Not all classroom discourse should be between the students and the teacher. When students interact with each other, the classroom becomes a community of learners. Teachers can also create a positive classroom climate by asking questions that promote creativity, critical thinking, and many different answers. For example, a teacher can ask, "How might we look at this differently?" or "Do you think there might be another way to solve this problem?" This type of climate invites students to take risks without fear of failure.

TABLE 11.1   Ways to Promote Positive Affective Factors

Affective factor	Teacher strategy
Curiosity about the world	• The teacher shows her own curiosity about what she reads in the paper by talking about it with the students.
Excitement about science	• The teacher brings in science books to read.
Enthusiasm to continue	• The teacher encourages students to keep trying to find an answer by pushing them to find other sources of information.
Ambition to investigate	• The teacher encourages students to ask "Why?" "What if?" and "Could we?" questions.
Sensitivity to others	• The teacher asks, "How could I help you with this problem?"
Disciplined thinking	• The teacher encourages students to analyze potential flaws in their thinking.
Respect for evidence	• The teacher says, "Do you think your conclusion is supported by your data?"
Willingness to change	• The teacher shows that the schedule can change if students are really interested in a topic.
Questioning attitude	• The teacher says, "I wonder if we could find an answer to this question."
Fascination with findings	• The teacher says, "Wow! I didn't realize that. Isn't that interesting!"
Responsibility to project	• The teacher asks students how they should play a role in completing the project.
Skepticism about results	• The teacher says, "You know, I am wondering if this is correct. Do you think we should try it again to see if we get the same results?"
Tolerance for change	• The teacher shows a willingness to change a classroom rule if it isn't working.
Confidence in self	• The teacher encourages a student by saying, "I know you can do this. You are really good at asking questions."
Open-mindedness	• The teacher shows a willingness to accept students' ideas.
Dependability	• The teacher demonstrates that he can be depended upon—for example, he brings something to school that he promised the students.
Independence	• The teacher encourages students to think for themselves.
Self-reliance	• The teacher says, "I think you can do this by yourself."
Willingness to compromise	• The teacher models compromise by saying, "I see your point. Let's compromise."
Willingness to cooperate	• The teacher encourages students to get along with others in the group.
Honesty in artifact	• The teacher stresses that students should be honest—it is not acceptable to lie about data just to create a good artifact.
Objectivity	• The teacher says, "Well, let's wait until we get some more evidence."
Flexibility with ideas	• The teacher encourages students to try different ideas.
Patience with others	• The teacher doesn't get frustrated when a student is having difficulty understanding something.
Precision	• The teacher encourages students to measure things accurately.
Thoroughness	• The teacher stresses thoroughness by asking students to take meticulous notes.
Satisfaction with artifacts	• The teacher tells students they should be proud of their product.
Self-discipline	• The teacher says, "I'm going to put mind over matter and finish this."
Methodicalness	• The teacher models being organized.
Persistence with a task	• The teacher says, "I think you could answer this if you just stick with it a little longer."

Ensuring Equality

One essential component of the current national reform efforts in the United States is the premise that *all* students need to be scientifically literate (NRC, 1996; Rutherford & Ahlgren, 1990). The No Child Left Behind Act of 2001 (http://www.nochildleftbehind.gov) is also designed to force each state to ensure that all students learn. Thus, teachers need to believe that all students can learn, and they must treat all students equitably. Research on equity issues finds, however, that girls and minorities still lag behind their white male counterparts in achievement and interest in science (National Council for Research on Women, 2001).

One way to measure your students' self-images when it comes to science is to have them draw pictures of scientists. The draw-a-scientist activity has been done with many populations of people, and the typical drawing is very stereotyped (Barman, 1996; Boylan, Hill, Wallace, & Wheeler, 1992; Huber & Burton, 1995; Rampal, 1992; Sumrall, 1995). The scientist is usually depicted as a white man who is either bald or has wild, unkempt hair. He is usually wearing a lab coat while working in a laboratory with chemicals and glassware. Often, he is pictured with a sinister look on his face. He wears eyeglasses, and he usually has a pocket protector filled with pens and pencils. Learning Activity 11.2 gives you the opportunity to test for yourself science-related self-images.

Although girls start elementary school with similar interests and abilities as boys and overall gender differences in science achievement are decreasing, the gap between high-achieving girls and boys is increasing. Girls do less well. Between 1978 and 1990, boys were the top-scoring students in school (National Council for Research on Women, 2001; Wilson, 1992), and the TIMSS study showed that boys continue to have significantly higher achievement in science literacy than do girls (Schmidt et al., 2001; TIMSS, 1998).

Middle-class girls now take about the same number of high school math and science courses as do middle-class boys. However, in college, these girls are much less apt to major in math, science, or engineering fields than are similarly talented boys. They also drop out at faster rates. Low-income girls have had access to far fewer programs than have middle-class girls (National Council for Research on Women, 2001). The end result is that women are still greatly underrepresented in fields like physical science, engineering, and technology (Pollina, 1995). White women and minorities continue to be underrepresented in science and engineering employment (National Council for Research on Women, 2001). Current workforce projections indicate that unless more women and minorities are attracted to science, the United States will not have the trained personnel necessary to meet its needs (National Council for Research on Women, 2001).

A number of different factors have been associated with the gender and racial differences in science. Some studies indicate that girls and women and minorities are socialized to believe that science is for white boys and men (Hardin & Dede, 1992). Television shows and movies that depict scientists as weird "geeks" continue to reinforce this stereotype. Patricia Campbell and Beatrice Chu-Clewell (2002) suggested that even the news reinforces the stereotype that boys are better than girls in science and mathematics:

Learning Activity 11.2

DRAW A SCIENTIST

Materials needed:

- Paper and drawing materials
- The following chart
- A group of students

A. Give children drawing paper and drawing materials and ask them to draw pictures of scientists. Do not give them any other directions.
B. Analyze the students' pictures according to the following list. How many students draw their pictures with these characteristics?

Gender:
Male _____
Female _____
Race or Ethnic Background:
White _____
Hispanic _____
Asian _____
African American _____
Other _____

Work Environment:
Laboratory _____
Office _____
Outdoors _____
Other _____
Personal Characteristics:
Frizzy/wild hair _____
Eyeglasses _____
Pocket protector with pens/pencils _____
Bald head _____
Mean or sinister look _____
Other _____

C. Ponder this: In 1957, Margaret Mead and Rhoda Metraux published a report entitled "Image of the Scientist Among High-School Students" in *Science* (vol. 126, p. 387). They found that:

The scientist is a man who wears a white coat and works in a laboratory. He is elderly or middle aged and wears glasses. He is small. He may be bald or may be unshaven or unkempt. He may be stooped or tired. He is surrounded by equipment: test tubes, Bunsen burners, flasks and bottles, a jungle gym of blown glass tubes and weird machines with dials. The sparkling white laboratory is full of sounds: the bubbling of liquids in test tubes and flasks, the squeaks and squeals of laboratory animals, the muttering voice of the scientist. He spends his days doing experiments. He pours chemicals from one test tube into another. He peers aptly through microscopes. He scans the heavens through a telescope (or a microscope!). He experiments with plants and animals, cutting them apart, injecting serum into animals. He writes neatly in black notebooks.

D. Has the image of a scientist changed since 1957? Why do you suppose it has or has not? Why are these images so hard to displace?
E. Where do you suppose these images come from?
F. Who benefits or suffers from the maintenance of these images? What effect do you think these images have on girls' and minorities' career aspirations? What might a classroom teacher do to counter these images?

Research that supports math and science stereotypes gets much more attention in the media than does work that challenges them. For example, when one researcher says boys' higher SAT scores mean that boys are biologically superior to girls in math, she gets invited to the "Today Show" and is written up in the *New York Times*. But when another researcher says that society and test bias cause boys' higher SAT scores, she gets to go home and make dinner. Her results aren't considered news. (Clu-Clewell & Campbell, 2002)

Parents may contribute to biases against science. It is not uncommon for parents to expect different behaviors from boys than from girls. Boys often are expected to be tough, play basketball, and be active in outdoor activities, while girls are expected to be soft, play with dolls, and stay neat and clean while playing indoors. Gender-biased toys still permeate television ads—boys are more likely to be shown playing with erector sets, microscopes, and chemistry sets, while girls are more commonly shown playing with dolls, makeup, and stuffed animals (Hardin & Dede, 1992).

School practices also tend to contribute to gender and racial differences in science. Kahle and Lakes (1983) found that teachers give boys more opportunity to engage in science and math tasks. Teachers speak to boys more often, ask boys more high-level questions, and favor boys' responses (Baker, 1988). They have boys elaborate more on their answers than they do girls, and they let boys dominate classroom conversations (Baker, 1988; Sadker & Sadker, 1992). Finally, Guzzetti and Williams (1996) found that boys dominate laboratory experiences by manipulating lab equipment, and girls are given the job of recording information in journals.

Textbook publishers have made concerted efforts to depict girls and women and minorities doing science, but Bazler and Simonis (1992) found that illustrations, photos, and text of boys and men far outnumber those of girls and women. Further, textbooks portray life science as a female interest and physical science as a male domain (Kahle & Lakes, 1983).

Some research studies suggest that curriculum connected to real life helps girls and women appreciate science (NSF, 2003). Connections to real life are a hallmark of project-based science. To help develop a project-based science classroom in which girls and minorities feel comfortable with science and are encouraged in science as much as their white male counterparts, teachers need to be cognizant of their curriculum and instructional practices (Lynch, 2000). Table 11.2 illustrates some of these practices (King, Hollins, & Hayman, 1997; Rosser, 1990; Sanders, 1994; Tobias, 1992; Wilson, 1992). In addition, teachers need to be aware of cultural influences on classroom behavior. Garrick-Duhaney (2000) pointed out that teachers can help prevent discipline problems and promote equitable learning by teaching children about cultural differences and being sensitive to cultural orientations and nuances. In addition, teachers need to create a responsive classroom with a strong sense of classroom community where *all* students are made to feel competent and connected to others in the class.

Learning Activity 11.3 will help you think about equitable classroom practices.

TABLE 11.2   Equitable Classroom Practices

- The teacher helps girls and minorities by making sure that they are provided with challenging activities and that they succeed in the activities.
- Girls and minorities are active participants in the lesson, rather than passive observers or recorders of information.
- The teacher has high expectations of all students (for example, he or she doesn't give the girls the answers and expects that they can complete the work in science).
- Girls use manipulative materials in science classes.
- The teacher fosters collaboration, so, in science, girls have equal roles to boys.
- The teacher chooses language carefully (doesn't use male pronouns only to refer to doctors, engineers, or scientists).
- If students are in mixed-gender groups, rules are established to make sure girls have a chance to participate equally with boys.
- The teacher exposes students to women and minority role models in science so that students of both genders and all backgrounds can think of science as a possible career area.
- Girls and minorities are encouraged to do well in science and mathematics (and to take these subjects in school).
- Girls use computers in the classroom as much as their male counterparts do.
- Girls and minorities are involved in science and mathematics competitions (such as MathCounts and Science Olympiad) and extracurricular science and math activities (visiting science museums, participating in science clubs, and reading science journals).
- Girls are called on in class as much as their male counterparts are.
- Girls' answers are elaborated upon (teachers don't simply say, "Okay," and move on).
- The teacher uses interdisciplinary curriculum materials.

Classroom Organization

Classroom organization is the second major topic of classroom management that we will examine. In this section, we discuss three aspects of **classroom organization**: planning the physical arrangement of a project-based science classroom, structuring the school schedule, and maintaining a safe classroom. In each area, we will consider some practical strategies for dealing with student behavior and maintaining a productive and safe classroom environment. Finally, we will review in each section management tools that can help teachers organize and maintain a project-based science environment.

Arranging the Classroom

In an inquiry science environment, the **physical arrangements** of the classroom need to be carefully planned to facilitate student investigation and collaboration and to minimize behavioral problems. To stimulate and encourage students to investigate the world around them, the classroom should offer science attractions for students to observe and manipulate, stimulating materials to read, and computers with telecommunication access. To facilitate investigation, science equipment should be readily available, and there should be areas where students can easily work on different aspects of projects.

Learning Activity 11.3

EQUITABLE CLASSROOM PRACTICES

Materials Needed:

- A classroom to visit or teach in
- Video recording equipment if you teach the lesson

A. Obtain permission to watch a teacher teaching a science lesson or film yourself teaching a science lesson. Interview the teacher about his or her views about equitable practices in schools or watch the video of your lesson.

B. Analyze the classroom interactions and interview for the following:
   - How did the teacher help girls and minorities be successful?
   - Were girls and minorities active participants in the lesson, rather than passive observers or recorders of information?
   - Did the teacher have high expectations of girls and minorities (didn't give them the answers and did expect that they could complete the work in science)?
   - Did girls and minorities use manipulative materials?
   - Did the teacher foster collaboration so that girls and minorities had equal roles to the boys?
   - Did the teacher choose language carefully (didn't use male pronouns exclusively to refer to doctors, engineers, or scientists)?
   - If students were in mixed-gender groups, were rules established to make sure girls had a chance to participate equally with boys?
   - Did the teacher expose students to women and minority role models in science so that students of both genders and all backgrounds could think of science as a possible career area?
   - Were the girls and minorities encouraged to do well in science and mathematics (and to take these subjects in school)?
   - Did girls and minorities use computers in the classroom as much as their male counterparts did?
   - Were girls and minorities involved in science and mathematics competitions (such as MathCounts and Science Olympiad) and extracurricular science and math activities (visiting science museums, participating in science clubs, and reading science journals)?
   - Were girls called on in class as much as their male counterparts?
   - Were girls' answers elaborated upon (teachers didn't simply say, "Okay," and move on)?
   - Did the teacher integrate science throughout the curriculum?

Depending on the project, the physical environment will change. For instance, the classroom might be overflowing for several months with plants while students are investigating the effect of fertilizers on plants; but, when they are studying pollution, the room might be filled with air- and water-quality testing equipment. Even though the physical organization of the room continually changes in a project-based science classroom, there are fundamental features that must be maintained.

Most teachers don't have the luxury of determining the size and shape of their classrooms, but teachers usually have a great deal of flexibility to move desks and tables, decorate, and arrange materials. A smoothly running classroom doesn't just happen. Teachers who have effective, productive, stimulating classrooms are very aware of how they have organized students' desks, equipment, and supplies (Haberman, 1995). Evertson and colleagues (1984, 2000) have identified four keys to good physical arrangement of a classroom: (a) keep high-traffic areas free of congestion, (b) be sure students can easily be seen by the teacher, (c) keep frequently used teaching materials and student supplies readily accessible, and (d) be certain students can easily see instructional presentations and displays (2000, pp. 4–5).

Keep High-Traffic Areas Free of Congestion

What are high-traffic areas? These can include, but are not limited to, doorways, the area around the trash can, the teacher's desk, the area around the pencil sharpener, sinks, bookshelves with literature or encyclopedias, investigation centers, and the drinking fountain. It is best to keep high-traffic areas away from each other. Keep aisles wide and clear around these areas so they are easy to get to. Be aware of maintaining accessibility for wheelchair users or students with mobility problems. All students should be able to reach necessary items during the course of the school day. It is also a good idea to keep quiet areas, like reading centers, away from the high-traffic areas. Classroom pets and aquariums should also be kept away from high-traffic areas, because animals can easily become stressed from too much noise or movement around their cages.

When people first start teaching, they often give little attention to the logistics of their rooms and, instead, arrange things so they look good, designing and putting up fancy bulletin boards. Although an attractive classroom is certainly part of classroom organization and climate, attractiveness is not the primary consideration; a well-conceived classroom can facilitate learning and help minimize behavior problems. One beginning teacher mistakenly placed a set of encyclopedias near a science learning center where students were supposed to work with magnets. While individual students were working in the science center, other students had to walk past to get encyclopedias from the shelf. Students at the center were continually distracted, often by passersby looking at what they were doing with magnets.

In a project-based classroom, different areas will become high-traffic areas at different times, depending on the project. For example, when students are investigating plants, the window area might become a high-traffic area; but when students are investigating wheels, they might never go near the windows. Areas that are typically high traffic, regardless of the project, are areas with resources (reference materials, children's literature, and children's magazines), computer areas, areas where laboratory equipment is housed, and office supply areas (where students might obtain construction paper, tape, and glue to construct artifacts). Since students often work at their desks or at tables in collaborative groups in a project-based environment, it is wise to try to arrange sets of desks away from other sets of desks. In this way, one collaborative group will not interfere with another.

Be Sure Students Can Be Seen Easily by the Teacher

Good instruction and good behavior management depend a great deal on teachers' abilities to see all of their students. This is especially true in a project-based science classroom in which students are moving about and collaborating with each other. There is nothing more annoying to a teacher than hearing students talking or misbehaving, but not being able to see who they are. To help solve this problem, teachers arrange their classrooms so they have a clear line of sight to all student desks, bookshelves, learning centers, table areas, and science storage facilities.

One beginning teacher arranged a free-standing literature bookshelf so the shelf created a private study area where there was a bean-bag chair. The bookshelf obstructed the teacher's view of any student sitting in the bean-bag chair, making it a prime site for misbehavior.

Project-based science teachers need to be able to see students in order to facilitate discussions, diagnose learning problems, and help students as needed. If teachers cannot see students, they cannot engage them in conversation or notice if they are struggling with reading material, a question, or an activity. In Scenario 4, the teacher sat down to grade papers while the students worked on multiple activities. This is when students became loud and disruptive. Moving about the room to help students, facilitating discussions, or keeping students engaged with prompts and questions will minimize misbehavior. Just being able to see students and the fact that students know they can be seen will go a long way to keeping the classroom running smoothly.

Another important reason for making sure that all areas of the room are visible is that even simple science materials, such as a glass jar, can become dangerous if dropped and broken. Teachers need to be able to keep an eye on students while they are using equipment to make sure they are not in danger of hurting themselves and to respond quickly when accidents happen.

Keep Frequently Used Teaching Materials and Student Supplies Readily Accessible

Keeping often-used materials and supplies readily accessible saves time for the teacher and students. When off-task time between activities and lessons is thus curtailed, invitations for students to misbehave are minimized. Frequently used teaching materials include student textbooks, science equipment, and office supplies (rulers, tape, glue, and paper).

Clear shoe boxes and zipper-closure bags make good storage containers for science equipment and supplies. They can be handled easily by students distributing materials, and they make collecting materials easy and fast. Because such storage containers are clear, the teacher and the students can easily see what is inside. They can be stored on shelves or in larger boxes. Rolling carts (the type that libraries use for returned books) make great locations for storing frequently used textbooks, reference materials, and magazines. The movable carts let teachers rearrange their classrooms as needed, and the carts can be moved out of the way when not in use.

In a project-based science classroom, it is useful to have materials for investigations and for making artifacts readily accessible. Keep magnifying lenses and simple microscopes accessible at all times. This accessibility promotes curiosity and inquiry since it enables students to examine objects at any time of the day. Other objects to keep readily available are measuring cups, measuring spoons, bowls or containers, bug cages, eye droppers or dropper bottles, balance scales, rulers, flashlights, mirrors, and tweezers. Materials that are commonly used to create artifacts include rulers, tape, glue, cardboard, tagboard, paper, stencils, crayons, markers, and construction paper.

Be Certain Students Can See Displays and See and Hear Instructional Presentations

The obvious reason for making presentations and displays accessible and visible is that students need to be able to see them in order to learn. The less-obvious reason is that students become bored or frustrated when they can't see what you and the other students are talking about, and boredom and frustration often lead to misbehavior.

Teachers sometimes erroneously blame students for misbehaving during presentations. In one classroom, a teacher was demonstrating a science technique (how to focus a microscope accurately) to approximately 25 students. The sole microscope was sitting on a desk in the front of the room where the teacher was demonstrating the techniques for adjusting the stage, adjusting the mirror, moving the knobs, and fine-tuning the view. Since the microscope was placed at the eye level of the students in the first row, the rest of the class could not see what was being demonstrated. Some students in the back began to misbehave, and the teacher became frustrated and angry with them. When students were given the microscopes to use, most did not know how to use them, and the teacher again became angry and frustrated. It did not occur to her that the primary cause of the problem was the position of the microscope during her demonstration.

In Chapter 8, we explored how lessons might include demonstrations, discrepant events, presentations (diagrams, pictures, graphs, movies, videos, and television shows), and guest lecturers. Make sure that students can easily see and hear all such presentations and guest speakers. Keep displays, such as diagrams, pictures, or graphs, in locations where students can see them. Position televisions, screens, and monitors so that all students in the class have a clear view of them. Make sure that students have a clear view and that other students' heads or classroom objects aren't in the way. Make sure that objects to be viewed are positioned at proper distances and heights. Check that glare from overhead lights or the sun doesn't interfere with viewing. In addition, make sure all students can hear the presentation. Particular attention needs to be given to making sure that students who are hearing impaired have a clear view of the teacher or person using sign language.

Table 11.3 presents a summary of key considerations to think about with regard to the physical arrangement of elementary and middle grade science classrooms. In Learning Activity 11.4, you will imagine that you are setting up a classroom. Try

TABLE 11.3   Arranging a Classroom

Keep high-traffic areas free of congestion
- Arrange high-traffic areas away from each other.
- Keep aisles wide and clear.
- Pay attention to accessibility for wheelchair users.
- Make sure students can easily get to and reach necessary items.
- Keep quiet areas away from the high-traffic areas.
- Keep classroom pets and aquariums away from high-traffic areas.
- Try to arrange sets of desks away from other sets of desks.

Be sure you can easily see students
- Arrange the classroom so you have a clear line of sight to all student desks, bookshelves, learning centers, table areas, and science storage facilities.
- Watch for potential safety problems.

Keep frequently used teaching materials and student supplies readily accessible
- Keep student textbooks, science equipment, and office supplies in readily accessible areas.
- Use clear shoe boxes and zipper-closure bags as storage containers.
- Use rolling carts.
- Keep materials used in investigations readily accessible.
- Keep materials used for making artifacts readily accessible.

Be certain students can see displays and see and hear instructional presentations
- Make sure that students can easily see and hear presentations and guest speakers, discrepant events, or demonstrations.
- Keep displays where students can see them.
- Position televisions, screens, and monitors so that all students have a clear view.
- Consider height, angle, distance, and glare.
- Make sure hearing-impaired students have a clear view of the sign language interpreter.

to think about the aspects of classroom organization just discussed as you design a classroom setting.

Structuring the School Day

Although a teacher may not be able to structure the entire school day, given that there are often fixed schedules for art, music, physical education, meetings, and assemblies, a teacher generally has a great deal of flexibility to structure the day within the classroom. The structure of the school day includes the number of minutes a teacher allocates to various subject areas, the delivery of science as a subject (discontinuous and presented as a separate subject or integrated and continuous), and the day's schedule of activities ("etched in stone" or flexible). In this section, we will discuss each of these aspects of structuring the school day.

The Number of Minutes Allocated to Subjects

The number of minutes allocated to a subject area is determined by many different factors: the students' knowledge, skills, and abilities; state and local curriculum

Learning Activity 11.4

ARRANGING A PROJECT-BASED SCIENCE CLASSROOM

Materials Needed:

- Drawing materials, such as pencils, rulers, and compass
- Paper or a computer with a drawing program*
- A classroom to observe

* You may also choose to design your classroom using the online Classroom Architect at http://classroom.4teachers.org/

A. Using the four principles of good classroom organization you just read about, design a 20-foot by 40-foot classroom so that it contains all of the following items:
  - 25 flat-topped (not slanted) student desks
  - A 12-foot by 3-foot table
  - 10 extra chairs
  - A freestanding bookshelf
  - A teacher's desk and chair
  - A round table with a 5-foot diameter
  - A file cabinet
  - An overhead projector
  - Two computers on carts
  - An aquarium with fish
  - A cage with gerbils
  - Four plants in pots
  - A pencil sharpener
  - Textbooks
  - Student belongings
  - The teacher's supplies

  Permanent features of the classroom are shown in Figure 11.1.

B. Describe how your classroom organization will facilitate project-based science learning. You may want to review chapters that discussed lessons, collaborative learning, and other techniques that facilitate learning science.

C. What specific types of learning centers, bulletin boards, and visual displays would you design to interest students in science and promote the development of driving questions and investigations?

D. Visit an elementary or middle grade classroom to determine whether the classroom follows the four guidelines for good organization. If not, what would you do to change it?

guidelines; and the driving question being investigated. These factors can sometimes dramatically alter a teacher's preplanned lesson. You might prepare a 40-minute lesson and take only 5 minutes to teach it. You might prepare a 40-minute lesson and take 3 days to teach it. The teacher in Scenario 2, for example, was surprised that her lesson took longer than planned. However, while plans always change, it is important

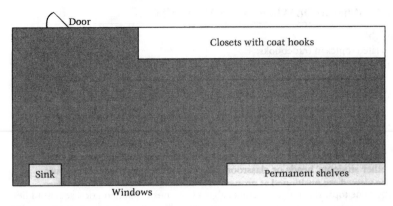

Figure 11.1 Permanent features of a classroom.

to plan, including determining the number of minutes a lesson should last. Without a plan, the day will likely become disorganized and chaotic.

Imagine that a group of first-grade students is investigating the driving question "What animals live in my neighborhood?" Because first graders are only beginning to learn to read, a teacher may spend more minutes teaching reading than teaching science. However, children can practice their reading skills by reading science trade-books. Because first graders are unlikely to know how to use a magnifying glass or a simple microscope, the teacher will need to spend a great deal of time developing these skills before students can investigate what animals are in a pond, for example.

With experience, allocating amounts of time for particular lessons becomes eas-ier for teachers. However, because sometimes even the most experienced teacher is fooled about the amount of time needed to cover a topic, develop a skill, or conduct an investigation, good teachers always think ahead to have contingency plans ready. Otherwise, classroom management problems can arise. In anticipation of lessons that go more quickly than expected, it is a good idea to have extra specimens for students to observe, science tradebooks available for students to look at, and computer-based activities or other activity centers set up to which students can move when they are finished with work. The teacher in Scenario 2 had students learn more about the topic of DDT through computer research when they were finished with their work. Table 11.4 lists activities for students to engage in when there is extra time.

Some state and local curriculum guidelines predetermine the number of minutes a teacher must allocate to a given subject area. For example, some schools may tell teachers that they need to spend 45 minutes a day on reading and writing and 30 minutes a day on science at the first-grade level. Good teachers understand their cur-riculum thoroughly enough that they can often blend science throughout the cur-riculum. They do not worry, as a result, about the exact number of minutes they spend on a subject. Martin Haberman (1995), in fact, pointed out that exemplary teachers working with children in poverty become masters at knowing how to work around "rules," such as time allotments, so that their children learn. They realize that a smoothly flowing school day facilitates good classroom management, and a smoothly flowing school day is not always characterized by a specific number of

TABLE 11.4   What to Do When There's Time Left Over

Students might:
- Look up related topics in tradebooks;
- Consult the encyclopedia for more information;
- Engage in silent reading for personal enjoyment;
- Use the World Wide Web to find more information on the topic;
- Explore additional examples of the same topic;
- Brainstorm related topics to investigate;
- Design an activity that classmates could participate in;
- Make an entry in a journal or portfolio;
- Make a drawing or model to explain what they learned today;
- Create a quiz for classmates;
- Explore the same topic through a different subject (art, music, mathematics, social studies, language arts, reading, or physical education);
- "Peer tutor" other students in the room;
- Play an educational game; or
- Go to activity centers to explore new topics or ideas.

minutes and a schedule set in concrete. Good teachers plan their days so that, ultimately, all subjects are covered and outcomes are reached. For example, on one day a teacher had students read about animals for an hour and plan an investigation for 15 minutes. The next day, students spent 45 minutes completing the investigation. Then students spent 15 minutes working with the teacher to write stories about their investigation. Next, the students *read* their investigation "stories" to others in the class for 15 minutes. On this second day, then, the class spent 45 minutes on science and 30 minutes on reading and writing. Over the course of 2 days, the students spent a total of 90 minutes on reading and writing and 60 minutes on science. This is equivalent to the recommended number of minutes of reading/writing (45 minutes per day) and science (30 minutes per day).

The driving question that is being investigated can dramatically influence the number of minutes allocated to science. The type of project, where the students are in the project, and the level of student motivation are all factors influenced by the driving question. For example, a project that investigates "What animals live in my neighborhood?" may take more time to prepare and finish than a project investigating "What do our classroom pets need to be healthy?" simply because it takes more time to get children ready to go outside, walk outside, investigate, and return to the school to resume other work than it takes to observe a pet eating in a cage in the classroom.

Where the students are in the project will also influence the time spent on different subjects. When students are beginning an investigation, the teacher may need to teach many science lessons to provide students with the knowledge and skills needed for their investigations. The science investigations themselves take time. As students finish their investigations and begin to make artifacts, the teacher may spend more time teaching writing skills.

Finally, student motivation will influence the amount of time spent on a topic. One driving question may motivate students so much that it is difficult to get them to stop their investigations. For other driving questions, it may be difficult to get students to finish their investigations.

Learning Activity 11.5 will work you through planning times for subjects during the school day.

Learning Activity 11.5

PREPLANNING THE NUMBER OF MINUTES ALLOCATED TO SUBJECTS

Materials Needed:

- Pencil and paper or a computer

A. Imagine that students in your third-grade class have decided (with your guidance) to investigate the driving question "How can magnets be used around school and home?" Think about the knowledge and skills that third graders would need to begin investigating this driving question and make a list of them, Then, think about lessons that might meet these needs and make a list of these as well.
B. Using these parameters, plan a week's worth of lessons focusing on the amount of time that you think it will take to cover various topics (make sure you include all subject areas: science, mathematics, social studies, and language arts/reading). Use Table 11.5 as your planning grid.
C. If possible, meet with a practicing classroom teacher to discuss your plans. Are they feasible? Where does the experienced teacher think you need more time and less time?
D. Find out what guidelines the teacher must follow with regard to the number of minutes per subject area. Does the state or local school district stipulate the number of minutes to allocate to various subjects? What is the teacher's experience with this in his or her classroom?

TABLE 11.5   Planning Grid

Monday	Tuesday	Wednesday	Thursday	Friday
9:00 School starts	9:00 School starts	9:00 School starts	9:00 School starts	9:00 School starts
11:00–12:00 Art class with art teacher				
12:00–12:45 Lunch	12:00–12:45 Lunch	12:00–12:45 Lunch	12:00–12:45 Lunch	12:00–12:45 lunch
		1:00–2:00 Physical education class with PE teacher		
			2:00–3:00 Music class with music teacher	
3:00 End of day	3:00 End of day	3:00 End of day	3:00 End of day	3:00 End of day

The Delivery of Science as a Subject

You probably discovered in Learning Activity 11.5 that the structure of the school day is influenced by how the teacher presents science and other subject areas. If the teacher presents each subject area separately, the school week might look like what is shown in Table 11.6.

In Table 11.6, each subject area is treated separately, and seldom is there a connection among the subject areas. If a teacher presents science as a continuous subject integrated with other subjects, the week will look very different. Imagine a teacher starts out the week by reading a 15-minute story about how magnets are used in everyday life. Next, the teacher explains to students that they will need to know something about magnets to explore the driving question "How can magnets be used around school and home?" In a lesson, the students spend about an hour learning that magnets have a north and a south pole, and they test the poles of the magnet to learn that opposite poles attract and like poles repel. The teacher explains that maps are used in everyday life and that maps have directional arrows on them to indicate north, south, east, and west. The teacher introduces the idea that the earth has a magnetic north that can be identified with a compass. The teacher spends about an hour showing the students how to use a compass. After lunch, the class discusses situations in which maps are used, and they identify objects on a map (such as cities, states, roads, streams, and lakes), using the compass. Finally, the students are given a home assignment requiring them to investigate how magnets are used in their homes.

The next day, students discuss the findings of their home assignment. They begin a portfolio where they gather pictures that illustrate how magnets are used in everyday life. To accompany the pictures, they create descriptions of how the magnets are used. The teacher notices that many of the descriptions in the students' portfolios contain incomplete sentences and lack end punctuation, so he includes a lesson that will teach students about complete sentences and end punctuation. Next, the teacher explains to students that they will be graphing information about magnets and that they may develop artifacts that display information in graph form. The teacher introduces graphing by having students sort objects that stick to a magnet and objects that don't stick to a magnet. Students make a bar graph to display data about the number of objects that attracted or did not attract the magnet. Students make two pie graphs to show what types of items (such as metal, paper, and plastic) attracted or did not attract the magnet.

On the third day, the teacher reviews how to use a compass for finding directions and contrasts this method with using the grids and coordinates on a map. Students practice finding various geographical features and sites on a map of their state. Next, the teacher holds a discussion about how students could use a compass if they were lost and did not have a map. The class goes outside and practices finding various things on the playground using a compass. After lunch, the teacher returns to the driving question "How can magnets be used around school and home?" The students discuss investigations that they might develop to answer this question. One group decides to investigate how a magnet could be used to keep things open or closed. Another group decides to investigate how magnets are used to keep things together. A third group decides to investigate how magnets are used in machines and other

TABLE 11.6   Sample Schedule

Monday	Tuesday	Wednesday	Thursday	Friday
9:00 School starts	9:00 School starts	9:00 School starts	9:00 School starts	9:00 School starts
9:00–10:00	9:00–10:00	9:00-10:00	9:00-10:00	9:00–10:00
Reading/writing: Read first chapter of *How to Eat Fried Worms* and discuss.	*Reading/writing:* Read second chapter of *How to Eat Fried Worms* and discuss.	*Reading/writing:* Write an opinion piece about whether you would be willing to eat fried worms.	*Reading/writing:* Read third chapter chapter of *How to Eat Fried Worms* and discuss.	*Reading/writing:* Read fourth chapter of *How to Eat Fried Worms* and discuss.
10:00–11:00	10:00–11:00	10:00–11:00	10:00–11:00	10:00–11:00
Science: Test whether opposite ends of a magnet attract or repel.	*Science:* Use iron filings to see evidence of magnetic fields at the end of a magnet.	*Science.* Test what magnets stick to and don't stick to.	*Science.* Plan an investigation to see how we use magnets in our daily lives.	*Science:* Carry out science investigation.
11:00–12:00	11:00–12:00	11:00–12:00	11:00–12:00	11:00–12:00
Art class with art teacher	*Mathematics:* Introduce graphing.	*Mathematics:* Make a graph showing how many students in the class have blue, brown, and green eyes.	*Mathematics:* Teach about types of graphs (bar, circle, and line).	*Mathematics:* Practice making one of each type of map using student eye color.
12:00–12:45 Lunch	12:00–12:45 Lunch	12:00–12:45 Lunch	12:00–12:45 Lunch	12:00–12:45 Lunch
12:45–1:00	12:45–1:00	12:45–1:00	12:45–1:00	12:45–1:00
Silent reading	Silent reading	Silent reading	Silent reading	Silent reading
1:00–2:00	1:00–2:00	1:00–2:00	1:00–2:00	1:00–2:00
Mathematics: Finish lesson on adding two-digit numbers and give end-of-chapter test.	*Social studies:* Introduce symbols on a map (keys, north, south, east, west).	*Physical education class with PE teacher*	*Social studies:* Learn about geographical features on maps and identify key features in state.	*Social studies:* Make a map of the neighborhood, including directional key and geographical features.
2:00–3:00	2:00–3:00	2:00–3:00	2:00–3:00	2:00–3:00
Social studies: Discuss how maps are used in everyday life. Examine a map.	*Health:* Introduce lesson on nutrition and food pyramid.	*Social studies:* Look at a map of the United States and locate regions based on directional keys.	*Music class with music teacher*	*Reading:* Write a summary of chapters read in *How to Eat Fried Worms.*
3:00 End of day	3:00 End of day	3:00 End of day	3:00 End of day	3:00 End of day

TABLE 11.7   Preliminary Investigation Plans

GROUP 1: HOW CAN MAGNETS OPEN OR CLOSE THINGS?
Plan: Using magnets of various sizes and shapes, investigate how magnets can open and close things (such as doors, boxes, bags, and folders). Find out if magnets are used this way in real life.

GROUP 2: HOW CAN MAGNETS KEEP THINGS TOGETHER?
Plan: Using one magnet, test things that can be kept together (such as papers, books, bags, and clothing). Find out if magnets are used this way in real life.

GROUP 3: HOW ARE MAGNETS USED IN MACHINES OR OTHER TECHNOLOGY?
Plan: Get on the Web and find out more about uses of magnets. Call an expert who uses machines. Take apart old, broken machines and computers.

GROUP 4: HOW ARE MAGNETS USED IN HOBBIES OR CRAFTS?
Plan: Visit a hobby or crafts store. Invite to class a parent who makes crafts. Try making a useful object out of magnets.

---

technology. A fourth group decides to investigate how magnets are used in hobbies and crafts.

The next day, each group works to come up with subquestions and a preliminary investigation plan. They write down the plans. Table 11.7 shows the groups' plans.

After lunch, the teacher helps students locate resources, supplies, and materials needed to conduct the investigations, and students begin their investigations.

On Friday, the morning begins with students continuing their investigations. The teacher notices that one group is having difficulty adding up a list of numbers of machines found that use magnets. This prompts him to take some time to teach a lesson about adding two-digit numbers. After the lesson, students are ready to continue with their investigation.

After lunch, the teacher shows the students how to summarize information, using a chart. At the tops of the charts, students write down statements that reflect the topics of their investigations. In the charts, they list all the things they found. Students use the topics at the tops of the charts to form the topic sentences of their summaries, and they use the items in the charts to make supporting sentences. The teacher has students summarize in writing what they have done so far in their investigations. Students share their summaries with classmates and discuss what else they will need to do to answer the driving question.

In a project-based science classroom, teachers are more likely to structure the school day in the integrated manner discussed in this second example. The delivery of science content in this example is continuous and integrated across the other subject areas. Connections are made for students to

CONNECTING TO NATIONAL SCIENCE EDUCATION STANDARDS

SCIENCE EDUCATION TEACHING STANDARDS

Teachers of science actively participate in the ongoing planning and development of the school science program. In doing this, teachers participate in decisions concerning the allocation of time and other resources to the science program (p. 51).

show them how the lessons fit with the driving question. For example, when students learn to graph, they understand that they need this skill in order to present information in their artifacts. When they learn about compasses, they know that they need this information to understand how some machines use magnets. Teachers who present science (and other subject areas) as separate entities run the risk that students won't see the relationships between knowledge and skills or transfer understanding and skills to new situations.

The Day's Schedule of Activities

Although planned, the daily schedule in an elementary project-based science classroom is flexible. In *some* middle schools organized around large blocks of time, the teacher has a great deal of flexibility, but in other middle schools, students switch classes every 50 minutes, and so the day is much less flexible. Regardless of the flexibility of school organizational structure, good teachers understand that they must be flexible, that some activities take more time or less time than anticipated, and that they must be willing to alter the day's schedule accordingly.

For example, the teacher may never get to the lesson on reading a compass on Monday because students' investigations take a different focus. These types of changes are expected and welcomed in a project environment. If it is necessary to teach the lesson on reading a compass, the teacher will introduce it on another day.

Because project-based science does not focus on correct answers, frequently there are multiple answers to a question. Therefore, lessons and activities evolve and flow with students' findings. Because learning is student directed, not all activities and lessons can be planned exactly as they will be executed. Imagine that students discover that magnets can be used to hold things (such as holding the lid of a can on a can opener) as well as close things (such as the door of a refrigerator). This discovery motivates students to invent a lunch box that can be opened and closed with a magnet. Although the hours needed to test this idea were not in the original plans, the teacher lets students complete the activity. It may even lead to a new lesson that she did not originally consider. For example, she may need to introduce a lesson on magnetic poles in magnets of different shapes—the poles of a bar magnet are on the ends, whereas the poles of a circular magnet are on opposite sides. Students may need this information if they opt to use a round magnet to open or close the lunch box.

Table 11.8 presents some key ideas to remember about structuring the school day.

TABLE 11.8 Structuring the School Day

Number of minutes allocated to subjects
- Preplan the number of minutes needed for a subject.
- Preplan the number of minutes a lesson will last.
- Have activities ready for situations in which too much time is left over.
- Become thoroughly aware of the curriculum so you know how to blend science throughout it.
- Don't worry about the exact number of minutes actually spent on a subject or lesson.
- Be concerned with meeting curriculum goals and lesson objectives.

Maintaining a Safe Classroom

Because students in a project-based science classroom are self-directed, they share a role in developing activities, locating resources, and conducting investigations. They frequently move about the room to collaborate with others, work on a computer, or engage in an activity. Often, different groups of students investigate different sub-questions. All of this activity raises unique **safety** concerns.

In general, teachers need to be aware of school district policies on all safety-related issues. For example, does the school have a policy about having animals in the classroom or the use of chemicals in a science class? Is there a policy on administering first aid? Who is supposed to be notified in case of an accident or injury? How are field trips arranged, and what are the legal liabilities?

> **CONNECTING TO NATIONAL SCIENCE EDUCATION STANDARDS**
>
> **SCIENCE EDUCATION TEACHING STANDARDS**
>
> Teachers of science design and manage learning environments that provide students with the time, space, and resources needed for learning science. In doing this, teachers ensure a safe working environment (p. 43).

In this section, we will discuss several issues related to maintaining a safe classroom. These include working with animals and plants in the classroom, planning for safe use of equipment and materials, dealing with fire and glassware, planning safe field trips, and being prepared to administer first aid. At the end of the section is a checklist to assist you in planning for a safe classroom.

Animals and Plants in the Classroom

Many teachers like to have animals and plants in the classroom. Not only are living things motivating for students (they might lead to driving questions), but they are needed in some investigations. However, there are a number of considerations involved in bringing plants and animals into the classroom.

It is essential that animals be healthy and well cared for and that they not harm the students in the class by biting, scratching, or passing on a disease. The National Science Teachers Association handbook on safety in the elementary classroom (Dean, Dean, Gerlovich, & Spiglanin, 1993) suggests that several rules be adopted regarding animals in the classroom:

- Don't let students bring in live or deceased wild animals, snapping turtles, snakes, insects, or arachnids that might be carrying a disease.
- Make sure animals are housed properly in clean and securely closed cages.
- Purchase animals from reputable stores or supply houses. Carolina Biological Supply Company (2700 York Road, Burlington, NC 27215; 1-800-334-5551) is one of the most-used suppliers of live animals and specimens for science classes. Other suppliers and resources can be found on the National Science Teachers Association Web page (http://www.nsta.org).
- If students bring personal pets to school as part of a class activity, make sure that the pets have proper housing during the day and that strangers don't handle them

because they may become frightened and bite or scratch and because it may injure them.

- Don't let students pick up or touch unfamiliar animals—even if they are other students' pets.
- Don't allow students to tease or poke at an animal in a cage, even if their intention is only to wake up a sleeping animal and not to hurt it.
- Teach students how to properly pick up different animals. For example, a rabbit should be picked up by the scruff of the neck. Make sure students wear gloves when picking up most animals and always wash their hands after handling animals.
- Familiarize yourself with the necessary care for and best way to handle the animals in your classroom. Check local libraries for books or call a local zoo or veterinarian for more information.

Some plants can be toxic (even fatal) to students. The National Science Teachers Association (Dean et al., 1993) provides several guidelines for using plants safely in the classroom:

- Teach students not to put plants in their mouth.
- If students are handling plants, have them wash their hands after the investigation so that sap or juice doesn't remain on their skin. Make sure they wash their hands before eating any food.
- Don't let students inhale smoke from any burning plant or pick any unknown flowers, seeds, or plants because some fumes or toxins can be very dangerous. For example, poison sumac and poison oak are poisonous to the skin, and foxglove or jimson weed are poisonous when eaten.

Equipment and Materials

Teachers should wear and require students to wear goggles to protect their eyes against impact or contact with liquids or fumes. Even simple activities, such as burning a candle, working with a mild acid like vinegar, or melting sugar in a test tube, have the potential to cause eye damage or irritation. Classrooms should have sets of safety goggles, and teachers need to teach students how to use them. Goggles should always be cleaned after each use, because eye diseases, such as conjunctivitis, are very contagious. Some activities involve blindfolding students. Once a blindfold has been used, it should not be used on another student.

Safety shields can protect students during demonstrations. For example, if you are conducting a lesson in which you demonstrate that hot air expands by heating a test tube with a cork in it, you should use a safety shield since the cork will pop out (and sometimes fly out) of the test tube.

Students in a project-based classroom are active participants in designing and carrying out investigations. This means that they may gather their own materials and supplies. Improperly stored materials are invitations for accidents. Make sure that the area where materials and supplies are stored has plenty of space, or limit the number of students allowed into the area. Make sure that the place where you store equipment and supplies has adequate shelving that is sturdy, deep enough so items don't easily fall off, and secured to a wall or floor so that it cannot tip over on students. Chemicals and glassware should be stored only on lower shelves so they are easy to get to. Keep supplies in small jars so that students can handle them easily. For example, rather than having students pour vinegar from a gallon jug, keep small dropper bottles of vinegar available for student use. Although the typical elementary classroom does not have many (if any) volatile liquids and acids, some middle school or junior high classrooms have these items. A popular middle grade life science book on the market uses nitric acid in one of its protein activities. Do not store volatile items together, near heat sources, or near electricity and keep these items in locked storage areas. Finally, make sure all items are labeled and keep handy a quick reference on precautions, antidotes, and proper disposal.

Always be vigilant when students are using electricity. During one teacher's first year of teaching, students were investigating batteries and bulbs (using size D dry cells, bell wire, and flashlight bulbs). She asked the students to figure out a way to make the bulb brighter (intended to be done by adding more batteries or using a different wattage flashlight bulb), and a student headed toward a wall outlet to plug in his little electrical setup. Fortunately, she caught the student before he got there and injured himself. Many elementary teachers let students connect too many batteries in a series circuit. The uninsulated wire in the overloaded circuit gets very hot, burning the students. Instruct children how to use electrical devices. Teach them not to touch items (such as hot plates) right after they have been turned off, because they will usually still be hot. Teach them to unplug electrical objects by pulling on the plug, rather than the wire. Avoid using electrical extension cords because they can be overloaded or students can easily trip over them in the classroom. Be very careful when using any small electrical device, such as a hot plate, small motor, fish aquarium pump, or fan, so students' hands or clothing don't come in contact with a hot surface or fast-moving part.

Fire

Many elementary teachers do not let students use fire in the classroom, but some middle schools and junior high schools use Bunsen burners, Sterno cans, matches, or other sources of flame. Teachers and students should know where fire alarms are positioned in a building and be aware of the quickest fire escape from the room. Accidental fires in a classroom often happen when clothing or hair come too close to an open flame. All classrooms should be equipped with fire extinguishers and fire blankets. There are different types of fire extinguishers (for electrical fire, flammable liquids, and so on), so most schools like to use multipurpose ABC fire extinguishers.

Glassware

Glassware presents particular problems in any classroom, but it is of great concern in an elementary classroom. Teachers need to think carefully about using glass jars, mirrors, thermometers, and prisms. Many teachers use glass only if it is absolutely necessary and a plastic alternative will not work. If glass is used, make sure that any sharp edges (such as on mirrors or prisms) are taped, painted, or ground smooth. If students are going to put glass tubing into a rubber or cork stopper, make sure that they use a lubricant and that they do not put the glass tubing into the palms of their hands to press in the stoppers (Dean et al., 1993).

Field Trips

Field trips are valuable in a project-based environment because they link the driving question and investigation to a broader community and the world outside of school. In fact, Haberman (1995) pointed out that exemplary teachers working with urban children and children of poverty use field trips and community resources much more frequently than do unsuccessful teachers. However, there are a number of precautions that should be taken when going on field trips. The National Science Teachers Association (Dean et al., 1993) provides several tips:

- Don't take anything for granted on the trip.
- Bring along a second adult approved by the school administration.
- Obtain parent permission before taking students away from school.
- Alert parents about proper clothing or supplies needed for the trip. For example, ask students to wear long-sleeved clothing if they will be outside and could come in contact with ticks or be scratched by branches.
- Always bring along a first aid kit on a field trip in case something happens to a student while away from the school building. Many teachers also bring along cell phones in case they need to make an emergency phone call for help.
- Take special precautions for any field trip near a body of water, a stream, or a river. Become familiar with animals or plants near or in the water. Use a "buddy system" if students are actually in the water. Learn CPR.
- If taking a field trip to a factory, laboratory, or business, make sure that the trip is well supervised and that someone from the site conducts the field trip. On-site guides will be the most familiar with potential danger areas for students and can alert participants to situations in which more caution is needed.

Additional information about field trips and community connections can be found in Robertson, W. C. (2001). *Community Connections for Science Education.* Arlington, VA: NSTA Press.

First Aid

First aid is meant primarily to protect, not treat. The American Red Cross is the best source of information for first aid procedures for staying calm, restoring breathing, stopping bleeding, and preventing shock, so we won't go into first aid procedures here. Become familiar with first aid procedures in case anything happens in the classroom or on field trips.

Safe Lessons

As a regular part of any lesson, teachers should think about potential safety precautions before beginning to teach. The checklist in Table 11.9 (modified from Dean et al., 1993) summarizes safety considerations and provides some guidelines for planning and teaching on a daily basis with safety in mind. Some teachers like to have students complete a safety contract whereby the student and their parents sign a contract regarding safe classroom procedures (Roy, 2006). This contract can include items from Table 11.9 rephrased for students. For example, under the section *Questions to Ask Regarding Safe Use of Chemicals,* number 3 reads: Do you forbid students from mixing acid and water? This item could be reworded on a student contract to state: Never touch or mix chemicals without the teacher's permission. Similar items listed on a contract can form the basis of a safety contract that can be signed by students and parents/guardians. Learning Activity 11.6 will work you through a safety audit of a classroom.

TABLE 11.9 Planning Safe Lessons

Questions to ask regarding general safety concerns	Answers
1. Do you have a copy of the federal, state, and local regulations that relate to school safety, as well as a copy of your school district's policies and procedures?	_____
2. Check your classroom. Are equipment and materials properly stored (in the right types of cabinets for chemicals, on sturdy shelving that won't tip over, and on deep shelving so that items won't fall off of easily)?	_____
3. Are you familiar with possible hazards involved in using the equipment and materials in your room?	_____
4. Do you know your school's policies and procedures in case of accidents?	_____
5. At the start of each science activity, do you instruct students regarding potential hazards and precautions?	_____
6. Is the number of students working together on an experiment limited to a number that can safely perform the experiment without causing confusion and accidents?	_____
7. Do students have sufficient time to perform the experiments and clean up and properly store the equipment and materials after use?	_____
8. Do you instruct students not to taste substances? Do you instruct students not to touch substances without first obtaining specific instructions from you?	_____
9. Are your students aware that all accidents or injuries—no matter how small— should be reported to you immediately?	_____
10. Do you instruct your students that it is unsafe to touch the face, mouth, eyes, and other parts of the body while they are working with plants, animals, or chemical substances and afterwards, until they have washed their hands and cleaned their nails?	_____
11. Does your classroom have safety goggles and a first aid kit? Do you know how to use these items? Do students use the safety goggles? Are the goggles cleaned and disinfected after each use?	_____
12. Are materials and supplies that students use stored in an area with plenty of space to avoid accidental collisions among students?	_____

Questions to ask regarding safe use of chemicals	Answers
1. Have you taught students that they must not mix chemicals "just to see what happens"?	_____
2. Have you taught students to never taste chemicals and to wash their hands after use?	_____
3. Do you forbid students from mixing acid and water?	_____
4. Do you keep combustible materials in a metal cabinet equipped with a lock?	_____
5. Do you store chemicals under separate lock in a cool, dry place, but not in a refrigerator?	_____
6. Do you store only a minimum amount of chemicals in the classroom? Do you give students only small amounts of materials to work with (such as a dropper bottle of vinegar, rather than a gallon jug)?	_____
7. Do you properly discard chemicals not used in a given period?	_____
8. Are all chemicals labeled?	_____
9. Do you keep handy a quick reference for precautions, antidotes, and proper disposal of all chemicals?	_____

Questions to ask regarding glassware   Answers

1. Do you use plastic instead of glass when possible? _____
2. Do students know how to use glassware? (For example, do they know how to insert glass tubing into a rubber stopper and how to heat hard glass test tubes—not from the bottom but tipped slightly and not in the direction of another student?) _____
3. Are a whisk broom and dust pan available for sweeping up pieces of broken glass? _____
4. Are students aware that they should not drink from glassware used for science experiments? _____
5. Are thermometers for use in the classroom filled with alcohol, not mercury? _____

Questions to ask regarding electricity   Answers

1. Are your students taught safety precautions for the use of electricity in all situations (not to touch an item recently turned off, not to pull out an electrical appliance using the cord, to keep fingers and clothing away from moving parts)? _____
2. Have you told students not to experiment with the electric current of home circuits? _____
3. Are you allowed to use extension cords in your building? Are they in good condition and plugged into the nearest outlet (so they won't cause a short circuit)? _____
4. Are students' hands dry when they touch electrical cords, switches, or appliances? _____

Questions to ask regarding fire   Answers

1. Do you know your school's fire regulations and evacuation procedures and the location of and use of fire-fighting equipment? _____
2. Does your room have an ABC fire extinguisher and a fire blanket? Do you know how to use them? _____
3. What extra cautions do you take when dealing with fire? What special instructions do you give students? Could you use a hot plate instead of fire? _____

Questions to ask regarding plants and animals   Answers

1. Are the animals healthy, well cared for, and in a suitable habitat? _____
2. Are your students aware that they are not allowed to bring live or deceased wild animals into the classroom? _____
3. Do you buy animals only from reputable stores or supply houses? _____
4. Do you instruct students not to pick up unfamiliar animals? Do they know not to poke at an animal in a cage? _____
5. Do you and your students know how to properly care for an animal and how to pick it up? _____
6. Are your students aware that they should never put a plant into their mouths? _____
7. Do students wash their hands after touching plants and animals? _____
8. Do you forbid students from inhaling smoke from a plant or picking up an unknown plant, flower, or seed? _____

Questions to ask regarding field trips   Answers

1. Are you thoroughly familiar with a field trip location before you take students there? _____
2. Do you have extra adult supervision for a trip? _____
3. Are your field trips approved by the proper school administration? _____
4. Do you secure written parent permission for taking students on a trip? _____
5. Are the students aware of proper clothing or supplies needed for a trip? _____
6. Do you have a first aid kit to take with you? Is a cellular telephone available to take with you? _____
7. Do you have a "buddy system" set up for students? _____
8. Do you have students' home phone numbers, medical records, and medications before leaving school? _____

Learning Activity 11.6

CONDUCTING A SAFETY AUDIT

Materials Needed:

- A copy of Table 11.9
- Permission to conduct a safety audit of a classroom

A. Obtain permission from proper school personnel (such as the principal and teacher) to conduct a safety audit of a classroom.
B. Study the features of the classroom. Observe a science lesson.
C. Audit the school and classroom for the safety concerns listed in Table 11.9. What recommendations would you make to the teacher or school to improve safety?

Management Strategies

Management strategies are the third major management concern and challenge that we will discuss. In this section, we will focus on three stages of **classroom management**: before instruction, during instruction, and after instruction. All management strategies are used to help develop welcoming classroom learning communities where students can learn effectively.

Before Instruction

Teachers need to take several steps before starting any classroom lesson to ensure a smoothly running classroom. First, teachers must establish classroom norms about what is acceptable behavior. Second, they must establish an atmosphere that fosters "good" behavior. Third, they must set up agreements, or contracts, with students to define objectives, choices, and consequences of behavior. Fourth, they must anticipate problems before they happen.

Establishing Norms for Acceptable Student Behavior

To establish acceptable student behavior, many teachers begin a school year by setting general class rules and ground rules for behavior during different types of lessons (such as investigations, reading time, or computer sessions). As discussed earlier in this chapter, an important question to ponder when setting rules is whether the teacher should determine acceptable classroom norms or whether the teacher should involve students in the development of those norms.

Kamii (1984, 1991) and Kamii, Clark, and Dominick (1994) argued that children need to invent ethical meanings. Katz (1984) suggested that getting children to comply with rules does not develop the primary goal of generating good children. What does is giving children choices and helping children learn for themselves in a caring classroom community. Haberman (1995) stressed that children in poverty quickly learn to rebel against mandates for compliance. In other words, teachers need to move beyond simply enforcing rules. Kohn (1996) wrote that it is far better to ask children to create rules—but even this strategy is not constructivist as students might come up with only the rules they think the teacher wants to hear or that they recall from previous years. He argues that it is better to get students to solve problems and engage them in discussion about what they want their class to be like. Kohn wrote, "The construction of meaning is an active process. It can't be done unless the learner has substantial power to make decisions.... If we are talking about learning to be a responsible, caring person, then the decisions include how to solve problems and get along with others" (p. 78).

Although the answer to the question of who should determine the rules may, in part, be determined by the age level of the students, rules tend to be easier to enforce if students in the class helped create them. When students have some ownership over the classroom environment, they are invested in making it work. In addition, when students help generate rules, they understand them better because they put them in the language that they use and that has meaning for them. For example, a word like *salty,* while meaningless to the teacher with respect to being a negative slang word, may be considered offensive by students and need to be put on the unacceptable language list.

To set general **classroom rules**, many teachers engage students in discussion about what their classroom should be like and what classroom rules might be established to make the classroom this way. Even very young students have experiences that help them accomplish this task. Frequently, students will come up with rules such as "treat others with respect," "be fair to others," "no hitting," and "no cheating." Many teachers post these rules on a chart in a prominent place in the classroom.

Having students create classroom rules may sound easy, but it is not always. Children may not be used to making decisions about their behavior, and they may act out during the transition period. They might "test" you to see if you are really going to let them have choices. You might experience outright resistance. Some children might be silent and not attempt to solve classroom problems. Children may just parrot "what they think you want to hear." To solve some of these transition problems, you can try several strategies:

- Try going slower; children may be overwhelmed with choices.
- Let students make some "bad" decisions and rules (as long as they are not emotionally or physically dangerous), and let them analyze and reflect on whether they are working. For example, if students say there is no reason to write their names on their papers, you might go along with this one time. If students don't quickly decide to change this rule, then approach them to help you solve the problem of not knowing who to give what grades to for written work.
- If students refuse to participate, saying it is your job to make the rules, reassure them that you will still make some rules, but that you can make some rules together.
- Find out why students are silent when it comes to making rules. Are they shy? Do they feel safe? The reason will determine your approach.

- If students are only parroting answers, push them to defend the answers they give. Ask, "Why is it important?" "What does it mean?" "Does everyone agree?" (Kohn, 1996, pp. 96–97).

Although these general rules are important for any classroom, there are some **ground rules** specific to a project-based science classroom. These ground rules establish acceptable student behavior during various aspects of project activities. For example, teachers might want students to be quiet when they are reading, but lively when critiquing each other's ideas. Table 11.10 illustrates some aspects of project-based science instruction for which specific rules might be needed. You can use the table with students to establish ground rules. It can also be used to communicate with parents about your expectations. Many teachers find the table most helpful before the start of a specific activity.

The teacher in Scenario 2 established rules before taking students outdoors to play a game. Another important time for establishing ground rules is when students are brainstorming ideas for a driving question. Before brainstorming, the teacher draws three columns on the board: *Activity, Questions,* and *Ground Rules.* Then the teacher might engage students in a discussion about the ground rules that need to be in place during the brainstorming activity. The resulting ground rules can be posted in the classroom and referred to at a later time.

Another useful technique for establishing acceptable student behavior is filling in a behavior tree. A **behavior tree** is a diagram in which the teacher and students establish the goals that need to be met during a specific activity and set ground rules that need to be followed to meet the goals. The behavior tree focuses students' attention on behavior needed for a specific task or activity. The behavior tree in Figure 11.2 illustrates acceptable behavior during an acid-rain activity. The teacher might say to the class, "Boys and girls, this afternoon we are going to be testing different samples of acid rain. You will be working in groups. What are some goals we have been working on as a class that might apply to group work during an investigation?" Students might respond, "Recording information accurately, completing the task, and giving everyone a chance to work." The teacher would record these three goals on the behavior tree, and students would generate the rules they believe applied to the desired goals.

Establishing an Atmosphere That Fosters "Good" Behavior

After classroom rules and ground rules for behavior during different situations have been set, an atmosphere that encourages the desired behaviors must be established. Helping students learn the required behaviors and norms often requires support. Also, many successful teachers encourage proper behavior by modeling it themselves. The teacher in Scenario 2 at the beginning of this chapter modeled for students how to come into the building quietly after being outdoors. Students are quick to notice when teachers say one thing, but do another. For example, when a teacher sits in a corner grading papers while students are working on their investigations, as in Scenario 4 at the beginning of this chapter, the message to students is that the teacher is not interested in what they are doing. In addition, when teachers are grading papers,

TABLE 11.10   Activities and Related Ground Rules

Project-based activity	Questions to ponder	Resulting ground rules
Brainstorming ideas for driving questions	How can we brainstorm ideas without getting off track? How can we brainstorm and not hinder good ideas? How can we value all students' ideas?	Keep to the topic. Don't put down people's ideas.
Collaboration with others	How can we make sure all people participate? How can we keep one person from taking over? How can we make sure there aren't "loafers" in the group? What voice level is acceptable?	Make sure everyone has a job to do in the group.
Critiquing each other's ideas	How can we make sure we don't insult someone? How can we nicely help someone make his or her project better?	Say good things about the project first.
Investigations	How can we safely use science equipment? How can we make sure we are exploring an idea fully? How can we make sure all students help with the investigation? Can we move around the classroom? Can we get our own equipment or must the teacher get it for us?	Don't run with science equipment.
Working on the computer	How do we find only information that is related to our topic? How do we use the computer safely? How does everyone get a chance to use the computer?	Let everyone in our group conduct one search on a topic.
Demonstrations or guest speaker visits	How do we treat a guest speaker? What is important to do during a demonstration? How do we sit and listen respectfully?	Don't talk when someone else is talking to the class.
Field trips	How do we behave in public?	Make our community proud of kids at our school.
Making artifacts	How can we make sure that all students' ideas are used in the artifact? What amount of noise is acceptable?	Keep a list of each person's contribution or idea. We should use talking voices.
Sharing artifacts	How do we give people new ideas without insulting them?	Say, "1 like your idea, but did your group consider...?

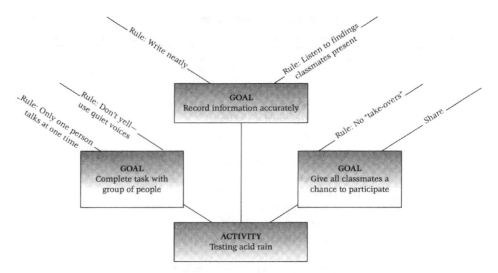

Figure 11.2   Behavior tree model.

they cannot help students think through questions, design investigations, and find resources; facilitate collaboration among students; and help students develop artifacts. Some teachers give lectures about wearing goggles during investigations, but fail to wear them themselves, or insist that students be quiet when working, but talk loudly with a colleague who walks by the room.

In addition to supporting learning about behavior and modeling classroom behaviors, teachers must reinforce good behavior. This **reinforcement** does not consist of tangible rewards, like stickers, points for free time, or certificates to local fast-food restaurants. Kohn (1996) argued that tangible rewards only accomplish temporary compliance, and they do nothing to help a child become kind or caring. He says that **rewards** also warp the relationship between student and teacher, because they don't create a caring alliance based on warmth and respect. Successful reinforcement consists instead of praise and, especially, recognition. Haberman (1995) noted that exemplary teachers working with children in poverty consistently reward and recognize *effort* more than right answers or achievement. Similarly, Campbell (1999) noted that teachers should promote collaboration over competition and improvement over ability. Students appreciate and respond well to the recognition of teachers, peers, and community members. A personalized note (that goes beyond a one-word comment like "good") by the teacher in a student's portfolio or a positive letter home to a student's parent(s) is very rewarding for most students. Having the results of a project published in the school or town newspaper, receiving a personal letter from a member of the community complimenting students' work, or hearing comments from peers about how a student's ideas helped the group complete a project are all effective forms of recognition that reinforce positive behavior. So are simple comments from the teacher that indicate that she is noticing a student's efforts. It is very easy to say to a student, "I noticed today that you seemed really interested in the discussion."

Comments like this go a long way. Remember that *all* students need to feel seen and appreciated.

Finally, a teacher can set an atmosphere for good behavior by focusing on what is positive, rather than what is negative. An enthusiastic teacher who shows a positive interest in students is much more successful than a teacher who is constantly correcting students or who does not seem to care about them as people. For example, rather than reprimanding the student who is talking out of place, recognize the student who is working well (ignoring the other one). However, make sure the recognition is genuine and not contrived to make other students in the class jealous. A teacher can sincerely recognize a student for her unique idea or question, creative way of solving a problem, multifaceted conclusions to an investigation, or insightful way to display the results of a project. Haberman (1995) stressed the importance of making all children feel important. Therefore, if a child misbehaves, it is important to focus less on the misbehavior than on the fact that the child is a vital and necessary member of the class or contributor to a project.

Setting Up Contracts With Students

Some teachers find that learning contracts effectively promote self-responsibility. The teacher in Scenario 3 used a contract with his students. Learning contracts, a written agreement made between a teacher and members of a group, accomplish several things. First, they force students to think about the expectations people have of them. They require them to delineate what type of project they will complete and what quality they expect to produce. Second, they outline specific ways students will meet objectives. Students can, for example, set parameters for work to be completed, artifacts to be created, or types of presentations to be given to the class. Third, contracts can specify how projects will be evaluated. Will the teacher evaluate it? Will peers judge its quality? What will be evaluated? Fourth, learning contracts can also establish procedures for group collaboration, such as making decisions equitably, getting all members of the group involved, and coming to decisions in a fair manner.

Essential features of contracts include, but are not limited to, the following:

- A description of the project (students list the driving question, the subquestions they are investigating, and the names of the students on their team);
- An outline of how students expect to accomplish tasks (students list what they will investigate, what materials they plan to obtain, who they intend to contact in the community, and what artifacts they expect to create); and
- An evaluation plan (students describe the quality of work they expect of themselves, how they want to be evaluated, such as with a rubric or with an interview by the teacher, and what aspects they want to be the assessment criteria, such as concepts learned and ability to collaborate together).

Figure 11.3 is an example contract that a group of fifth graders might complete for an acid-rain project.

Names of team members: _____

Date: _____

Description of project:

We are trying to answer questions about acid rain. We are trying to find out if there is acid rain in our environment. Our group's driving question is "Can acid rain hurt plants in our neighborhood?"

How we're going to accomplish this project:

We are going to set up four geranium plants in pots. We are going to put all of them in the window in the room and give them the same amount of water. But to the water for one plant, we will add 1/2 teaspoon of vinegar. For the second plant, we will add 1 teaspoon of vinegar to the water, and for the third plant we will add 2 teaspoons of vinegar. The fourth plant will get plain water. We will measure the pH of the water with pieces of universal indicator paper that the teacher showed us how to use when we learned about acids and bases. We might call a nursery or a farmer to interview him or her about the effects of acid rain on plants.

Evaluation plan:

We will take photographs of our plants as they grow for 2 weeks. We will look up *acid rain* in the encyclopedia or on the Web to see what we can find out. We will measure the plants and see if they look healthy. The height and color of the plants will be recorded each day. The pictures, a graph of height versus time, and descriptions of how they look will be our artifacts. The teacher can grade this, but we will tell how well we think we did in our portfolio. We expect to get along with everyone in our group so we can answer this question.

Signatures: _____

Figure 11.3   Learning contract.

Anticipating Problems

Effective classroom teachers think about problems that might occur in their classroom, and they plan accordingly. In a project-based science classroom, teachers must consider the fact that different groups of students might be working on different tasks at the same time, compounding potential problems. In general, potential problems fall into several general categories:

- Giving directions
- Materials and supplies
- Student behavior
- Timing
- Complexity of task
- Sustaining interest over time

Cultural differences can also be associated with problems. For example, Brown (2005) reported that Caucasian teachers oftentimes give directions that are vague, such as "It is about time we get started." He reports that African American students, particularly in urban environments, are more accustomed to direct directions. An example in science might be, "Sit down at your desk and take out your science notebook." Similarly, Brown reported that Caucasian teachers expect students to engage in discussions by sitting quietly and raising their hands, while African American and Chicano students are more accustomed to loud, emotional discussions where it is acceptable to yell out responses. These types of cultural differences can cause unanticipated problems for teachers who are not aware of the cultural differences.

Some teachers find it useful to think through potential problems using a fish-bone model. A **fish-bone model** is a diagram that lists tasks leading to a goal. On each "bone," teachers list possible problems. Figure 11.4 is a fish-bone model representing potential problems during four different tasks related to learning about acids and bases. Learning Activity 11.7 will work you through the creation of a fish-bone model.

During Instruction

We will discuss five factors to consider during instruction. These are the distributing of materials and supplies, making transitions, dealing with disturbances, handling multiple instances of the same activities, and handling multiple instances of different activities at the same time. On first sight, project-based science classrooms may seem chaotic. However, effective project-based classrooms are actually organized and efficient—and the reason is that the teachers have very clear management strategies for instruction.

Distributing Materials and Supplies

Whether it is the teacher or students who distribute materials and supplies in a science class, the distribution portion of a lesson is often the time when students misbehave or accidents occur. In part, this is because distribution represents downtime—a time when students are not actively participating in instruction or investigation. It is a time when students are simply waiting for something to happen, and so it often does. For this reason, it is important to distribute materials and supplies in an efficient manner to limit the amount of downtime between tasks. There are several efficient ways to distribute materials and supplies.

One way is to keep all materials and supplies in handy clear plastic boxes, plastic tubs, or zipper-closure bags that can be located quickly and easily passed around a table area to students.

Some teachers like to place all of the needed materials for an investigation or lesson in a single box or bag. For example, to investigate circuits, students need a dry cell, wires, bulb holders, a flashlight bulb, and a dry cell holder. Rather than pass around five different boxes with these different items, the teacher could put one of

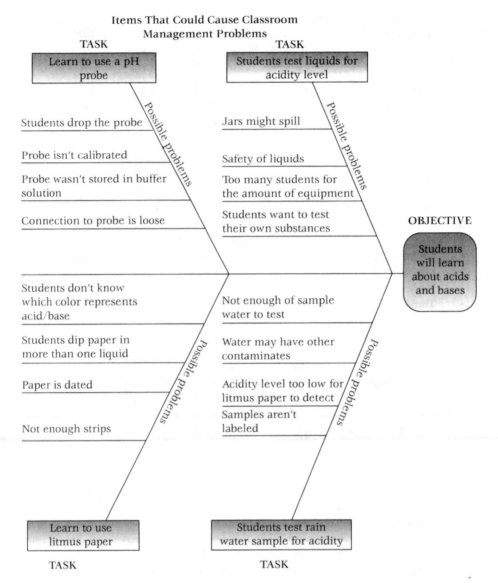

Figure 11.4 Fish-bone model.

each of the items into a zipper-closure bag before class and hand each table area one bag with all of the materials the students at that table would need for the lesson. This method is especially useful for young children with short attention spans who don't tolerate downtime well.

It takes teachers a considerable amount of time to set up and organize materials in advance. Many teachers consider this part of their professional lives and spend time gathering materials and setting up before school, during lunch, and after school. Other teachers solicit help from parents, grandparents, and retirees in the community; older students in the school; or students in the class who volunteer to be lab assistants.

Learning Activity 11.7

ANTICIPATING PROBLEMS

Materials Needed:

- Figure 11.4
- Something to write with

A. Using the fish-bone diagram in Figure 11.4, determine what the teacher would need to do to avoid the potential problems.
B. Plan to teach a lesson. Complete a fish-bone diagram of the potential problems in the lesson.
C. How could the potential problems be resolved?

Many teachers find that once students have materials and supplies in their possession, they are eager to begin the activity or investigation. At this point, it is usually very difficult to give directions to students. For this reason, it is most often a good idea to give clear and complete directions *before* materials are distributed for a lesson or other whole-class activity.

In a project-based science classroom, students will frequently obtain their own supplies and materials because they are investigating different subquestions related to the driving question, and they are making different artifacts. For example, students using pillbugs to study ecosystems might investigate separate subquestions about the habitats of the crustaceans, the desired temperature preferred by the crustaceans, and the amount of light preferred by the crustaceans. Although these are related activities, they require different materials: bug cages, different soils, thermometers, sources of light, and light probes. For this reason, teachers in a project-based science classroom face some unique problems regarding distribution of materials and supplies. Students need to know where to obtain necessary supplies, and they need to have some parameters for doing this. Some teachers require students to submit to them lists of needed materials, materials that the teacher then gathers and distributes. Others teach students where materials and supplies are stored and allow them to get items for themselves. With this approach, it is important to make sure the supplies and materials are stored in an area that is safe for students. It is also important to teach students how to use materials. For example, students need to learn how to carry expensive microscopes and focus the lenses without scratching them.

Making Transitions

Transitions are the few minutes that elapse between investigations, discussions, readings, and other activities. Transitions, like distribution of materials, are points of departure for students. Teachers handle transitions in different ways, but always try to limit the amount of time between activities. Some teachers like to establish a standing rule for transitions, such as requiring students to read silently or make entries in their journals during transitions. Some teachers like to make transitions fun and interesting. For instance, they might pose an interesting science question for students to ponder during this time. Some like to engage students in quick games or songs during transitions. Finally, teachers can engage students in discussions about what they think can be done during transitions; students may have good ideas for handling downtime.

Dealing With Disturbances

Many potential disturbances are eliminated in a project-based environment because students are genuinely interested in what is happening. Haberman (1995) asserted that projects are the most effective way to capture the attention of children living in poverty and, thus, eliminate many behavior problems. In addition, if there is a positive classroom climate, if there are established ground rules for behavior, if good behavior is modeled, and if downtime is minimized, the potential for disturbance is already lessened. Nevertheless, disturbances will still occur. For example, one group of students might be working at the computer while another group is conducting an investigation. These multiple instances of classroom activity create opportunities for one group of students to disturb another group, often unintentionally.

It is impossible within the scope of this book to discuss all potential disturbances and all possible teacher responses. However, there are several general strategies that are recommended in most cases that we can discuss.

First, it is usually best to *have students try to solve their own problems*. For example, when a disturbance occurs, you might ask the student involved to think about how he or she could solve the problem, or you might ask all the students in the class to come up with solutions. This technique fits with constructivist theories discussed earlier.

Second, *establish consequences* in advance for disturbances that are severe. Established consequences are *not* threats or punishments. Threats and punishments are usually only temporarily effective. They don't teach students to solve their problems, and they frequently cause more problems. Kohn (1996) argued that bribes and threats do three things: disrupt a lesson, warp the relationship between punisher and punished, and impede moral or ethical development. Threats and punishments teach children not to think about how the behavior affects others, but, rather, about what might happen if they get caught. **Consequences**, on the other hand, are the predetermined outcomes of student choice. For example, if a student chooses to disrupt work in her group, she knows that the consequence is that she will have to take a time out. Consequences can be established when class rules are set, or, if absolutely necessary, students can be informed of consequences by the teacher.

Third, *be consistent* in reacting to disturbances. Do not let students get away with a certain behavior sometimes but enforce consequences on other instances. For example, if calling a classmate a bad name is considered inappropriate during science class, it should also be considered inappropriate on the playground.

Fourth, *be fair*. Some teachers treat girls less harshly than boys. For example, they might let girls talk in class but scold boys for doing the same thing. Similarly, teachers sometimes let more academically oriented students get away with behaviors that they would not tolerate from other students in the class. For example, they might let the brightest student in the class turn in a late paper, but insist on deadlines for the other students. Students will be quick to spot unfair treatment, and perceiving that they are being treated unfairly may induce them to further misbehave.

Fifth, *be a good role model.* It is important that teachers not display the very behaviors they are trying to eliminate. Hardin and Harris (2000) pointed out that teachers must be good role models by not having tantrums, making insults, raising their voices, berating children, or being rude toward children. These authors stressed that crisis prevention begins with treating all students with respect.

For more information about classroom management and different viewpoints on discipline (sometimes conflicting viewpoints), see Figure 11.5.

Albert, L. (1992). *An introduction to cooperative discipline* [Videotape]. Circle Pines, MN: American Guidance Service.

Albert, L. (1989). *A teacher's guide to cooperative discipline: How to manage your classroom and promote self-esteem.* Circle Pines, MN: American Guidance Service.

Canter, L., & Canter. M. (1992). *Lee Canter's assertive discipline: Positive behavior management for today's classroom.* Santa Monica, CA: Lee Canter & Associates.

Child Development Project. (1996). *Ways we want our class to be: Class meetings that build commitment to kindness and learning.* Oakland, CA: Developmental Studies Center.

DeVries, R., & Zan, B. (1994). *Moral classrooms, moral children: Creating a constructivist atmosphere in early education.* New York: Teachers College Press.

Dreikurs, R., Grunwald, B., & Pepper, F. C. (1982). *Maintaining sanity in the classroom: Classroom management techniques* (2nd ed.). New York: HarperCollins.

Emmer, E. T., & Evertson, C. M. (1981). Synthesis of research on classroom management. *Educational Leadership, 38,* 342–347.

Evertson, C. M., Emmer, E. T., Clements, B. S., Sanford, J. P., & Worsham, M. E. (1984). *Classroom management for elementary teachers.* Englewood Cliffs, NJ: Prentice Hall.

Evertson, C. M., Emmer, E. T., Clements, B. S., & Worsham, M. E. (2000). *Classroom management for elementary teachers.* Boston: Allyn and Bacon.

Freiberg, J. H. (Ed.). (1999). *Beyond behaviorism: Changing the classroom management paradigm.* Boston: Allyn and Bacon.

Kamii, C. (1984). Obedience is not enough. *Young children, 39,* 11–14.

Kamii, C. (1991). Toward autonomy: The importance of critical thinking and choice making. *School Psychology Review, 20,* 382–388.

Kohn, A. (1996). *Beyond discipline: From compliance to community.* Alexandria, VA: Association for Supervision and Curriculum Development.

National Association for the Education of Young Children. (1986). *Helping children learn self-control: A guide to discipline.* Pamphlet. Washington, DC: Author.

Purkey, W. W., & Strahan, D. B. (1986). *Positive discipline: A pocketful of ideas.* Columbus, OH: National Middle School Association.

Wolfgang, C. H., Bennett, B. J., & Irvin, J. L. (1999). *Strategies for teaching self-discipline in the middle grades.* Boston: Allyn and Bacon.

Figure 11.5 Classroom management resources.

Handling Multiple Instances of the Same Activities

Even in project-based classrooms, teachers frequently facilitate the same activity for all students. This occurs frequently in a lesson on a concept or skill that all students will need in order to complete a project. The teacher described in Scenario 3 was handling multiple instances of the same activity to teach about parts per million. Several problems can occur when all students are participating in the same activity. Some students will finish earlier than others. Some students will "drop out."

The problem of early and late finishers can be caused by a diversity of abilities among students. More academically advanced or skilled students frequently finish activities much sooner than do their less academically able peers. Students with learning problems can have related behavioral problems. Gifted and talented students might become disruptive because they are unchallenged and bored. When there are differences among early and later finishers, it is a good idea to set classroom rules for finishing early, such as "Always take out a book to read silently when finished before others." Some teachers have students who finish early provide peer tutoring for classmates who are still struggling with an activity. Some teachers set up learning centers (locations in the room with enrichment activities) where students can go to work if they finish early. If learning centers are used, teachers need to be careful that the activities in these locations aren't more motivating than the primary classroom activity, or students may rush through their primary work to get to the enrichment activities.

"Dropping out" can become a problem when all students in the class are completing the same activity. **Dropouts** are students who lack interest in an activity or who fail to be motivated to complete it. Teachers need to monitor carefully students' interests and plan activities that are motivating. One way to monitor students' interests is to take time to talk with each student periodically. Ask questions to find out what the student is interested in, if he or she is unchallenged, frustrated, or bored. Another way is to have students keep dialogue journals (in which they write to each other or in which you and students correspond). These journals provide you with powerful insight into concerns students have about classroom activities.

Handling Multiple Instances of Different Activities at the Same Time

The apparent chaos found in a project-based science class is compounded when there are different activities going on at the same time. Scenario 4 illustrates a classroom with different activities going on at the same time. In such situations, teachers must be able to orchestrate students' working on different activities. They must be able to make different materials, supplies, and equipment available for different activities and keep different groups of students focused on the main project.

Having materials, supplies, and equipment available for different activities at the same time is no easy task. In fact, because of the evolving nature of projects, students don't even always know what they will need before the start of an investigation. To make this job easier, teach students how to get many of their own science supplies, materials, and equipment. In addition, rely on parent and community volunteers to help with materials distribution. Structure your lessons so that students submit supply lists to you before a break, lunch period, or classes taught by other teachers so that you have time to gather materials.

To keep different groups of students focused on the project, a teacher must show students the relationships among their separate activities and to the driving question. Although one group of students may be on the computer, looking up information about current weather conditions throughout the state, and another group may be measuring current weather conditions outside the classroom, these two activities may be tied together by the driving question "How does weather affect our environment?" The activities are also related to each other since both are looking at current weather conditions. Students don't always see these connections, but knowing that everyone's work is connected can keep students focused and eliminate some of the management problems associated with working on different activities at the same time.

One way to emphasize this connection is to have students share ideas and discoveries with each other throughout a project. Groups can share their ongoing investigations with the whole class, or small groups can be paired to discuss their projects with each other. Another strategy is to have students draw concept maps and Venn diagrams or complete charts like the one shown in Table 11.11.

After Instruction

Just because ground rules and procedures have been established in a classroom does not mean they will be followed. The teacher needs to continually reinforce classroom norms; revisit classroom norms; and evaluate progress *before, throughout*, and *after* the project. In addition, after instruction students and teachers should *reflect* on classroom norms, on the effectiveness of instruction, and on student behavior. *The National Science Education Standards* (NRC, 1996) promotes such reflective practices in science education: "Teachers of science engage in ongoing assessment of their teaching and of student learning. In doing this, teachers use student data, observations of teaching, and interactions with colleagues to reflect on and improve teaching practice" (p. 37). It adds, "They use self-reflection and discussion with peers to

TABLE 11.11   Projects Are Linked

Group 1 activity	Group 2 activity	Relationship to driving question "How does weather affect our environment?"
We're using the Web to find current weather conditions in Ohio.	We're taking the temperature outside of the school.	We're both trying to find out what the weather is right now.
We're doing an investigation to see if more molds grow in our environment when weather conditions are wet. We're growing mold in wet and dry conditions.	We're growing different grass seeds under different conditions (wet, dry, hot, and cold) to see if the weather will affect plants in our area.	We're both investigating how weather conditions affect the growth of things.

understand more fully what is happening in the classroom and to explore strategies for improvement" (NRC, 1996, p. 42).

Reinforcing Classroom Norms

Teachers need continually to ask students to *reflect* on their objectives and encourage behaviors that help students meet these objectives. In a project-based environment, this is frequently accomplished by reminding students of the driving question they are trying to answer. This type of reminder focuses their attention on the purpose of their activities, and it can guide them in thinking about appropriate behaviors needed to accomplish their goals.

Teachers also need to *reinforce* established classroom rules. The teacher in Scenario 1 reminded students about classroom rules regarding talking when a reading assignment was given. Some teachers like to have students complete self-evaluation checklists or T-charts (like those discussed in Chapter 7) at the end of a lesson to assess whether or not they are following established classroom norms. Some teachers like to have debriefing sessions or class meetings in which students discuss their progress toward following classroom norms (Croom & Davis, 2006). Finally, some teachers like to send home "progress reports" or letters to report on students' behavior and progress in school.

Revisiting Classroom Norms

Classroom rules or norms are not established only once during a school year. Good teachers continually revisit and reexamine the established rules and norms. They have students analyze whether the rules make sense and if new rules should be made to replace those no longer appropriate. Not only does revisiting classroom norms limit problems that might occur throughout the school year, it teaches students many of the problem-solving skills that project-based science seeks to foster. Students analyze whether rules are effective and solve the resulting problems if they are not working.

Evaluating Progress

One way to evaluate progress is to hold students accountable for their work. In a project-based environment, however, students must be accountable not only to the teacher, but also to one another. Each individual becomes accountable for his or her own learning, and the group is held accountable for learning and working well together.

Individual Accountability. Individual **accountability** is the responsibility that individuals in a group have for their own learning. Teachers make it clear to students that individual members of a group cannot coast along on the work of others and cannot disturb the learning of others. Each person must learn the assigned material with the support of the group. For example, while completing a project, all students

TABLE 11.12   How Did I Do?

Name: _____

Rate yourself on how well you think you collaborated with your teammates on this activity.

Criteria	5 Very well	4 Somewhat well	3 Somewhat	2 Somewhat poorly	1 Poorly
I accepted ideas from my classmates.	_____	_____	_____	_____	_____
I listened carefully to the people on my team.	_____	_____	_____	_____	_____
I provided constructive criticism to help my team.	_____	_____	_____	_____	_____
I followed the rules our class generated.	_____	_____	_____	_____	_____

in a group contribute to the creation of an artifact and to the group's presentation. This individual accountability must be assessed. Students might keep journals or complete scoring rubrics to reflect their personal contributions to the project investigation, ideas, or artifacts. A rubric can also measure each person's collaborative skills, such as accepting ideas from others, listening carefully to others, interacting constructively with others, and following classroom rules. Individual students might be required to pass a test on certain material, or each student might keep a portfolio to demonstrate his or her own personal progress toward mastering learning concepts. A sample rubric that students could use to evaluate their collaborative skills is pictured in Table 11.12.

Group Accountability. We often think of accountability as the ability to demonstrate one's skill and knowledge to the teacher. In project-based classrooms, groups also have accountability. Group accountability most often falls into one of two types—collaboration and acquisition of certain knowledge or mastery of skills. In order for a group to demonstrate its ability to work well, many teachers have students complete questionnaires designed to ascertain whether interpersonal skills are being acquired and maintained, whether the classroom environment fosters sharing and respect, and whether students feel comfortable sharing and critiquing each other's work. Students can also complete a group scoring rubric that evaluates whether everyone participated, solved problems without arguing, talked in appropriate voices, treated one another with respect, stayed on task, and negotiated solutions. Another option is to assign to one student the task of evaluating with a rubric his or her group's collaborative skills. A sample group-scoring rubric is shown in Table 11.13. The teacher may also assess the group's progress and have a conference with each team about its collaborative skill development.

To evaluate group members' acquisition of knowledge or mastery of skills, a teacher may have students ask one another questions about what they learned. Some teachers issue a final grade on a project based on a student's individual achievement and his or her accomplishment as part of a group. The grade could be the student's individual score plus bonus points from a group score. Teachers must decide what

TABLE 11.13   How Did Our Team Do?

Name: _____

Names of members of my team: _____

Rate how well you think you worked together on this activity.

Criteria	5 Very well	4 Somewhat well	3 Somewhat	2 Somewhat poorly	1 Poorly
Everyone participated.	____	____	____	____	____
We solved problems without arguing.	____	____	____	____	____
We talked in appropriate voices.	____	____	____	____	____
We treated each other with respect.	____	____	____	____	____
We stayed on task.	____	____	____	____	____
We negotiated to solve problems.	____	____	____	____	____

weights individual achievement and group accomplishment have in determining grades. Some teachers give all students in a group the same grade, the average of all students' work. Other teachers collect only one paper (sometimes randomly) from a group and assign the grade from that paper to all members of the group. Whatever method is used, it is important to establish assessment guidelines before the start of a project. Before a collaborative project is assigned, students should understand what weight will be given to individual and group progress. They should also understand how grades will be determined. If students recognize that collaboration is important and that they will not be harmed in the assessment process, there will be group cohesion, and potential problems will be eliminated.

Using Technology Tools to Facilitate Classroom Management

Sometimes teachers benefit from the help of others to improve their classroom management strategies. Other teachers may have encountered similar problems and have ideas for handling specific situations. The process of collaborating with peers to discuss problems faced in the classroom can also be a great mentoring experience for new teachers. Even veteran teachers sometimes need a "shoulder to cry on," and peers can provide them needed support to help them cope with the stresses related to teaching.

Electronic communication with peers through such sites as the one maintained by the National Science Teachers Association (http://www.nsta.org) can be the tool needed to obtain and share classroom management ideas. Teachers can talk with colleagues who have used project-based science through the Web-based Inquiry Science Environment (WISE).

The Web provides many up-to-date sites that deal with safety issues that have application to classrooms. A good one is the American Red Cross (http://www.redcross.org/services/hss/). Online tools, such as the Classroom Architect (http://classroom.4teachers.org), also help with management concerns, such as setting up a classroom.

Chapter Summary

This chapter introduced three key components of the basic management of a project-based science classroom: classroom climate, organization, and management strategies. In the classroom climate section, we considered a constructivist framework for establishing a positive classroom environment. In this framework, teachers try to have students participate fully in helping solve classroom problems and forming norms or ground rules. The teacher serves as a role model who promotes a positive attitude toward science and a positive classroom climate. The teacher develops affective factors, such as curiosity about the world, excitement about science, and enthusiasm to continue activities, and generates worthwhile curriculum materials that are of interest to students. To develop a positive classroom climate in which students help solve problems, there must be a balanced relationship between the teacher and the students. Finally, in a positive classroom climate all students—boys and girls, minorities and nonminorities—feel capable of learning science.

In the section on classroom organization, we discussed aspects of physical facilities, the structure of the school day, and safety. We considered specific strategies for managing a project environment.

In the section on management strategies, we focused on before-teaching considerations, such as establishing rules, using contracts, or anticipating problems; during-instruction considerations such as distributing materials, making transitions, dealing with disturbances, and working with multiple instances of activities; and after-instruction considerations, such as reinforcing classroom norms, revisiting and reflecting upon norms, and evaluating progress. We also touched on ways that teachers might collaborate with others to discuss classroom management strategies.

Chapter Highlights

- Many unwelcome classroom behaviors can often be traced to the larger classroom context and the curriculum.
- A constructivist classroom environment asks students to reflect on what they should do and to solve problems together.
- In a positive classroom climate:
  - Students have real choices.
  - Teachers are role models for good behavior and science inquiry.
  - Instructional planning focuses on good curriculum.
  - Teachers exude a positive attitude toward science.
  - Teachers promote positive affective factors, such as curiosity, excitement, persistence, enthusiasm, skepticism, and open-mindedess.

- Maintaining a balanced teacher-student relationship helps create a classroom learning community where students feel safe.
- Teachers ensure equality by focusing on classroom practices that help all students learn.
- A well-planned classroom arrangement supports project-based science through:
  - Creating an effective physical arrangement.
  - Keeping high-traffic areas free of congestion.
  - Ensuring that students can be seen by the teacher.
  - Keeping frequently used materials and supplies readily accessible.
  - Being certain students can see displays, as well as see and hear presentations.
- The structure of the school day influences classroom management through:
  - The number of minutes allocated to a subject area,
  - The delivery of science as a subject area, and
  - The daily schedule of activities or lessons.
- A safe classroom helps ensure that all students learn.
- In a safe classroom:
  - Animals and plants are healthy and well cared for and do not harm students.
  - Goggles are used in science classrooms.
  - Shelving, supplies, and materials are safely stored.
  - Students learn the safe use of electricity.
  - Sources of fire are eliminated or controlled.
  - Glassware is replaced with plastic or its use is carefully monitored.
  - Field trips are well planned to ensure safety and learning by all students.
  - Teachers are familiar with first aid procedures.
- Management strategies are used to help develop welcoming classroom learning communities where students can learn effectively.
- Teachers need to focus on classroom management before instruction, during instruction, and after instruction.
- To create a positive classroom before instruction:
  - Establish norms for acceptable student behavior.
  - Establish an atmosphere that fosters "good" behavior.
  - Reward students by focusing on effort.
  - Use caution when using "rewards" because they are viewed by some to be coercive.
  - Set up contracts with students to promote self-responsibility.
- Effective classroom teachers anticipate problems before they occur and plan accordingly.
- Several factors to consider during instruction are:
  - The distribution of materials and supplies.
  - Making transitions.
  - Dealing with disturbances.
  - Handling multiple instances of the same activities.
  - Handling multiple instances of different activities at the same time.
- Teachers need to continually reinforce classroom norms; revisit classroom norms; and evaluate progress before, throughout, and after the project.
- After instruction, students and teachers should reflect on classroom norms, on the effectiveness of instruction, and on student behavior.
- Teachers can collaborate with their peers to develop good classroom management strategies.

Key Terms

Accountability	Flexible
Affective factors	Ground rules
Atmosphere	High-traffic areas
Attitudes	Learning contracts
Behavior tree	Norms
Classroom management	Physical arrangements
Classroom organization	Real choices
Classroom rules	Reinforcement
Climate	Rewards
Consequences	Role models
Downtime	Safety
Dropouts	Safety contract
Fish-bone model	Transitions

References

Ajzen, I., &. Fishbein, M. (1980). *Understanding attitudes and predicting social behavior.* Englewood Cliffs, NJ: Prentice Hall.

Baker, D. (1988, April). Teaching for gender differences. *Research matters to the science teacher.* Columbus, OH: National Association for Research in Science Teaching.

Bandura, A. (1986). *Social foundations of thought and action: A social cognitive theory.* Englewood Cliffs, NJ: Prentice Hall.

Barman, C. (1996). How do students really view science and scientists? *Science and Children, 34,* 30–33.

Bazler, J. A., & Simonis, D. A. (1992). Are women out of the picture? Sex discrimination in science texts. In M. Wilson (Ed.), *Options for girls: A door to the future.* Austin, TX: ProEd.

Boylan, C. R., Hill, D. M., Wallace, A. R., & Wheeler, A. E. (1992). Beyond stereotypes. *Science Education, 76*(5), 465–476.

Brooks, J. G., & Brooks, M. G. (1993). *The case for constructivist classrooms.* Alexandria, VA: Association for Supervision and Curriculum Development.

Brown, D. F. (2005). The significance of congruent communication in effective classroom management. *Clearing House, 79*(1), 12–15.

Campbell, J. (1999). *Student discipline and classroom management.* Springfield, IL: Charles C. Thomas.

Chu-Clewell, B., & Campbell, P. B. (2002). Taking stock: Where we have been, where we are, where we're going. *Journal of Women and Minorities in Science and Engineering, 8,* 255–284.

Croom, L., & Davis, B. H. (2006, Spring). It's not polite to interrupt, and other rules of classroom etiquette. *Kappa Delta Pi Record,* 109–113.

Dean, R., Dean, M. M., Gerlovich, J. A., & Spiglanin, V. (1993). *Safety in the elementary science classroom.* Washington DC: National Science Teachers Association.

Evertson, C. M., Emmer, E. T., Clements, B. S., Sanford, J. P., & Worsham, M. E. (1984). *Classroom management for elementary teachers.* Englewood Cliffs, NJ: Prentice Hall.

Evertson, C. M., Emmer, E. T., Clements, B. S., & Worsham, M. E. (2000). *Classroom management for elementary teachers.* Boston: Allyn and Bacon.

Garrick-Duhaney, L. M. (2000). Culturally sensitive strategies for violence prevention. *Multicultural Education, 7*(4), 10–17.

Guzzetti, B. J., & Williams, W. O. (1996). Gender, text, and discussion: Examining intellectual safety in the science classroom. *Journal of Research in Science Teaching, 33*, 5–20.

Haberman, M. (1995). *Star teachers of children in poverty.* West Lafayette, IN: Kappa Delta Pi.

Hardin, C. J., &. Harris, E. A. (2000). *Managing classroom crises.* Bloomington, IN: Phi Delta Kappa Educational Foundation.

Hardin, J., & Dede, C. J. (1992). Discrimination against women in science education: Even Frankenstein's monster was male. In M. Wilson (Ed.), *Options for girls: A door to the future.* Austin, TX: ProEd.

Horsch, P., Chen, Q., & Nelson, D. (1999). Rules and rituals: Tools for creating a respectful, caring learning community. *Phi Delta Kappan, 81*(3), 223–227.

Huber, R. A., & Burton, G. M. (1995). What do students think scientists look like? *School Science and Mathematics, 95*(7), 371–76.

Kahle, J., & Lakes, M. (1983). The myth of equality in science classrooms. *Journal of Research in Science Teaching, 20*(2), 131–140.

Kamii, C. (1984). Obedience is not enough. *Young Children, 39*, 11–14.

Kamii, C. (1991). Toward autonomy: The importance of critical thinking and choice making. *School Psychology Review, 20*, 382–388.

Kamii, C., Clark, F. B., & Dominick, A. (1994). The six national goals: A road to disappointment. *Phi Delta Kappan, 75*, 672–677.

Katz, L. G. (1984). The professional early childhood teacher. *Young Children, 39*, 3–9.

King, J. E., Hollins, E. R., &. Hayman, W. C. (1997). *Preparing teachers for cultural diversity.* New York: Teachers College Press.

Kohn, A. (1996). *Beyond discipline: From compliance to community.* Alexandria, VA: Association for Supervision and Curriculum Development.

Lynch, S. J. (2000). *Equity and science education.* Mahwah, NJ: Lawrence Erlbaum.

McEwan-Landau, B., & Gathercoal, P. (2000). Creating peaceful classrooms: Judicious discipline and class meetings. *Phi Delta Kappan, 81*(6), 450–454.

Mead, M., & Metraux, R. (1957). Image of the scientist among high school students. *Science, 126*, 387.

National Council for Research on Women. (2001). *Balancing the equation: What we know and what we need.* New York: Author. www.ncrw.org/research/scifacts/htm

National Research Council. (1996). *National science education standards.* Washington, DC: National Academy of Sciences.

National Science Foundation. (2003). *New formulas for America's workforce: Girls in science and engineering.* Arlington, VA: Author.

Nelson, B. H., Weiss, I. R., & Capper, J. (1990). *Science and mathematics education briefing book, Volume II.* Chapel Hill, NC: Horizon Research.

Nelson, B. H., Weiss, I. R., & Conaway, L. E. (1992). *Science and mathematics education briefing book, Volume III.* Chapel Hill, NC: Horizon Research.

Pajares, F. M. (1992). Teacher's beliefs and educational research: Cleaning up a messy construct. *Review of Educational Research, 62*(3), 307–32.

Pollina, A. (1995). Gender balance: Lessons from girls in science and mathematics. *Educational Leadership, 53*(1), 30–33.

Rampal, A. (1992). Images of science and scientists: A study of school teachers' views. Characteristics of scientists. *Science Education, 76*(4), 415–436.

Rosser, S. V. (1990). *Female friendly science.* New York: Pergamon Press.

Roy, K. (2006, March). Safety contracts: Let it be written, let it be done. *Science Scope,* 10–11.

Rutherford, F. J., &. Ahlgren, A. (1990). *Science for all Americans.* New York: Oxford University Press.

Sadker, M., & Sadker, D. (1992). Sexism in the classroom: From grade school to graduate school. In M. Wilson (Ed.), *Options for girls: A door to the future.* Austin, TX: Pro.Ed.

Sanders, J. (1994). *Lifting the barriers: 600 strategies that really work to increase girls' participation in science, mathematics, and computers.* Port Washington, NY: Jo Sanders Publications.

Schmidt, W. H., McKnight, C. C., Houang, R. T., Wang, H. C., Wiley, D. E., Cogan, L. S., & Wolfe, R. G. (2001). *Why schools matter: A cross-national comparison of curriculum and learning.* San Francisco: Jossey-Bass.

Sumrall, W. J. (1995). Reasons for the perceived images of scientists by race and gender of students in grades 1–7. *School Science and Mathematics, 95*(2), 83–90.

Taylor, P. C. S., Fraser, B. J., &. White, L. R.. (1994, March). *A classroom environment questionnaire for science educators interested in the constructivist reform of school science.* Paper presented at the annual meeting of the National Association for Research in Science Teaching, Anaheim, CA.

Third International Mathematics and Science Study. (1998). http://timss.bc.edu

Tobias, R. (1992). *Nurturing at-risk youth in math and science: Curriculum and teaching considerations.* Bloomington, IN: National Educational Service.

Weiss, I. (1978). *Report of the 1977 National Survey on Science, Mathematics, and Social Sciences.* Research Triangle Park, NC: Center for Educational Research and Evaluation, Research Triangle Institute.

Weiss, I. (1987). *1985–86 National survey of science and mathematics education.* Research Triangle Park, NC: Research Triangle Institute.

Wilson, M. (Ed.). (1992). *Options for girls: A door to the future.* Austin, TX: ProEd.

Chapter 12

Planning a Project-Based Curriculum

Introduction

This chapter focuses on the various ways teachers can plan lessons to help students develop meaningful understandings in a project-based science environment. To implement a project, teachers develop a series of lessons around a driving question. You may have questions about implementing project-based science in a classroom: *How do I create lesson plans? How do I plan a project? How do I modify existing curriculum materials to fit a project approach? What resources are needed in a project environment? Where can I get help in finding resources? How do I help all students learn science?* This chapter focuses on answering these types of questions and walks you through the process of planning a project-based curriculum using an insect project as an illustration of the process. We will discuss how to develop a project, including deciding on concepts and learning performances, developing a driving question, writing lessons, designing assessments, and sequencing lessons. Next, we will discuss how to select and obtain resources from a variety of sources, including commercial suppliers, noncommercial suppliers, local stores, and community resources. Because a project-based curriculum is not limited to science, we will discuss how project-based science supports curriculum integration. We will consider concept mapping as

Chapter Learning Performances

- *Create a lesson plan and teach a lesson.*
- *Explain the important factors teachers must consider when planning a project.*
- *Justify why concept mapping is critical to planning.*
- *Describe how a teacher can use a variety of resources to plan a project.*
- *Create a project-based unit of study.*
- *Explain the value of integrating science across the curriculum.*

a way to plan science integration with other subject areas. We will begin by observing three teachers as they plan an insect project.

Scenario: Planning the Curriculum

Imagine three teachers planning a unit on insects. Scattered in front of them are their district's curriculum framework, the textbook adopted by the school, the *National Science Education Standards* (NRC, 1996), *Science for All Americans* (Rutherford & Ahlgeen, 1990), *Benchmarks for Science Literacy* (AAAS, 1993), *Atlas of Science Literacy* (AAAS, 2001), and other curriculum resources.

"Well, our curriculum framework requires us to cover information about insects. I think we should cover this chapter in the spring when the insects reappear," says Holly.

Martino, looking at the district's curriculum framework, says, "Yeah. We could teach about metamorphosis, insect classification, and predator–prey relationships. Butterflies and other insects would be hatching in the spring."

Judy says, "To show metamorphosis, the kids could raise their own butterflies."

"No, we can just use this chapter. It has lots of nice photographs of butterfly metamorphosis," replies Martino. "If you really want, we could go to the zoo's insect exhibit for the kids to see insects at different stages."

Holly exclaims, "Shouldn't we worry about what the children should learn? What do we want them to be able to do?"

Martino agrees and states, "Right, we should begin by clearly defining what we want students to learn. How about, 'Students, describe the life cycle of a butterfly'?"

Judy then replies, "But that is too low level of a learning performances. How about, 'Draw and explain the life cycle of a butterfly'?"

Holly exclaims, "I like that! What else do we want students to be able to do? Maybe we should identify four of the concepts?"

Judy says, "Well, how are we going to meet our inquiry objectives?"

Holly says, "Right. We have to worry about that."

Judy leans forward in her chair. "I've been learning about project-based science in this course I'm taking, and I think we can cover many of our learning performances using this approach. We just turn this insect chapter into an interesting driving question."

"But how can we do that?" asks Holly.

"Yeah and what's a driving question?"

"Well, it's a question, like 'What kinds of insects live in our neighborhood?' We could get kids interested in studying insects by taking them outside on a walk and looking for and collecting bugs. This way, kids could become really interested in the topic and learn lots about it."

Holly still isn't sure. "But aren't we supposed to use the school's adopted book? This textbook is what I think we're supposed to use."

"Sure. We could assign sections of the book for kids to read. And you're right; it has some nice pictures that kids could refer to. One of the things I learned in this course is that we need to cover basic science content through some structured lessons. We could use this chapter to cover metamorphosis and animal classification."

Martino is worried. "But we have lots of other curriculum objectives to cover, and those objectives are going to be covered on the proficiency tests. I have to worry about the low scores of some of my students."

"Well, remember, like I said before, one of our curriculum objectives that is on the proficiency exam has to do with kids doing inquiry. By using a project approach, the kids can work together to investigate insects in the neighborhood. They can make posters, presentations, and maybe videos about insects. These things can demonstrate what the kids are learning. They can show that they know how to collect and analyze data, communicate information, and make conclusions. Those are objectives we're supposed to cover, too. And in my class, I learned that girls and minorities traditionally left behind respond more positively to this approach."

"Yeah, but what do we use for these investigations?" Martino is still feeling anxious.

Judy explains how project investigations work: "In a project approach, students are working on many different subquestions, so we can draw from lots of different resources. This STC kit on *The Life Cycle of Butterflies* (National Science Resources Center, 1992, 1994) is very good for teaching about metamorphosis. We can get butterfly eggs, raise caterpillars, and watch them grow into butterflies. This would make a great lesson to teach the objectives in our curriculum. The kids will also come up with some of their own ideas for the investigations in the project."

"Well, I think I'd like to try this. It sounds like a good way to teach these objectives," says Holly.

Martino says, "Well, I'm not sure."

Judy takes this as a green light, "Okay, let's start by concept mapping out our content and inquiry learning performances. I think you'll see that we can also get in some math and language arts stuff, too."

Planning Lessons

Although the best planned lessons don't always succeed, unplanned lessons rarely succeed. Planning is critical for a successful lesson. As a beginning teacher (or even an experienced teacher trying project-based science for the first time), you will spend much time planning your lessons and thinking through each step. You will need to ask and answer a series of questions: *What am I teaching? What are the learning and performance outcomes? How is what I am teaching related to the driving question? What materials are needed for the lesson? How will I proceed with the lesson? How will I evaluate students?* With practice, you will find that you are asking yourself many

> ## CONNECTING TO NATIONAL SCIENCE EDUCATION STANDARDS
>
> ### SCIENCE TEACHING STANDARDS
>
> In determining the specific science content and activities that make up a curriculum, teachers consider the students who will be learning the science. Whether working with mandated content and activities, selecting from extant activities, or creating original activities, teachers plan to meet the particular interests, knowledge, and skills of their students and build on their questions and ideas (pp. 30–31).

of these questions automatically. However, it will still be essential for you to make plans to structure your lessons so that you can help students learn what you are trying to teach them.

Developing Learning Performances

To begin planning, teachers need to first specify what students will learn. We prefer the term **learning performances** to the commonly used *objectives* for specifying student learning outcomes. The term *learning performances* specifies what achievement we expect of the students (Perkins, Crismond, Simmons, & Unger, 1992; Wiggins & McTighe, 1998). *Objectives,* on the other hand, usually are vague and only specify what students will know at the conclusion of an activity, project, or lesson. They don't tell us *how* we will know that students have learned or *how* we will know what they have learned. They don't tell us what students will *do* to demonstrate understanding. Without identifying the learning performances we expect, it is difficult to design a lesson and assess deeper conceptual understanding. Do we want students to recall information or do we want them to apply principles to explain what might happen? A range of learning performances can be identified that require different cognition from students. For instance, we might want to have students define the term *ecosystem,* or we might want them to explain the relationship among different species in an ecosystem. Either performance will make a different cognitive demand of students. The first asks for simple recall of a definition, whereas the second requires students to understand connections among principles.

The cells in Table 12.1, which was modified from Anderson and Krathwohl (2001), illustrate various learning performances. Each row shows different knowledge dimensions: factual knowledge, conceptual knowledge, procedural knowledge, and metacognitive knowledge. **Factual knowledge** is knowledge of facts and details. For example, a student uses factual knowledge to recall the definition of *photosynthesis.* **Conceptual knowledge** is knowledge of interrelationships, principles, theories, and models. For instance, a student uses conceptual knowledge to explain the relationships between geologic time and the theory of evolution. **Procedural knowledge** is knowledge of how to do something, conduct inquiry, or use a skill. A student uses procedural knowledge of technological probes when she uses a light probe to measure the amount of light a plant receives. **Metacognitive knowledge** is self-knowledge or knowledge of one's own cognition. Students use metacognitive knowledge when they think about their own learning and record their ideas in a science journal.

TABLE 12.1 Learning Performances*

The knowledge domain	The cognitive process dimension					
	Remember	Understand	Apply	Analyze	Evaluate	Create
	Definition: To recall something from memory	*Definition:* To comprehend and be able to explain the meaning of a concept or process	*Definition:* To develop solutions to familiar or new problems	*Definition:* To examine a concept or process in detail to learn more about it	*Definition:* To make judgments about value, quality, importance, or condition	*Definition:* To produce something
Factual knowledge *Definition:* Knowledge of details and facts	*Example:* State that matter is made up of atoms	*Example:* Use the idea of conservation of atoms to explain how the mass after a chemical reaction compares to before the reaction	*Example:* Carry out an experiment to illustrate the conservation of mass	*Example:* Given a table of data, be able to discriminate the relevant from irrelevant data needed to answer a question related to conservation of mass in a chemical reaction	*Example:* Judge if appropriate conclusions were drawn regarding a set of data collected from a series of chemical reactions	*Example:* Design an experiment to illustrate the conservation of mass
Conceptual knowledge *Definition:* Knowledge of interrelationships, principles, theories and models	*Example:* Define "flow of energy" in an ecosystem	*Example:* Explain how sunlight is transformed through photosynthesis to provide food for plants, herbivores, and carnivores	*Example:* Implement an experiment to show the influence of sunlight on the growth of plants	*Example:* Analyze an ecosystem to show how extinction of a small omnivore might influence the population of a herbivore	*Example:* Evaluate how a new government policy might influence various species in ecosystem	*Example:* Create an alternative EPA policy to improve the ecosystem of the Great Lakes

(Continued)

TABLE 12.1 Continued

The knowledge domain	Remember	Understand	Apply	Analyze	Evaluate	Create
Procedural knowledge *Definition:* Knowledge about how to do something, conduct inquiry, or use a skill	*Example:* Recognize the parts of a microscope	*Example:* Clarify to a fellow student how to focus a microscope	*Example:* Use a microscope to observe single-celled pond organisms	*Example:* From observations of single-celled organisms, organize a data chart to distinguish the organisms	*Example:* Determine whether conclusions drawn from observations of single-celled organisms are appropriate	*Example:* Devise a procedure to determine if the number of single-celled organisms in a pond sample changes over time
Metacognitive knowledge *Definition:* Self-knowledge, knowledge of one's own cognition	*Example:* Recall that you find writing in a science journal challenging	*Example:* Summarize which tasks you find most difficult in science classes and why they are difficult for you	*Example:* Implement a procedure to help you study for a science test	*Example:* Breakdown the components of a science test that you will need to focus on to be successful	*Example:* Critique whether you have improved your own test-taking skills	*Example:* Construct a portfolio and organize it in a manner that would be most compatible with your own learning style to show what you have learned

* Modified from L. W. Anderson & D. R. Krathwohl, 2001.

In Table 12.1, each column shows categories and cognitive processes: remember, understand, apply, analyze, evaluate, and create. Students **remember** when they simply recall something from memory. For example, students might remember the names and symbols on a periodic table of elements. When students **understand**, they are able to comprehend and explain. For example, a student understands how to use a light probe if she can explain to a fellow student how to use it to measure the light reaching a plant. An elementary student will **apply** knowledge when he uses it to solve problems or develop solutions. A young student who uses her

> ### CONNECTING TO NATIONAL SCIENCE EDUCATION STANDARDS
>
> #### SCIENCE TEACHING STANDARDS
>
> Teachers plan activities that they and the students will use to assess the understanding and abilities that students hold when they begin a learning activity (p. 32).

knowledge of magnets to create a way to pick up paper clips scattered on the floor is applying her knowledge to solve a problem. Students **analyze** when they examine a concept or process in detail to learn more about it. If a student breaks down a table of data to determine which factors are affecting plant growth and which are not, he is analyzing the data. Students **evaluate** when they use their knowledge to make judgments about the value, quality, importance, or condition of something. A student who critically examines the town's policy on recycling and writes a letter to the mayor pointing out problems is evaluating the recycling process. Finally, students create when they produce something. A young student creates when she devises a new procedure for observing changes in the weather.

Creating a Lesson Plan

Think of a lesson plan as a road map. You wouldn't plan a trip to an unknown region without consulting a map. Similarly, you shouldn't teach a lesson without first constructing a lesson plan. When planning, it is a good idea to use a lesson plan format. Lesson plan formats vary, but we introduce two formats that are useful for science teachers in a project-based classroom.

Basic Lesson Plan Format

- *Student learning performances*: What do I hope to accomplish in the lesson? What concepts or inquiry skills will students develop? Will students develop background experiences for the project? What is it that I want students to be able to do?
- *Relationship to the driving question:* How is the lesson related to the driving question of the project?
- *Materials*: What materials will I or the students need?

- *Instructional strategies*: What strategies or learning activities will I use to help students reach the learning performance? Will I use demonstration or discussion, for example?
- *Time required*: How much time will it take to complete the lesson?
- *Cautions*: Are there any dangerous or hazardous components of the activities associated with the lesson? What precautions need to be taken? (We discuss safety more thoroughly in Chapter 11.)
- *Instructional sequence*: How will I proceed through this lesson?
  - *Introducing the lesson*: How will I introduce the lesson to the students? How will I motivate and capture students' attention? How will I find out about students' prior knowledge?
  - *Representing the content*: How am I representing the content students will learn? What explanations or learning activities am I using? How will I connect the ideas to students' prior knowledge?
  - *Establish links to the driving question*: How is the activity related to the driving question? How will I point out to students how what they are learning is related to the driving question of the project?
  - *Evaluating learning*: How will I determine if students met the learning objectives? What is the relationship between this lesson and student products or artifacts?

Sample Basic Lesson Plan

Let's look at a specific plan for a lesson that helps middle grades students develop the idea that, when chemicals react, they often give observable evidence. The lesson is situated in a project with the driving question "What chemicals are in my home?" When developing a lesson, you need to ask yourself a series of questions based on the lesson plan format. The answers given here pertain to the chemical reaction lesson:

- Student learning performances:
  - Based on the concept map you developed for the project (see Figure 12.1), you decide that students need to explain and give an example of what is meant by a chemical reaction.
- Relationship to the driving question:
  - The driving question was selected so that the study of basic chemical ideas could be introduced to students. This is necessary to help them identify chemical changes that occur in their homes.
- Materials:
  - You will need about 10 tablespoons of sulfur (this can be purchased through any science catalog); steel wool; a magnet; a Pyrex test tube; and a propane torch, Bunsen burner, or Sterno can.
- Instructional strategies:
  - You will use a teacher demonstration.
- Time required:
  - At least 40 minutes will be required.
- Cautions:
  - Good safety precautions should be followed throughout the demonstration. You and the students should wear safety goggles. This will both protect your eyes

and model good laboratory practice. Do not point the mouth of the test tube toward the children. Finally, the test tube will get hot. Do not let the children touch it. Wear heat-resistant gloves or use hot pads. Break the test tube open by wrapping it in an old cloth that you can then discard. After wrapping the test tube, gently tap it with a heavy object like a hammer. Always practice the demonstration before doing it in class. Make sure the room is well ventilated.

- Instructional sequence:
  1. Students should sit in groups of four. At each table there should be a sample of steel wool and sulfur.
  2. Ask students to describe what the steel wool and sulfur look and feel like.
  3. Ask the students to test the magnetic properties of steel wool with the magnet (the magnet will attract to the steel wool).
  4. Have groups share their observations.
  5. Ask students to come to a class consensus about the descriptions.
  6. Ask students to predict what will happen if you heat the sulfur and steel wool together. Also ask them to give justifications for their predictions.
  7. Next, have students share their predictions and justifications with the class. Write predictions and justifications on the board or newsprint paper.
  8. Now place about two tablespoons of sulfur and a small amount of steel wool in a test tube. Heat the two materials using a propane torch, a Bunsen burner, or a Sterno can. Caution: Wear safety glasses and don't point the mouth of the test tube toward the children. Children should also wear safety glasses for this demonstration. The room should be ventilated.
  9. Once all of the sulfur has burned off or reacted with the steel wool, walk around the classroom, showing the students the material in the test tube. Ask them to describe what they see. Caution: The test tube will be very hot. Wait a few minutes to let it cool down and pick it up using an insulated test tube holder or hot pad.
  10. Once the test tube has cooled, pass out a small sample to each group. Note: You may have to break the test tube to get the sample out of it. Caution: To break the test tube, first wrap it in an old cloth that can be discarded. Again ask students to describe what the material looks and feels like.
  11. Have them test the magnetic properties of this material (it is no longer magnetic).
  12. Ask them to compare what they observed with their predictions.
  13. Have them try to explain why their predictions differ from their observations.
  14. Have the students share their observations with the class.
  15. Ask students to discuss in their groups what occurred.
  16. Ask them if the material that was in the tube is the same as the sulfur and steel wool (iron).
  17. Ask them to justify their reasons.
  18. Have the groups share their ideas with the class.
  19. Share with the class a verbal explanation of *chemical reaction*. The material sulfur and the material iron reacted (came together) to form a new substance called *iron sulfide*. This new substance has different properties than the original substances. It looks different. It is not magnetic.
  20. Tell students that this is an example of a chemical reaction: Starting materials come together to form a new material. Ask them how they know a new material was formed.

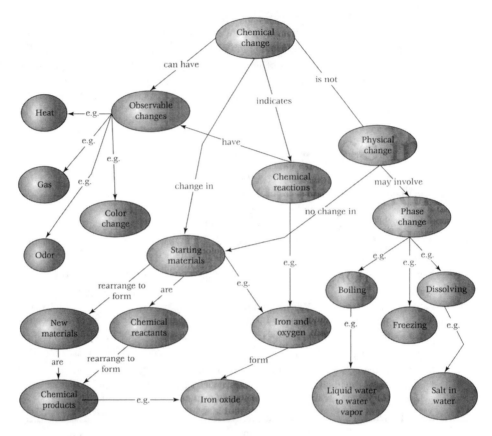

Figure 12.1 Concept map of chemical change.

This demonstration could be followed by another showing that chemical reactions usually give evidence. For instance, you could break water up into hydrogen and oxygen through electrolysis (running electricity through it with a 9- or 12-volt battery). With a little practice, this demonstration is not difficult to do. Another idea is to follow up this demonstration with a student activity. For instance, you could have students mix baking soda and vinegar and then starch and iodine together.

5-E Model Lesson Planning Format

The **5-E Model**, developed by the Biological Science Curriculum Study (BSCS, http://www.bscs.org/library/BSCS_5E_Model_Executive_Summary2006.pdf) organization, is a five-stage lesson planning model, which includes the stages of engagement, exploration, explanation, elaboration, and evaluation. The 5-E model is shown in Figure 12.2. The model is supported by constructivist theory, because they take into consideration children's prior experiences, and it enables students to construct understanding from activities in which they are actively engaged. Recent research suggests that these types of models are significantly better than expository methods for teaching concepts (Odom & Kelly, 2001).

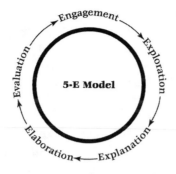

Figure 12.2   The 5-E model.

Engagement. In the **engagement** stage, the teacher tries to make some type of connection between students' past experiences and the present lesson. The purpose is to focus students' thinking on the current concept or skill to be learned. The engagement stage is consistent with the constructivist model because it provides the teacher a chance to activate students' prior understandings and experiences before starting the exploration stage. There are many ways to connect past experiences with the present lesson. The teacher might provide the students with an advance mental organizer (such as telling them they are about to read about electricity), tell a story, create an analogy, recall a previous experience, ask about students' experiences, remind students of a lesson or investigation done earlier in the year. This can also be a useful stage for connecting the current lesson to the driving question for the project.

We will illustrate the 5-E model lesson planning format, using examples from magnet lessons and an electricity project framed around the driving question "How do electrical appliances work?" In a lesson used to teach about electrical circuits, conductors, and insulators, a teacher might read a passage from the story *Dear Mr. Henshaw* by Beverly Cleary (Cleary, 1983). The passage is about a boy named Leigh Botts who keeps having his dessert stolen out of his lunch. He decides he wants to build a burglar alarm so he can catch the dessert thief. The teacher asks students how they think Leigh might build this burglar alarm. This sets the stage for learning about circuits, conductors, and insulators; and it enables the teacher to find out about students' prior knowledge about electricity.

Exploration. In the exploration stage, students have concrete experiences with phenomena that are essential to developing understanding of a concept. Most students have at one time or another memorized information with little meaning, such as the words to a song in a school play or vocabulary words for a test. For example, a little boy who knew his baby sitter was studying to become a teacher asked her on several occasions who Richard Stands was. Each time she said that she did not know. One day in frustration the child yelled, "You're going to be a teacher, so you better find out who Richard Stands is!" When she asked him where he had learned about "this Richard Stands guy," the boy said that they pledged to him every day—"I pledge allegiance to the flag of the United States of America and to the Republic for Richard Stands, one Nation, under God, invisible [as opposed to indivisible] with liberty and justice for all." This type of confusion also happens in science. Students sometimes encounter concepts that are foreign to them. Unless they have the opportunity to explore the concepts, develop understanding of them, put words to them, and expand upon them, the concepts have little real meaning. For instance, many of us may have memorized the names of various types of rocks without first observing and classifying the rocks in our environment. The exploration stage provides children with the opportunity to create meaning from phenomena.

Imagine a child about the age of 7 who is learning about magnets. She places the magnet near a paper clip, and it is picked up. She places it near a pile of paper clips, and they are all picked up. She slowly moves the magnet across the top of a table, and, to her surprise, a paper clip jumps over to the magnet before the magnet actually touches it. She tries this again. She holds the magnet about an inch above the paper clips, and several jump into the air to reach it. She places the magnet below the table and moves it around under the pile of paper clips. They follow the magnet across the table. She thinks about the refrigerator magnets used to hold up her schoolwork, so she places a sheet of paper over the paper clips and places the magnet over the paper. Sure enough, the magnet lifts both the paper and the paper clips.

What has this child discovered? She knows that the paper clips are attracted to the magnets, and she knows that the magnet can "go through" air, paper, and desk "to attract" the paper clips. Does she know that this phenomenon is called the magnetic field? Does she know that the magnetic field can pass through nonmagnetic materials? Does she know that magnets attract to materials, such as iron and nickel? Conceptually, she understands the answers to many of these questions, but she simply has not yet put the words to them.

Contrast her situation with that of the child whose teacher gives him the definitions of magnetic field and magnetic attraction. Does the child have any real understanding of the concepts this way? Maybe, but it is unlikely that a 7-year-old really understands these concepts very well after having only been given definitions. Most children will not really understand such concepts unless they have had prior experiences with them. The child who thought about the refrigerator magnet holding up her work was developing new understandings (of the magnet being able to attract paper clips through air) with those she already had (about magnets going through paper) and is now in a position to learn more formal ideas about magnets.

Similarly, in an electricity lesson, the students use dry cells, wires, and bulbs to try to construct their own circuits to light the bulbs. They test various materials in the circuit pathway, such as rubber bands, pop can tabs, coins, erasers, wires, and paper clips, to see if they will light the bulb.

Explanation. In the **explanation** stage, the teacher introduces formal vocabulary or students verbalize understanding about the explorations in which they have been engaged. The intention is to focus the students mentally on the concepts they are exploring and tie them to the project.

In the magnet lesson, the teacher might gather the children to discuss what they observed. Students would share their findings about the magnet being able to "pass through" air, paper, and desk to pick up a paper clip. The teacher would introduce related concepts to the students, explaining that these findings showed that the magnet's *magnetic field* could pass through *nonmagnetic* materials. Because the students had experience with these phenomena, they are able to use the related scientific terminology with much less confusion. The scientific words most likely merely substitute for the words they have been seeking. A child who has been searching for a word to explain the magnet's ability to "go through" things can accommodate the term *magnetic field* easily into his or her mental structure or schema.

In an electricity lesson, the teacher directs students at this stage to notice that there are two types of circuits: series and parallel. She formally introduces the word *conductor* for the items that let the bulb light when they are in the circuit pathway and the word *nonconductor* to those items that do not let the bulb light when they are in the circuit pathway.

Elaboration. During **elaboration**, students can gain a deeper understanding of the concept by engaging in additional activities related to the concept. For example, the teacher might engage students in another investigation in which they explore whether the magnet will attract through plastic, glass, aluminum, copper, and other materials. Students expand their current understanding to encompass the idea that magnetic fields will pass through nonmagnetic materials but not other magnetic fields. The teacher might also show a video or an educational television program or read related text passages.

In an electricity lesson, students make a simple switch that, when connected, will set off a buzzer. This extends students' learning about circuits (the circuit is closed when the switch is on and it is open when the switch is off) and conductors (the metal in the switch conducts electricity when the switch is on).

Evaluation. The last stage in the 5-E instructional lesson planning model is **evaluation**. The evaluation stage is in accord with the constructivist theory that advocates embedded assessment. This also gives the teacher an opportunity to assess learning occurring in the context of the project.

In an electricity lesson, students are evaluated on their knowledge of circuits, conductors, and insulators when they complete the following task: Figure out a way to "alarm" a lunch box so that, when it opens, the buzzer goes off. Students must have working knowledge of circuits, conductors, and insulators to be able to figure out how to make the buzzer come on when the lunch box opens. Some students construct the lunch boxes so that the electrical circuit is connected when a piece of metal (a conductor) touches a loose metal end of the buzzer when the lid of the lunch box is pulled up (thus closing the circuit). Note: To learn more about this electricity lesson and ways to integrate science with other disciplines (mathematics, reading, and social studies), see Sandmann, Weber, Czerniak, and Ahern (1999).

The 5-E model, however, must not be viewed as a prescriptive sequence or a rigid five-step lesson plan model. Although we have examined it as a five-step cycle, the steps are flexible and iterative. For example, a series of lesson plans might have several explorations before a formal explanation is introduced. Similarly, all five steps could occur in a single day's lesson, or it could take a week or more to get through all five steps.

At this point in the chapter, you have learned why lesson plans are a critical component of a project-based classroom, and you have learned some strategies for plan-

CONNECTING TO NATIONAL
SCIENCE EDUCATION STANDARDS

SCIENCE EDUCATION
TEACHING STANDARDS

The plans of teachers provide opportunities for all students to learn science. Therefore, planning is heavily dependent on the teacher's awareness and understanding of the diverse abilities, interests, and cultural backgrounds of students in the classroom. Planning also takes into account the social structure of the classroom and the challenges posed by diverse student groups. Teachers plan activities that they and the students will use to access the understanding and abilities that students hold when they begin a learning activity (p. 32).

ning a lesson. The remainder of this chapter will explore in more detail how an entire project is developed.

Developing a Project

As the scenario with Judy, Martino, and Holly illustrated, instructional planning is never an easy task, and planning project-based instruction is especially challenging. Teachers must do more than plan individual investigations, artifacts, and classroom activities; they must also interrelate these instructional components in such a way that they complement each other and build toward assessment goals for all students. There are many different planning styles; we will examine one method for developing a project.

Identifying Concepts, Specifying Learning Performances, and Matching to Curriculum Objectives

The decision to help students meet various learning performances and cover certain concepts at a particular grade level is usually not one that teachers make by themselves. Day-to-day lessons are usually guided by the need to cover basic concepts and curriculum objectives that a school district has for its students. At the national level, the *National Science Education Standards* were published in 1996 (NRC), and the American Association for the Advancement of Science established guidelines for school curriculum in *Benchmarks for Science Literacy* (AAAS, 1993). Many school districts adopted these ideas and formed curricula based on these sets of standards. Some states established state-mandated curricula that spell out essential knowledge and skills. Therefore, although curriculum objectives may vary from state to state or from school district to school district, you will find a great deal of similarity among concepts and objectives covered at most grade levels around the country.

What learning performances and concepts do you plan students to learn by engaging in a particular project? To answer this question, start by examining local and state guidelines, and national guidelines such as The *National Science Education Standards* (NRC, 1996), *Benchmarks for Science Literacy* (AAAS, 1993), and *Atlas of Science Literacy* (AAAS, 2001, 2007). Determine which local, state, or national objectives you plan to meet.

Imagine that the grade level you teach is required by your state to learn about various concepts related to insects. This is a common standard in the primary grades. The *National Science Education Standards* (NRC, 1996) lists "life cycles of organisms" as a content standard for grades K–4. At this point, you might want to clarify your own understanding of this standard. You might read what the *National Science Education Standards* or the *Science for All Americans* say about this standard. You could consult various trade books and textbooks. Your reading will help clarify the important concepts and ideas involved in meeting this standard. Make a list of important science concepts associated with this and other standards. Once you have identified and made a list of concepts, teachers need to organize the lessons in an effective way. You will notice that the teachers described in the opening scenario were using a variety of resources, including the *National Science Education Standards* and their curriculum framework. They also started by creating a concept map and thinking about the needs of their own students.

Organizing the Lessons in a Project

We suggest three ways to determine the organization of lesson plans: the use of concept maps (Novak & Gowin, 1984), observing and listening to students during a project, and the KWL strategy (Ogle, 1986).

Concept maps are visual representations of the relationships among concepts. As such, concepts maps are an external representation of ideas an individual holds. Concept maps (Novak & Gowin, 1984) enable learners (students and teachers) to link ideas and concepts, helping them construct integrated understanding. By developing concept maps of a project, a teacher can identify the major concepts students will need to understand to complete the project and plan lessons accordingly.

Teachers also plan lessons by observing and listening to children during a project. As teachers observe students during project work, they notice what concepts and skills students need to complete the project. They also become acutely aware of students' interests so they can make instruction relevant to their lives (Haberman, 1995).

A third technique is a strategy frequently used in reading instruction, the **KWL strategy** (Ogle, 1986). In this technique, the *K* stands for "what you already know," the *W* stands for "what you want to know," and the *L* stands for "what you learned." *K* and *W* are typically determined prior to beginning a topic, investigation, or lesson. *L* is determined after instruction or the investigation. Frequently, *L* is used as an assessment technique.

Using Concept Maps

The various concepts drawn on a concept map can help you identify potential lessons. We illustrate the procedure using a project with the driving question "Where do animals get their food?" In Figure 12.3, *food chain* is the uppermost concept. A teacher might want to have a lesson in which the children in the class create a drawing of a

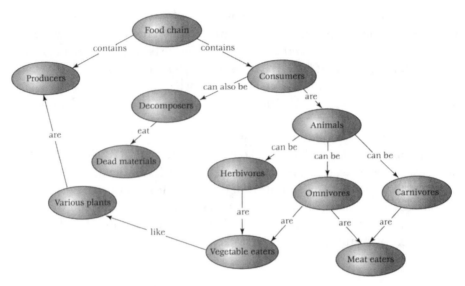

Figure 12.3 Concept map of food chain.

food chain, showing what animals eat what other animals and plants. Other lessons might teach students about producers, consumers, and decomposers.

You may need to redo or expand a concept map during a project. Like the children in a class, a teacher will develop understanding as a project progresses. When you revise a concept map, you may identify additional lessons. For example, students might become interested in the nutritional value of being a vegetarian. This might lead to new lessons on various aspects of health, such as cholesterol, vitamins, and fiber. A concept map on insects might look like the one pictured in Figure 12.4.

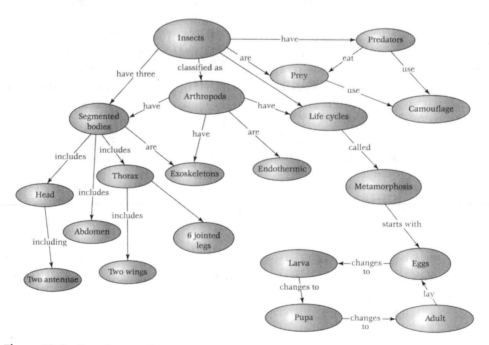

Figure 12.4 Insect concept map.

Using Observations and Listening

Teachers can also identify necessary lessons by observing and listening to students. During a project, you might notice that students are having particular trouble understanding or applying a concept, employing a laboratory technique (like using a microscope), or perhaps following a procedure (like calculating the mean of a list of numbers). Based on your observations, you could plan a lesson for the next day. Or you might change the plans for the day and carry out an impromptu lesson. You might hold an immediate discussion to help clarify a concept that students are having difficulty with at the moment.

> ### CONNECTING TO NATIONAL SCIENCE EDUCATION STANDARDS
> #### SCIENCE CONTENT STANDARDS
>
> Students need opportunities to present their abilities and understanding and to use the knowledge and language of science to communicate scientific explanations and ideas. Writing, labeling drawings, completing concept maps, developing spreadsheets, and designing computer graphics should be a part of science education (p. 144).

Using KWL

The **KWL strategy** (Ogle, 1986) is another good strategy for determining what lessons to use during project activities. As already mentioned, the *K* stands for "what you already know," the *W* for "what you want to know," and the *L* for "what you learned."

Asking students what they already know and what they want to learn helps teachers get to better know their students and determine what lessons to use. The *K* assesses students' prior knowledge and, so, helps teachers better understand what concepts and skills will be needed in the project. The *W* helps teachers develop lessons based upon students' interests. The *L* can be used to develop new investigations—once you have an understanding of what students have learned, you also have an idea of what they still need to learn. Some teachers like to add a "how," or *H,* section to the KWL strategy—KWHL—so that students focus on how they will acquire knowledge before they report what they learned.

> ### CONNECTING TO NATIONAL SCIENCE EDUCATION STANDARDS
> #### SCIENCE TEACHING STANDARDS
>
> At all stages of inquiry, teachers guide, focus, challenge, and encourage student learning. Successful teachers are skilled observers of students, as well as knowledgeable about science and how it is learned (p. 33).

Imagine that students are exploring the project question "When do various insects appear on our playground?" A teacher might start this project by creating three columns on the board: *Know, Want to know,* and *Learned.* The teacher asks the students,

TABLE 12.2   KWL Chart 1

Know	Want to know	Learned
Insects are dangerous.	_____	_____
Insects come from eggs.	_____	_____
Insects come out in the spring.	_____	_____

"What do you *know* about insects?" The teacher uncritically accepts the students' answers; she allows students to generate as many ideas as possible. Some of the students' responses are not accurate, but the teacher doesn't try to correct them at this point. Students might say, "Insects are dangerous," "Insects come from eggs," and "Insects come out in the spring." These responses inform the teacher that students need lessons on insect characteristics, metamorphosis, insect defense mechanisms, and seasons. The KWL chart the teacher is creating so far looks like the one in Table 12.2.

Next, the teacher asks, "What do you *want* to know about the insects?" and records responses. The teacher lets students brainstorm questions for the list. Students might generate questions like "How are insects born?" "What insects are dangerous?" and "Where do insects live in the winter?" These questions help the teacher plan lessons on metamorphosis and insect behavior. The chart now looks like the one in Table 12.3.

To further explore insects, the teacher has students observe insects on different 3-foot by 3-foot plots on the playground. Once this activity is completed, the teacher has the students list their observations. Throughout the project, the teacher has students update this list and the two columns of the chart, correcting what they thought they knew or adding to what they want to learn.

CONNECTING TO
NATIONAL SCIENCE
EDUCATION STANDARDS

SCIENCE TEACHING
STANDARDS

The actions of teachers are deeply influenced by their understanding of and relationships with students (p. 29).

At the end of the project, the teacher refers again to the chart and asks the students, "What did you learn?" Students might say, "We learned that insects go through several stages called *metamorphosis*," "Insects lay eggs that stay dormant during the

TABLE 12.3   KWL Chart 2

Know	Want to know	Learned
Insects are dangerous.	I want to know how insects are born.	_____
Insects come from eggs.	I want to know what insects are dangerous.	_____
Insects come out in the spring.	I want to know where insects live in the winter.	_____

TABLE 12.4   KWL Chart 3

Know	Want to know	Learned
Insects are dangerous.	I want to know how insects are born.	We learned that insects go through several stages called metamorphosis.
Insects come from eggs.	I want to know what insects are dangerous.	Insects lay eggs that stay dormant during the winter and hatch in the spring.
Insects come out in the spring.	I want to know where insects live in the winter	Insects can camouflage themselves to hide from predators.

winter and hatch in the spring," and "Insects can camouflage themselves to hide from predators." Answers like these help the teacher develop new lessons, such as one on defense mechanisms other than camouflage. The final KWL chart looks like the one in Table 12.4.

Creating Learning Performances and Matching to Curriculum Objectives

After you have identified and organized the concepts and skills to be taught, create learning performances you plan for students to learn through engaging in the project. For instance, you might write learning performances that state: Students will draw the stages of the life cycle of a butterfly. Students will explain the life cycle of a butterfly. Students will describe the parts of an insect's body. Sometimes, your local school district's objectives might be written in a way that specifies learning performances. For example, the concept "life cycles" may be listed in a school's curriculum objectives as "Each fourth-grade student will describe the stages of the life cycle of a butterfly in correct sequence." However, your school district's objectives may be written in terms of "Students will know …" or "Students will understand …" In any case, once you have written learning performances, match them up with your local or state curriculum objectives. Identifying concepts, specifying learning performances, and matching them to curriculum objectives is actually a very iterative process. For example, you might ask yourself, "What concepts will students be learning if they explore this curriculum objective?" or "What curriculum objective will I cover if students learn this concept?" Learning Activity 12.1 will get you started planning a project by having you select concepts, specify learning performances, and match objectives.

Developing the Driving Question

Often after identifying concepts, learning performances, and curriculum objectives, teachers brainstorm several driving questions that will allow students to explore the concepts selected. Sometimes, however, teachers will brainstorm driving questions before even starting the concept map.

Developing good driving questions takes time and thought. Don't worry if a good question doesn't pop into your mind immediately. Driving questions can always be

Learning Activity 12.1

IDENTIFYING CONCEPTS, SPECIFYING LEARNING PERFORMANCES, AND MATCHING TO CURRICULUM OBJECTIVES

Materials Needed:

- The curriculum objectives from a local school district
- A copy of the *National Science Education Standards* (NRC, 1996)
- A copy of *Benchmarks for Science Literacy* (AAAS, 1993) and *Atlas for Science Literacy* (AAAS, 2001, 2007)

A. Select a topic to teach to elementary or middle grade students.
B. Consult the *National Science Education Standards* (NRC, 1996), *Atlas for Science Literacy* (AAAS, 2001, 2007), or *Benchmarks for Science Literacy* (AAAS, 1993) to identify the concepts related to this topic.
C. Consult the curriculum from a local school district to find out what objectives need to be covered related to this topic.
   - Using this list, create a concept map.
   - Using the concept map, specify learning performances your students will accomplish.
   - Match the concepts and learning performances to the school's curriculum objectives.
D. Keep your concept map, your learning performances, and your comparison of objectives. You will be using them throughout this chapter.

modified as project development progresses. As we discussed in Chapter 3, there are many sources for driving questions: personal experiences, students' interests and culture, hobbies and personal interests, the newspaper or television, other teachers, textbooks and other curriculum materials, and the World Wide Web.

Here are some initial examples of driving questions related to insects:

- What insects live on our playground? (The idea for this question came from a fellow teacher.)
- What are the names of common insects? (The idea for this question came from a commercial curriculum publisher.)
- How do insects grow and change? (The idea for this question came from a school's curriculum.)
- What insects did you see this morning on the way to school? (The idea for this question came from personal experiences.)
- Where do insects go in the winter? (The idea for this question came from a child.)
- Why are they spraying insecticide to kill gypsy moths? (The idea for this question came from a news story reported in the local paper.)

Of these possible driving questions, the best for elementary students is most likely "What insects live on our playground?" This question meets all of the features of a good driving question (discussed in Chapter 3):

- *Feasibility.* Students can design and perform investigations about insects that live on the playground. For example, they can plot out a section of the playground to study the variety of insects that live on it.
- *Worth.* Students can learn science concepts related to the district's curriculum standards, such as *metamorphosis, classification, predator and prey,* and *camouflage.*
- *Contextualization.* The question is a real-world question for students because it relates to their own physical environment—their playground.
- *Sustainability.* Students can learn about insects on the playground throughout the school year by studying subquestions, such as "Which insects are around in the fall?" "Where do they go in the winter?" and "Which reappear earliest in the spring?"
- *Ethical.* Students can observe insects without harming the insects or themselves.
- *Meaning.* Insects are interesting and exciting to most children because they have strange defense mechanisms, use camouflage in fascinating ways, and have interesting life cycles.

Each of the other questions lacks fundamental features of a good driving question. For example, "What are the names of common insects?" is not sustainable over time. Students can learn the names of insects in a very short period of time, and then they will likely become bored. Learning the names of insects is also unlikely to be meaningful to most children. "How do insects grow and change?" is feasible—students can design and perform investigations to watch insects grow and change. However, the question may not be contextualized in students' lives. Studying the growth of an insect students never see in their own environment might cause some children to lose interest in the topic. This question also can be subsumed as a subquestion under a broader driving question. "What insects did you see this morning on the way to school?" is contextualized, but it may not be worthwhile because important concepts in the curriculum are unlikely to be covered in a study of insects seen on the way to school one day. This question, too, can be subsumed under a broader driving question. "Where do insects go in the winter?" might seem sustainable since it covers a longer period of the year. However, the question may not be feasible because students may not be able to design an investigation that they can conduct outdoors in the winter. Finally, "Why are they spraying insecticide to kill gypsy moths?" may be contextualized because it is related to students' lives; but for safety reasons, children should not be handling insecticide and, therefore, there is no investigation students can conduct related to the question. In Learning Activity 12.2, you develop a driving question for your project.

Developing Lessons

After teachers have identified concepts, learning performances, and a driving question, they begin to develop lessons. Don't expect to develop lessons from scratch. Original ideas are rare. You might come up with some original ideas, but more likely you will modify activities that have already been developed to meet the needs of your own students and to ensure that *all* students in your class learn. As you learned in Chapter 8, you will select instructional strategies that meet the needs of your students

Learning Activity 12.2

TURNING CONCEPTS, LEARNING PERFORMANCES, AND CURRICULUM OBJECTIVES INTO DRIVING QUESTIONS

Materials Needed:

- The list of concepts, learning performances, and curriculum objectives generated in Learning Activity 12.1

A. Take the list of concepts, learning performances, and curriculum objectives generated in Learning Activity 12.1 and brainstorm several driving questions from it.
B. Share this list with a colleague or classmate to try to generate additional driving questions.
C. Evaluate each question for:
   - Feasibility
   - Worth
   - Contextualization
   - Sustainability
   - Meaning
   - Ethical

D. Select one driving question that you will develop into a project throughout the rest of this chapter. Write an argument about why you are selecting this question over others you brainstormed.

and the educational goals. For our insect project, there are several good resources that we could use to develop lessons.

The FOSS *Insects* kit is valuable for teaching about the structure and life cycle of several different insects: mealworms, waxworms, milkweed bugs, silkworms, and butterflies (Lawrence Hall of Science, 1993).The FOSS kit *Environments* has several lessons about insects. In one lesson, students investigate the environments (moisture and darkness) preferred by isopods and beetles. In another lesson, students investigate the level of salinity needed for brine shrimp to hatch (Lawrence Hall of Science, 1993).

Another useful curriculum is the Science and Technology for Children (National Science Resources Center, 1992) curriculum *The Life Cycle of Butterflies*. Intended for second graders, this material can generate several lessons. Students raise painted lady butterflies to learn about metamorphosis; they use observation and recording skills to raise caterpillars and butterflies; they learn about caterpillars' basic needs for air, water, food, and shelter; and they learn about insect body parts and function.

The Ranger Rick series entitled *Incredible Insects* (Braus, 1989) is filled with interdisciplinary activities related to insect classification, metamorphosis, habitats, and adaptations for survival and the influence of insects on our lives. One activity that comes from *Incredible Insects* is a butterfly metamorphosis role play.

The Lawrence Hall of Science GEMS series contains two useful books: *Hide a Butterfly* (Echols, 1986) and *Ladybugs* (Echols, 1993). *Hide a Butterfly*, designed for preschool through kindergarten, focuses on skills (observing, communicating, comparing, and matching), concepts (camouflage and predator/prey), and themes (systems and interactions, models and simulations, scale, structure, evolution, and diversity and unity). *Ladybugs*, which also focuses on skills, concepts, and themes, covers body structure, life cycles, defense mechanisms, predatation, and the environmental role of ladybugs.

Guest lecturers and field trips also make good lessons. An entomologist from a local university could provide students with useful information and resources. A field trip to the zoo's insect exhibit would provide information and expose students to a variety of insects.

Children's literature is another rich resource for a lesson. The following books teach about insect metamorphosis and insect lore:

Livo, L. J., McGlathery, G., & Livo, N. J. (1995). *Of bugs and beasts*. Englewood, CO: Teacher Ideas Press.
Carle, E. (1969). *The very hungry caterpillar*. New York: Philomel Books.
Ryder, J. (1989). *Where butterflies grow*. New York: Lodestar Books, E. P. Dutton.

Using these various resources, we can create a lesson for second graders:

- *Learning performances*: Each student will describe that most insects change from egg to adult through a process called *metamorphosis*.
- *Relationship to the driving question*: The driving question "What insects live on our playground?" will be addressed through studying metamorphosis because students may find different stages of insects on the playground at any given time (egg, larvae, pupa, or adult). In order to understand that these different stages are forms of the same insect, students must understand metamorphosis.
- *Materials needed*: J. Ryder's (1989) *Where Butterflies Grow*.
- *Instructional strategies:*
  A. Reading a children's literature book
  B. Role-playing butterfly metamorphosis
  C. Drawing pictures for artifacts
- *Time required*: 4 to 5 days
- *Cautions*: Make sure students have a clear understanding of behavioral expectations during the role-playing activity.
- *Instructional sequence*:
  A. Introducing the lesson:
    1. Read *Where Butterflies Grow*.
    2. Discuss what it might be like to change from a caterpillar to a butterfly.
    3. Discuss the stages that a butterfly goes through.
  B. Representing the content:
    1. Show students real butterfly eggs and an adult butterfly (if not available, show photographs of these stages).
    2. Ask the children if they know how the egg changes to a butterfly. Discuss the students' ideas.

    3. Have students role play the stages of metamorphosis that a butterfly goes through. *Ranger Rick's Incredible Insects* has a good example of a play in which children can participate.
  C. Establishing links to the driving question: See if students can identify other insects on their playground that are simply in different stages of metamorphosis. For example, moths, ants, flies, and bees also go through complete metamorphosis.
  D. Evaluating learning:
    1. Give students a set of photographs or drawings. Have them place them in the correct order of the metamorphosis.
    2. Have collaborative groups of students explain to others how their drawings illustrate complete metamorphosis.

In the sample lesson plan above, the teacher used direct and indirect instructional strategies: role playing and children's literature. We might design a class investigation (an experiential instructional strategy) based on the FOSS *Insects* (Lawrence Hall of Science, 1993) curriculum and the STC *The Life Cycle of Butterflies* unit (National Science Resources Center, 1992). One option for an investigation is to have student groups observe the development of a butterfly egg into an adult butterfly. As with the teacher-directed lessons, to plan this investigation we would have to think about the learning performances, relationship to the driving question, materials needed, instructional strategies, time required, instructional sequence, and cautions.

- *Learning performances*:
  A. Each student will observe and describe butterfly metamorphosis.
  B. Each student will record butterfly metamorphosis by drawing a picture and noting the date of each stage change.
  C. Each student will explain how most insects change from egg to adult through a process called *metamorphosis*.
- *Relationship to the driving question*: The driving question "What insects live on our playground?" will be addressed through studying metamorphosis because students may find different stages of insects on the playground at any given time (egg, larvae, pupa, or adult). In order to understand that these different stages are forms of the same insect, students must understand metamorphosis.
- *Materials needed*:
  A. Butterfly eggs (can be ordered from animal supply houses, such as Carolina Biological)
  B. Bug cages
  C. Drawing paper
  D. Drawing supplies (colored pencils, markers, or crayons)
  E. Magnifying glasses
  F. Caterpillar food supply (varies, depending upon the stage the butterfly is in and the type of butterfly; for example, monarch caterpillars eat milkweed)
- *Instructional strategies*:
  A. Investigation to raise butterflies
  B. Drawing pictures as artifacts
- *Time required*: 4 to 6 weeks
- *Instructional sequence*:

A. Have each group of students set up a bug cage with butterfly eggs and caterpillar food.

B. Each day, have students observe the butterfly's stage of metamorphosis, using a magnifying glass. Clean the cage daily. Add food if needed.

C. Have students draw pictures of the stages and record the date. Have them keep these drawings in their science portfolios.

D. After students have observed and recorded the complete metamorphosis of the butterfly—egg, larvae (caterpillar), pupa (chrysalis), and adult (butterfly)—have them explain how their butterfly egg changed throughout the last several weeks.

E. Ask students how this investigation helps them answer the question "What insects live on our playground?" Remind them that, when they found a caterpillar and a butterfly, these were actually the same insect. To answer the driving question, they needed to understand that the caterpillar and adult were just in different stages.

- *Cautions*: Always have children wash their hands after touching insects being raised in the classroom. Bring only known species into the room (such as painted ladies, swallowtails, or monarchs).

After students complete the class investigation, we might want groups of students to ask their own questions and design their own investigations. For instance, one group might be interested in finding out what kinds of food caterpillars eat. To answer this subquestion, students would need to plan and conduct an experiment. Students could observe other insects feeding on the playground, or they could research what insects eat. Then they could place monarch caterpillars into different containers holding different foods such as milkweed, carrot tops, grass, and other leaves. Students could determine which foods the caterpillars ate the most by inspecting the leaves or by weighing the leaves before and after the caterpillars had eaten for a day. This investigation could be tied to important concepts of metamorphosis (what animals eat during various life cycle stages), and it would teach basic skills (observation, measurement, recording, and making conclusions). Other groups of students might be interested in the life cycle of other insects, such as mealworms and waxworms. Through a series of investigations, these students could compare and contrast the life cycles of various insects.

Developing Assessments in Lesson Plans

Assessments linked to classroom activities and investigations can be used to gain a complete view of student learning. In Chapters 9 and 10, we discussed many types of assessment strategies that teachers can use to measure student understanding. For our insect project, the following assessment strategies would be effective:

- Discussion-based observations: Discuss with students what they have noticed about the changes their butterflies are going through.
- Anecdotal records: Make a notation about which students seem to be having difficulty recording daily information about their insect's growth.

- Checklists: Check off which students have used a magnifying glass to observe insects up close.
- Clinical interviews: Talk with students about what they have noticed in their investigation of the foods caterpillars eat.
- Concept maps: Have students develop maps to illustrate connections among concepts being studied in the insect project.
- Performance-based assessment: Have students design an investigation to explore life cycles of various insects.
- Student writing samples: Have students write stories about what it might feel like to be a butterfly.
- Daily journals: Have students make daily records of their observations of the life cycle of a butterfly and conclude what the pattern of cycles is for a butterfly.
- Physical products: Have students create models illustrating the life cycle of a butterfly.
- Drawings: Have students make daily drawings of each day in the life cycle of a butterfly.
- Music: Have students adapt songs to describe butterfly metamorphosis.
- Videos: Have students make a video explaining their investigation of the foods caterpillars eat and summarizing their findings.
- Multimedia documents: Have students create electronic documents illustrating the stages of butterfly growth, including graphs and photographs, of the changes a butterfly goes through.

CONNECTING TO NATIONAL SCIENCE EDUCATION STANDARDS

SCIENCE EDUCATION TEACHING STANDARDS

Teachers of science engage in ongoing assessment of their teaching and of student learning (p. 37).

Each of these assessment items can be embedded into instruction and can help teachers determine whether students have met science standards. For example, most of the aforementioned assessments could be used to show whether students know the K–4 AAAS benchmark on understanding life cycles. Many high-stakes assessments, such as state achievement tests, require students to use higher level thinking and processing skills, such as writing, analyzing data, and synthesizing information.

For example, fourth graders in one state are expected to be able to "describe the duration and timing of a pattern given a repetitive pattern in nature." Students who can make daily records of observations of the life cycle of a butterfly and conclude what the pattern of cycles is have used important skills, such as writing and synthesizing information, that can be used on the fourth-grade tests in that state.

In Learning Activity 12.3, you will develop a lesson.

Developing a Calendar of Activities

You learned earlier that teachers can use such techniques as concept mapping, listening to students, and KWL to organize lessons. Concept maps, for example, create a

Learning Activity 12.3

PLANNING, TEACHING, AND EVALUATING A LESSON

Materials Needed:

- Access to a variety of curriculum materials and resources
- Paper and pencil or computer
- A video camera
- Teaching supplies and equipment as planned

A. Using one of the lesson plan formats presented earlier, prepare a short lesson, no more than 10 or 15 minutes long, related to the driving question, concepts, learning performances, and curriculum objectives generated in Learning Activity 12.2.
B. Teach the lesson to a small group of your peers. Arrange to have a video made of your lesson.
C. For the project topic you have chosen, identify several possible assessments that could be used to measure student learning.
D. Design plans for carrying out one of the assessments. What would be included? How would it be evaluated?
E. How does the assessment help students show what they have learned about the driving question?
F. Watch yourself on video and analyze your lesson:
  1. How did you help students understand the concepts or skills?
  2. What kinds of questions did you ask?
  3. What did your classmates think of the lesson?

hierarchy that helps define the sequence of concepts students need to learn in the lessons. It is a good idea to develop a calendar to guide the sequence of lessons and orchestrate the various components in a time frame. The calendar is an *estimate* of when each lesson will take place. It almost always gets changed during the course of implementation. Table 12.5 is a sample calendar of activities for our insect project. In Learning Activity 12.4, you make a calendar for your project.

Selecting and Obtaining Resources

One important aspect of lesson and project planning involves selecting and obtaining resources to use during instruction. **Resources** include a variety of printed and electronic resources, such as textbooks, workbooks, kits of materials, slides, transparencies, charts and maps, games, audiotapes, filmstrips, videos, television, computers, CD-ROMs, DVDs, and other technologies, such as iPods. Most communities also offer a wealth of resources through local businesses, organizations, zoos, community centers, and museums. Each type of resource is used for a different reason, offers advantages, and includes disadvantages. You might think of all the available

TABLE 12.5 Insect Project Calendar of Activities

Monday	Tuesday	Wednesday	Thursday	Friday
2 Introduce driving question "What insects live on our playground?" Read Ryder's story, "Where Butterflies Grow."	**3** Set up insect cages with butterfly eggs and record daily observation of butterflies.	**4** Get on Web to find information on monarch butterflies. Record daily observation of butterflies.	**5** Take field trip to playground to observe insects. Record daily observation of butterflies.	**6** Read Carle's book, *The Very Hungry Caterpillar.* Record daily observation of butterflies. Take photographs or make drawings of insects on the playground in order to identify them.
9 Record daily observation of butterflies and weigh the gain in mass of the caterpillars. Research information about insects found on the playground.	**10** Have students design investigations to help answer "What insects live on our playground?" Record daily observation of butterflies.	**11** Set up student investigations. Record daily observation of butterflies.	**12** Record daily observation of butterflies. Listen to guest speaker on pest control.	**13** Record daily observation of butterflies. Take trip to insect exhibit at the zoo.
16 Record daily observation of butterflies. Role-play the stages the butterfly has gone through.	**17** Record daily observation of butterflies. Work on investigations. Look on Web for more information about butterflies.	**18** Record daily observation of butterflies. Listen to guest speaker—an entomologist from the local university.	**19** Record daily observation of butterflies. Work on insect computer program. Use Web to find information related to groups' investigations.	**20** Record daily observation of butterflies. Write a song to depict metamorphosis. Try to find insects in different life-cycle stages on the playground.
23 Record daily observation of butterflies. Arrange drawings of butterfly stages in portfolio.	**24** Record daily observation of butterflies. Revisit playground for additional observations.	**25** Record daily observation of butterflies. Revisit playground for observations.	**26** Record daily observation of butterflies. Prepare for multimedia presentation (artifact).	**27** Record daily observation of butterflies. Prepare for multimedia presentation (artifact).
30 Record daily observation of butterflies. Prepare for multimedia presentation.	**31** Give presentation to parents.	**1** Give presentation to parents.	**2**	**3**

Learning Activity 12.4

DEVELOPING A CALENDAR OF ACTIVITIES

Materials Needed:

A. planning calendar

Using the ideas you collected on your topic throughout this chapter, plan a calendar of activities for the project you are designing.

resources as a tool kit for teaching. Several issues come into play when planning which resources to use. In this section, we will discuss these issues.

Selecting Resources

Given the variety of resources available to a teacher, it is important to develop or identify criteria for selecting materials to use in a science classroom. Several organizations have published guidelines for selecting resources that can help teachers make good judgments about materials (see: the National Science Resources Center, Tuomi, 1993; the National Association for the Education of Young Children, Bradekamp, 1987; the National Middle School Association, 1982; Alexander & George, 1981; Wiles & Bondi, 1981; and Barnes, Shaw, & Spector, 1989). Table 12.6 summarizes the common features of these recommendations.

The FOSS *Insects* curriculum meets almost all of the criteria for selecting resources for our insect project. It spans several months, giving students the opportunity to explore the topic in-depth. Children work directly with concrete materials (insects, such as mealworms, waxworms, and butterflies) to make their own observations and conclusions. They engage in inquiry to learn about the metamorphoses of various insects. Integration is incorporated in the form of suggested children's literature, writing assignments, and illustrations the children keep. Ongoing assessments occur throughout the unit—children keep observations of their insects, teachers use checklists that focus on process skills, and alternative assessments (such as making models and drawing life cycles) are suggested. Although this unit does not use technology, it is an excellent example of a resource that stresses exploration and depth over coverage of information.

Obtaining Resources

Many types of resources are available to help teachers develop a project-based curriculum. A good classroom contains a mixture of inexpensive, everyday household

TABLE 12.6　Summary of Criteria for Selecting Resources

Criteria	Rating		
Allows children to explore a science topic in depth	Exceptional 1	Moderate 2	Poor 3
Presents the topic in a manner relevant to students' everyday lives	Exceptional 1	Moderate 2	Poor 3
Lets students engage in direct, purposeful experiences in which they can make their own observations and conclusions	Exceptional 1	Moderate 2	Poor 3
Allows children to work collaboratively	Exceptional 1	Moderate 2	Poor 3
Presents accurate information	Exceptional 1	Moderate 2	Poor 3
Actively engages students in their learning through experiences with concrete materials	Exceptional 1	Moderate 2	Poor 3
Promotes inquiry, problem solving, and critical thinking	Exceptional 1	Moderate 2	Poor 3
Uses technology as a tool to enhance learning	Exceptional 1	Moderate 2	Poor 3
Helps children learn how to learn (establishes a foundation for lifelong learning)	Exceptional 1	Moderate 2	Poor 3
Develops children's self-esteem, sense of competence, and positive feelings toward learning science	Exceptional 1	Moderate 2	Poor 3
Responds to children's individual differences in ability, development, and learning styles (the curricula are novel and varied and use a variety of instructional strategies)	Exceptional 1	Moderate 2	Poor 3
Integrates science across subject areas	Exceptional 1	Moderate 2	Poor 3
Engages children in discussions and conversation that challenge their thinking and help them construct understanding	Exceptional 1	Moderate 2	Poor 3
Stresses such skills as observing, measuring, hypothesizing, and predicting	Exceptional 1	Moderate 2	Poor 3
Uses portfolios, practical assessment, or other forms of alternative assessment	Exceptional 1	Moderate 2	Poor 3
Provides opportunities for physical movement	Exceptional 1	Moderate 2	Poor 3
Stresses exploration and depth over coverage of information	Exceptional 1	Moderate 2	Poor 3

materials, as well as commercial and noncommercial materials, magazines, and books. Although there are numerous sources for obtaining materials, there are a few ideal places to start your search. In the following sections, we will discuss in more detail how to obtain a variety of resources, how to use the community as a resource, and how to use the Web to search for resources.

By exploring the Eisenhower National Clearinghouse, you can identify both commercial and noncommercial resources. This clearinghouse, funded by the U.S. Department of Education, serves to improve access to mathematics and science resources for teachers, students, and parents. The Clearinghouse collects the most up-to-date materials and catalogs them. The materials can be accessed on the Web at http://www.goENC.com/. The Clearinghouse is a free resource, and it can save you time.

Commercial Suppliers

The National Science Teachers Association publishes an annual supplement to its journals entitled *NSTA Science Education Suppliers*. This publication is packed with information about commercial companies that sell equipment and supplies (such as microscopes, slides, dissecting kits, tuning forks, and hand generators), software (such as CD-ROMs and DVDs), media (such as maps, games, and kits), curriculum programs (such as FOSS and STC), and tradebooks (such as children's literature). The publication includes addresses, telephone numbers, Web pages, and e-mail addresses of suppliers. It summarizes the type of equipment carried by each company and categorizes it in a number of ways, such as by science subject and grade level. You can obtain this publication by either subscribing to the NSTA journals (*Science and Children, Science Scope,* or *Science Teacher*); contacting NSTA at the National Science Teachers Association, 1840 Wilson Boulevard, Arlington, VA 22201-3000; or browsing the resource section on their Web site (http://www.nsta.org).

There is no shortage of ideas for teaching science. Teachers' resource books are packed with ideas for hands-on activities that make good lessons. Many include pages that can be duplicated for children. For example, teacher resource books, such as *The Pillbug Project* (Burnett, 1999), can be purchased online at the National Science Teachers Association Web page. This book focuses on the study of the pillbug or sowbug and crustaceans and provides a useful resource to contrast crustaceans with insects.

Noncommercial Suppliers

Noncommercial suppliers include nonprofit groups, such as the Audubon Society, businesses, such as utility companies, and science parks, such as Sea World. These types of suppliers frequently distribute free or inexpensive materials. Most of these materials are educationally sound and useful resources for the classroom. However, some businesses create materials for the purpose of conveying the value of their particular products or services. For example, a local utility company might publish free materials on the benefits of nuclear energy. Most of these materials are accurate, but you should watch out for materials that present biased opinions. Sometimes materials can be void of valid scientific evidence, so teachers need to help students recognize the quality and validity of these materials. One example of a high quality Web site moni-

TABLE 12.7   Criteria for Using Noncommercial Resources

Criteria	Rating		
Furthers my learning performances	Exceptional 1	Moderate 2	Poor 3
Is free of objectionable advertising, propaganda, or bias	Exceptional 1	Moderate 2	Poor 3
Is scientifically accurate	Exceptional 1	Moderate 2	Poor 3
Is interesting	Exceptional 1	Moderate 2	Poor 3
Is intellectually stimulating	Exceptional 1	Moderate 2	Poor 3
Is beneficial for children and classroom use	Exceptional 1	Moderate 2	Poor 3
Meets criteria for selecting resources (Table 12.6)	Exceptional 1	Moderate 2	Poor 3

tored by scientists is *Earth and Sky: A Clear Voice for Science* (http://www.earthsky.org/). This Web site has numerous features, including daily radio broadcasts; a blog; and links to numerous topics, including animals, body and mind, weather and climate, earth, oceans, plants, and space. One way that teachers can help students evaluate free materials is by examining the sponsors. In the case of *Earth and Sky: A Clear Voice for Science*, the sponsors are respected organizations, such as the USDA Forest Service, National Science Foundation, National Space Grant Foundation, National Fish and Wildlife Foundation, US Fish and Wildlife Service, National Park Service, National Oceanic and Atmosphere Administration, National Aeronautics and Space

TABLE 12.8   National Noncommercial Suppliers

Organization	Resources provided
National Wildlife Federation	*Nature Scope* series of books *Ranger Rick* magazine *My Big Backyard* magazine
Project Wild (5430 Grosvenor Lane, Bethesda, MD 20814, 301-493-5447)	*Project WILD Aquatic* *Project Learning Tree*
NASA Spacelink	Slides, photographs, videotapes, lesson plans
Project WET (201 Culbertson Hall, Montana State University, Bozeman, MT 59717-0057, 406-994-5392)	*Project WET*
TERC	*Hands On!* magazine Papers Innovative projects

Learning Activity 12.5

SELECTING GOOD RESOURCES

Materials Needed:

- Table 12.6 and Table 12.7

A. Find resources that you might use for lesson ideas for the project topic that you have been developing throughout this chapter.
B. Using the criteria listed in Tables 12.6 and 12.7, rate the quality of the resources.
C. Keep track of items you could use and their costs.

Administration, United States Geological Survey, and the Environmental Protection Agency. Table 12.7 shows criteria for evaluating free and inexpensive materials provided by noncommercial suppliers.

Table 12.8 lists some nationally recognized noncommercial suppliers. You may also want to contact local utility companies, fire departments, police departments, zoos, aquariums, botanical gardens, museums, departments of natural resources, environmental organizations, and businesses for science-related information and materials in your region.

Everyday Household Materials

Everyday household materials, such as cotton balls, cups, vinegar, and baking soda, are used in numerous elementary and middle grade investigations and activities. For example, for our insect unit, food supplies and bug cages can be purchased at pet stores. Hardware stores sell nylon screen and wood for building simple habitats. Grocery stores sell jars and food supplies that can be used for investigations on insects. Learning Activity 12.5 asks you to identify local resources you could use in your project.

The Community as a Resource

Community resources provide valuable sites for field trips, are sources of educational materials, and create opportunities for experiences. Besides providing a change of pace, community resources show children how science is relevant to their daily lives. For example, for our insect project, a local public health official or nurse would be a good resource for talking about health hazards (such as Lyme disease) associated with insect bites. Science museums provide students with hands-on displays, interactive experiences, and special programs. Other community resources are local zoos,

CONNECTING TO
NATIONAL SCIENCE
EDUCATION
STANDARDS

SCIENCE EDUCATION
TEACHING STANDARDS

Teachers consider their own strengths and interests and take into account available resources in the local environment (p. 31).

science centers, parks, hospitals, police stations, courts, radio stations, universities, and businesses. Many teachers find that they can use all of their children's parents as community resources. A parent doesn't need to be a doctor or professional to be a resource. A mother who grows houseplants may be an expert in this area as much as any other community member. For the insect project, we could visit the zoo to see insects that are nonindigenous to our region.

Technology has, in many ways, expanded our definition of community. Numerous World Wide Web sites let us visit sites and talk with people around the world. Google Earth (http://earth.google.com/), for example, combines satellite imagery and maps to put the world's geographic information at your fingertips. The GLOBE Project (Global Learning and Observations to Benefit the Environment) is an exciting hands-on project for elementary and secondary students (see http://www.globe.gov) that gives students opportunities to take scientific measurements, report data to others on the Web, publish their projects, and collaborate with scientists and others around the world, thereby forming a large community.

The community can also be a wonderful network for providing teachers with free materials. Although schools usually provide teachers with small budgets for purchasing materials, many school budgets are very tight and teachers frequently need to identify creative alternatives for stocking their classrooms. Many schools have joined a trend toward collaborative alliances and partnerships with local businesses. These partnerships provide many advantages: Local businesses become engaged in education, businesses provide schools with needed supplies and materials, and students see the relevance of what they are learning to local businesses. If your school does not have a partnership with local businesses, you might want to consider asking local businesses for donations. Many teachers do this by writing simple letters. A personal follow-up in the form of a telephone call or visit can often secure a donation. Check to make sure your school district's policies allow you to seek donations. Figure 12.5 shows a sample letter requesting donations from businesses.

Integrated Curriculum

Curriculum integration has become popular among educators. However, even at the elementary level, it is common for students to move from subject to subject and learn topics in a fragmented, disconnected fashion that bears little resemblance to real life. This frequently leads students to be bored by and overloaded with information. It seems only logical that subject areas should not be separated in schools because they are not separated in the world. Paul DeHart Hurd (1991), in an article entitled "Why We Must Transform Science Education," wrote,

Dear Local Grocery Store Owner:

I am a fifth-grade teacher at Kenwood Elementary School, and I am very interested in improving the science skills of my students. As a member of our community and a business leader, you are, I am sure, also concerned about the quality of our students' science education. Most science activities in the elementary grades require only simple household items. However, with limited financial resources at our school, we cannot purchase these items. I am asking that you consider donating to our school the items listed below. These materials will be used in a series of lessons to teach students about insects.

10 aluminum pans
5 lbs. of flour
rolls of netting
10 plastic jars
10 sets of mixing spoons
10 measuring cups
2 bags of potting soil
1 bag of cornmeal
5 flashlights
100 Styrofoam cups

I will call you in a week to see if you are interested in meeting to discuss this donation. Thank you for your interest in our students' educational needs.

Sincerely,

Teacher's Name

Figure 12.5   Letter requesting donations from local businesses.

> Science today is characterized by some 25,000 to 30,000 research fields. Findings from these fields are reported in 70,000 journals, 29,000 of which are new since 1978. Traditional disciplines have been hybridized into such new research areas as biochemistry, biophysics, geochemistry, and genetic engineering.... These changes in the way modern science is organized have yet to be reflected in science courses. There is little recognition that in recent years the boundaries between the various natural sciences have become more and more blurred and major concepts more unified.

Hurd stressed the need for greater integration of school subjects and the integration of science with social issues, technology, and other school subjects.

Consistent with Hurd's (1991) recommendations, most national reform efforts currently stress the need to integrate the curriculum (Czerniak, 2007; International Reading Association, 1996; National Council of Teachers of English, 1996; National Council for the Social Studies, 1994; National Council of Teachers of Mathematics, 1989, 2000; National Research Council, 1996). Integration is a major focus in such science reform initiatives as *Science for All Americans* (Rutherford & Ahlgren, 1989) and the *National Science Education Standards* (NRC, 1996). Curriculum integration is also stressed by the National Association for the Education of Young Children

(NAEYC), an organization that specializes in instructional practices appropriate for the education of the young child, and the National Middle School Association (NMSA), an association that focuses on young and early adolescents.

Howard Gardner (1983, 1993, 1999) challenged our conventional thinking about intelligence by identifying eights forms of intelligence, or **multiple intelligences**: mathematical–logical, linguistic, spatial, bodily–kinesthetic, musical, interpersonal, intrapersonal, and naturalist. In classrooms, learners use multiple intelligences to execute complex tasks. Armstrong (1994) suggested that students become active learners when they use most of their intelligences. Teachers now use Gardner's multiple intelligence theory to support multidimensional instruction, integrated curriculum, and assessments that match students' strengths. Gardner described the eight intelligences as follows:

- **Linguistic intelligence** is characterized by the capacity to use language (written or oral). Students with linguistic intelligence learn best by speaking, hearing, and seeing. Science teachers capitalize on this by using such strategies and techniques as debates, writing, taperecordings, and books.
- **Logical-mathematical intelligence** uses numbers and logic. Students with this intelligence solve puzzles, explore patterns and relationships, work with numbers and data, use logical reasoning, classify, and use time sequencing. Science teachers have these students manipulate numbers, find patterns in data, and make conclusions based on data.
- People with **visual-spatial intelligence** think in terms of physical space. They use graphic images, drawings, and pictures. They learn best by visualizing objects and working with colors, photographs, visual diagrams, concept maps, charts, and 3D models.
- Students with **musical intelligence** are sensitive to sound, pitch, and rhythm. They like to sing, hum, listen to music, and play an instrument. Teachers help them learn science through songs and jingles, by writing songs, and by listening to the radio and CDs.
- **Body-kinesthetic intelligence** involves the use of movement in learning. They like to make things, touch objects, talk, use body language, dance, act, and participate in sports. Teachers help them learn science through acting and role plays, demonstrating to others, and using physical games.
- **Interpersonal intelligence** is the ability to understand and interact with other people. People with interpersonal intelligence tend to have many friends, talk a great deal, and understand others. Science teachers engage these students in collaborative learning, interviews, and activities in which they can write to or talk with others.
- **Intrapersonal intelligence** is the ability to understand oneself. Students with strong intrapersonal intelligence like to work alone and explore their own interests. They are focused inward on dreams and goals. They work well on self-paced, individualized projects. Science teachers frequently have them keep personal journals or work on independent=study projects.
- **Naturalist intelligence** is a person's ability to identify and classify patterns in nature. People who are sensitive to changes in weather patterns or who are good at seeing relationships and patterns among plants and animals may be expressing naturalist intelligence abilities.

Many schools have adopted integrated curriculum techniques in response to Howard Gardner's theory of multiple intelligences. The process of blending a topic across

TABLE 12.9 Strategies to Adapt Teaching Activities to Meet Students' Multiple Intelligences

Linguistic intelligence	Logical–mathematical intelligence	Visual–spatial intelligence	Musical intelligence	Body–kinesthetic intelligence	Interpersonal intelligence	Intrapersonal intelligence	Naturalist intelligence
Tape-recordings	Formulas	Drawings and paintings	Songs, raps, jingles	Field trips; Team tasks	Reflective journals	Examining patterns in nature	
Biographies	Puzzles	Photos	Choral reading	Role playing	Collaborative learning	Self-evaluation	Outdoor education
Poetry	Timelines	Storyboards	Writing songs	Sports and physical games	E-mail and chat-room discussions	Independent studies	Nature walks
Books	Venn diagrams	Props and scenes for plays	Tone and rhythmic patterns	Simulations	Group projects	Personal goal setting	Photographing nature
Journals	Data analysis	Posters	Background music for meditation or concentration	Dances	Group and class discussions	Problem solving	Outdoor field trips; trips to the zoo, museums, etc.

TABLE 12.10   Sample Multiple Intelligences Lessons on Insects

Linguistic intelligence	Logical–mathematical intelligence	Visual–spatial intelligence	Musical intelligence	Body–kinesthetic intelligence	Interpersonal intelligence	Intrapersonal intelligence	Naturalist intelligence
Read a book about insects.	Use measurements and data to compare and contrast insects found in different environmental areas.	Draw pictures of various insects.	Listen to a song about insects or go to the children's section of a local music store for songs.	Take a field trip to learn more about insects.	Work with team member to learn more about insects.	Keep a reflective journal about insect findings.	Examine patterns found in insect environments.
Discuss a book with others.	Design a puzzle about insect taxonomy.	Photograph insects in their natural habitats.	Participate in a choral reading about insects.	Role play insect movement.	Design a collaborative learning investigation on insects.	Conduct a self-evaluation on what has been learned about insects.	Learn how insects are adapted to survive in their environment.
Write a play about insects.	Keep a timeline documenting insect activity.	Create a storyboard depicting insect metamorphosis.	Write lyrics about insects to the tune of a popular song.	Design and play a physical game to depict an insect concept.	Share data on insects with others over e-mail.	Design and complete an independent study to learn more about a particular insect.	Take a nature walk around the school to look for insects.
Listen to an entomologist guest speaker.	Compare and contrast insects with other animal classifications.	Design props for a play about insects.	Note rhythmic patterns in nature.	Simulate insect metamorphosis.	Work with others on a group project on insects.	Set a personal goal for overcoming fear of insects.	Photograph insects in nature.
Write a report on insects.	Analyze data collected on insect observations throughout the school year.	Design a poster to display findings related to an insect investigation.	Play background music while students are investigating insects.	Create a dance to teach a concept related to insects.	Have a whole-class discussion about insect investigation.	Figure out how to safely control insects in a garden.	Go to the zoo exhibit on insects.

different subject areas, such as science, art, music, and drama, helps teachers engage students' multiple intelligences. In addition, technology tools facilitate the use of many different intelligences. When students create a multimedia document, for example, they might use logical thinking to organize the document, visual-spatial skills to design the project, music to add special effects, charts and data to present an argument, and written words to tell a story.

Table 12.9 illustrates how a teacher might plan activities on the topic of insects, using Howard Gardner's multiple intelligences theory. Table 12.10 shows how a teacher might adapt such activities to a project on insects.

Research on females and minorities in science indicates that an integrated focus helps spark interest in science (Bianchini, Cavazos, & Helms, 2000). Teachers need to emphasize how technology and science can be integrated with other fields with a human interest, such as ecology, health, and sociology (National Council for Research on Women, 2001). Haberman (1995) also stressed the importance of an integrated project approach related to real life for working with children in poverty. Barton and Yang (2000) stressed that teachers need to present science in such a way that encourages all students to participate in science. Teachers need to integrate science into students' daily lives and help students learn science in nontraditional ways that meet their interests and cultures (Barton, 1998). For example, Western science does not traditionally recognize traditional folk medicine, yet long before the discovery of aspirin in 1853, Hispanic families knew that that tea made from the flower of the daisy fleabane could be used to treat aches and pains (Meyer-Monhardt, 2000). These flowers contain acetylsalicylic acid (aspirin). Educators also stress the importance of culturally relevant curriculum for American Indian students. Davison and Miller (1998) wrote that curriculum can be made culturally relevant for American Indian students by presenting it in an integrated fashion. An integrated approach that focuses on cultural understanding can help minority students become engaged in science.

Before we continue discussing the integrating of school subject areas, think about your beliefs about curriculum integration with the help of Learning Activity 12.6.

> ## CONNECTING TO NATIONAL SCIENCE EDUCATION STANDARDS
>
> ### SCIENCE EDUCATION TEACHING STANDARDS
>
> Schools must restructure schedules so that teachers can use blocks of time, interdisciplinary strategies, and field experiences to give students many opportunities to engage in serious scientific investigation as an integral part of their science learning (p. 44).

The Definition of Integration

Sometimes educators use the terms *integrated, interdisciplinary,* and *thematic* synonymously. Lederman and Niess (1997) defined **integrated** as a blending in which the separate parts are not discernible. They use the metaphor of tomato soup; you cannot discern the tomatoes in the soup. They define **interdisciplinary** as a mixture of subjects connected, but still identifiable. The metaphor they use is chicken noodle

Learning Activity 12.6

INVESTIGATING YOUR BELIEFS ABOUT CURRICULUM INTEGRATION

Materials Needed:

- An elementary or middle grade science textbook
- Pencil and paper

A. Examine a fourth-grade science textbook for a selected topic. A traditional textbook will usually contain such topics as the human body, electricity, magnetism, sound, light, animals, plants, machines, the earth's crust, volcanoes, earthquakes, and the solar system.

B. Form a team with four other classmates and assign yourselves the roles of particular subject matter specialists (science teacher, mathematics teacher, social studies teacher, language/reading teacher, and art teacher). Take the topic selected and, together, design a set of integrated lesson ideas for teaching the topic.

C. After you have finished planning, individually critique the lessons, using the following criteria. This should establish what you personally believe about integrated planning. Then, have your group critique the lesson ideas using the same criteria. Make sure your group comes to a consensus.

1. What prior knowledge and experiences do students need before engaging in integrated lessons?
2. Will important content and inquiry objectives be met? If not, what is missing? Explain.
3. If a teacher teaches this way for the entire school year, will there be important topics that are not covered? Standards that are not met?
4. Are the learning performances watered down or less meaningful in these lessons? Explain.
5. Should the topic stay as it is in the textbook (taught separately by subject matter)? Why or why not?
6. Is curriculum integration beneficial? Why or why not?
7. Do you know enough about each subject area to teach this way? Explain.

soup; it is a soup, but you can still recognize the broth, chicken, and noodles. Similarly, Jacobs (1989) defined *interdisciplinary* as "a knowledge view and curriculum approach that consciously applies methodology and language from more than one discipline to examine a central theme, issue, problem, topic, or experience." Finally, Lederman and Niess (1997) defined **thematic** as a unifying topic or subject transcending traditional subject boundaries. For purposes of this book, we use the word *integrated* to define the crossing of subject matter boundaries.

Beane (1995) suggested that curriculum integration, like project-based science, begins with "problems, issues, and concerns posed by life itself" (p. 616). Integrated curriculum, according to Beane (1996), has four characteristics: (a) it is organized around problems and issues that have personal and social significance in the real world, (b) it uses pertinent knowledge in the context of topic without regard for subject lines, (c) it is used to study current problems, rather than for a test or grade level outcome, and (d) it emphasizes projects and activities with real application of knowledge and problem solving. Hopkins (1937, as cited in Beane, 1996) defined *integration* similarly—as cooperatively planned, problem centered, and integrated knowledge. These definitions of curriculum *integration* are consistent with our view of project-based learning as crossing curricular subject areas.

How Project-Based Science Supports Curriculum Integration

Rakow and Vasquez (1998) wrote, "Project-based integration may be the most authentic form of cross-curricular integration because it involves students in real-world learning experiences. In project-based integration, students investigate real issues in real contexts." Project-based science's key features (driving questions, student engagement in investigations, communities of learners collaborating together, use of technology, and creation of artifacts) are all congruent with curriculum integration. As students answer driving questions, they develop deeper understandings because they make connections among the central concepts of a variety of subject areas. Driving questions

> ### CONNECTING TO NATIONAL SCIENCE EDUCATION STANDARDS
>
> ### SCIENCE EDUCATION CONTENT STANDARDS
>
> The standard for unifying concepts and processes is presented for grades K–12, because the understanding and abilities associated with major conceptual and procedural schemes need to be developed over an entire education, and the unifying concepts and processes transcend disciplinary boundaries (p. 104).

are contextualized; they are anchored in the lives of learners and deal with important, real-world issues. Real-world questions are not separated into different subject areas.

Integration builds understanding of concepts. When science topics involve people, as they usually do, social studies concepts, such as economics, politics, culture, and history, come into play. Teachers can use children's literature of all types (realistic fiction, historical fiction, fantasy, plays, newspapers, science fiction, traditional or classical literature, poetry, tradebooks, and biographies) to teach about science topics. Many science investigations make use of the arts: painting, sketching, drawing, collages, sculpture, drama, role playing, pantomime, charades, skits, movies, puppets, improvisation, music, songs, instruments, and jingles. Integrating mathematics into science allows students to use computation, measurement, ratios, proportion, graphing, and geometry in their investigations. For example, students studying the question "Where does all the garbage go?" will study such science concepts as decomposition and pollution. However, the question does not fit neatly into the school subject of science. The question flows over to social studies because it involves laws, regulations, ethics, value judgments, and decisions. Mathematics comes into play when the amount of garbage is extrapolated for every human on earth over the next few years at present rates. Language, reading, and communication are used to find information, debate issues, discuss possible solutions to our garbage problem, and communicate findings.

When students engage in investigations in project-based science, they use many skills that cross the curriculum. For example, they ask questions, look for relationships, organize procedures, consider multiple factors, take notes, use reference materials, record data, summarize information, interpret data, formulate conclusions, and communicate results. These skills are also used in other subject areas, such as mathematics, language arts, and social studies.

Project-based science emphasizes communities of learners collaborating together. Such collaboration allows students to integrate understandings from a variety of careers and subject areas. For example, students investigating the topic of insects might interact with an entomologist from a local university to learn about biological research on insects, cultural norms regarding insects, and mathematical models associated with insect population control.

Use of technology is an integral component of project-based science. Technology is rarely separated along subject-area lines. Software programs frequently include digital photographs, sound, and music, for example. Many provide students with opportunities to read and write. Some include mathematics concepts and social studies concepts. For example, a popular software program called *Great Ocean Rescue* (Tom Snyder Productions, 1994) integrates the sciences and tackles social issues related to oceans, such as pollution and coral-reef destruction. The Web is not separated into subject areas; topics are organized, as they are in the real world, around themes. For example, if you were to conduct a Web search on acid rain, you would uncover a host of subjects: coal mining, air pollution laws, industrial regulations, sulfur in coal, and mutations in frogs.

CONNECTING TO NATIONAL SCIENCE EDUCATION STANDARDS

SCIENCE EDUCATION TEACHING STANDARDS

Individual and collective planning is a cornerstone to science teaching; it is a vehicle for professional support and growth. In the vision of science education described in the *Standards*, many planning decisions are made by groups of teachers at grade and building levels to construct coherent and articulated programs within and across grades (p. 32).

As students create artifacts, they use all types of knowledge and skills from other subject areas, such as music, art, mathematics, and language arts. Students might develop posters or make models from recycled materials. They might write songs, write reports of their results, or give oral presentations. Students might mix written work, sound, video, and graphics in hypermedia artifacts.

A Word of Warning

Lonning and DeFranco (1997) argued that integration can be justified only when connecting subjects enhances the understanding of them. In other words,

TABLE 12.11 Rubric for Evaluating Integrated Curriculum

Project topic: _____
Rating scale   4 = Strong   3 = Adequate   2 = Weak   1 = No evidence

National standards in science are followed (for grades 5–8)

4 3 2 1   The activities support unifying concepts and processes (systems, order, and organization; evidence, models, and explanation; constancy, change, and measurement; evolution and equilibrium; and form and function).

4 3 2 1   The activates support scientific inquiry (abilities to do scientific inquiry and understand scientific inquiry).

4 3 2 1   The activities develop understanding in physical science (properties and changes of properties in matter, motions and forces, and transfer of energy), life science (structure and function in living systems, reproduction and heredity, regulation and behavior, populations and ecosystems, diversity and adaptations of organisms), and/or earth-space science (structure of the earth system, earth's history, and earth in the solar system).

4 3 2 1   The activities connect science with technology (students have understandings about science and technology, and they have abilities of technological design).

4 3 2 1   Science is presented as it relates to personal and societal perspectives (personal health; populations, resources, and environments; natural hazards; risks and benefits; and science and technology in society).

4 3 2 1   The history and nature of science are presented to students (science as a human endeavor, nature of science, and history of science).

National standards in mathematics are followed

4 3 2 1   The activities support mathematical communications.
4 3 2 1   The activities support mathematical connections.
4 3 2 1   The activities support mathematical problem solving.
4 3 2 1   The activities support mathematical reasoning.

National standards in language arts/reading are followed

4 3 2 1   Students have opportunities to read a variety of print and nonprint materials.
4 3 2 1   Students are able to write for a variety of purposes.
4 3 2 1   Students are able to adjust their spoken language for a variety of audiences.
4 3 2 1   Students use the language arts—reading, writing, listening, and speaking—to nurture their learning through research.

National standards in social studies are followed

4 3 2 1   The activities provide for the study of culture and cultural diversity.
4 3 2 1   The activities provide for the study of time, continuity, and change.
4 3 2 1   The activities provide for the study of people, places, and environments.
4 3 2 1   The activities provide for the study of individual development and identity.
4 3 2 1   The activities provide for the study of individuals, groups, and institutions.
4 3 2 1   The activities provide for the study of power, authority, and governance.
4 3 2 1   The activities provide for the study of production, distribution, and consumption.
4 3 2 1   The activities provide for the study of science, technology, and society.
4 3 2 1   The activities provide for the study of global connections.
4 3 2 1   The activities provide for the study of civic ideas and practices.

Lessons include connections across the curriculum

4 3 2 1   The lessons integrate science, mathematics, social studies, and/or language arts.
4 3 2 1   The learning of concepts and skill is enhanced because of the connections made across the curriculum.
4 3 2 1   The unit allows students to see one subject from the viewpoint subject (multiple perspectives).

Source: This rubric is modified from various professional standards (NRC, 1996; NCSS, 1994; NCTE-IRA, 1996; & NCTM, 1989, 2000) and was used in a Professional Development Grant entitled PRISM-CLASS (Project for Integrating Science and Mathematics Curriculum with Language Arts and Social Studies) at the University of Toledo.

teachers should not force integration for the sake of integration. We have seen teachers create integrated units of study around such topics as "teddy bears." Although this topic may encourage reading and writing in language arts, it provides little opportunity for students to engage in meaningful science learning. Good driving questions will help you to integrate curriculum in a meaningful way.

Another way to judge the worth of curriculum integration is to use the rubric in Table 12.11, which lists standards of the *National Science Education Standards* (NRC, 1996), the National Council for the Social Studies (NCSS, 1994), the National Council of Teachers of English/International Reading Association (1996), and the National Council of Teachers of Mathematics (1989, 2000).

An Example of Curriculum Integration

The driving question "What kinds of insects live in our neighborhood?" offers many opportunities for teachers to integrate the curriculum:

- In science, students can study camouflage, metamorphosis, classification, body structure, and predator–prey relationships.
- Mathematics can come into play in several ways. Students might investigate the area of an insect's territory. They might graph the number of insects found in different areas of the playground or graph the weight of the insects during different stages of development. They might use ratio and proportion to contrast the weight of an ant with the mass it can carry. Students might calculate the length of tunnels that ants build. They might explore the geometric shapes found in nature's insect populations or find the average number of days it takes for insects to complete metamorphosis.
- Social studies are involved when students learn about the historical impact of insects (such as locusts) and the development of insecticides to limit insect populations. Students might study the cultural behaviors (such as eating insects) of people in other countries.
- Language arts are involved when students write reports, communicate findings, give each other feedback, and read and discuss insect-related stories. Some stories that could be used in an insect science–language arts activity project include Livo et al.'s *Of Bugs and Beasts*, Carle's *The Very Hungry Caterpillar*, and Ryder's *Where Butterflies Grow*.

Concept Mapping to Plan Integrated Projects

Earlier in this chapter, we discussed concept mapping as a way to identify and organize important concepts in a project. Concept mapping can also be used to organize the integration of a project across the curriculum.

A teacher in a *self-contained classroom* (with one group of students all day) can use concept mapping to develop lessons across subject areas. Teachers in **departmentalized** (separated by subject area) schools can use concept mapping to plan together to integrate a theme or topic across their disciplines. Frequently intermediate and middle schools (usually grades 4–8) purposefully place teachers in teams so they can plan

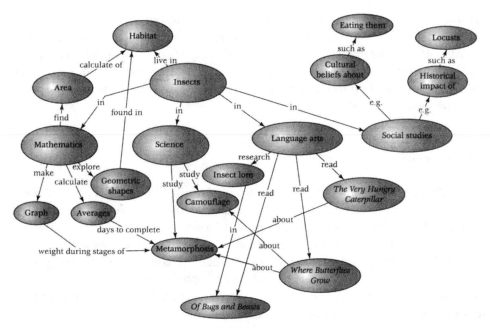

Figure 12.6   Concept map for science integration.

integrated lessons together. Regardless of how you plan integrated lessons, you will find that using an integrated approach provides an exciting way to teach science that is consistent with current reform efforts in science education. However, few people

---

Learning Activity 12.7

DEVELOPING INTEGRATED CURRICULUM

Materials Needed:

- The National Council of Teachers of Mathematics (2000) standards
- The National Research Center (1996) science standards
- The International Reading Association (1996) standards
- The National Council of Teachers of English (1996) standards
- The National Council for the Social Studies (1992) standards
- The AAAS (2001, 2007) *Atlas of Science Literacy*

A. Compare and contrast the science concepts and skills that need to be covered in the project you have been developing throughout this chapter with important content in the NCTM, NRC, IRA, NCTE, and NCSS standards.
B. What reading, language arts, mathematics, and social studies content might be covered in this project?
C. To the concept map you developed earlier, add integrated reading, language arts, social studies, and mathematics content. You might also add art, music, and drama content to your map.
D. Outline several lesson ideas that can be integrated with mathematics, social studies, reading, and language arts.

are knowledgeable in all subject areas. An excellent resource to help teachers grow professionally by learning more about topics that can be included in a concept map is the *Atlas of Science Literacy* (AAAS, 2001, 2007). Figure 12.6 is a sample concept map for the integration of science with language arts, reading, mathematics, and social studies in our insect project. Learning Activity 12.7 asks you to integrate the project you have developed with other disciplines.

Chapter Summary

In this chapter, we explored lesson and project planning. Planning is a critical aspect of developing a project environment. This chapter focused on several steps that teachers typically take to plan lessons and a project-based curriculum.

Throughout the chapter, we developed an insect project. First, we identified concepts and selected learning performances. We matched concepts and learning performances to local, state, and national standards. We used concept mapping to organize the concepts. Next, we discussed ways that teachers can develop a driving question, lessons, assessments, and sequence of activities.

Second, we examined some ways to locate commercial, noncommercial, household, and community resources because they are important in any project. We also presented techniques to judge the quality of resources.

Finally, we discussed the reasons for the natural support of curriculum integration in a project-based environment. Once again, we used concept mapping as a planning tool for integrating the curriculum.

Chapter Highlights

- Lessons must be carefully planned to be successful.
- Lesson plans are a road map to teaching.
- The 5-E model is a constructivist model for planning lessons.
- Teachers decide the organization of lessons by:
  - Using concept maps to evaluate what students understand.
  - Observing and listening to students.
  - Using the KWL strategy to probe what students know, want to know, and have learned.
- Many teachers like to use a concept map to organize a project's activities, because the hierarchy in the concept map helps define the sequence of concepts students need to learn in lessons.
- There are many factors that influence teachers' decisions about the concepts and learning performances to be covered in a project:
  - State and national standards.
  - Adopted curricula.
  - Students' interests and culture.
- After identifying concepts and selecting learning performances, teachers brainstorm several driving questions that will allow students to explore the concepts selected.

- Teachers develop lessons to help students learn important questions that fit with the driving question.
- Various assessments that are linked to classroom activities and investigations can be used to gain a complete view of student learning.
- A calendar, although only an estimate of time, helps to guide the sequence of lessons and orchestrate the various components in a timeframe.
- Teachers use many resources to help plan lessons:
  - Commercial suppliers
  - Noncommercial materials
  - Everyday household materials
  - Community resources
- Curriculum integration is the crossing of curricular subject areas.
- Curriculum integration has become a popular reform movement for several reasons:
  - It resembles real life more closely.
  - National standards are calling for curriculum integration.
  - Integration helps teachers teach to children's multiple intelligences.
  - Curriculum integration has been found to help girls and minorities learn science.
- Project-based science supports curriculum integration because its key features (driving questions, student engagement in investigations, communities of learners collaborating together, use of technology, and creation of artifacts) are all congruent with curriculum integration.
- Teachers do not integrate for the sake of integrating. Rather, they integrate across subjects to enhance learning for all students.
- Concept mapping is a useful tool for planning an integrated curriculum project.

Key Terms

Analyze
Apply
Body–kinesthetic intelligence
Conceptual knowledge
Concept maps
Curriculum integration
Departmentalized
Elaboration
Engagement
Evaluate
Evaluation
Explanation

Exploration
Factual knowledge
5-E model
Integrated
Interdisciplinary
Interpersonal intelligence
Intrapersonal intelligence
KWL strategy
Learning performances
Lesson plan
Linguistic intelligence
Listening

Logical–mathematical intelligence
Metacognitive knowledge
Multiple intelligences
Musical intelligence
Naturalist intelligence
Observing
Procedural knowledge
Resources
Remember
Thematic
Understand
Visual–spatial intelligence

References

Alexander, W. M., & George, P. S. (1981). *The exemplary middle school*. New York: Holt, Rinehart and Winston.

American Association for the Advancement of Science. (1993). *Benchmarks for science literacy*. New York: Oxford University Press.

American Association for the Advancement of Science. (2001). *Atlas of science literacy*. Washington, DC: American Association for the Advancement of Science and National Science Teachers Association Press.

American Association for the Advancement of Science. (2007). *Atlas of science literacy, Volume 2*. Washington, DC: Author.

Anderson, L. W., & Krathwohl, D. R. (Eds.). (2001). *A taxonomy for learning, teaching, and assessing: A revision of Bloom's taxonomy of educational objectives*. New York: Longman.

Armstrong, T. (1994). *Multiple intelligences in the classroom*. Alexandria, VA: Association for Supervision and Curriculum Development.

Barnes, M. B., Shaw, T. J., & Spector, B. S. (1989). *How science is learned by adolescents and young adults*. Dubuque, IA: Kendall/Hunt.

Barton, A. C. (1998). *Feminist science education*. New York: Teachers College Press.

Barton, A. C., & Yang, K. (2000). The culture of power and science education: Learning from Miguel. *Journal of Research in Science Teaching, 37*(8), 871–889.

Beane, J. (1995). Curriculum integration and the disciplines of knowledge. *Phi Delta Kappan, 76*, 616–622.

Beane, J. (1996). On the shoulders of giants! The case for curriculum integration. *Middle School Journal, 28*, 6–11.

Bianchini, J. A., Cavazos, L. M., & Helms, J. V. (2000). From professional lives to inclusive practice: Science teachers and scientists' views of gender and ethnicity in science education. *Journal of Research in Science Teaching, 37*(6), 511–547.

Bradekamp, S. (1987). *Developmentally appropriate practice in early childhood programs serving children from birth through age 8*. Washington, DC: National Association for the Education of Young Children.

Braus, J. (Ed.). (1989). *Ranger Rick's nature scope: Incredible insects*. Washington, DC: National Wildlife Federation.

Burnett, R. (1999). *The pillbug project: A guide to investigation*. Washington, DC: National Science Teachers Association.

Cleary, B. (1983). *Dear Mr. Henshaw*. New York: Harper Collins.

Czerniak, C. M. (2007). Interdisciplinary science teaching. In S. K. Abell & N. G. Lederman (Eds.), *Handbook of research on science education*. Mahwah, NJ: Lawrence Erlbaum.

Davison, D. M., & Miller, K. W. (1998). An ethnoscience approach to curriculum issues for American Indian students. *School Science and Mathematics, 98*(5), 260–265.

Echols, J. C. (1986). *Hide a butterfly*. Berkeley, CA: Lawrence Hall of Science.

Echols, J. C. (1993). *Ladybugs*. Berkeley, CA: Lawrence Hall of Science.

Gardner, H. (1983). *Frames of mind: The theory of multiple intelligence*. New York: Basic Books.

Gardner, H. (1993). *Multiple intelligence: The theory into practice*. New York: Basic Books.

Gardner, H. (1999). *The disciplined mind*. New York: Simon & Schuster.

Haberman, M. (1995). *Star teachers of children in poverty*. West Lafayette, IN: Kappa Delta Pi.

Hurd, P. D. (1991). Why we must transform science education. *Educational Leadership, 49*(2), 33–35.

Jacobs, H. H. J. (1989). *Interdisciplinary curriculum: Design and implementation.* Alexandria, VA: Association for Supervision and Curriculum Development.

Lawrence Hall of Science. (1993). *Full-option science system (FOSS).* Chicago: Encyclopedia Britannica Educational Corporation.

Lederman, N. G., & Niess, M. L. (1997). Integrated, interdisciplinary, or thematic instruction? Is this a question or is it questionable semantics? *School Science and Mathematics, 97*(2), 57–58.

Lonning, R. A., & DeFranco, T. C. (1997). Integration of science and mathematics: A theoretical model. *School Science and Mathematics, 97*(4), 212–215.

Meyer-Monhardt, R. (2000). Fair play in science education: Equal opportunities for minority students. *The Clearing House, 74*(1), 18–22.

National Council for Research on Women. (2001). *Balancing the equation: What we know and what we need.* www.ncrw.org/research/scifacts/htm

National Council for the Social Studies. (1994). *Curriculum standards for social studies.* Washington, DC: Author.

National Council of Teachers of English/International Reading Association. (1996). *Standards for English language arts.* Newark, DE: Author.

National Council of Teachers of Mathematics. (1989). *Curriculum and evaluation standards for school mathematics.* Reston, VA: Author.

National Council of Teachers of Mathematics. (2000). *Principle and standards for school mathematics.* Reston, VA: Author.

National Middle School Association. (1982). *This we believe.* Columbus, OH: Author.

National Research Council. (1996). *National science education standards.* Washington, DC: National Academy Press.

National Science Resources Center. (1992). *The life cycle of butterflies.* Burlington, NC: Carolina Biological.

National Science Resources Center. (1994). *STC: Science and technology for children.* Washington, DC: National Academy of Sciences.

Novak, J. D., & Gowin, D. B. (1984). *Learning how to learn.* Cambridge, England: Cambridge University Press.

Odom, A. L., & Kelly, P. V. (2001). Integrating concept mapping and the learning cycle to teach diffusion and osmosis concepts to high school biology students. *Science Education, 85*(6), 615–635.

Ogle, D. (1986). A teaching model that develops active reading of expository text. *The Reading Teacher, 39*(2), 564–570.

Perkins, D., Crismond, D., Simmons, R., & Unger, C. (1992). Inside understanding. In D. Perkins, J. Schwartz, M. West, & M. Wiske (Eds.), *Software goes to school: Teaching for understanding with new technologies* (pp. 70–87). New York: Oxford University Press,.

Rakow, S. J., & Vasquez, J. (1998). Integrated instruction: A trio of strategies. *Science and Children, 35*(6), 18–22.

Rutherford, J., & Ahlgren, A. (1989). *Science for All Americans: Project 2061.* New York: Oxford University Press.

Sandmann, A., Weber, W., Czerniak, C., & Ahern, J. (1999). Coming full circuit: An integrated unit plan for intermediate and middle grade students. *Science Activities, 36*(3), 13–20.

Tom Snyder Productions. (1994). *The great ocean rescue.* Watertown, MA: Author.

Tuomi, J. (1993, Fall). *Effective materials for science instruction. NSRC Newsletter.* Washington, DC: National Science Resources Center.

Wiggins, G., & McTighe, J. (1998). *Understanding by design.* Alexandria, VA: Association for Supervision and Curriculum Development.

Wiles, J., & Bondi, J. (1981). *The essential middle school.* Columbus, OH: Merrill.

Chapter 13

Next Steps

Introduction

Even though you have learned so much about project-based science, you probably still have questions about this approach: *What challenges will I face when I first implement project-based science? What are the benefits of project-based science? How will I improve my science teaching?* This chapter summarizes some of the benefits of this approach. Project-based science also presents challenges, so we will discuss ways to overcome a number of them. Since one way to overcome challenges is to continue your professional growth, we will examine some strategies for becoming a professional teacher who is a lifelong learner.

Benefits of Project-Based Science

Teaching is hard work. Nevertheless, you will find there are many benefits to running a project-based classroom:

- Because your students are pursuing solutions to important and meaningful questions, you too will find the work interesting and motivating. Your teaching will

Chapter Learning Performances

- *Explain the important benefits of using project-based science.*
- *Describe how a teacher can continue his or her professional development.*
- *Create a professional development plan.*
- *Compare and contrast the challenges to implementing a project-based curriculum and specify strategies that can be used to overcome challenges.*

vary each year since you will be exploring new projects with each new group of students. Even within a project, students are exploring several subquestions, making daily lessons novel and varied. Finally, because students are interested in and excited about the questions they are investigating, you will find teaching a pleasurable experience.

- Often, as your students pursue answers to questions, they explore interesting content. A second benefit of teaching in a project-based manner is that you continually learn new ideas about how the world works. This makes you a lifelong learner.

- In project-based science, classroom management is facilitated. Because students are interested and motivated, discipline problems are minimized. This doesn't mean that you don't need good classroom management skills. Good classroom management is essential. However, when students are interested and their minds are active, they are less likely to cause discipline problems. The best way to manage classrooms is to keep children intellectually engaged in the work.

- In a project-based classroom, your teaching and student learning will match the reform efforts in science education including *the National Science Education Standards* (National Research Council, 1996), *Project 2061: Science for All Americans* (Rutherford & Ahlgren, 1989), and *Benchmarks for Science Literacy* (AAAS, 1993). For example, the National Research Council (1996) argued that "there needs to be new emphasis placed on inquiry-based learning focusing on having students develop a deep understanding of science embedded in the everyday world." Many states have structured their objectives and high-stakes tests to correspond with these national standards. For example, many state tests now ask students to make conclusions from data, write responses to situations, and demonstrate skills on performance-based components.

- Through engaging in project-based science, learners develop deep, integrated understanding of content and process. Learners build meaningful relationships with and connections between ideas that they can use to understand their world. Students also learn the process of science, which can be used to solve problems in a variety of contexts. For example, students can apply the ability to ask good questions, plan and design investigations, and draw conclusions about everyday problems.

- In project-based science, students learn to work together to solve problems. In order to be successful in the real world, students need to know how to work with people from different backgrounds. Collaborative skills, such as being able to listen to others, debate ideas respectfully, and share ideas, are abilities that can be used in many contexts at all ages. Moreover, these are important abilities that employers seek in order to create a productive workforce.

- Project-based science promotes responsibility and independent learning. As students come up with their questions and design investigations, they learn to take responsibility for their own learning. In an investigation, students learn to find information from a variety of different sources, follow through on data collection procedures, and take responsibility for gathering necessary materials. Teachers encourage students to engage in self-evaluation strategies, such as assessing progress in their portfolios, which encourages reflection and self-improvement. Group and individual accountabilities are enhanced by collaboration.

- Project-based science is sensitive to the needs of a diverse group of students. Project-based science meets the needs of male and female students of varied cultures, races, and academic abilities by focusing on issues and questions important in their lives. Because project-based science actively engages students in different types of

> *Learning Activity 13.1*
>
> ### BENEFITS OF PROJECT-BASED SCIENCE
>
> *Materials Needed:*
>
> - A pencil and paper
>
> A. Of the eight benefits of project-based science listed, which are the most important to you. Why?
> B. What additional benefits do you see for project-based teaching?
> C. Share your ideas with a classmate. How are your ideas similar or different?

tasks, it meets the varied learning needs of many different students. Multidimensional assessment techniques allow students to demonstrate their understandings in a variety of formats, formats that work for them. Collaboration teaches students to work together despite their differences—in fact, project-based science puts these differences to work.

In Learning Activity 13.1 you will further explore the benefits of project-based science.

Challenges of Project-Based Science

Elementary and middle school teachers face a number of **challenges** when implementing project-based science for the first time. In previous chapters, we discussed some challenges of implementing project-based science and some strategies for dealing with these challenges. For example, obtaining resources can be challenging. Some of the strategies for dealing with this challenge are using free and inexpensive materials, such as 2-liter soda bottles, and finding a local company or university that is willing to donate equipment. Another challenge is getting students to debate ideas. One strategy for dealing with this challenge is building collaborative skills and building trust. Conflict management strategies are also helpful.

You may be confronted by additional challenges not addressed in this book. For example, you might feel that your science background is not strong enough, and you might feel uncomfortable with the subject matter involved. You might feel uncomfortable helping your students carry out an investigation if you have never carried out one yourself. Although these are critical issues, there are a number of positive steps you can take to overcome them.

We will discuss a number of challenges—teachers' content knowledge and knowledge of the process of science, limited student experience, lack of time, and real or perceived external pressures. We will consider a number of recommendations for meeting these challenges. However, don't expect to meet all potential challenges your first year of teaching in a project environment. Teaching involves lifelong learning. Even after 20 years of teaching, you will still wrestle with how to improve many aspects of your teaching, each year showing more improvement. Any ideas we consider here are just a start. You will need to continue your professional development throughout your career. For this reason, we will also discuss strategies for continuing your professional development.

Teacher Discomfort With Content Knowledge

Numerous reports claim that teachers' lack of content knowledge limits the teaching of science at the K–8 level (Weiss, 1978, 1987). Certainly teachers need to understand what they are teaching, and feeling uncomfortable with content is a major reason that science is cut out of the curriculum. Although you might feel a little shaky during your first attempts to teach science, as we all did, there are ways to increase your science knowledge and increase your comfort level with science:

- Scientific knowledge expands daily. Neither you nor anyone else can know everything. On many occasions, we have been confronted by students' questions to which we did not know the answers. It is okay to say, "I don't know," and then to model how to find the answer. Showing your students that you are interested in learning more and how to learn is more important than demonstrating that you know everything. Demonstrate to students how you can use reference books, the Web, and local experts to find answers to questions.
- Like your students, use each project to learn more about a science topic and about engaging in the process of science. If you take this approach, you will gradually become more comfortable with science content. This gradual, "learn as you go" approach is less intimidating than trying to learn a great deal of new information at one time.
- Develop the habit of reading science magazines and books. Reading is a wonderful way to learn. Magazines, such as *Science Scope* and *Science and Children* published by the National Science Teachers Association and *Nature Scope* published by the National Wildlife Federation, are excellent, nonthreatening choices. Often, the articles in them give additional sources of reading materials. Also, develop your own personal library of books. For instance, *The Pillbug Project* by Robin Burnett will give you some good background knowledge of pillbugs, as well as a sense of how to carry out investigations. Finally, the National Science Foundation distributes a number of books that you will find valuable both for your learning and your classroom teaching. Table 13.1 gives the addresses of several professional organizations that publish journals.
- Attending science classes; workshops; and local, regional, and national science conferences will help you learn new science content. Local school districts, a local university, a local environmental group, or a state organization may offer some workshops. You will find these types of classes and conferences extremely valuable; you will not only get new ideas for teaching and learn some science, but you will also meet others who are as interested as you are in improving teaching and learning. (We will discuss more thoroughly how to join professional organizations in the section on professional development.) Once you join a professional development organization, you will receive in the mail information about workshops and conferences.

- Community members and parents can be guest speakers who provide science content expertise. You and your students can learn valuable science knowledge when guest speakers come to your class. For example, if your class is carrying out a project on water quality and some of your students want to perform chemical tests that you don't understand, there is likely to be a parent or other community member who would be very willing to volunteer time to talk with your class and help with a demonstration.
- The World Wide Web is a valuable source of science information. Although you might have to do a little digging, you will find that the Web is filled with a number of very useful pages that help you and your students learn content. For instance, the Great Lakes Information Network (http://www.great-lakes.net) provides a wealth of information about the Great Lakes region and its economy, environment, tourism, news, events, and weather. Volcano World (http://volcano.und.nodak.edu) contains all the information you could want about all aspects of volcanoes. You can even send a question to a volcanologist via the Web site. Finally, The National Science Foundation (http://nsf.gov/news/classroom) offers many online documents with up-to-date science information for classroom teachers and students.

> **CONNECTING TO NATIONAL SCIENCE EDUCATION STANDARDS**
>
> **STANDARDS FOR PROFESSIONAL DEVELOPMENT**
>
> Professional development for teachers of science requires learning essential science content through the perspectives and methods of inquiry (p. 59).

Teacher Discomfort With the Process of Science

Many of us have performed "cookbook" activities, following steps to complete science-related tasks, but these are not investigations. Unfortunately, few of us have had experiences conducting investigations. Like a lack of content knowledge, a lack of experience and understanding of how to carry out investigations can inhibit your ability and willingness to help students engage in the process of investigation. The best way to learn how to carry out an investigation is to do it—ask questions, make observations, manipulate variables, and analyze data. If you take an active role in your students' projects, you will soon develop a level of comfort. As in baking a cake, you don't become good at it unless you do it.

Belonging to professional organizations also will help you become comfortable with the process of science. At conferences, speakers present sessions that focus on inquiry. These presentations can help you learn how to support students in the process of investigations. Further, at conferences, expert teachers frequently model inquiry activities they use with their own students; you may be able to talk with these teachers after such sessions.

Another very valuable mechanism to improve your teaching is observation of an expert teacher carrying out an investigation with his or her class. Notice how a knowledgeable teacher supports students in the process of inquiry. Discuss the lesson with the teacher after class to learn the reasons behind the approach.

It is perhaps more important to invite an expert teacher in to observe and critique your teaching. This will be difficult to do at first. We all are hesitant to invite others to critique us, but commentary from an experienced teacher can be very valuable. If you can establish a long-term mentor arrangement with an expert teacher, you will profit even more. Another way to get feedback on your teaching is to record it. You can show the video to a colleague or critique it yourself. How are you supporting students? In what ways are you promoting understanding of science?

The Web also provides opportunities for teachers and students to learn more about the process of science. For example, Annenberg Media (http://www.learner.org/) provides videos, assessment tools, and other resources to help teachers learn about quality science teaching. Knowledge Networks On the Web (KNOW) is a professional development tool for use by teachers who are using curriculum materials or technologies developed by the Center for Highly Interactive Computing in Education (hi-ce) at the University of Michigan (see http://know.umich.edu). Resources like these can help you learn more about the process of science and generate confidence-building relationships with others who are conducting investigations.

Finally, field trips and internships in science settings can help you learn about the process of science as it is actually carried out in the real world. Field trips to governmental and commercial research laboratories, local university science laboratories, and medical hospitals can provide valuable insight into the process of science. Sometimes these institutions offer summer paid or volunteer internships.

Limited Student Experience

Carrying out an investigation is difficult cognitive work. You are challenging your students with such activities. Many students see the purpose of school as learning content knowledge. As students advance through grade levels, they become more regimented in their learning. At-risk children are frequently exposed to lots of "kill and drill" strategies that focus on memorizing low-level factual knowledge. Many students develop good strategies for memorizing information, and they may resist new ways of learning that require them to think in more cognitively challenging ways. Although memorizing is an important learning strategy, finding information, analyzing data, making plans, carrying out investigations, developing concept maps, and making products that synthesize understanding cannot be done through memorization.

Children will sometimes ask, "Why can't we just read from the book?" or "Why can't we just answer the questions at the end of the chapter?" These children have likely learned strategies that require very little effort. They read from the book, look for reading cues (such as bold-print words, topic sentences, or summary sentences), and easily identify the answers to questions at the end of the chapter. Although reading for information is a central component of project-based science, it can be easier for some students to find answers at the end of a chapter than to develop a design for an investigation.

You can take several positive measures to overcome this resistance to learning in a more active manner:

- Make your expectations clear to your class. Stress to children that, although they may have done activities differently in other years, this is the way your classroom works.
- Explain to children that they are working the way scientists work. Scientists find information, ask questions, analyze data, and collaborate with others. Explain that just reading and finding answers distorts what science is all about. Science is about inquiry, and that is what students will do in your class. Interactions with scientists, parents, and others in the community will reinforce this spirit of inquiry.
- Point out to your students how the work they are doing in class helps find solutions to real problems and issues. Most children, especially pre- and young adolescents, have a strong desire to engage in meaningful activities.
- Encourage students to take risks, ask questions, and plan designs, even if they don't work. Young inquirers will go down a number of wrong paths, as do real scientists, and students need to know this is okay.

Lack of Time

Having students learn science through inquiry takes time. There is no way around this. It will take your time, and it will take classroom time. We are interested in creating useful science knowledge, knowledge that students will retain; and investigations and projects, as time-consuming as they are, foster such knowledge. The National Science Education Standards (NRC, 1996, p. 219) suggest that time is a major resource that must be allotted in the school day. A few suggestions for finding time in the day to teach science are described below.

Often there are downtimes during an investigation—while the plants are growing, material is decomposing, or metal is rusting. Science teachers should use these downtimes productively, slipping additional lessons into them. For example, while plants are growing, the class could be learning how to take measurements of plant growth using rules or calipers. These downtimes are also periods in which elementary teachers can teach lessons in other subject areas.

Another way to make maximal use of time is to integrate the curriculum. Earlier we discussed how project-based science supports curriculum integration. Curriculum integration can be a time-saving device, because it allows many academic objectives to be reached at once. For example, curriculum integration would blend the following objectives into a project designed to answer "Where does all of our garbage go?": (a) students will be able to discern why people make laws (in social studies, students could learn why recycling laws were created in their state); (b) students will be able to discern fact from opinion (in reading, students could debate whether it is a fact or an opinion that recycling makes a difference); and (c) students will be able to find the volume of a container (in mathematics, students can calculate the amount of garbage that fills a trash dumpster outside of the school building during a week's time). Coordinating subject areas to save time is also suggested in the National Science Education Standards (NRC, 1996, p. 214).

One common complaint is how time-consuming it is to set up investigations and clean up after them. Don't deny your students the opportunity to learn through inquiry because of these problems. One solution is to assign class jobs. Some teachers have students help pass out materials at the beginning of class, and they save 5 minutes or so at the end of every class session by having the entire class help clean up. Teachers of young elementary students frequently rely on the help of volunteer parents or older students in the school.

Another technique that has worked for many schools is block scheduling (creating longer class periods of time, such as 90-minute blocks instead of 45-minute segments) or flexible scheduling, which creates extended time periods for students to work on the projects. Since much class time is devoted to setting up and cleaning up project work, longer work periods save the time of setting up and cleaning up for numerous shorter periods. In an elementary school where you teach several subjects, it will be easier for you to build in extended periods of time to work on projects.

At the middle school level, where subject matter classes are from 45 to 55 minutes long, you might take a proactive approach, as some teachers have, to lead efforts to shift your school to a block schedule. In block scheduling, classes meet for double periods for 2 or 3 days each. If your students move from class to class as a group, which is common in middle schools, you might also arrange to trade classes with a teacher who teaches in the time slot before or after you so you can have a longer period of time for a science investigation.

Real or Perceived External Pressures

Because project-based science is built around meaningful, hands-on science investigations that extend over time, it can create tension between breadth (superficial coverage of many topics and objectives) and depth (extensive coverage of a few topics). Meaningful, in-depth learning of limited but question-relevant content is preferable to superficial surveys of a wide body of content, which is the approach of most science survey texts. Doing science is more important than being exposed to a wide body of meaningless science content that is likely to be forgotten. This position is supported by the Third International Mathematics and Science Study (TIMSS, 1997, 1998), which suggests that students would learn more if the curriculum covered fewer topics in a more in-depth manner. Engaging learners in investigations is now strongly supported by a number of prominent national organizations, like the National Research Council (1996) and the American Association for the Advancement of Science (1993). These organizations have taken strong "less is more" stances and advocate sustained, project-oriented science teaching. A number of local school districts and state education agencies are becoming more sensitive to these new ideas in education and are beginning to stress cognitive strategies, such as planning and analyzing data, on state examinations.

Unfortunately, curriculum in the United States has, for many years, taken the breadth approach, and it has only been in the last few years that educators have been moving toward covering fewer subjects in a school year. As a result, many principals, parents, and school board members are more familiar with learning science the old way, reading about numerous topics in a textbook, than they are with learning a few science topics in-depth through long-term inquiry. This difference in viewpoint can cause problems for a teacher trying to implement a project-based approach.

Parents, administrators, colleagues, and even the janitor may question what you are doing. This is only natural, since you are doing something new. To counteract this questioning, take positive steps:

- Gain the support of administrators and colleagues. Talk to your principal and administrators about how you plan to do science and explain the project-based science approach. You might share a copy of the National Science Education Standards (NRC, 1996) with them to show them how this approach is supported in national reform efforts. Extend an open invitation to your principal to visit your class. Share with your colleagues your methods of teaching science and invite them into your classroom as well.
- Hold a science open house to inform parents and members of the community of your science practices. Create a school or class newsletter that includes information about your science class, write information sheets to parents about specific projects students are working on, and extend open invitations to parents to visit your class. Some teachers have even made videos of their teaching and shared it with parents to illustrate how the approach motivates students to learn and how the teaching objectives match state-mandated tests or curriculum standards. You could also lend a copy of *Every Child a Scientist: Achieving Literacy for All* to parents. This is a useful resource for parents, showing them how they can help implement science standards. It is available from the National Research Council and can be ordered by calling 1-800-624-6242 or 202-334-3313, or it can be ordered online at the Web site of National Academy Press. Teachers who have used these techniques have found them to be very successful.
- You need to inform people in your school community of your teaching approach and educate them about the importance of in-depth coverage of topics through long-term inquiry. Several resources can help you educate others about your position:
  - The *National Science Education Standards* (NRC, 1996) support project-based science. This book can be ordered from the National Academy Press, 2101 Constitution Ave., NW, Box 285,Washington, DC, 1-800-624-6242 or 202-334-3313. Information about the *National Science Education Standards* can also be found online at the Web site of National Academy Press.
  - Other excellent resources that can be used to educate others about effective science teaching are the National Science Teachers Association position papers. Position papers are available on a number of topics, such as the use of computers, elementary science, multicultural science education, research in science education, and parent involvement in science education. These position papers are available online at the National Science Teachers Association Web site.
  - The Eisenhower National Clearinghouse contains a wealth of information, including interesting articles, standards frameworks, findings from educational research, and information about reform efforts.
  - The National Association for Research in Science Teaching (NARST) publishes a series entitled *What Research Says to the Science Teacher* that can be obtained through NARST. These papers provide useful information about a variety of topics, such as gender equity in science, inquiry, and constructivism. NARST and the National Science Teachers Association (NSTA) have also partnered to publish research-based information for classroom teachers.

Learning Activity 13.2 asks you to further explore each of these challenges.

Learning Activity 13.2

YOUR CHALLENGES

Materials Needed:

- A paper and pencil

A. Of the challenges listed in this chapter, which one is the most critical for you to overcome? Why?
B. What other challenges do you think you might experience as you implement project-based science?
C. How might you resolve these challenges?
D. Work with a few colleagues. What are their suggestions for resolving these challenges?

Continuing Your Professional Growth

Teaching presents challenges, but it also gives rewards. Perhaps one of the best rewards is becoming part of a community that cares about the learning of children. As a teacher, you will need to continue your professional growth by joining professional organizations, attending conferences, subscribing to journals, and getting information from the World Wide Web. Many teachers use these types of resources to help them become National Board certified teachers. The National Board for Professional Teaching Standards (NBPTS) developed professional standards that define what accomplished teachers should know and be able to do. NBPTS administers National Board Certification, a voluntary assessment program that certifies educators who meet those standards (http://www.nbpts.org).

Joining Professional Organizations

Joining **professional organizations** is essential to being a professional. It is one of the best ways to continue your development as a teacher. Joining national, state, and local organizations will allow you to build connections with other teachers who may have many of the same questions and issues that you do. These organizations, through their publications, conferences, newsletters, and Web sites, are also sources of information on content and instruction. They can also inform you about national policy and upcoming events.

At the national level, join the National Science Teachers Association (NSTA), which focuses on teaching science to children at all levels. Many states have science education organizations affiliated with NSTA. You can find out about them by contacting NSTA. State Departments of Education employ science consultants who are good sources of information about professional organizations in your state. If you live in a large urban area, there may be local science teacher organizations that you can join. If your school district has a curriculum director, this person should be knowledge-able about local organizations. Principals and other school administrators are also help-ful sources of information. Table 13.1 lists national organizations you may want to join.

Attending Conferences

Membership in professional organizations, such as those Listed in Table 13.1, will give you opportunities to attend their **confer-ences**. Conferences are exciting because they connect you with other educators. In the pro-cess, you get to learn about what others are doing in their teaching. You can also attend presentations by nationally known sci-entists and educators. At most conferences, you can attend the exhibit hall, where you can see some of the newest commercially available materials. Free materials are often distributed at these conferences. As you become comfortable in your own teaching, you might want to share some of your project ideas with others by presenting at pro-fessional conferences yourself. Many teachers find it more comfortable to first present at local or state conferences to gain more experience and confidence.

Subscribing to Publications

Many professional organizations also create associated **publications**. For example, NSTA offers a number of important publications, including *Science and Children* (for elementary teachers) and *Science Scope* (for middle grade teachers). These publica-tions are filled with articles written by educators, science teaching ideas, informa-tion about science education reform, reviews of curriculum materials and software, and conference information. State and local organizations usually publish newslet-ters with helpful teaching hints and important local information. Table 13.2 lists the publications associated with each organization.

Using Technology Tools For Professional Development

As we have discussed throughout this book, the World Wide Web can be used as a source of both content and instructional resources. It also can serve as a source of **professional development**, because it is filled with numerous sites that can give you helpful information regarding various education and science topics. For example, the Eisenhower National Clearinghouse contains information about science and math-ematics reform, research findings in education, lesson plans, and curriculum mate-rials. The National Science Teachers Association has also developed many online

TABLE 13.1   Professional Science Education Organizations

Name of organization	Mailing address	Web page	Focus
National Science Teachers Association	1840 Wilson Blvd. Arlington, VA 22201-3000 703-243-7100	www.nsta.org	Science education at all ages
School Science and Mathematics Association	SSMA Central Office The Ohio State University 238 Arps Hall 1945 North High Street Columbus, OH 43210-1172 Phone: 614-292-8061 Fax: 614-292-7695 E-mail: white.32@osu.edu	www.ssma.org	Science and mathematics education; integration of science and mathematics
AIMS Education Foundation	1595 S. Chestnut Ave. Fresno, CA 93702 209-225-4094 Fax: 209-255-6396	www.aimsedu.org	Activities for integration of science and mathematics at the elementary level
American Association of Physics Teachers	AAPT Membership Dept. One Physics Ellipse College Park, MD 20740 Fax: 301-209-0845	www.aapt.org	Physical science teaching and learning
Chemical Educational Foundation	1525 Wilson Blvd. Suite 750 Arlington, VA 22209 703-527-6223 Fax: 703-527-7747	www.chemed.org	Chemistry education
National Association of Biology Teachers	11250 Roger Bacon Drive 19 Reston, VA 20190-5202 703-471-1134 800-406-0775 Fax: 703-435-5582	www.nabt.org	Biology teaching and learning
Geological Society of America	P.O. Box 9140 Boulder, CO 80301-9140 303-447-2020 Fax: 303-447-1133	www.geosociety.org	Geology and geology education
National Association for Research in Science Teaching	NARST National Office 12100 Sunset Hills Road, Suite 130 Reston, VA 20190 Phone: 703-234-4138 Fax: 703-435-4390	http://www.narst.org	Research in science teaching and learning

TABLE 13.2   Professional Publications

Name of organization	Mailing address	Web page	Title of journal
The National Science Teachers Association	1840 Wilson Blvd. Arlington, VA 22201 –3000 703-243-7100	www.nsta.org	*Science and Children* *Science Scope* *Science Teacher*
School Science and Mathematics Association	Journal Editor and Office, 2007–2012 Gerald Kulm Texas A&M University College Station, TX 77866 Phone: 979-862-8100 E-mail: gkulm@coe. tamu.edu	www.ssma.org	*School Science and Mathematics*
National Association for the Education of Young Children	1509 16th Street NW Washington, DC 20036	www.naeyc.org	*Young Children*
National Middle School Association	2600 Corporate Exchange Dr., 370 Columbus, OH 43231 10800-528-NMSA	www.nmsa.org	*Middle School Journal* *Middle Ground*
Association for Supervision and Curriculum Development	1250 N. Pitt Street Alexandria, VA 22314-1453 1-800-933-ASCD FAX: 703-299-8631	www.ascd.org	*Educational Leadership*
Phi Delta Kappa	408 N. Union P.O. Box 789 Bloomington, IN 47402	www.pdkintl.org/ kappan/kappan/htm	*Phi Delta Kappan*

professional development opportunities for teachers. Table 13.3 lists the Web addresses of several organizations that can provide you with information that will help you continue your professional development.

Inquiry Into Your Teaching

One of the best ways to improve your teaching is to reflect on and to engage in inquiry about your own teaching. In this book, you have often been asked to reflect on your beliefs and past experiences. You can continue to use *reflection* throughout your professional career to improve your teaching. This book has also stressed the importance of inquiry to learn about the

CONNECTING TO NATIONAL SCIENCE EDUCATION STANDARDS

STANDARDS FOR PROFESSIONAL DEVELOPMENT

Science learning experiences for teachers must introduce teachers to scientific literature, media, and technological resources that expand their science knowledge and their ability to access further knowledge (p. 59).

TABLE 13.3   Useful Science Professional WWW Addresses

Name	Web page	Focus
The National Science Teachers Association	www.nsta.org	Science teaching and learning at all ages
Eisenhower National Clearinghouse	http://www.goENC.com	Information about science and mathematics education
The National Science Foundation	http://nsf.gov/news/classroom	Science research and funding
National Center for Education Statistics	http://nces.ed.gov/timss/	Findings from the Third International Mathematics and Science Study
The American Association for the Advancement of Science	www.aaas.org	Advancement of science; publisher of *Science for All Americans* and *Benchmarks for Science Literacy*
National Academy of Science	www.nas.edu www.nap.edu	Advice on scientific issues; author of the National Science Education Standards

world. You can use a form of inquiry, called *action research,* to learn about your own teaching.

Reflection can be defined as a voluntary effort to share and critique ideas about teaching, assess one's teaching and students' learning, formulate aims and goals about the curriculum through collaboration, and take responsibility for actions and the consequences of actions (Baird, 1992; Barnes, 1992; Putnam & Grant, 1992). The *National Science Education Standards* (NRC, 1996) promotes reflective practices in science education: "Teachers of science engage in ongoing assessment of their teaching and of student learning. In doing this, teachers use student data, observations of teaching, and interactions with colleagues to reflect on and improve teaching practice" (p. 37).

Action research, like the inquiry process, involves asking questions, making plans, carrying out the plans, and analyzing and making use of what was learned. The process is similar to the investigation web described in Chapter 4; the main difference is that action research is not for the purpose of generating findings to share with others, but to generate information to put into practice in your own teaching. For example, a teacher might want to investigate whether collaborative learning increases motivation among students in her class. She would make a plan to answer this question, carry out an investigation, and analyze whether collaboration changed motivation levels. Put differently, teachers are engaging in project-based work to investigate their own teaching. Dick Arends (1991), in his book *Learning to Teach,* providesd great detail about inquiry into and reflection on one's own teaching.

The following strategies are part of reflection and action research:

- *Keep a journal.* Keeping a journal and taking notes of what works and what doesn't work is one form of action research. For example, you might want to keep notes on the types of driving questions that students seem to find meaningful. In what types of questions do children seem to be interested?

- *Make a video of your teaching.* Another way you can examine your own teaching is to video it. Many people dislike seeing themselves on video, but making a video of your teaching and then analyzing it is an excellent way of improving your teaching—athletes watch videos of themselves constantly to learn how they can maximize their performance. Keeping a record also can show how you change over time.

- *Collaborate with colleagues.* Colleagues can be excellent sources of help for reflection and action research. They can serve as mentors who can provide helpful suggestions to improve your teaching, and they can be good role models. Visiting the classrooms of others can provide you with new ideas and practical suggestions for improving your teaching. Letting others critique your teaching by visiting your classroom or watching you on video can provide you with valuable insight, insight that you might not have on your own.

- *Have students fill out questionnaires.* Many teachers ask their students to fill out questionnaires about their teaching. These questionnaires can be simple open-ended statements, like "The thing I liked most about this lesson was ..."or "One thing I would change about this project is ...." Questionnaires can also be simple scales: "The acid-rain project was 1 (excellent), 2 (good), 3 (satisfactory), 4 (bad)." Teachers of younger students often use "smiley faces" on their questionnaires—students circle a picture of a smile, a neutral face, or a frowning face next to pictures or statements, such as "growing plants," "using a magnifying glass," and "recording our results."

Looking back at what we have done can also lead to improvements in teaching. In Learning Activity 13.3, you will reexamine some of your initial ideas about teaching and learning. In Learning Activity 13.4, you will conduct a case study of an elementary or middle grade classroom to see for yourself whether classrooms in your region are meeting the fundamental national goals for science and how many of the elements of project-based science are evident in these classrooms.

> **CONNECTING TO NATIONAL SCIENCE EDUCATION STANDARDS**
>
> **STANDARDS FOR PROFESSIONAL DEVELOPMENT**
>
> Professional development activities must provide opportunities for teachers to learn and use various tools and techniques for self-reflection and collegial reflection, such as peer coaching, portfolios, and journals (p. 68).

Chapter Summary

This book has presented a new approach to teaching elementary and middle school science, an approach called *project-based science.* We started this last chapter by considering some of the benefits of this approach. This approach can present a number of challenges for the teacher. Some teachers are not comfortable with the science content knowledge necessary to carry out projects. Some do not know how to conduct

Learning Activity 13.3

REVISITING INITIAL IDEAS ABOUT TEACHING SCIENCE

Materials Needed:

- Answers to Learning Activity 1.3

A. In Learning Activity 1.3, you identified your science teaching goals. Have these goals changed since you completed this activity? How so?

B. Examine your response to Learning Activity 1.2. How have your beliefs about why children should learn science changed since the beginning of this book?

C. Throughout this book, you have encountered a number of ideas about teaching science to children. Examine your beliefs now about these ideas.
   - What are your beliefs about the role of investigations now? How have they changed since you read the book?
   - What are your views on collaboration? Have they changed? How? What will you do differently in your classroom in the future?
   - What were your views about covering basic science content and skills before exploring the ideas in this book? Have they changed? How will this affect your teaching?
   - What were your views about assessment before reading this book? Have your views changed? How will this affect your teaching in the future?
   - What were your beliefs about classroom management before reading this book? Have they changed? How so? How will they affect your teaching in the future?
   - What did you think about curriculum integration prior to exploring the ideas in this book? Explain your beliefs now. How will your teaching be affected?

D. How has this book informed your teaching practices? What will you try to change? What will you keep about your current teaching?

an investigation with students. Time is always a problem in teaching, but it can be a bigger problem for teachers who are trying to carry out long-term investigations. Finally, because using a project-based approach may be new to parents, colleagues, administrators, and community members, this approach may be met with resistance. These real or perceived pressures can affect a teacher who is trying to implement this new teaching approach.

In this chapter, we discussed these challenges and ways to overcome them. One way to become a better teacher who can implement a project approach is to continue your professional development. We discussed several ways to do this: join professional organizations, attend conferences, read professional publications, and access information on the Web. Finally, we examined the notion of inquiry into your teaching.

We hope you will continue to grow as a teacher, and we wish you luck as you begin using a project approach in your teaching.

Learning Activity 13.4

CASE STUDY OF AN ELEMENTARY OR MIDDLE GRADE CLASSROOM

Materials Needed:

- An elementary or middle school to study

A. Obtain permission from the appropriate officials (usually a school district superintendent, curriculum director, or principal and classroom teacher) at an elementary or middle school to conduct a small case study at the school.
B. Research how the school's course of study, curriculum, instructional practices, and assessment techniques compare with the guidelines proposed by national organizations (such as AAAS and NRC).
C. In your opinion, how well is this school meeting suggested goals? In what ways is it falling short?
D. How well is the school implementing the features of project-based science? In what ways is it failing to implement project-based science ideas?
E. What obstacles might you face trying to implement project-based science in this environment?

Chapter Highlights

- One of the most rewarding ways to teach science is a method known as *project-based science.*
- Project-based science has many benefits:
  - Teachers find the work interesting and motivating.
  - Teachers become lifelong learners.
  - Classroom management is facilitated.
  - Project-based science matches the reform efforts in science education.
  - Learners develop deep, integrated understanding of content and process.
  - Students learn to work together to solve problems.
  - Project-based science promotes responsibility and independent learning.
  - Project-based science is sensitive to the needs of a diverse group of students.
- Teachers can face a number of challenges when implementing project-based science for the first time:
  - Discomfort with science content knowledge
  - Discomfort with the process of science
  - Limited student experience, which might cause some students to resist project work
  - The length of time required
  - External pressures that can create tension
- Professional growth is a continuous process throughout a teacher's career.
- Teachers can continue their professional growth by:
  - Joining professional organizations.
  - Attending conferences.
  - Subscribing to journals.

- Getting information from the World Wide Web.
- Teaching can be improved by reflecting on and engaging in inquiry about one's own teaching.

Key Terms

Action research
Challenges
Conferences
Professional development

Professional organizations
Project-based science
Publications
Reflection

References

American Association for the Advancement of Science. (1993). *Benchmarks for science literacy.* New York: Oxford University Press.

Arends, R. (1991). *Learning to teach.* New York: McGraw-Hill.

Baird J. R. (1992). Collaborative reflection, systematic inquiry, better teaching. In T. Russell & H. Munby (Eds.), *Teachers and teaching from classroom to reflection.* Bristol, PA: The Falmer Press.

Barnes, D. (1992). The significance of teachers' frames for teaching. In T. Russell & H. Munby (Eds.), *Teachers and teaching from classroom to reflection.* Bristol, PA: The Falmer Press.

National Research Council. (1996). *National science education standards.* Washington, DC: National Academy of Sciences.

Putnam, J., & Grant, S. S. (1992). Reflective practice in the multiple perspective program at Michigan State University. In L. Valli (Ed.), *Reflective teacher education cases and critiques.* New York: State University of New York Press.

Rutherford, J., & Ahlgren, A. (1989). *Science for All Americans: Project 2061.* New York: Oxford University Press.

Third International Mathematics and Science Study. (1997). http://nces.ed.gov/ TIMSS/

Third International Mathematics and Science Study. (1998). http://nces.ed.gov/ TIMSS/

Weiss, L. (1978). *Report of the 1977 national survey on science, mathematics, and social sciences.* Research Triangle Park, NC: Center for Educational Research and Evaluation, Research Triangle Institute.

Weiss, L. (1987). *1985–86 national survey of science and mathematics education.* Research Triangle Park, NC: Research Triangle Institute.

Index

Page numbers in italics refer to Figures or Tables